Zoology: A Comprehensive Approach

Zoology: A Comprehensive Approach

Editor: Celeste Stewart

R CALLISTO REFERENCE

www.callistoreference.com

Callisto Reference,
118-35 Queens Blvd., Suite 400,
Forest Hills, NY 11375, USA

Visit us on the World Wide Web at:
www.callistoreference.com

ISBN: 978-1-64116-123-7 (Hardback)

Cataloging-in-Publication Data

Zoology : a comprehensive approach / edited by Celeste Stewart.
    p. cm.
Includes bibliographical references and index.
ISBN 978-1-64116-123-7
1. Zoology. 2. Animals. 3. Biology. I. Stewart, Celeste.
QL45.2 .Z66 2019
590--dc23

Table of Contents

Preface VII

Chapter 1 A new species of *Hirudo* (Annelida: Hirudinidae): historical biogeography
of Eurasian medicinal leeches 1
Naim Saglam, Ralph Saunders, Shirley A. Lang and Daniel H. Shain

Chapter 2 Sex-specific patterns in body mass and mating system in the Siberian
flying squirrel 13
Vesa Selonen, Ralf Wistbacka and Andrea Santangeli

Chapter 3 LemurFaceID: a face recognition system to facilitate individual identification
of lemurs 23
David Crouse, Rachel L. Jacobs, Zach Richardson, Scott Klum, Anil Jain,
Andrea L. Baden and Stacey R. Tecot

Chapter 4 *Caenorhabditis monodelphis* sp. n.: defining the stem morphology and
genomics of the genus *Caenorhabditis* 37
Dieter Slos, Walter Sudhaus, Lewis Stevens, Wim Bert and Mark Blaxter

Chapter 5 Tracking the migration of a nocturnal aerial insectivore in the Americas 52
Philina A. English, Alexander M. Mills, Michael D. Cadman, Audrey E. Heagy,
Greg J. Rand, David J. Green and Joseph J. Nocera

Chapter 6 Rediscovering a forgotten canid species 63
Suvi Viranta, Anagaw Atickem, Lars Werdelin and Nils Chr. Stenseth

Chapter 7 Survey of avifauna of the Gharana wetland reserve: implications for
conservation in a semi-arid agricultural setting on the Indo-Pakistan border 72
Pushpinder S. Jamwal, Pankaj Chandan, Rohit Rattan, Anupam Anand,
Prameek M. Kannan and Michael H. Parsons

Chapter 8 Shoaling promotes place over response learning but does not facilitate
individual learning of that strategy in zebrafish (*Danio rerio*) 81
Claire L. McAroe, Cathy M. Craig and Richard A. Holland

Chapter 9 Whole mitochondrial genomes provide increased resolution and indicate
paraphyly in deer mice 89
Kevin A. M. Sullivan, Roy N. Platt II, Robert D. Bradley and David A. Ray

Chapter 10 White-nose syndrome fungus, *Pseudogymnoascus destructans*, on bats
captured emerging from caves during winter in the southeastern
United States 95
Riley F. Bernard, Emma V. Willcox, Katy L. Parise, Jeffrey T. Foster and
Gary F. McCracken

Chapter 11   **Effects of agrochemicals on disease severity of *Acanthostomum burminis* infections (Digenea: Trematoda) in the Asian common toad, *Duttaphrynus melanostictus*** 106
Uthpala A. Jayawardena, Jason R. Rohr, Priyanie H. Amerasinghe, Ayanthi N. Navaratne and Rupika S. Rajakaruna

Chapter 12   **High-resolution monitoring from birth to sexual maturity of a male reef manta ray, *Mobula alfredi*, held in captivity for 7 years: changes in external morphology, behavior, and steroid hormones levels** 116
Ryo Nozu, Kiyomi Murakumo, Rui Matsumoto, Yosuke Matsumoto, Nagisa Yano, Masaru Nakamura, Makio Yanagisawa, Keiichi Ueda and Keiichi Sato

Chapter 13   **Indigenous house mice dominate small mammal communities in northern Afghan military bases** 124
Christoph Gertler, Mathias Schlegel, Miriam Linnenbrink, Rainer Hutterer, Patricia König, Bernhard Ehlers, Kerstin Fischer, René Ryll, Jens Lewitzki, Sabine Sauer, Kathrin Baumann, Angele Breithaupt, Michael Faulde, Jens P. Teifke, Diethard Tautz and Rainer G. Ulrich

Chapter 14   **Demography of a small, isolated tiger (*Panthera tigris tigris*) population in a semi-arid region of western India** 138
Ayan Sadhu, Peter Prem Chakravarthi Jayam, Qamar Qureshi, Raghuvir Singh Shekhawat, Sudarshan Sharma and Yadvendradev Vikramsinh Jhala

Chapter 15   **Species-level divergences in multiple functional traits between the two endemic subspecies of Blue Chaffinches *Fringilla teydea* in Canary Islands** 151
Jan T. Lifjeld, Jarl Andreas Anmarkrud, Pascual Calabuig, Joseph E. J. Cooper, Lars Erik Johannessen, Arild Johnsen, Anna M. Kearns, Robert F. Lachlan, Terje Laskemoen, Gunnhild Marthinsen, Even Stensrud and Eduardo Garcia-del-Rey

Chapter 16   **Body condition scoring of Bornean banteng in logged forests** 170
Naomi S. Prosser, Penny C. Gardner, Jeremy A. Smith, Jocelyn Goon Ee Wern, Laurentius N. Ambu and Benoit Goossens

Chapter 17   **Active parental care, reproductive performance, and a novel egg predator affecting reproductive investment in the Caribbean spiny lobster *Panulirus argus*** 178
J. Antonio Baeza, Lunden Simpson, Louis J. Ambrosio, Nathalia Mora, Rodrigo Guéron and Michael J. Childress

Chapter 18   **The cellular basis of bioadhesion of the freshwater polyp *Hydra*** 193
Marcelo Rodrigues, Philippe Leclère, Patrick Flammang, Michael W. Hess, Willi Salvenmoser, Bert Hobmayer and Peter Ladurner

Permissions

List of Contributors

Index

Preface

This book was inspired by the evolution of our times; to answer the curiosity of inquisitive minds. Many developments have occurred across the globe in the recent past which has transformed the progress in the field.

Zoology is a branch of biology that studies the anatomical, physiological, behavioral and evolutionary characteristics of animals. It also explores the habits and distribution of animals as well as their interactions with their environment. All studies of living and extinct species of animals are covered by this discipline. Zoology branches out into diverse taxonomical fields such as mammalogy, entomology, ornithology and herpetology. The development of modern techniques of genetic engineering like DNA fingerprinting, has aided the understanding of animal populations. This book contains some path-breaking studies in the field of zoology. It outlines the processes and applications of this field in detail. Students, researchers, zoologists and all associated with the area of zoology will benefit alike from this book.

This book was developed from a mere concept to drafts to chapters and finally compiled together as a complete text to benefit the readers across all nations. To ensure the quality of the content we instilled two significant steps in our procedure. The first was to appoint an editorial team that would verify the data and statistics provided in the book and also select the most appropriate and valuable contributions from the plentiful contributions we received from authors worldwide. The next step was to appoint an expert of the topic as the Editor-in-Chief, who would head the project and finally make the necessary amendments and modifications to make the text reader-friendly. I was then commissioned to examine all the material to present the topics in the most comprehensible and productive format.

I would like to take this opportunity to thank all the contributing authors who were supportive enough to contribute their time and knowledge to this project. I also wish to convey my regards to my family who have been extremely supportive during the entire project.

Editor

A new species of *Hirudo* (Annelida: Hirudinidae): historical biogeography of Eurasian medicinal leeches

Naim Saglam[1], Ralph Saunders[2], Shirley A. Lang[3] and Daniel H. Shain[2*]

Abstract

Background: Species of *Hirudo* are used extensively for medicinal purposes, but are currently listed as endangered due to population declines from economic utilization and environmental pollution. In total, five species of *Hirudo* are currently described throughout Eurasia, with Turkey being one of the major exporters of medicinal leech, primarily *H. verbana*.

Results: To define the distribution of *Hirudo* spp. within Turkey, we collected 18 individuals from six populations throughout the country. Morphological characters were scored after dorsal and ventral dissections, and Maximum Likelihood (ML) and Bayesian Inference (BI) analyses resolved phylogenetic relationships using mitochondrial cytochrome *c* oxidase subunit I (COI), 12S ribosomal RNA (rRNA), and nuclear 18S rRNA gene fragments. Our results identify a new species of medicinal leech, *Hirudo sulukii* n. sp, in Kara Lake of Adiyaman, Sülüklü Lake of Gaziantep and Segirkan wetland of Batman in Turkey. Phylogenetic divergence (e.g., 10–14 % at COI), its relatively small size, unique dorsal and ventral pigmentation patterns, and internal anatomy (e.g., small and pointed atrium, medium-sized epididymis, relatively long tubular and arc formed vagina) distinguish *H. sulukii* n. sp. from previously described *Hirudo* sp.

Conclusions: By ML and BI analyses, *H. sulukii* n. sp. forms a basal evolutionary branch of Eurasian medicinal leeches. Phylogeographic interpretations of the genus identify a European *Hirudo* "explosion" during the upper Miocene followed by geological events (e.g., Zanclean flood, mountain building) that likely contributed to range restrictions and regional speciation of extant members of the clade.

Background

Hirudinid leeches are parasitic to a variety of vertebrates leading many to regard them with distaste, but their medicinal utility is well established. For centuries, *Hirudo medicinalis* and related species (e.g., *H. verbana, H. troctina*) were prescribed to treat virtually every human ailment from arthritis to yellow fever, most without efficacy. In 1830, during their peak usage, a Paris hospital employed more than five million medicinal leeches [30]. Consequently, populations of *H. medicinalis* in Central Europe were depleted, and non-sustainable collecting led to their extinction in many areas. Pollution and habitat drainage further

added to their decline, forcing Europe to import medicinal leeches from the Ottoman Empire (Anatolia), North Africa and Russia [31] to meet demand. By the late 1900's, the advent of "modern" medicine drastically reduced clinical demand for leeches, allowing some threatened populations to rebound.

Leech therapy languished for most of the 20th century, considered "quackery" by mainstream medical practitioners [66], but the discovery of various bioactive compounds in leech saliva [27, 39], and recognition of the leech's superior ability to relieve venous congestion (e.g., [58]), has led to renewed interest in clinical applications. Current fields of employment include reconstructive microsurgery, hypertension, and gangrene treatment [24]. In light of 19th century threats to medicinal leech populations as demand increased, considerable conservation steps were

* Correspondence: dshain@camden.rutgers.edu
[2]Biology Department, Rutgers The State University of New Jersey, 315 Penn Street, Camden, NJ 08102, USA
Full list of author information is available at the end of the article

implemented to ensure their continued availability. Pursuant to these efforts, much confusion resulted regarding the taxonomic status of different morphological forms [18, 28, 56, 65]. Phylogenetic analysis of nuclear and mitochondrial DNA sequences suggest that the genus *Hirudo* is monophyletic [60], and that species or morphological varieties can be readily identified by coloration patterns. Molecular studies have shown that European medicinal leeches, although usually marketed as *H. medicinalis*, comprise a complex of at least three species: *H. orientalis*, the commonly sold *H. verbana* and the relatively rare *H. medicinalis* [4, 37, 54, 55, 60]. Kutschera and Elliott [36] analyzed the behavior of adult *H. medicinalis*, but could not find differences with respect to its sister taxon *H. verbana*. Morphological and molecular data demonstrate that commercially available medicinal leeches are generally not *H. medicinalis* [35, 56, 60], but rather specimens belonging to the Eastern phylogroup *H. verbana* [61, 62], which is predominantly bred in leech farms and used as a modern 'medicinal' stock.

Turkey is rich in wetlands and known to support at least two species of medicinal leech, *H. medicinalis* and *H. verbana*. Prior to ~2000, it was believed that medicinal leeches from Turkey's wetlands were only *H. medicinalis* [21, 31]. Molecular characterization of Turkish leeches was not performed until the turn of the century, however, and leeches from the Kızılırmak and Yesilirmak Deltas on the Black Sea coast, comprising the majority of leech specimens destined for export, have proven to be to *H. verbana* [4, 51, 55].

Mapped localities of all *Hirudo* species show extensive, belt-shaped ranges extending from east to west. To establish the distribution of *Hirudo* species in Turkey, one of the major exporters of medicinal leeches worldwide, we sampled broadly in three representative localities within the western, eastern and southeastern regions of Turkey. Our data identifies a new species for the genus, *H. sulukii* n. sp., that forms a basal evolutionary branch among European medicinal leeches and sheds light on the evolutionary history of the genus.

Methods

Specimen collection and maintenance

Leech specimens collected throughout Turkey (Kara Lake, Beyaz Cesme Marsh, Uluabat Lake, Segirkan wetland, Balik Lake, Sülüklü Lake) were transported to Fırat University, Fisheries Faculty (Elazig, Turkey) and maintained in separate 600 L fiberglass tanks based on collection location. Tank bottoms were elevated with peat soil ~10 cm on one side to create a terrestrial to aquatic continuum. Leeches were fed one adult frog (e.g., *Pelophylax ridibunda*) blood meal per month (others have utilized mammalian blood), and typically survived 2+ years in the laboratory. Specimens were fixed in 70 % ethanol for molecular analysis and some were fixed with 10 % formaldehyde in PBS for dissection. External traits of live specimens were observed by stereomicroscopy. Preserved specimens were dissected dorsally and ventrally, with representative sketches of internal morphology derived directly from the type specimen.

DNA extraction

Tissue samples from live specimens were obtained by placing the leech in a 10 % ethanol sedating solution until it was unresponsive to touch. Approximately half of the caudal sucker was removed with a scalpel, and tissue cuttings were immediately processed using the E.Z.N.A.™ Tissue DNA kit (Omega Bio-Tek) following the manufacturer's instructions. Whenever possible, tissue from postmortem specimens was taken from the caudal sucker to avoid contamination from gut contents.

DNA sequence amplification of target genes

Nuclear 18S rRNA, mitochondrial 12S rRNA and partial cytochrome *c* oxidase subunit 1 (COI) DNA fragments were amplified from genomic DNA using the polymerase chain reaction (PCR). All 12S sequences were obtained under conditions described by Borda and Siddall [8]. PCR amplification protocols were conducted as described by Wirchansky and Shain [67] employing primers listed in Table 1. PCR products were purified using the Wizard SV Gel and PCR Clean-Up System kit (Promega, Inc.) according to the manufacturer's protocol.

Table 1 Primers used for PCR amplification and DNA sequencing

Gene	Primer name	Primer sequence	Reference
18S rDNA	C	5'- CGGTAATTCCAGCTCCAATAG -3'	Apakupakul et al. (1999) [4]
	Y	5'-CAGACAAATCGCTCCACCAAC -3'	Apakupakul et al. (1999) [4]
12S rDNA	12S-A	5'-AAACTAGGATTAGATACCCTATTAT-3'	Palumbi, 1996 [44]
	12S-B	5'-AAGAGCGACGGGCGATGTGT-3'	Simon et al. [57]
CO1	LCO1490	5'-GGTCAACAAATCATAAAGATATTGG-3'	Folmer et al. [20]
	HCO2198	5'-TAAACTTCAGGGTGACCAAAAAATCA-3'	Folmer et al. [20]

DNA sequencing and editing

Purified PCR products were shipped to GeneWiz, Inc. (South Plainfield, NJ) for Sanger DNA sequencing using an ABI 3730xl DNA analyzer. Each PCR product was sequenced in both directions using amplification primers, and sequence chromatograms were viewed and manually adjusted in ChromasPro (Technelysium, Queensland, Australia) or BioEdit [26]. Sequence alignments were made with MUSCLE [17] or CLUSTAL W [29, 38]. Accession numbers for all CO1, 12S and 18S sequences are listed in Suppl. Data (Table 1).

Phylogeny

Maximum-likelihood (ML) analyses were performed for all DNA comparisons, using the pipeline sequence MEGA 7 [34] to align corresponding sequences from multiple individuals or homologous DNA across species, Gblocks [9] for alignment curation, PhyML [25] for tree building and TreeDyn [11] for tree drawing, as configured in the Phylogeny.fr platform [14]. The aLRT statistical test (approximation of the standard Likelihood Ratio Test; [3]) embedded in PhyML determined branch support values. Default settings were used for all parameters.

Bayesian Inference (BI) analysis was performed on the combined data set (morphological parameters, 18S, 12S, COI in Nexus format) in MrBayes v. 3.2.1x64 [48, 49]. Data were partitioned for 18S and 12S, and by codon position for COI. ModelTest [47] via FindModel was used to determine the optimal model of evolution for each gene under the Akaike Information Criterion (AIC; [46]). The general time reversible (GTR) model with a gamma distributed rate parameter was used for COI, 12S and 18S. Two analyses were run simultaneously with all parameter sets unlinked by partition for two million generations each, sampling every 100 generations, with a burn-in achieved by <50,000 generations. Setting the burn-in to 500,000 generations left a total of 7413 trees sampled for assessment of posterior probabilities. Gaps were treated as missing data, and default settings were used for all other parameters.

Results

Specimens of Hirudo were collected from multiple locations in Turkey (Fig. 1; Tables 2 and 3). These localities are separated by 1312 km (Uluabat Lake to Kara Lake), 1306 km (Uluabat Lake to Beyaz Cesme Marsh) and 289 km (Kara Lake to Beyaz Cesme Marsh). Leeches were typically found in muddy bottoms, as well as underwater and in aquatic/terrestrial vegetation (typically reedbeds), with banks of water proving the most prevalent habitat.

Specimens were scored for morphological characters according to Borda and Siddall [8], Utevsky and Trontelj [65], Klemm [33], Sawyer [53], Nesemann and Neubert [42], Saglam [50] and Govedich et al. [23], Elliott and Dobson [19] (Additional file 1). By these criteria, 10 leeches were identified as H. verbana, while

Fig. 1 Locations of field sites (*small circles*) in Turkey from where *Hirudo* specimens were collected. See Tables 2 and 3 for geographic coordinates

Table 2 Collection field sites in Turkey and specimen descriptions. Depositions in the Academy of Natural Sciences, Philadelphia, PA (ANSP) and Cukurova University Parasitology Museum, Adana, Turkey (CUPM)

Locality	Province	Designation	Catalogue number	Type	Coordinates	Elev.
Hirudo sulukii n. sp.						
Kara Lake	Adiyaman	HS1	CUPM-HIR/2016-1	Para	37°59'35"N 38°48'52"E	1233 m
		HS2	CUPM-HIR/2016-2	Para		
Sülüklü Lake	Gaziantep	HS3	CUPM-HIR/2016-3	Para	37°18'12" N 37°14'53"E	877 m
		HS4	ANSP G1 19489	Para		
		HS5	ANSP G1 19488	Holo		
Segirkan wetland	Batman	HS6	CUPM-HIR/2016-4	Para	37°51'46"N 41°01'00"E	525 m
		HS7	CUPM-HIR/2016-5	Para		

six specimens did not match characters described for any known *Hirudo* species. Specifically, external pigmentation was unique, along with internal distinctions of the epididymis and vagina (see below).

Hirudo sulukii n. sp

Based on morphological and genetic criteria, we formally propose the new species designation, *Hirudo sulukii* n. sp. (LSID: *urn:lsid:zoobank.org:act:C338A26A-A205-4894-AB01-AA012293DD25*), for leech specimens collected near Adiyaman, Batman and Gaziantep in southeastern Anatolia (Tables 2 and 3). The name "sulukii" is derived from the Turkish word "sülük" in reference to "leech". Description based on holotype (specimen HS5 from Sülüklü Lake, catalogue ANSP G1 19488 in the Academy of Natural Sciences, Philadelphia, PA, USA). Paratypes deposited in the Academy of Natural Sciences (ANSP G1 19489) and Cukurova University Parasitology Museum, Adana, Turkey (CUPM-HIR/2016-1). Description: adult 64.06 ± 23.06 mm (27–105 mm) mean long, 6.71 ± 2.61 mm (4–12 mm) mean wide, mean width of anterior sucker 3.36 ± 1.10 mm (2–5.2 mm), mean width of posterior sucker 4.53 ± 1.33 mm (2–7 mm) (Fig. 2). Dorsum (Figs. 3 and 4a) pigmentation variably olive green, two orange paramedian stripes thin, two orange paramarginal stripes broad and encompassing black, segmentally-arranged united ellipsoid and elongated spots, dorsal lateral margins of body

with yellow stripes encompassing zigzagged black longitudinal; covered with numerous papillae of body surface; background pigmentation of ventral (Figs. 3 and 4b) surface light greenish and covered with small number irregular dark markings. With classic *Hirudo* arc eyespot pattern [53], containing five pairs bilateral eyespots. Eyespots, five pairs on II, III, IV a1, V a1 and VI a2, forming a parabolic arc (Fig. 5). Number of annuli per somite: I-II-III: one, IV-V: two, VI-VII: three, VIII: four, IX: five (b1, b2, a2, b5, b6). Gonopores situated in furrow between annuli, separated by five annuli, male pore in the furrow XI b5/b6, female pore in the furrow XII b5/b6. Jaws trignathous, monostichodont, papillated.

Male reproductive apparatus notably large, with thick muscular penis sheath terminating in a bulbous prostate, located at ganglion in segment XI. Epididymis medium-sized, spherical, more than twice size of pearlescent-sheened ejaculatory bulb, tightly packed masses of ducting standing upright on either side of the atrium. Testisacs ovoid and larger than ovisacs, located posterior to ganglion in segment XIII. Female reproductive system relatively coiled tubing. The pearlescent-sheened vagina long and upright, evenly bowed tube entering directly into ventral body wall. Oviducts a thin duct forming several coiled and covered with a thick layer of glandular tissue, bi-lobed ovaries. Ovisacs globular ovoid or small bean seed-shaped (Fig. 6).

Table 3 Collection field sites in Turkey and specimen descriptions. Depositions in the Academy of Natural Sciences, Philadelphia, PA (ANSP) and Cukurova University Parasitology Museum, Adana, Turkey (CUPM)

Locality	Province	Designation	Coordinates	Elev.
Hirudo verbana				
Beyaz Cesme Marsh	Elazig	HV	38°59'51"N 39°55'48"E	1225 m
Uluabat Lake	Bursa	HV1, HV2, HV3, HV4, HV5, HV6, HV7	40°10'23"N 28° 37'26"E	4 m
Balik Lake	Samsun	HV19, HV20a	41°34'48"N 36° 04'30"E	0 m

Fig. 2 Dorsal (**a**) and ventral (**b**) views of *Hirudo sulukii* n. sp. Holotype HS5 from Sülüklü Lake, Turkey (catalogue ANSP G1 19488)

Remarks

Despite similarities between *Hirudo sulukii* n. sp. and other *Hirudo* species, the former can be distinguished from its closest relatives using internal and external features. *Hirudo sulukii* n. sp. differs from *H. medicinalis* and *H. orientalis* by the form of black spots on the dorsal, paramedian stripes of the body. *Hirudo sulukii* n. sp. has black, segmentally-arranged united ellipsoid and elongated spots, and dorsal lateral margins of body a pair of zigzagged black dorsolateral longitudinal stripes (Fig. 4a). The ventral coloration pattern of *H. sulukii* n. sp. has a variable number of irregular spots (Fig. 4b); *H. orientalis* has black, dorsal rounded or quadrangular spots while *H. medicinalis* has elongated spots. The marginal spots of *H. medicinalis* are fused to form distinct black stripes. The ventral of *H. medicinalis* has an irregular dark mesh-like pattern while that of *H. orientalis* is more regular, formed by segmentally-arranged pairs of light markings on a predominantly black background. *Hirudo verbana* has broad, diffuse paramedian stripes orange in color. The ventral pattern of *H. verbana* is unicolored greenish to yellow, bounded by a pair of

black ventrolateral stripes. *Hirudo troctina* has a pair of zigzag-shaped, black ventrolateral longitudinal stripes [65]. Hechtel and Sawyer [28] considered external pigmentation to be not only the most useful, but also arguably the best character to distinguish species of *Hirudo*.

In this study we used the approach of Hechtel and Sawyer [28] and Utevsky and Trontelj [65] regarding the size of the epididymis in relation to the ejaculatory duct. The epididymes of *Hirudo sulukii* n. sp. (Fig. 6) and *H. orientalis* are medium-sized. In contrast, the epididymes of *H. verbana* are relatively small, whereas *H. troctina* and *H. medicinalis* have massive epididymes [65]. The vagina of *Hirudo sulukii* n. sp. is relatively long tubular and arc formed (Fig. 6), while in *H. orientalis* the vagina is tubular and evenly curved. The former two species do not show the central swelling and sharp folding typical for *H. verbana*. In *H. medicinalis*, the vagina can have two conditions: straight and tubular, or terminally curved [65]. *Hirudo troctina* has a bulbous vagina [28].

Moquin-Tandon [40] described at least five species of *Hirudo* including *H. verbana* and *H. medicinalis*, but later concluded that they were all varieties of the same leech species. The medicinal leech, *H. sulukii* n. sp., considered here was determined to be morphologically different than all species described by Moquin-Tandor [40, 41].

Phylogenetic analyses

To determine the relationship of specimens to other *Hirudo* species, we subjected them to the comparative analysis of CO1 (cytochrome c oxidase subunit 1) and 12S rRNA from mitochondria, and nuclear 18S rRNA. Combined COI, 12S and 18S rRNA analysis contained 13 terminals with 1514 aligned characters. Maximum Likelihood of the combined data set yielded five equally parsimonious trees with 500–1000 steps (Fig. 7; Additional file 1); concordant trees were generated independently with COI data (Fig. 8; Additional file 1). Collectively, *H. sulukii* n. sp., formed a basal branch among European medicinal leeches with strong bootstrap support, while resolution among *H. medicinalis*, *H. orientalis* and *H. verbana* lineages was ambiguous, as noted in previous studies [45, 56]. Population structure was shallow among the collected specimens (<2 % divergence at CO1; Table 4), suggesting recent invasions into field sites sampled in the current study (see Fig. 1). The Asian species, *H. nipponia*, fell outside the *Hirudo* clade in combined sequence analyses (Fig. 7), suggesting a deep ancestral split with European species, and calling into question the designation of *H. nipponia* within the *Hirudo* phylogroup. Interestingly, *H. nipponia* was equidistant to European *Hirudo* species (~22–25 % at CO1), the latter of which were approximately equidistant to each other (i.e., ~10–14 % at CO1; Table 4). Inferring a divergence rate of ~2 % per million years at the CO1

Hirudo specimens	Dorsal view	Ventral view
HS1		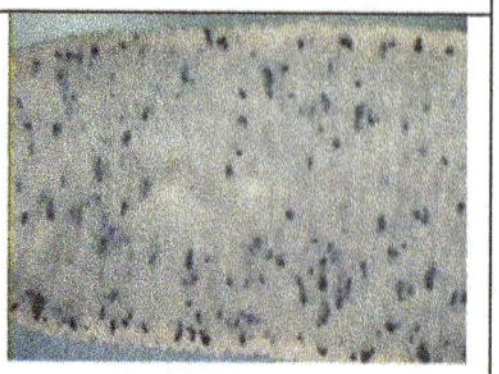
HS2		
HV		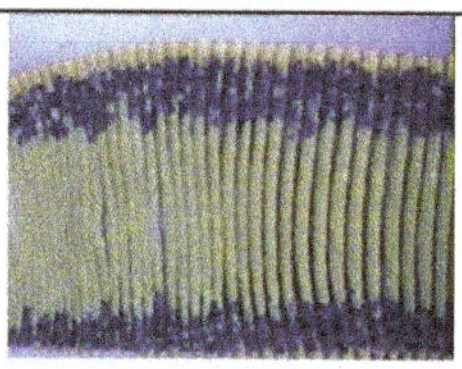
HV1	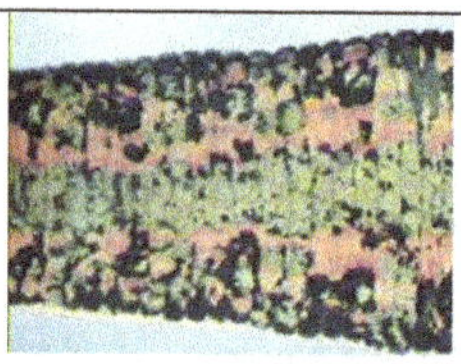	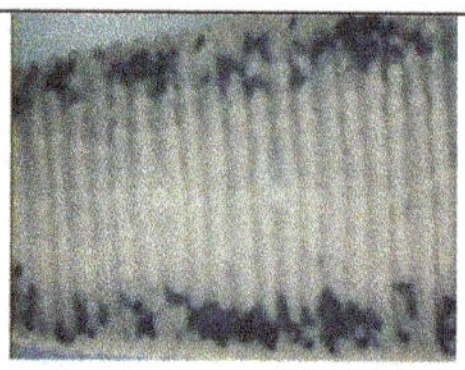
HV4	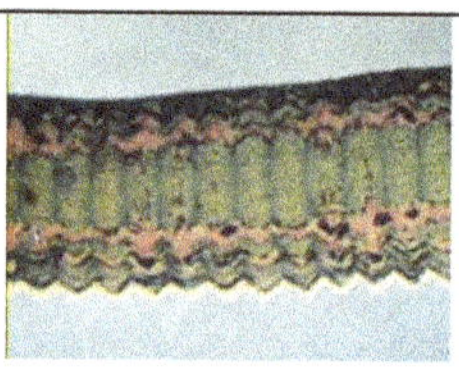	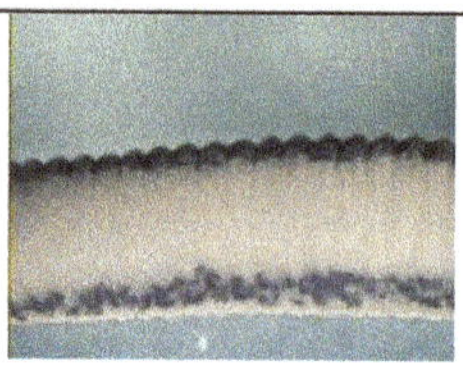

Fig. 3 Pigmentation patterns of representative *Hirudo* specimens. HS1, 2 – *H. sulukii n. sp.*; HV, HV1, 4 – *H. verbana*. See Tables 2 and 3 for specimen descriptions

locus based on combined geological and molecular data within Oligochaeta [10, 15, 67], we estimate a lower Miocene split between lineages leading to *H. nipponia* and European *Hirudo sp.*, and radiation of the latter species during the upper Miocene. Branch patterns of remaining species were consistent with those reported previously [45].

Discussion

Maximum Likelihood and Baysian Inference analyses yielded trees with concordant topologies and strong support for *H. sulukii* as a basal branch of the European medicinal leeches. Relationships between *H. medicinalis*, *H. verbana* and *H orientalis* were less conclusive, consistent with confusion regarding their morphological identification [45, 56]. The relatively small size of *H. sulukii*, unique dorsal and ventral pigmentation patterns,

and internal anatomy (e.g., small and pointed atrium, medium-sized epididymis, relatively long tubular and arc formed vagina) are distinguishing features of this previously undescribed leech. Note that *H. sulukii* has thus far been collected only from relatively high elevation field sites (i.e., Kara Lake-Adiyaman 1233 m, Sülüklü Lake-Gaziantep 877 m, and Segirkan wetland- Batman 525 m), and its small size in comparison with other *Hirudo* species may reflect an adaptation to this environment (e.g., reduced foraging season/food supply), as suggested for other annelid species (e.g., [15]).

Previously, only two medicinal leeches were thought to occur in Turkey, *H. verbana* and *H. medicinalis*, while a total of five are currently described throughout Eurasia. The range of *H. verbana* occurs to the south of *H. medicinalis* in an almost parapatric fashion with little overlap [5, 32, 42, 43, 51]. The former is subdivided into

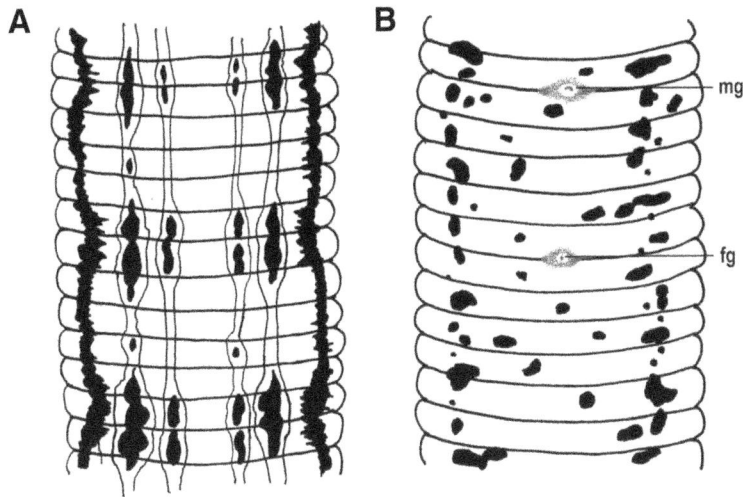

Fig. 4 *Hirudo sulukii* n. sp. Dorsal view (**a**) and ventral view (**b**); mg, male gonopore; fg, female gonopore. Based on holotype HS5 from Sülüklü Lake, Turkey (catalogue ANSP G1 19488)

an Eastern (southern Ukraine, North Caucasus, Turkey and Uzbekistan) and Western phylogroup (Balkans and Italy) that do not overlap, suggesting distinct postglacial colonization from separate refugia [61, 64]. Easternmost records are from Samarqand Province in Uzbekistan [61, 64, 65], resulting in an east-to-west extent of ~4600 km. Leeches supplied by commercial facilities belong to the Eastern phylogroup, originating mostly from Turkey and the Krasnodar Territory in Russia, two leading areas of leech export.

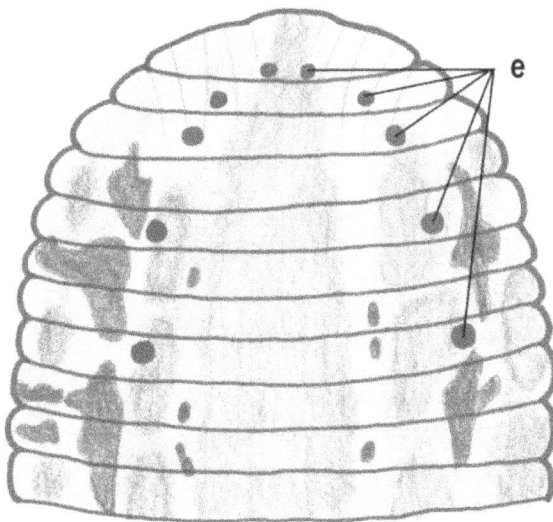

Fig. 5 View of anterior-dorsal and eyes of *Hirudo sulukii* n. sp. e, eyes. Based on holotype HS5 from Sülüklü Lake, Turkey (catalogue ANSP G1 19488)

Hirudo medicinalis is distributed from Britain and southern Norway to the southern Urals and probably as far as the Altai Mountains, occupying the deciduous arboreal zone [6, 12, 16, 21, 22, 31, 43, 51, 52, 59, 63, 68]. *Hirudo orientalis* is associated with mountainous areas in the sub-boreal eremial zone and occurs in Transcaucasian countries, Iran and Central Asia, while *H. troctina* has been found in northwestern Africa and Spain in the Mediterranean zone [64]. *Hirudo verbana* and *H. medicinalis* have recently experienced range expansions while *H. orientalis* has remained geographically isolated within arid and alpine areas of Central Asia and Transcaucasia [61].

By molecular clock inference using divergence estimates at the CO1 locus [10, 15, 67], our data suggest a deep, ancestral split between European and Asian (i.e., the lineage leading to *H. nipponia*) medicinal leeches somewhere in the lower Miocene, followed by an "explosion" of *Hirudo* species upon their putative arrival to the European continent during the upper Miocene, 5–10 mya (Fig. 9). The possible misclassification of *H. nipponia* does not affect this evolutionary scenario since it represents a basal, sister branch to the European *Hirudo* phylogroup (see Fig. 8). This evolutionary timeline is supported by tree topologies and relative genetic distance of European *Hirudo* species to each other at the COI locus (i.e., 10–14 % divergence; see Table 4). The time frame of these events suggest the presence of an open habitat corresponding with, for example, formation of Levantine land bridges, which may have facilitated mammalian-based, passive dispersal of an ancestral *Hirudo* archetype throughout Europe. Thereafter, tectonic activity at the onset of the Pliocene ~5.3 mya broke the

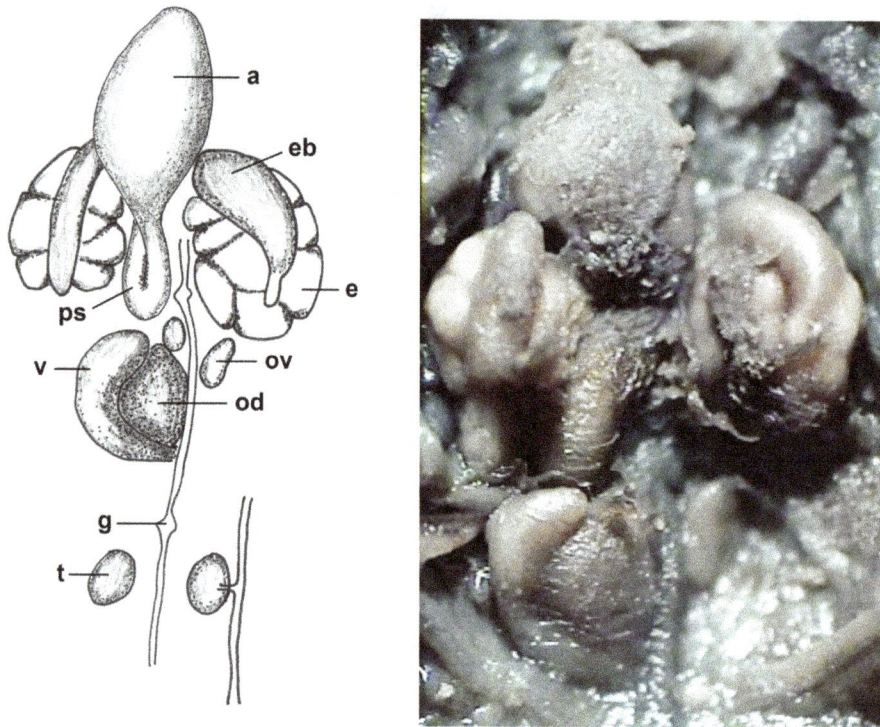

Fig. 6 *Hirudo sulukii* n. sp. Dorsal view of reproductive system. *a*, Atrium; *e*, epididymis; *eb*, ejaculatory bulb; *g*, ganglion; *o*, ovisac; *od*, oviduct; *ps*, penis sheath; *t*, testisac, *v*, vagina. Based on holotype HS5 from Sülüklü Lake, Turkey (catalogue ANSP G1 19488)

land bridge between Morocco and Spain causing the Zanclean Flood that filled the Mediterranean basin, and in combination with mountain building throughout the European continent [7], appears to have restricted panmixia among extant *Hirudo* lineages, leading in part to their speciation and current geographic ranges. For instance, concurrent with the closing of the Tethys Sea by continental drift of the African and Arabian plates, mountain building events occurred in Southern Turkey forming the Taurus Mountain chain [13]. At present, the

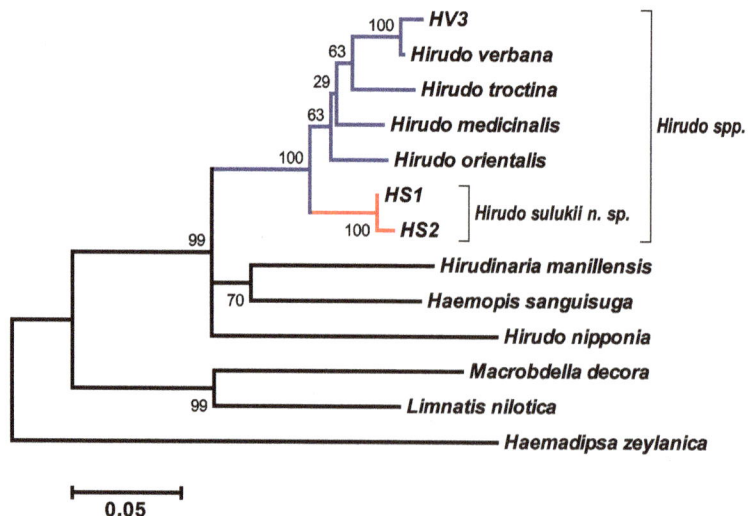

Fig. 7 Maximum Likelihood analysis of a combined COI, 12S and 18S data set (1514 total positions). Bootstrap values are indicated. European *Hirudo* species form a distinct clade with *H. sulukii* n. sp. as a basal member

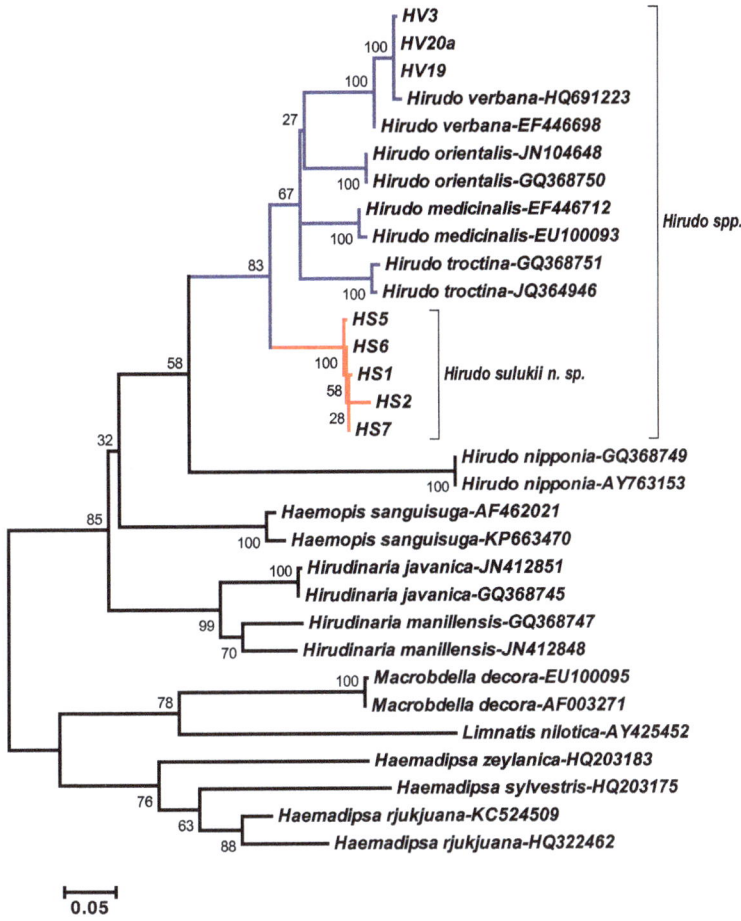

Fig. 8 Maximun Likelihood analysis of COI mtDNA haplotypes (542 total positions). The tree is drawn to scale, with branch lengths measured in the number of substitutions per site

Table 4 Pairwise distance matrix of *Hirudo* specimens

		1	2	3	4	5	6	7	8	9	10	11	12	13
1	HS1	-												
2	HS2	0.0145	-											
3	HS5	0.0036	0.0182	-										
4	HS6	0.0036	0.0182	0.0036	-									
5	HS7	0.0018	0.0163	0.0054	0.0018	-								
6	HV3	0.1238	0.1373	0.1240	0.1259	0.1237	-							
7	HV19	0.1238	0.1373	0.1240	0.1259	0.1237	0.0018	-						
8	HV20a	0.1238	0.1373	0.1240	0.1259	0.1237	0.0018	0.0000	-					
9	H. verbana-HQ691223	0.1237	0.1372	0.1238	0.1258	0.1235	0.0072	0.0054	0.0054	-				
10	H. medicinalis-EU100093	0.1053	0.1142	0.1054	0.1053	0.1031	0.1059	0.1059	0.1059	0.1101	-			
11	H. orientalis-JN104648	0.1153	0.1285	0.1154	0.1153	0.1130	0.0979	0.0979	0.0979	0.1001	0.0875	-		
12	H. troctina-JQ364946	0.1220	0.1330	0.1221	0.1220	0.1197	0.1097	0.1097	0.1097	0.1139	0.0957	0.0918	-	
13	H. nipponia-AY763153	0.2366	0.2495	0.2363	0.2366	0.2394	0.2342	0.2341	0.2341	0.2393	0.2273	0.2513	0.2202	-

Numbers are divergence values within a 560 bp fragment of the cytochrome c oxidase subunit I (COI) locus

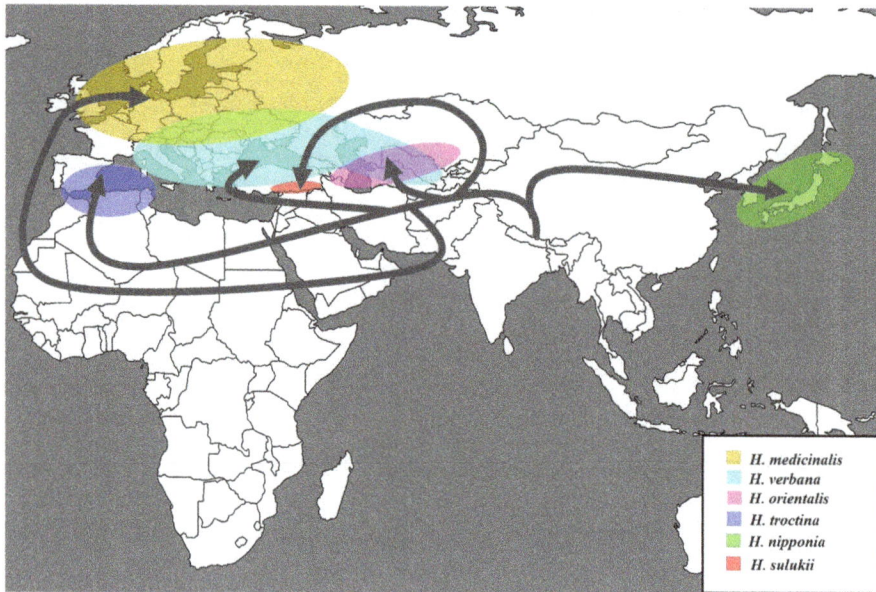

Fig. 9 Proposed biogeographical history of Eurasian *Hirudo* species based on phylogeny and current distribution patterns. Colored regions indicate reported geographic ranges of respective *Hirudo* species; the arrow topology reflects relationships between species based on Maximum Likelihood and Baysian phylogenies

Anatolia region has been isolated completely by the southeastern Taurus Mountains [1], likely isolating *H. sulukii* from other European *Hirudo* populations.

Species of *Hirudo* have had broad applications in medicine, ranging from reconstructive surgeries (e.g., facial, finger reattachment, ear flap) to anticoagulants/analgesics secreted from salivary glands [2, 24]. Thus the discovery of a new *Hirudo* species, particularly a basal member of this phylogroup, has considerable value in the context of medical potential. Specifically, natural variants of known bioactive factors (e.g., hirudin, antistasin, etc.) are logical candidates to explore for their potentially enhanced or novel pharmaceutical properties. The current study has prompted a more systematic survey of *Hirudo* throughout Turkey and surrounding regions with the collective aims of refining the evolutionary history of the genus, facilitating conservation efforts, and identifying species that may expand the repertoire of medicinal applications for this important Hirudinid genus.

Conclusions

By phylogenetic and morphological criteria, specimens collected from Kara Lake of Adiyaman, Sülüklü Lake of Gaziantep and Segirkan wetland of Batman in Turkey comprise a new species, *Hirudo sulukii*. Geographic isolation by the Taurus Mountain chain has likely contained *H. sulukii* within the regional sampling area. By ML and BI analyses, *H. sulukii* n. sp. forms a basal evolutionary branch of Eurasian medicinal leeches, preceded by a

deeper ancestral split with the Asian medicinal leech. *H. nipponia*. Phylogeographic interpretations of the genus identify a European *Hirudo* "explosion" during the upper Miocene followed by geological events (e.g., Zanclean flood, mountain building) that likely contributed to range restrictions and regional speciation of extant members of the European clade.

Acknowledgements
We thank Mark Siddall for helpful comments.

Funding
Supported by The Scientific and Technological Research Council of Turkey (TUBITAK) to NS, and Busch Biomedical and GAIA grants to DHS.

Authors' contributions
NS collected leech specimens, conducted experimental analyses including specimen dissections, and contributed to writing the manuscript; RS conducted DNA analyses and contributed to writing; SAL assisted with DNA analyses, phylogenies and writing; DHS oversaw experimental analyses and writing of the manuscript. All authors read and approved the final manuscript.

Competing interests
The authors declare that they have no competing interests.

Endnotes
Not applicable

Author details
[1]Department of Aquaculture and Fish Diseases, Fisheries Faculty, Firat University, 23119 Elazig, Turkey. [2]Biology Department, Rutgers The State University of New Jersey, 315 Penn Street, Camden, NJ 08102, USA. [3]Rowan University Graduate School of Biomedical Sciences at SOM, Stratford, NJ 08084, USA.

References

1. Altinli E. Geology of Eastern and Southeastern Anatolia. Bull Mineral Res Explor Inst Turk. 1966;66:35–76.

2. Abdualkader AM, Ghawi AM, Alaama M, Awang M, Merzouk A. Leech Therapeutic Applications. Indian J Pharm Sci. 2013;75:127–37.

3. Anisimova M, Gascuel O. Approximate likelihood-ratio test for branches: A fast, accurate, and powerful alternative. Syst Biol. 2006;55:539–52.

4. Apakupakul K, Siddall ME, Burreson EM. Higher level relationships of leeches (Annelida: Clitellata: Euhirudinea) based on morphology and gene sequences. Mol Phylogenet Evol. 1999;12:350–9.

5. Balık S, Ustaoğlu MR, Sarı HM, Özdemir Mis D, Aygen C, Taşdemir A, Yıldız S, Topkara ET, Sömek H, Özbek M, İlhan A. A preliminary study on the biological diversity of Bozalan Lake (Menemen- İzmir). E.U. J Fish Aquat Sci. 2006;23:291–4.

6. Bat L, Akbulut M, Culha M, Sezgin M. The macrobenthic fauna of Sırakaraağaçlar Stream flowing into the Black Sea at Akliman, Sinop. Turk J Mar Sci. 2000;6:71–86.

7. Blondel J, Aronson J, Bodiou JY, Boeuf G. The Mediterranean region, biological diversity in space and time. 2nd ed. New York: Oxford University Press Inc; 2010. p. 376.

8. Borda E, Siddall ME. Arhynchobdellida (Annelida: Oligochaeta: Hirudinida): phylogenetic relationships and evolution. Mol Phylogenet Evol. 2004;30: 213–25.

9. Castresana J. Selection of conserved blocks from multiple alignments for their use in phylogenetic analysis. Mol Biol Evol. 2000;17(4):540–52.

10. Chang CH, Lin SM, Chen JH. Molecular systematics and phylogeography of the gigantic earthworms of the Metaphire formosae species group (Clitellata, Megascolecidae). Mol Phylogenet Evol. 2008;49:958–68.

11. Chevenet F, Brun C, Banuls AL, Jacq B, Chisten R. TreeDyn: towards dynamic graphics and annotations for analyses of trees. BMC Bioinformatics. 2006;7:439.

12. Demirsoy A, Kasparek M, Akbulut A, Durmus Y, Emir Akbulut N, Çaliskan M. Phenology of the medicinal leech, Hirudo medicinalis L. in north-western Turkey. Hydrobiologia. 2001;462:19–24.

13. Dercourt J, Zonenshain LP, Ricou LE, Kazmin VG, Le Pichon X, Knipper AL, Grandjacquet C, Sbortshikov IM, Geyssant J, Lepvrier C, Pechersky DH, Boulin J, Sibuet J-C, Savostin LA, Sorokhtin O, Westphal M, Bazhenov ML, Lauer JP, Biju Duval B. Geological evolution of the Tethys belt from the Atlantic to the Pamirs since the Lias. Tectonophysics. 1986;123:241–315.

14. Dereeper A, Guignon V, Blanc G, Audic S, Buffet S, Chevenet F, Dufayard JF, Guindon S, Lefort V, Lescot M, Claverie JM, Gascuel O. Phylogeny.fr: robust phylogenetic analysis for the non-specialist. Nucleic Acids Res. 2008;36: W465–9.

15. Dial CR, Dial RJ, Saunders R, Lang SA, Tetreau MD, Lee B, Wimberger P, Dinapoli MS, Egiazarov AS, Gipple SL, Maghirang MR, Swartley-McArdle DJ, Yudkovitz SR, Shain DH. Historical biogeography of the North American glacier ice worm, Mesenchytraeus solifugus (Annelida: Oligochaeta: Enchytraeidae). Mol Phylogenet Evol. 2012;63:577–84.

16. Duran M, Akyıldız GK, Özdemir A. Gökpınar Çayı'nın Büyük Omurgasız Faunası ve Su Kalitesinin Değerlendirilmesi. Türk Sucul Yaşam Dergisi. 2007; 5:577–83.

17. Edgar RC. MUSCLE: multiple sequence alignment with high accuracy and high throughput. Nucleic Acids Res. 2004;32(5):1792–7.

18. Elliott JM, Kutschera U. Medicinal leeches: historical use, ecology, genetics and conservation. Freshw Rev. 2011;4:21–41.

19. Elliott JM, Dobson M. Freshwater Leeches of Britain and Ireland. Keys to the Hirudinea and a Review of their Ecology. Freshwater Biological Association Scientific Publication No: 69. 2015. p. 1–108.

20. Folmer O, Black M, Hoeh W, Lutz R, Vrijenhoek R. DNA primers for amplification of mitochondrial cytochrome c oxidase subunit I from diverse metazoan invertebrates. Mol Mar Biol Biotechnol. 1994;3:294–9.

21. Geldiay R. Çubuk Barajı ve Emir Gölünün Makro ve Mikro Faunasının Mukayeseli İncelenmesi. Ankara Üniv Fen Fak Mecm. 1949;2:106.

22. Geldiay R, Tareen IU. Bottom Fauna of Gölcük Lake. 1. Population study of Chironomids, Chaoborus and Oligochaeta. İzmir: E.Ü.F.F. İlmi Raporlar Serisi No:137; 1972. p. 15.

23. Govedich FR, Bain BA, Moser WE, Gelder SR, Davies RW, Brinkhurst RO. Annelida (Clitellata): Oligochaeta, Branchiobdellida, Hirudinida, and Acanthobdellida. In: Thorp JH, Covich AP, editors. Ecology and classification of North American freshwater invertebrates 3rd edition, Academic press. 2009. p. 385–410.

24. Gödekmerdan A, Arusan S, Bayar B, Sağlam N. Tıbbi Sülükler ve Hirudoterapi. Turkiye Parazitol Derg. 2011;35:234–9. doi:10.5152/tpd.2011.60.

25. Guindon S, Gascuel O. A simple, fast, and accurate algorithm to estimate large phylogenies by maximum likelihood. Syst Biol. 2003;52(5):696–704.

26. Hall TA. BioEdit: a user-friendly biological sequence alignment editor and analysis program for Windows 95/98/NT. Nucleic Acids Symp Ser. 1999;41:95–8.

27. Haycraft JB. On the action of secretion obtained from the medicinal leech on coagulation of the blood. Proc R Soc Lond. 1884;36:478.

28. Hechtel FOP, Sawyer RT. Toward a taxonomic revision of the medicinal leech Hirudo medicinalis Linnaeus, 1758 (Hirudinea: Hirudinidae): re-description of Hirudo troctina Johnson, 1816 from North Africa. J Nat Hist. 2002;36(11):1269–89.

29. Higgins D, Thompson J, Gibson T, Thompson JD, Higgins DG, Gibson TJ. CLUSTAL W: improving the sensitivity of progressive multiple sequence alignment through sequence weighting, position-specific gap penalties and weight matrix choice. Nucleic Acids Res. 1994;22:4673–80.

30. Kaestner A. Invertebrate zoology, vol. I. New York: Interscience; 1967. p. 597.

31. Kasparek M, Demirsoy A, Akbulut A, Emir Akbulut N, Çaliskan M, Durmus Y. Distribution and status of the medicinal leech (Hirudo medicinalis L.) in Turkey. Hydrobiologia. 2000;441:37–44.

32. Kazancı N, Ekingen P, Türkmen G. A study on Hirudinea fauna of Turkey and habitat quality of the species. Rev Hydrobiol. 2009;1:81–95.

33. Klemm DJ. The leeches (Annelida: Hirudinea) of North America. Cincinnati: Aquatic Biology Section, Environmental Monitoring and Support Laboratory, United States Environmental Protection Agency; 1982.

34. Kumar S, Stecher G, Tamura K. MEGA7: molecular evolutionary genetics analysis version 7.0 for bigger datasets. Mol Biol Evol. 2016;33:1870–4. doi:10.1093/molbev/msw054

35. Kutschera U. Leeches underline the need for Linnaean taxonomy. Nature. 2007;447:775.

36. Kutschera U, Elliott JM. The European medicinal leech Hirudo medicinalis L.: morphology and occurrence of an endangered species. Zoosyst Evol. 2014;91:271–80. doi:10.3897/zse.90.8715.

37. Kvist S, Oceguera-Figueroa A, Siddall ME, Erseus C. Barcoding, types and the Hirudo files: using information content to critically evaluate the identity of DNA barcodes. Mitochondrial DNA. 2010;21:198–205. doi:10.3109/19401736. 2010.529905.

38. Larkin MA, Blackshields G, Brown NP, Chenna R, McGettigan PA, McWilliam H, Valentin F, Wallace IM, Wilm A, Lopez R, Thompson JD, Gibson TJ, Higgins DG. ClustalW and ClustalX, version 2. Bioinformatics. 2007;23(21):2947–8.

39. Markwardt F. Untersuchungen über Hirudin. Naturwissenschaften. 1955;52:537.

40. Moquin-Tandon A. Monographie de la Famille des Hirudinees. Montpellier: Maison de Commerce; 1827.

41. Moquin-Tandon A. Monographie de la Famille des Hirudinees. Paris: Balliere; 1846.

42. Nesemann H, Neubert E. Annelida: Clitellata: Branchiobdellida, Acanthobdellea, Hirudinea. In: Süßwasserfauna von Mitteleuropa, 6/2. Heidelberg, Berlin: Spektrum Akad Verl; 1999.

43. Özbek M, Sarı HM. Batı Karadeniz Bölgesi'ndeki Bazı Göllerin Hirudinea (Annelida) Faunası. E.Ü. Su Ürünleri Dergisi. 2007;24:83–8.

44. Palumbi SR. Nucleic acids II: the polymerase chain reaction. In: Hillis DM, Moritz C, Mable BK, editors. Molecular systematics. Sunderland: Sinauer & Associates Inc.; 1996. p. 205–47.

45. Phillips AJ, Siddall ME. Poly-paraphyly of Hirudinidae: many lineages of medicinal leeches. BMC Evol Biol. 2009;9:246. doi:10.1186/1471-2148-9-246.

46. Posada D, Buckley TR. Model selection and model averaging in phylogenetics: advantages of Akaike Information Criterion and Bayesian approaches over likelihood ratio tests. Syst Biol. 2004;53:793–808.

47. Posada D, Crandall KA. MODELTEST: testing the model of DNA substitution. Bioinformatics. 1998;14:817–8.

48. Ronquist F, Huelsenbeck J, Teslenko M. Draft MrBayes version 3.2 manual: tutorials and model summaries. 2011. p. 172.

49. Ronquist F, Huelsenbeck JP. MrBayes 3: Bayesian phylogenetic inference under mixed models. Bioinformatics. 2003;19:1572–4.

50. Saglam N. Key of freshwater and marine leeches. Elazığ, Turkey: Fırat Üniversitesi Basım Evi; 2004. p. 38.

51. Saglam N. Protection and sustainability, exportation of some species of medicinal leeches (*Hirudo medicinalis* L., 1758 and *Hirudo verbana* Carena, 1820). J FisheriesSciencescom. 2011;5(1):1–15.

52. Saglam N, Dorucu M, Ozdemir Y, Seker E, Sarieyyupoglu M. Distribution and economic importance of medicinal leech, *Hirudo medicinalis* (Linnaeus, 1758) in Eastern Anatolia/Turkey. Lauterbornia. 2008;65:105–18.

53. Sawyer RT. Leech biology and behavior. New York: Oxford University Press; 1986.

54. Siddall ME, Apakupakul K, Burreson EM, Coates KA, Erséus C, Gelder SR, Källersjö M, Trapido-Rosenthal H. Validating livanov: molecular data agree that leeches, branchiobdellidans, and acanthobdella peledina form a monophyletic group of oligochaetes. Mol Phylogenet Evol. 2001;21:346–51.

55. Siddall ME, Burreson EM. Phylogeny of leeches (Hirudinea) based on mitochondrial cytochrome c oxidase subunit I. Mol Phylogenet Evol. 1998;9: 156–62.

56. Siddall ME, Trontelj P, Utevsky SY, Nkamany M, Macdonald KS. Diverse molecular data demonstrate that commercially available medicinal leeches are not *Hirudo medicinalis*. Proc R Soc B Biol Sci. 2007;274:1481–7.

57. Simon C, Paabo S, Kocher TD, Wilson AC. Evolution of mitochondrial ribosomal RNA in insects as shown by the polymerase chain reaction. In: Clegg M, Clark S, editors. Molecular evolution, U.C.L.A. Symposia on molecular and CellularBiology, New series, vol. 122. New York: Alan R. Liss, Inc; 1990. p. 235–44.

58. Soucacos PN, Beris QE, Malizos KN, Xenakis TA, Georgoulis A. Successful treatment of venous congestion in free skin flaps using medical leeches. Microsurgery. 1994;15(7):496–501.

59. Taşdemir A, Yıldız S, Topkara ET, Özbek M, Balık S, Ustaoğlu MR. Benthic fauna of Yayla Lake (Buldan-Denizli). Türk Sucul Yaşam Dergisi. 2004;2:182–90.

60. Trontelj P, Utevsky SY. Celebrity with a neglected taxonomy: molecular systematics of the medicinal leech (genus *Hirudo*). Mol Phylogenet Evol. 2005;34:616–24.

61. Trontelj P, Utevsky SY. Phylogeny and phylogeography of medicinal leeches (genus *Hirudo*): Fast dispersal and shallow genetic structure. Mol Phylogenet Evol. 2012;63:475–85.

62. Trontelj P, Sotler M, Verovnik R. Genetic differentiation between two species of the medicinal leech, *Hirudo medicinalis* and the neglected *H. verbana*, based on random-amplified polymorphic DNA. Parasitol Res. 2004;94:118–24. doi:10.1007/s00436-004-1181-x.

63. Ustaoğlu MR, Balık S, Özbek M, Sarı HM. The Freshwater leeches (Annelida-Hirudinea) of the Gediz catchment area (İzmir region). Zool Middle East. 2003;29:118–20.

64. Utevsky S, Zagmajster M, Atemasov A, Zınenko O, Utevska O, Utevsky A, Trontelj P. Distribution and status of medicinal leeches (genus *Hirudo*) in the Western Palaearctic: anthropogenic, ecological, or historical effects? Aquat Conserv Mar Freshwat Ecosyst. 2010;20:198–210.

65. Utevsky SY, Trontelj P. A new species of the medicinal leech (Oligochaeta, Hirudinida, *Hirudo*) from Transcaucasia and an identification key for the genus *Hirudo*. Parasitol Res. 2005;98:61–6.

66. Whitaker IS, Rao J, Izadi D, Butler PE. Historical Article: Hirudo medicinalis: ancient origins of, and trends in the use of medicinal leeches throughout history. Br J Oral Maxillofac Surg. 2004;42:133–7.

67. Wirchansky BA, Shain DH. A new species of *Haemopis* (Annelida: Hirudinea): Evolution of North American terrestrial leeches. Mol Phylogenet Evol. 2010; 54:226–34.

68. Yıldırım N. Fırnız Çayı (Kahramanmaraş)'nın Fiziko-Kimyasal ve Bazı Biyolojik (Bentik makroinvertebrat) Özellikleri. In: Fen Bilimleri Enstitüsü, Biyoliji Ana Bilim Dalı. Kahramanmaraş: Kahramanmaraş Sütçü İmam Ünivrsitesi; 2006. p. 32.

2

Sex-specific patterns in body mass and mating system in the Siberian flying squirrel

Vesa Selonen[1*], Ralf Wistbacka[2] and Andrea Santangeli[3]

Abstract

Background: Reproductive strategies and evolutionary pressures differ between males and females. This often results in size differences between the sexes, and also in sex-specific seasonal variation in body mass. Seasonal variation in body mass is also affected by other factors, such as weather. Studies on sex-specific body mass patterns may contribute to better understand the mating system of a species. Here we quantify patterns underlying sex-specific body mass variation using a long-term dataset on body mass in the Siberian flying squirrel, *Pteromys volans*.

Results: We show that female flying squirrels were larger than males based on body mass and other body measures. Males had lowest body mass after the breeding season, whereas female body mass was more constant between seasons, when the pregnancy period was excluded. Male body mass did not increase before the mating season, despite the general pattern that males with higher body mass are usually dominant in squirrel species. Seasonal body mass variation was linked to weather factors, but this relationship was not straightforward to interpret, and did not clearly affect the trend in body mass observed over the 22 years of study.

Conclusions: Our study supports the view that arboreal squirrels often deviate from the general pattern found in mammals for larger males than females. The mating system seems to be the main driver of sex-specific seasonal body mass variation in flying squirrels, and conflicting selective pressure may occur for males to have low body mass to facilitate gliding versus high body mass to facilitate dominance.

Keywords: Sexual size dimorphism, Climate change, Scramble competition mating system, Female defence

Background

Reproductive strategies and evolutionary pressures differ between sexes and often lead to sex differences in body size, in body mass as well as in bone measurements [1–3]. The same factors may also lead to sex-specific seasonal variation in body mass, depending on energy expenditure and condition of individuals [4, 5]. Studies on sex-specific seasonal patterns of body mass are relatively scarce [6–9], but may contribute to understand the mating system of a species.

In mammals, lactation is one of the main factors contributing to the seasonal difference in energy expenditure between sexes. In addition, the levels of intra-sexual competition during the breeding season typically differ between

sexes, depending on the mating system of a species [3, 10–13]. An interesting mating system to study intra-sexual patterns in body mass is the so called scramble competition mating system [14]. This mating system has been frequently reported for insects [15], but is poorly studied in mammals, although it may occur, for example, in arboreal squirrels (subfamily Sciurinae; [16, 17]). In a scramble competition mating system, females are solitary and males move to visit different females which may be in oestrus during different days. Within this context, selection may favour males that are effective in locating female territories scattered across the landscape [16, 18].

Mating system and sexual selection are, however, not the only factors shaping differences in body mass between sexes [19]. For example, the seasonality of resource availability (such as food) and weather conditions (e.g. harsh winter periods) may also affect body mass difference between sexes beyond the more commonly recognised

* Correspondence: vessel@utu.fi
[1]Department of Biology, Section of Ecology, FI-20014 University of Turku, Turku, Finland
Full list of author information is available at the end of the article

effect of reproductive strategies [20, 21]. Consequently, climate change can have sex-specific effects on body mass [22], potentially creating temporal changes also in sexual body mass dimorphism of the species.

In this study, we broadly aim to quantify patterns underlying sex-specific body mass variation between and within years in the Siberian flying squirrel, *Pteromys volans* (Linnaeus, 1758) in Central-Western Finland. We used long-term datasets on body mass of individuals spanning 22 years and also a short-term dataset on body measurements. In flying squirrels, the locomotor system of gliding likely places a unique set of selective forces related to body mass in this group [23, 24]. Female flying squirrels live in isolation and males need to rapidly move between females during the mating season [17]. In addition, Siberian flying squirrels, like tree squirrels [25], perform mating chases, whereby a few males glide and run one after another from tree to tree following a female [26]. These behaviours may promote high movement abilities of males in competition with each other, and may favour fast gliders with low body mass. However, within the mating system of tree and flying squirrels, the defence of females by large dominant males may promote high body mass necessary for dominating mating opportunities and prevent smaller subdominant males from reproducing [17, 25, 27]. Earlier studies indicate that flying squirrels may have female-biased sexual dimorphism, with females being larger than males based on body mass [17] and body measurements of few museum specimens [28]. However, these studies are based on a limited number of individuals. In addition, the links between the mating system and the seasonal variation in body mass are still poorly understood in squirrels. For example, conflicting pressures between a defence mating system and a scramble competition mating system [16, 18] may affect seasonal variation in body mass in male flying squirrels. In addition, seasonal sexual dimorphism in body mass may be related to different seasonal patterns in energy expenditure between the sexes.

We predict (i) that seasonal patterns in body mass underlie the specific mating system: if males have highest body mass before the breeding season and lowest body mass after the breeding season, then the female defense mating systems is dominant in this species [2, 9]. Instead, if the scramble competition mating system is operative, then we predict that males should not have the highest body mass before the mating season in order to be fast in mating chases and in locating females. We also predict (ii), against the general pattern in mammal species, that in flying squirrels the females will have a larger body mass (measured outside of the breeding season) and other body measurements than males. Finally, we predict (iii) that weather conditions affect seasonal variation in body mass. Spring weather, which corresponds to the start of the breeding season of flying squirrels and, thus, may have sex-specific

effects on body mass, has significantly warmed in Finland during our study period [29, 30]. Thus, climate change has the potential to affect the 22 year trend in body mass of the two sexes in flying squirrels.

Methods
Study species
The Siberian flying squirrel is a nocturnal arboreal rodent which nests in tree cavities, nest-boxes and dreys (twig nests) in spruce-dominated boreal forests. Flying squirrels feed on deciduous trees within spruce-dominated forests. The mating season starts in mid-March and the first litter is born in late April [26]. Females can sometimes have a second litter which is born in late June. Females are territorial and live in non-overlapping home ranges (on average 4 ha in size), whereas males have much larger home ranges (average size of 60 ha) that can overlap with several other male and female home ranges [17]. The movement activity of males increases during the mating season when males actively move between territories of different females. Females come into oestrus, albeit not synchronously, within a short period starting from mid-March [17, 26].

Study areas and data collection
The study was carried out in two areas: Luoto (63°49'N, 22°49'E) and Vaasa (63°3'N, 22°41'E). In Luoto, flying squirrels were studied between 1993 and 2014 within an area of 44 km². The main land-uses in Luoto are shoreline spruce-dominated mixed forests, clear-cuts, and cultivated Scots pine plantations. The Vaasa study area is located about 90 km southwest of Luoto. The marking of flying squirrels started in 1992 in Vaasa within an area of 4 km², after the year 2000, the area was expanded to cover 25 km². Vaasa is covered by spruce-dominated forest patches, clear-cuts, and agricultural fields (for more information see [31, 32]).

The studied populations bred in nest-boxes. Nest-boxes were placed in forest patches of various sizes in groups of 2 to 4 nest-boxes per site, on average 2 nest-boxes per mature spruce forest hectare. The nest-boxes were made from a piece of aspen or spruce trunk, so that they resembled natural cavities. No known differences are apparent in behaviour or reproductive output, e.g. in number of offspring produced or communal nesting patterns, between individuals living in nest-boxes and those living in dreys or natural cavities (unpublished data; [32]), nor are we aware of significant differences in predator communities between sites. The nest-boxes have an entrance-hole diameter of 4.5 cm. This diameter is the same of that of the entrance of cavities made by the Great spotted woodpecker, *Dendrocopus major*, which represent the most common natural nesting site for flying squirrels in our study area. This size of the entrance-hole prevents main predators (e.g. the pine marten, *Martes martes*, and also large owls) from entering the nest-box. The same size of the entrance-hole, as well as the cavity size, between nest boxes and natural cavities makes the former readily accepted by flying squirrels for breeding and resting.

In total 489 male and 562 female flying squirrels were captured by hand from nest-boxes, sexed, weighed, and marked with ear-tags (Hauptner 73850, Hauptner, Germany). The main nest-box checking session was in June, and sites found occupied were checked again in August. In addition, in the years 1992–2003 part of the nest boxes were checked also between September and March, but for the following years there were only sporadic observations during these months. In total there were 1812 observations in June and August and 284 observations for October-March (for the 1051 studied individuals). The same observer (R. Wistbacka) was responsible for measurements taken in Luoto study area and also in Vaasa after 2001. Before 2001, weighing in Vaasa was made by R. Wistbacka and A. Mäkelä. The same weighing scale type was used (Pesola) and the scale was calibrated with a similar scale used in another flying squirrel population [33]. Thus, biases due to observer error between study areas were reduced.

We knew the age of 182 out of 489 males and 239 out of 562 females, because those individuals had been previously captured and marked as juveniles. Recapturing probability of individuals was high, above 0.8 for females and 0.7 for males [31], and we can conclude that new unmarked adult individuals located within our study area were very likely new recruits to our study system. Recruits arrive during the natal dispersal period, typically concentrated in September, whereas breeding dispersal is rare in flying squirrels [34, 35]. In other words, within our study areas the likelihood of an unmarked adult individual of being 1 year old is very high ([31], unpublished data). We used age based on this assumption in our models, because it was better aiming to control for the possible effect of age than leaving age out of the analyses (the same approach was used by [16]). The results for age were similar when only individuals captured as juveniles were used (Additional file 1: Figure S1).

Female body mass is affected by pregnancy between the start of mating season in mid-March and the birth of summer litters. Based on our data, second litters are born by mid-July at the latest. Thus, for the analysis of the effect of age and year on body mass, we excluded female observations recorded between 15 March and end of July. Nevertheless, the energy expenditure during the breeding season may have carry over impacts on body mass of females still in August.

For analysing sexual dimorphism in size, in addition to body mass, we measured skull length of 72 individuals, femur length of 60 individuals, and tail length of 56 individuals, all adults, during June 2014 and 2015 (the same

individuals were used for measurements of skull, femur and tail, but femur and tail length were missing for some individuals). All these measurements were taken by the same observer (A. Santangeli) to avoid observer-induced measurement bias.

Weather variables

We used weather information from the weather station (maintained by the Finnish Meteorological Institute) nearest to each study area. For Vaasa, the closest weather station was located within our study area, and for Luoto it was 10 km southeast of the study area. Weather recording stations were at the same altitude as the study areas. We used monthly average temperature and precipitation indices between December and June. Early winter weather was averaged for December and January, late winter was February and March. Spring season was represented by April and May, and June described the summer weather. For the spring weather, we also used the starting date of the tree growing season. Temperatures consistently above +5 °C indicate the beginning of tree growing season (Finnish Meteorological Institute; http://en.ilmatieteenlaitos.fi/seasons-in-finland), and have also been shown to correlate with birch bud burst in Finland [36].

Statistical analyses

Models on sexual differences in body measurements

We first built three generalized linear mixed models (GLMM) using as a response the skull, femur and tail length (n = 56–72). Because the main rationale for these models was to quantify sex differences in the abovementioned three measurements, we used sex as a categorical predictor, and territory identity nested within year in the random part to account for multiple observations from different individuals in the same territory (e.g. measurements of male and female from the same territory). We then used all observations where body mass was measured (n = 695) during the months from January to March, and from August to November to explore sex differences in body mass. Thus we excluded the months when females were pregnant in order to make the comparison in body mass between the two sexes. Additionally, we fit a GLMM with similar structure as those explained above, but now also with age and month controlled for in the model.

Models on seasonal and weather effects on body mass

We used GLMMs to investigate the relationship between body mass of flying squirrels (separately for sex and for two seasons, see below) in relation to environmental, life-history and temporal predictors. Specifically, we ran four GLMMs using in turn the body mass of adult females or males separately within two different seasons (winter and summer) as the response variable. Here we considered as winter all measurements of body weight collected between

January and March from males and females (n = 120 and 99, respectively), and as summer the measurements collected between June and November for males (n = 804), and between August and November for females (n = 284). We excluded female measurements collected in June from all analyses as these are affected by the breeding state (pregnancy). We run separate sex-specific models for the winter and summer season because data for the winter season were only collected up to the year 2003, whereas those from the summer period spanned until the year 2014 (see above).

In each of the four models, we included the individual identity nested within the study area (Luoto or Vaasa) in the random part of the model to account for pseudo-replication (i.e. multiple measurements collected on the same individual in the same study area over the years). We then included, in each of the four models, the month when body mass was measured (as a categorical variable), the age of the individual and the year (both as continuous variables). Finally, for the winter models we also included as predictors the average temperature and average precipitation during December and January, and for the summer models, the average temperature and precipitation in December and January combined, February and March combined, as well as average temperature and precipitation in May, April and June separately. In the summer models we also included the starting date of the tree growth season, with the rationale that an early start of the growth season would result in higher body mass later in the summer. We also tested the effect of age squared (to fit non-linear trends) by including this variable in each full model, and removing it if non-significant. We assessed the significance of each level combination within the categorical variables (e.g. between body mass in January and February within the month variable) by means of post-hoc comparisons adjusted for multiple testing using the Tukey method.

Before fitting the models we checked for collinearity using variance inflation factor (VIF) analyses. All variables had a VIF value lower than 2.5, indicating low collinearity levels and no need for excluding any of them from the models. We then built the four full models (i.e. the ones with all candidate predictor variables), one for each sex-class and season combination (see above). Next we applied model selection based on the Akaike's information criterion (AIC), followed by multi-model inference and averaging [37] using the MuMin package in R [38]. We derived averaged model coefficients and p-values for each variable from across the set of best ranked models (i.e. with ΔAIC < 4; listed in Additional file 1: Table S1).

Finally, we tested whether there was any temporal trend in body mass during winter (from early January to

15th of March), i.e. before the start of the mating season. This model was similar to the above models for body mass of males and females, but month was replaced with the day of the year so that 1st January got a value of 0 and 15th of March a value of 105. For this latter analysis we did not have repeated measurements from the same individual, therefore there was no need to include the individual identity as a random effect, whereas study area was included as a class variable.

All analyses were performed in R software v. 3.0.3 [39].

Results

Sexual differences in body measurements

Females body mass was on average 12 g higher than that of males (using data from the 22 years study period: t = -14.61, p < 0.001). Moreover, based on data from the years 2014 and 2015 only, females appear to have longer femur than males (t = -3.10, p = 0.01), whereas skull (t = 0.30, p = 0.77) and tail length (t = -1.32, p = 0.22) were similar between the two sexes (Fig. 1).

Seasonal and weather effects on body mass

We found considerable model uncertainty when running all possible combinations of body mass sub-models for the four separate analyses (see Additional file 1 for the list of 10 best supported models for males and females). This underscores the need for multi-model averaging, from which results are shown below.

Body mass of adult flying squirrels did not vary significantly between the different winter months (Fig. 2) and there was no change in body mass over the period preceding the start of the mating season for male or female flying squirrels. This was tested with the correlation between date, from January to mid-March, and body mass: males: n = 104, $F_{1,\ 95.6}$ = 0.35, p = 0.55; females: n = 93, $F_{1,\ 21.5}$ = 0.81, p = 0.38. However, we show that male body mass declined after the breeding season, amounting to about 10 % loss of weight from the winter body mass (Fig. 2). This decline was observable right after the start of the mating season as the pattern seems clear already in April (i.e. the decline from March to April was on average 5 g from the raw data, n = 32 measurements in April). Conversely, the body mass of females did not vary between seasons and summer months (Fig. 2).

For both sexes, low temperatures in late winter, as well as the early start of the tree growth season, resulted in increased body mass during the following summer season (Fig. 3, Table 1). Moreover, for male flying squirrels higher temperature in spring was associated to lower body mass in summer, whereas increased rain in June was related to lower body mass of females. Winter body mass was not related to any weather variable (Tables 1 and 2). The only significant temporal trend observed over the years of study was an increasing trend in female weight measured in

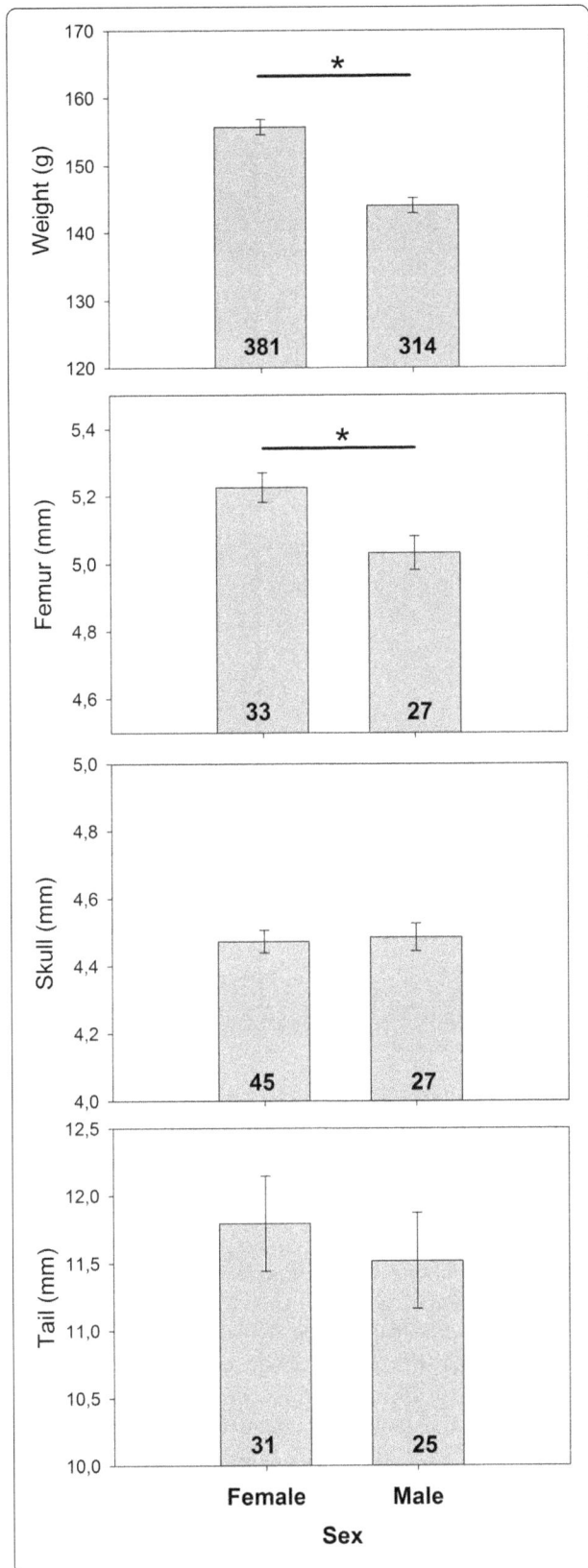

Fig. 1 Predicted mean values (and standard errors) for weight, femur, skull and tail length of male and female adult flying squirrels. Sample sizes are given with values within the column bars, whereas the * depicts significant differences between sexes

winter. Moreover, body mass increased with age in both sexes (Tables 1 and 2).

Discussion

We did not observe changes in male body mass before the breeding season. This was against our prediction for the female defence mating system. The only observable seasonal pattern we found was a decrease in male body mass after the breeding season. As we predicted, female flying squirrels were larger than males, but there were no detectable seasonal variation in female body mass outside of the pregnancy period. Unexpectedly, cold winter and cold spring conditions were linked to an increase in body mass of males and females in the following summer. The only temporal trend observed over the 22 year study period was a slight increase in winter body mass of females. Thus, no measurable temporal trends that might be related to sexual-size dimorphism were detected.

The observed female-biased sexual size dimorphism in the Siberian flying squirrels of this study is consistent with patterns observed for the southern flying squirrel *Glaucomys volans* [24]. In our study, the female-biased body mass was largest after the breeding season, because female body mass did not vary much within the year (outside of the pregnancy period). Similar observations of a stable female body mass between seasons have been noted for North American red squirrels [40]. We observed no difference in skull size between sexes. Conversely, femur length was longer in females than in males. The explanation for this may be the need for females to secure gliding potential when pregnant, and, thus, increase the aerofoil area compared to males [23]. For example, it is known that for flying mammals (i.e. bats) pregnancy restricts the ability to move [41]. Larger female body mass, as compared to that of males in flying squirrels may be related to reproductive benefits from large size due to the territoriality of females ([17, 24], see also [42]). Alternatively, it may also be due to the benefits for males to be smaller and thus more vagile within a scramble competition mating system.

We predicted changes in male body mass before the start of the mating season, but did not find any indication for these changes, within winter or between autumn and winter. Thus, there was no clear pattern in male body mass that would support the female defence mating system. For example in North American red squirrels [40] and grey-headed flying-foxes, *Pteropus poliocephalus* [9], body mass of males increased before the start of the

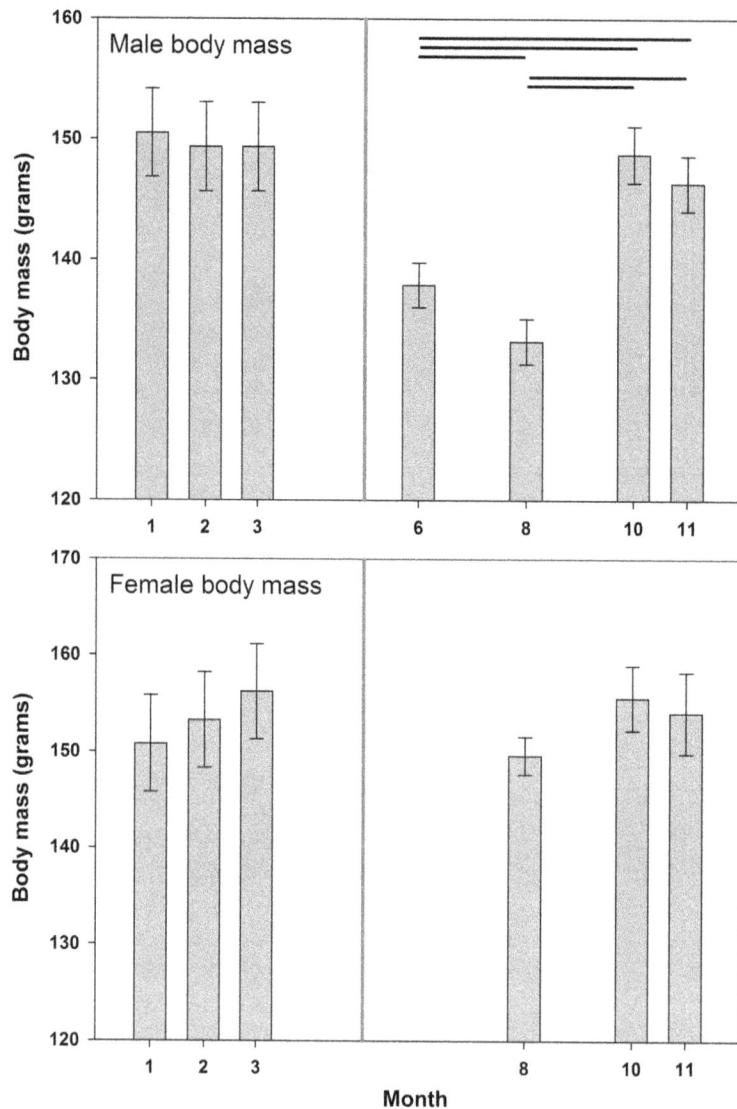

Fig. 2 Male and female body mass (least square means and standard errors) for each month from the winter model (1 January to 3 March) and for the summer model (6 June to 11 November). For females, June was omitted because they may still be pregnant in June. Lines above the bars join months for which body mass was significantly different after post-hoc testing (adjusted for multiple testing using the Tukey method). For females, there were no significant differences in body mass between the months of the winter season and between the months of the summer season

mating season and declined rapidly then after. However, it is known that in the Siberian flying squirrel the male body mass is positively related to reproductive success [17]. Thus, some aspects of the female defence mating system may be operative in flying squirrels, as is the case for tree squirrels [25, 27]. However, the stable body mass of males during winter seems to fit the hypothesis that extra weight just before breeding may not enhance fast movement to locate females. This fits the scramble competition mating system. Perhaps in the case of the Siberian flying squirrel, the selective forces for increased gliding potential versus dominance (i.e. low versus high body mass, respectively) at some level cancel each other out.

Body mass before the mating season is likely affected by environmental factors during the preceding winter. However, food availability for flying squirrels during the winter months is relatively stable, since the main winter food, catkins of deciduous trees, develop already during the autumn and persist on the trees during winter. In other words, food availability does not vary within the winter season [30]. Only after bud burst in spring do flying squirrels start consuming leaf material, but this is after the start of the mating season (from mid-May onwards in our study area). In addition, we did not find any weather variable to have an effect on body mass in winter, which is in line with a previous study

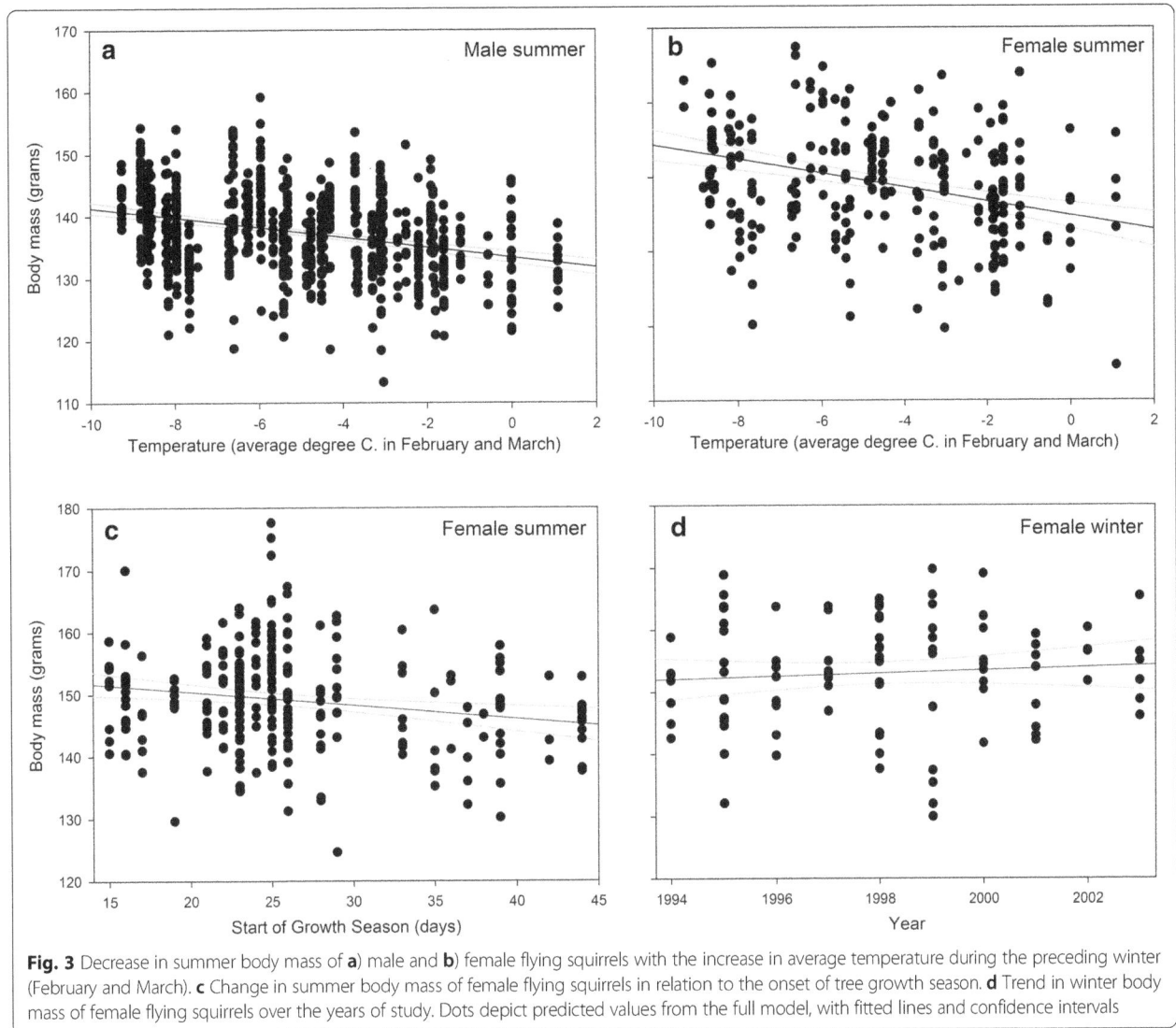

Fig. 3 Decrease in summer body mass of **a)** male and **b)** female flying squirrels with the increase in average temperature during the preceding winter (February and March). **c** Change in summer body mass of female flying squirrels in relation to the onset of tree growth season. **d** Trend in winter body mass of female flying squirrels over the years of study. Dots depict predicted values from the full model, with fitted lines and confidence intervals

where no effect of winter temperature on communal nesting behaviour of flying squirrels was found [32].

Surprisingly, male body mass did not increase during summer, from June to August, although the breeding season for male flying squirrels ends already in the spring after they have fertilised the females. Males had even lower body mass in August than in June. Instead, for some other species male mass is observed to decline rapidly during the mating season, but also increased rapidly during the summer [8, 43, 44]. In August, male flying squirrels frequently move between forest patches occupied by different females [45], perhaps to increase social links with new females which might affect future reproductive success. This may increase the energy expenditure of male flying squirrels and explain the decline in body mass observed during the late summer months. Moreover, another possible explanation for the observed

monthly variation in body mass might also relate to the food available to flying squirrels in the different periods, because in early summer flying squirrels eat mainly leaves [46]. However, food would affect body mass in both sexes. In addition, whether the nutritional value of leaves differs from that of catkins for flying squirrels is still unknown. In any case, food obviously is related to observed increase in male body mass from August to October, as squirrels likely prepare for winter by increasing body mass in late autumn.

Body mass increased with the age of Siberian flying squirrels in both sexes, which is in line with earlier studies reporting body mass increases with age in adult squirrels [25, 27, 47]. It should be noted that our analysis did not separate mortality and individual growth curves and is not suitable for determining growth patterns related to senescence [48, 49]. Nevertheless, age, body mass, dominance

Table 1 Generalized Linear Mixed Model results for adult male body mass measurements

	Variable	Estimate	SE	Z value	p-value
a)	(Intercept)	11,29	414,58	0,03	0,978
	Age	**3,01**	**0,59**	**4,89**	**<0,001**
	T_DecJan	0,27	0,32	0,81	0,419
	Year	0,24	0,34	0,66	0,507
	R_DecJan	−0,04	0,07	0,54	0,592
b)	(Intercept)	137,35	54,18	2,53	0,011
	Age	**0,80**	**0,22**	**3,70**	**<0,001**
	R_June	−0,02	0,01	1,72	0,086
	SGS	**−0,12**	**0,05**	**2,43**	**0,015**
	T_April	**−0,76**	**0,27**	**2,81**	**0,005**
	T_DecJan	−0,16	0,09	1,77	0,076
	T_FebMar	**−0,38**	**0,11**	**3,54**	**<0,001**
	T_May	**−0,58**	**0,20**	**2,90**	**0,004**
	R_DecJan	0,02	0,02	1,11	0,267
	R_May	−0,01	0,01	1,00	0,318
	R_FebMar	−0,02	0,02	0,89	0,376
	R_April	0,01	0,02	0,68	0,498
	T_June	−0,10	0,19	0,52	0,605
	Year	0,02	0,07	0,33	0,745

a) Winter (years 1992–2003) and b) summer (years 1992–2014). Statistics show the model averaged results from across the best supported models (listed in Additional file 1: Table S1). T depicts Temperature, R precipitation, DecJan the months of December and January combined, FebMar is February and March combined. SGS indicates the start of the growth season of trees. Significant variables are highlighted in bold font. Results for the month variable (which was set as categorical) are omitted here and are shown in Fig. 2

Table 2 Generalized Linear Mixed Model results for adult female body mass measurements

	Variable	Estimate	SE	Z value	p-value
a)	(Intercept)	−1448,00	1212,00	1,19	0,235
	Age	**14,30**	**3,49**	**3,90**	**<0,001**
	age*age	**−1,48**	**0,47**	**3,03**	**0,002**
	Year	**1,03**	**0,48**	**2,06**	**0,039**
	T_DecJan	−1,02	0,55	1,79	0,074
	R_DecJan	0,01	0,10	0,12	0,902
b)	(Intercept)	83,69	173,40	0,48	0,631
	Age	**3,97**	**0,65**	**6,04**	**<0,001**
	R_June	**−0,07**	**0,03**	**2,22**	**0,027**
	SGS	**−0,23**	**0,10**	**2,23**	**0,026**
	T_FebMar	**−0,69**	**0,26**	**2,56**	**0,010**
	R_DecJan	0,05	0,05	1,01	0,311
	R_April	0,05	0,05	0,98	0,329
	T_May	−0,43	0,50	0,85	0,398
	Year	0,14	0,14	0,96	0,335
	T_April	0,87	0,76	1,13	0,260
	R_FebMar	−0,03	0,06	0,43	0,671
	T_June	0,18	0,47	0,38	0,703
	T_DecJan	−0,14	0,28	0,50	0,620
	R_May	0,00	0,03	0,00	0,998

a) Winter (years 1992–2003) and b) summer (years 1992–2014). Statistics show the model averaged results from across the best supported models (listed in Additional file 1: Table S2). T depicts Temperature, R precipitation, DecJan the months of December and January combined, FebMar is February and March combined. SGS indicates the start of the growth season of trees. Significant variables are highlighted in bold font

hierarchies and reproductive success are linked in squirrels, as older individuals are usually heavier, higher in dominance hierarchy, and consequently have increased reproductive success [17, 25, 47].

Our results do not suggest any impact of climatic trends on body mass dimorphism in flying squirrels. The only temporal pattern observed during our study period was a slight increase in female body mass during winter. Our temporal data for winter months was limited (winter body mass data ended in year 2003) and it remains unclear what process is behind the observed temporal trend, because winter body mass was not linked to winter temperatures. In addition, the observed trend was against the expectation that climate warming would result in smaller body size ([50, 51]; but see [52]). The increasing trend in early spring temperatures over our study period ([29], for more detailed analysis on weather changes in our study area, see [30]) indicates potential for temporal changes in sexual dimorphism in body mass. However this was not the case, as we did not observe any trends in summer body mass of males or females.

Weather in winter had unexpected effects on summer body mass of flying squirrels. The summer body mass of both sexes increased when the preceding winter and spring seasons were colder. The reasons for this pattern remain unknown, but a possible explanation could be that warm weather at the beginning and during the breeding season may increase the intensity of mating season and result in high energy expenditure and low body mass after a successful reproduction. Lower body mass in summer was also associated with delayed tree growth in spring. This correlation might indicate a poor food situation in the late spring, because after the flowering of catkins in early spring leaves represent the main food for flying squirrels. Late start of the tree growing season indicates delay in leave growth. We also found a negative correlation between increased rain in June and female body mass in August, which may indicate that rainy conditions in summer have negative impacts on female body mass. In earlier studies, weather has been observed to affect timing of reproduction and reproductive success in Siberian flying squirrels [30] and red squirrels [53, 54]. In addition, for example in ungulates, body mass interacts strongly with weather and food availability in spring and summer [55, 56].

Conclusions

Our study supports the general view that tree and flying squirrels often deviate from the general pattern in mammals for males being larger than females [24, 57, 58]. Seasonal body mass patterns of male flying squirrels support the hypothesis that competition during the breeding season is the main driver of seasonal body mass variation in males. Instead, female body mass was more constant between seasons, when the effects of pregnancy are not considered. However, we also suggest that, in the case of Siberian flying squirrels, competing forces may play a role in selecting for male size that would represent an optimal balance between fast gliding (i.e. low body mass) and dominant (i.e. high body mass), with these competing pressures potentially masking some of the seasonal variation (before breeding season) in male body mass. However, verifying this hypothesis would require further studies. Our results also support the view that the effects of temperature on body mass may be complex and in correlative studies not necessarily straightforward to interpret [52]. Nevertheless, no indication that climate change had affected sexual body mass dimorphism was detected.

Additional file

Additional file 1: Table S1. The best ranked models explaining body mass of male Siberian flying squirrels measured in summer and winter (upper and lower panel respectively). Columns show the variables included in each model (see main text for an explanation of the variable names), the degrees of freedom of each model, the Akaike value corrected for small sample size (AICc) and the difference in AIC between best ranked and the target model (ΔAICc) as well as the AIC weight of each model. List is restricted to 10 best ranked models. Table S2. The best ranked models explaining body mass of female Siberian flying squirrels measured in summer and winter (upper and lower panel respectively). Columns show the variables included in each model (see main text for an explanation of the variable names), the degrees of freedom of each model, the Akaike value corrected for small sample size (AICc) and the difference in AIC between best ranked and the target model (ΔAICc) as well as the AIC weight of each model. List is restricted to 10 best ranked models. Figure S1. Male and female body mass (least square means and standard errors) for each month from the winter model (1 January to 3 March) and for the summer model (6 June to 11 November). Only observations where the exact age was known have been used for this supporting analysis (see methods). For females, June was omitted because they may still be pregnant in June. Lines above the bars join months for which body mass was significantly different after post-hoc testing (adjusted for multiple testing using the Tukey method). For females, there were no significant differences in body mass between the months of the winter season and between the months of the summer season. (DOCX 127 kb)

Acknowledgements
We thank all of the field workers Timo Hyrsky, Rune Jakobsson, Antero Mäkelä and Markus Sundell, who have assisted during data gathering. Three referees in peerage of science are thanked for their useful comments that helped to improve the manuscript.

Funding
The study was financially supported by Academy of Finland (grant number 259562 to VS), Kone Foundation (to AS), Oskar Öflunds stiftelse (to RW), Societas Pro Fauna et Flora Fennica (to RW), Svensk-Österbottniska samfundet (to RW), and Vuokon luonnonsuojelusäätiö (to RW).

Authors' contributions
RW collected data, VS and AS analysed data and wrote the manuscript. All authors read and approved the final manuscript.

Competing interests
The authors declare that they have no competing interests.

Author details
[1]Department of Biology, Section of Ecology, FI-20014 University of Turku, Turku, Finland. [2]Department of Biology, FI-90014 University of Oulu, Oulu, Finland. [3]The Helsinki Lab of Ornithology, Finnish Museum of Natural History, University of Helsinki, Helsinki, Finland.

References
1. Ralls K. Mammals in which females are larger than males. Quart Rev Biol. 1976;51:245–76.
2. Andersson M. Sexual Selection. Princeton: Princeton University Press; 1994.
3. Clutton-Brock TH. Sexual selection in males and females. Science. 2007;318: 1882–5.
4. Lindstedt SL, Boyce MS. Seasonality, fasting endurance, and body size in mammals. Am Nat. 1985;125:873–8.
5. Gittleman JL, Thompson SD. Energy allocation in mammalian reproduction. Am Zool. 1988;28:863–75.
6. Soderquist TR. Ontogeny of sexual dimorphism in size among polytocous mammals: tests of two carnivorous marsupials. J Mammal. 1995;76:376–90.
7. Boratynski Z, Koteja P. Sexual and natural selection on body mass and metabolic rates in free-living bank voles. Funct Ecol. 2010;24:1252–61.
8. Rughetti M, Festa-Bianchet M. Seasonal changes in sexual size dimorphism in Northern chamois (Rupicapra rupicapra). J Zool. 2011;284:257–64.
9. Welbergen JA. Fat males and fit females: sex differences in the seasonal patterns of body condition in grey-headed flying-foxes (Pteropus poliocephalus). Oecologia. 2011;165:629–37.
10. Mitchell B, McCowan D, Nicholson IA. Annual cycles of body weight and condition in Scottish red deer, Cervus elaphus. J Zool. 1976;180:107–27.
11. Michener GR, Locklear L. Differential costs of reproductive effort for male and female Richardson's ground squirrels. Ecology. 1990;71:855–68.
12. Wolff JO. Breeding strategies, mate choice, and reproductive success in American bison. Oikos. 1998;83:529–44.
13. Korine C, Speakman J, Arad Z. Reproductive energetics of captive and free-ranging Egyptian fruit bats (Rousettus aegyptiacus). Ecology. 2004;85: 220–30.
14. Ims RA. Spatial clumping of sexually receptive females induces space sharing among male voles. Nature. 1988;335:541–3.
15. Thornhill R, Alcock J. The evolution of insects mating systems. Cambridge: Harvard University Press; 1983.
16. Lane JE, Boutin S, Gunn MR, Coltman DW. Sexually-selected behaviour: red squirrel males search for reproductive success. J Anim Ecol. 2009;78:296–304.
17. Selonen V, Painter JN, Rantala S, Hanski IK. Mating system and reproductive success in the Siberian flying squirrel. J Mammal. 2013;94:1266–73.
18. Schulte-Hostedde AI, Millar JS. "Little chipmunk" syndrome? Male body size and dominance in captive yellow-pine chipmunks (Tamias amoenus). Ethology. 2002;108:127–37.
19. Isaac JL. Potential causes and life-history consequences of sexual size dimorphism in mammals. Mammal Rev. 2005;35:101–15.

20. Chan-McLeod AC, White RG, Russell DE. Comparative body composition strategies of breeding and nonbreeding female caribou. Can J Zool. 1999; 77:1901–7.

21. Beck CA, Bowen WD, Iverson SJ. Sex differences in the seasonal patterns of energy storage and expenditure in a phocid seal. J Anim Ecol. 2003;72:280–91.

22. Sheriff MJ, Richter M, Buck CL, Barnes BM. Changing seasonality and phenology of free-living arctic ground squirrels; the importance of sex. Phil Trans R Soc B. 2013;368:20120480.

23. Fokidis HB, Risch TS. Does gliding when pregnant select for larger females? J Zool. 2008;275:237–44.

24. Fokidis HB, Risch TS, Glenn TC. Reproductive and resource benefits to large female body size in a mammal exhibiting female-biased sexual size dimorphism. Anim Behav. 2007;73:479–88.

25. Koprowski JL. Alternative reproductive tactics and strategies of tree squirrels. In: Wolff JO, Sherman PW, editors. Rodent Societies: an Ecological and Evolutionary Perspective. Chicago: University of Chicago Press; 2007. p. 86–95.

26. Hanski IK, Mönkkönen M, Reunanen P, Stevens PC. Ecology of the Eurasian flying squirrel (Pteromys volans) in Finland. In: Goldingay R, Scheibe J, editors. Biology of gliding mammals. Furth: Filander; 2000. p. 67–86.

27. Wauters L, De Vos R, Dhondt AA. Factors affecting male mating success in red squirrels (Sciurus vulgaris). Ethol Ecol Evol. 1990;2:195–204.

28. Nandini R-R. The evolution of sexual size dimorphism in squirrels. PhD dissertation. Alabama: Auburn University; 2011.

29. Mikkonen S, Laine M, Mäkelä HM, Gregow H, Tuomenvirta H, Lahtinen M, Laaksonen A. Trends in the average temperature in Finland, 1847–2013. Stoch Env Res Risk Assessm. 2015;29:1521–9.

30. Selonen V, Wistbacka R, Korpimäki E. Food abundance and weather modify reproduction of two arboreal squirrel species. J Mammal. 2016; doi:10.1093/jmammal/gyw096.

31. Lampila S, Wistbacka A, Mäkelä A, Orell M. Survival and population growth rate of the threatened Siberian flying squirrel (Pteromys volans) in a fragmented forest landscape. Ecosci. 2009;16:66–74.

32. Selonen V, Hanski IK, Wistbacka R. Communal nesting is explained by subsequent mating rather than kinship or thermoregulation in the Siberian flying squirrel. Behav Ecol Socio. 2014;68:971–80.

33. Koskimäki J, Huitu O, Kotiaho J, Lampila S, Mäkelä A, Sulkava R, Mönkkönen M. Are habitat loss, predation risk and climate related to the drastic decline in a Siberian flying squirrel population? A 15 year study. Popul Ecol. 2014;56:341–8.

34. Hanski IK, Selonen V. Female-biased natal dispersal in the Siberian flying squirrel. Behav Ecol. 2009;20:60–7.

35. Selonen V, Hanski IK. Condition-dependent, phenotype-dependent and genetic-dependent factors in the natal dispersal of a solitary rodent. J Anim Ecol. 2010;79:1093–100.

36. Rousi M, Heinonen J. Temperature sum accumulation effects on within-population variation and long-term trends in date of bud burst of European white birch (Betula pendula). Tree Physiol. 2007;27:1019–25.

37. Burnham KP, Anderson DR. Model Selection and Multimodel Inference: A Practical Information-Theoretic Approach. 2nd ed. New York: Springer; 2002.

38. Bartoń K. Package "MuMIn" - Multi-Model Inference. 2014.

39. R Core Development Team. R: A language and environment for statistical computing. Version 3.0.3. 2013.

40. Koprowski JL. Annual cycles in body mass and reproduction of endangered Mt. Graham Red Squirrels. J Mammal. 2005;86:309–13.

41. Hayssen V, Kunz TH. Allometry of litter mass in bats: maternal size, wing morphology, and phylogeny. J Mammal. 1996;77:476–90.

42. Bondrup-Nielsen S, Ims RA. Reversed sexual size dimorphism in microtines: are females larger than males or are males smaller than females? Evol Ecol. 1990;4:261–72.

43. Falls JB, Falls EA, Fryxell JM. Fluctuations of deer mice in Ontario in relation to seed crops. Ecol Monogr. 2007;77:19–32.

44. Hoogland JL. Sexual dimorphism of prairie dogs. J Mamm. 2003;84:1254–66.

45. Hanski IK, Stevens P, Ihalempiä P, Selonen V. Home-range size, movements, and nest-site use in the Siberian flying squirrel, Pteromys volans. J Mammal. 2000;81:798–809.

46. Mäkelä A. Liito-oravan (Pteromys volans L.) ravintokohteet eri vuodenaikoina ulosteanalyysin perusteella (diet of flying squirrel, in Finnish). WWF Finland Rep. 1996;8:54–8.

47. Wauters L, Dhondt AA. Lifetime reproductive success and its correlates in female Eurasian red squirrels. Oikos. 1995;72:402–10.

48. Nussey DH, Coulson TN, Delorme D, Clutton-Brock TH, Pemberton JM, Festa-Bianchet M, Gaillard JM. Patterns of body mass senescence and selective disappearance differ among three species of free-living ungulates. Ecology. 2011;92:1936–47.

49. Mumby HS, Chapman SN, Crawley JAH, Mar KU, Htut W, Thura Soe A, Aung HH, Lummaa V. Distinguishing between determinate and indeterminate growth in a long-lived mammal. BMC Evol Biol. 2015;15:214.

50. Millien V, Lyons SK, Olson L, Smith FA, Wilson AB, Yom-Tov Y. Ecotypic variation in the context of global climate change: revisiting the rules. Ecol Lett. 2006;9:853–69.

51. Blois JL, Feranec RS, Hadly EA. Environmental influences on spatial and temporal patterns of body-size variation in California ground squirrels (Spermophilus beecheyi). J Biogeogr. 2008;35:602–13.

52. Teplitsky C, Millien V. Climate warming and Bergmann's rule through time: is there any evidence? Evol Appl. 2014;7:156–68.

53. Williams CT, Lane JE, Humphries MM, McAdam AG, Boutin S. Reproductive phenology of a food-hoarding mast-seed consumer: resource- and density-dependent benefits of early breeding in red squirrels. Oecologia. 2014;174:777–88.

54. Studd EK, Boutin S, McAdam AG, Krebs CJ, Humphries MM. Predators, energetics and fitness drive neonatal reproductive failure in red squirrels. J Anim Ecol. 2015;84:249–59.

55. Mysterud A, Yoccoz NG, Langvatn R, Pettorelli N, Stenseth NC. Hierarchical path analysis of deer responses to direct and indirect effects of climate in northern forest. Proc R Soc B. 2008;363:2357–66.

56. Herfindal I, Saether B-E, Solberg EJ, Andersen R, Hdga KA. Population characteristics predict responses in moose body mass to temporal variation in the environment. J Anim Ecol. 2006;75:1110–8.

57. Don BAC. Home range characteristics and correlates in tree squirrels. Mammal Rev. 1983;13:123–32.

58. Lurz PWW, Gurnell J, Magris L. Sciurus vulgaris. Mamm Species. 2005;769:1–10.

3

LemurFaceID: a face recognition system to facilitate individual identification of lemurs

David Crouse[1,7†], Rachel L. Jacobs[2*†], Zach Richardson[1], Scott Klum[1,8], Anil Jain[1*], Andrea L. Baden[3,4,5] and Stacey R. Tecot[6]

Abstract

Background: Long-term research of known individuals is critical for understanding the demographic and evolutionary processes that influence natural populations. Current methods for individual identification of many animals include capture and tagging techniques and/or researcher knowledge of natural variation in individual phenotypes. These methods can be costly, time-consuming, and may be impractical for larger-scale, population-level studies. Accordingly, for many animal lineages, long-term research projects are often limited to only a few taxa. Lemurs, a mammalian lineage endemic to Madagascar, are no exception. Long-term data needed to address evolutionary questions are lacking for many species. This is, at least in part, due to difficulties collecting consistent data on known individuals over long periods of time. Here, we present a new method for individual identification of lemurs (LemurFaceID). LemurFaceID is a computer-assisted facial recognition system that can be used to identify individual lemurs based on photographs.

Results: LemurFaceID was developed using patch-wise Multiscale Local Binary Pattern features and modified facial image normalization techniques to reduce the effects of facial hair and variation in ambient lighting on identification. We trained and tested our system using images from wild red-bellied lemurs (*Eulemur rubriventer*) collected in Ranomafana National Park, Madagascar. Across 100 trials, with different partitions of training and test sets, we demonstrate that the LemurFaceID can achieve 98.7% ± 1.81% accuracy (using 2-query image fusion) in correctly identifying individual lemurs.

Conclusions: Our results suggest that human facial recognition techniques can be modified for identification of individual lemurs based on variation in facial patterns. LemurFaceID was able to identify individual lemurs based on photographs of wild individuals with a relatively high degree of accuracy. This technology would remove many limitations of traditional methods for individual identification. Once optimized, our system can facilitate long-term research of known individuals by providing a rapid, cost-effective, and accurate method for individual identification.

Keywords: Animal biometrics, Conservation, *Eulemur rubriventer*, Linear discriminant analysis, Mammal, Multiscale local binary pattern, Pelage, Photograph, Primate

Background

Most research on the behavior and ecology of wild animal populations requires that study subjects are individually recognizable. Individual identification is necessary to ensure unbiased data collection and to account for individual variation in the variables of interest. For short-term studies, researchers may rely on unique methods for identification based on conspicuous natural variation among individuals at the time of data collection, such as differences in body size and shape or the presence of injuries and scars. These methods may or may not allow for identification of individuals at later dates in time. To address many evolutionary questions, however, it is necessary to collect data on known individuals over long periods of time [1]. Indeed, longitudinal studies are essential for characterizing life history parameters, trait heritability, and fitness effects (reviewed in [1]). Consequently, they are invaluable for identifying the demographic and evolutionary processes influencing wild animal populations [1].

* Correspondence: rachel_jacobs@gwu.edu; jain@cse.msu.edu
†Equal contributors
2Department of Anthropology, Center for the Advanced Study of Human Paleobiology, The George Washington University, Washington, DC, USA
1Department of Computer Science and Engineering, Michigan State University, East Lansing, MI, USA
Full list of author information is available at the end of the article

Unfortunately, longitudinal monitoring can be challenging, particularly for long-lived species. One of the primary challenges researchers face is establishing methods for individual identification that allow multiple researchers to collect consistent and accurate demographic and behavioral data over long periods of time (in some cases several decades). Current methods for individual identification often involve either capturing and tagging animals with unique identifiers, such as combinations of colored collars and/or tags [2–5], or taking advantage of natural variation in populations (e.g., scars, skin and pelage patterns) and relying on researchers' knowledge of individual differences [6–9]. The former method (or a combination of the two methods) has been used in some of the best established long-term field studies, such as the St. Kilda Soay Sheep and Isle of Rum Red Deer Projects [2, 3], as well as the Wytham Tit and Galápagos Finch Projects [4, 5]. Because they have long-term (multi-generation) data on known individuals, these projects have contributed substantially to the field of evolutionary biology by documenting how and why populations change over time (e.g., [10–13]).

Similar methods involving capturing and collaring have been used in many longitudinal studies of wild primates, such as owl monkeys [14], titi monkeys [15], colobines [16], and in particular, many Malagasy lemurs [17–20]. Through the long-term monitoring of individuals, many of these studies have provided important data on longevity, lifetime reproductive success, and dispersal patterns [15, 17, 18, 20–23].

Despite its utility for many longitudinal studies, the tagging process might sometimes be inappropriate or otherwise impractical. Tagging often requires that study subjects be captured via mist netting or in nest boxes (for birds) [4, 5], trapping (e.g., Sherman traps or corrals for some mammals) [2, 3, 24], and, in the case of some larger mammals, including many primates, darting via blow gun or air rifle [10, 25–27]. Capturing has several advantages, such as enabling data to be collected that would otherwise be impossible (e.g., blood samples, ectoparasites), but it can also be expensive, often making it unfeasible for studies with large sample sizes and/or those conducted over large spatial and temporal scales. Furthermore, capturing and tagging may pose additional risks to already threatened species. For example, such methods have been shown in some cases to cause acute physiological stress responses [16], tissue damage [28] and injury (e.g., broken bones, paralysis) [29], as well as disrupt group dynamics, and pose risks to reproduction, health, and even life [29–32].

An alternative method for individual identification relies on researcher knowledge of variation in individual appearances. It is less invasive and removes some of the potential risks associated with capturing and tagging.

Such methods have been successfully used in long-term studies of elephants, great apes, and baboons (among others) and have provided similarly rich long-term datasets that have been used to address demographic and evolutionary questions [6–9]. However, this method is more vulnerable to intra- and inter-observer error and thus can require substantial training. Moreover, for research sites involving multiple short-term studies in which researchers may use different methods for individual identification, it can be difficult to integrate data [33]. Additionally, long-term research is often hindered by disruptions to data collection (e.g., between studies, due to lack of research funds, political instability [1]). These breaks can result in lapses of time during which no one is present to document potential changes to group compositions and individual appearances, which can also complicate integrating data collected at different time points.

Under such circumstances, projects would benefit from a database of individual identifications, as well as a rapid method for identifying individuals that requires little training and can be used across different field seasons and researchers. The field of animal biometrics offers some solutions [34]. For example, some methods that have shown promise in mammalian (among other) research, including studies of cryptic animals, combine photography with computer-assisted individual identification programs to facilitate long-term systematic data collection (e.g., cheetahs: [35]; tigers: [36]; giraffes: [37]; zebras: [38]). These methods use quantifiable aspects of appearances to identify individuals based on probable matches in the system [34]. Because assignments are based on objective measures, these methods can minimize intra- and inter-observer error and facilitate integrating data collected across different studies [34]. At the same time, in study populations with large sample sizes, researchers might be limited in the number of individuals known on-hand. Computer-assisted programs can facilitate processing data to rapidly identify individuals when datasets are large, which reduces the limitations on sample size/scale imposed by the previous methods [34].

Despite their potential utility, such methods have not been incorporated in most studies of wild primates, and, particularly in the case of wild lemur populations, even with several drawbacks, capture and collar methods remain common [17–20]. As a result, multi-generation studies of lemur populations that incorporate individual identification are limited.

Here we present a method in development for non-invasive individual identification of wild lemurs that can help mitigate some of the disadvantages associated with other methods, while also facilitating long-term research (Table 1). Our system, called LemurFaceID, utilizes

Table 1 Individual identification methods

Method	Advantages	Disadvantages
Tagging/Collaring	Systematic across studies; opportunities to collect data that require animal to be in hand; precise location of animal known at all times (using GPS collar)	Invasive; poses risks to animals; expensive; less feasible for studies requiring large sample sizes; individual IDs may be unknown with loss of tag/collar
Manual identification based on physical variation	Non-invasive, low cost	Substantial training required; IDs may differ across studies/researchers; prone to intra- and inter-observer error; time-consuming for large sample sizes when individuals are not recognized instantly (e.g., manual comparisons of photographs are required)
Face recognition	Systematic across studies; non-invasive; minimal user training; reduces time to make identifications when datasets are large allowing for increased sample size/scale	Requires large dataset for development; currently requires partial knowledge of individual IDs; individual IDs may be unknown to the researcher if the system is unavailable for use

computer facial recognition methods, developed by the authors specifically for lemur faces, to identify individual lemurs based on photographs collected in wild populations [39].

Facial recognition technology has made great strides in its ability to successfully identify humans [40], but this aspect of computer vision has much untapped potential. Facial recognition technology has only recently expanded beyond human applications. While there has been limited work with non-human primates [41, 42], to our knowledge, facial recognition technology has not been applied to any of the >100 lemur species. However, many lemurs possess unique facial features, such as hair/pelage patterns, that make them appropriate candidates for applying modified techniques developed for human facial recognition (Fig. 1).

We focus this study on the red-bellied lemur (*Eulemur rubriventer*). Males and females in this species are sexually dichromatic with sex-specific variation in facial patterns ([43]; Fig. 2). Males exhibit patches of white skin around the eyes that are reduced or absent in females. In addition, females have a white ventral coat (reddish-brown in males) that variably extends to the neck and face. Facial patterns are individually variable, and the authors have used this variation to identify individuals in wild populations, but substantial training was required. Since the 1980s, a population of red-bellied lemurs has been studied in Ranomafana National Park, Madagascar [44–47], but because researchers used different methods for individual identification, gaps between studies make it difficult to integrate data. Consequently, detailed data on many life history parameters for this

Fig. 1 Examples of different lemur species. Photos by David Crouse (*Varecia rubra*, *Eulemur collaris*, and *Varecia variegata* at the Duke Lemur Center), Rachel Jacobs (*Eulemur rufifrons* in Ranomafana National Park), and Stacey Tecot (*Hapalemur griseus*, *Eulemur rubriventer* in Ranomafana National Park; *Propithecus deckenii* in Tsingy de Bemaraha National Park; *Indri indri* in Andasibe National Park)

Fig. 2 Red-bellied lemurs. The individual on the right is female, and the individual on the left is male

species are lacking. A reliable individual identification method would help provide these critical data for understanding population dynamics and addressing evolutionary questions.

In this paper we report the method and accuracy results of LemurFaceID, as well as its limitations. This system uses a relatively large photographic dataset of known individuals, patch-wise Multiscale Local Binary Pattern (MLBP) features, and an adapted Tan and Triggs [48] approach to facial image normalization to suit lemur face images and improve recognition accuracy.

Our initial effort (using a smaller dataset) was focused on making parametric adaptations to a face recognition system designed for human faces [49]. This system used both MLBP features and Scale Invariant Feature Transform (SIFT) features [50, 51] to characterize face images. Our initial effort exhibited low performance in recognition of lemur faces (73% rank-1 recognition accuracy). In other words, for a given query, the system reported the highest similarity between the query and the true match in the database only 73% of the time. Examination of the system revealed that the SIFT features were sensitive to local hair patterns. As matting of hair changed from image to image, the features changed substantially and therefore reduced match performance. The high dimensionality of the SIFT features also may have led to overfitting and slowing of the recognition process. Because of this, the use of SIFT features was abandoned in the final recognition system.

While still adapting methods originally developed for humans, LemurFaceID is specifically designed to handle lemur faces. We demonstrate that the LemurFaceID system identifies individual lemurs with a level of accuracy that suggests facial recognition technology is a potential useful tool for long-term research on wild lemur populations.

Methods
Data collection
Study species
Red-bellied lemurs (*Eulemur rubriventer*) are small to medium-sized (~2 kg), arboreal, frugivorous primates, and they are endemic to Madagascar's eastern rainforests [46, 52] (Fig. 3a). Despite their seemingly widespread distribution, the rainforests of eastern Madagascar have become highly fragmented [53], resulting in an apparent patchy distribution for this species. It is currently listed by the IUCN as Vulnerable with a decreasing population trend [54].

Study site
Data collection for this study was concentrated on the population of red-bellied lemurs in Ranomafana National Park (RNP). RNP is approximately 330 km² of montane rainforest in southeastern Madagascar [22, 55] (Fig. 3b). Red-bellied lemurs in RNP have been the subjects of multiple research projects beginning in the 1980s [44–47].

Dataset
Our dataset consists of 462 images of 80 red-bellied lemur individuals. Each individual had a name (e.g., Avery) or code (e.g., M9VAL) assigned by researchers when it was first encountered. Photographs of four individuals are from the Duke Lemur Center in North Carolina, while the remainder are from individuals in RNP in Madagascar. The number of images (1–21) per individual varies. The dataset only includes images that contain a frontal view of the lemur's face with little to no obstruction or occlusion. The dataset comprises images with a large range of variation; these include images with mostly subtle differences in illumination and focus (generally including subtle differences in gaze; ~25%), as well as images with greater variation (e.g., facial orientation, the presence of small obstructions, illumination and shadows; ~75%). Fig. 4 contains a histogram of the number of images available per individual. Amateur photographers captured photos from RNP using a Canon EOS Rebel T3i with 18–55 and 75–300 mm lenses. Lemurs were often at heights between 15–30 m, and photos were taken while standing on the ground. Images from the Duke Lemur Center were captured with a Google Nexus 5 or an Olympus E-450 with a 14–42 mm lens. Lemurs were in low trees (0–3 m), on the ground, or in enclosures, and photos were taken while standing on the ground.

The majority of images taken in Madagascar were captured from September 2014 to March 2015, though some individuals had images captured as early as July 2011. Images from the Duke Lemur Center were captured in July 2014. Due to the longer duration of the image collection in Madagascar, there was some difficulty establishing whether certain individuals encountered in

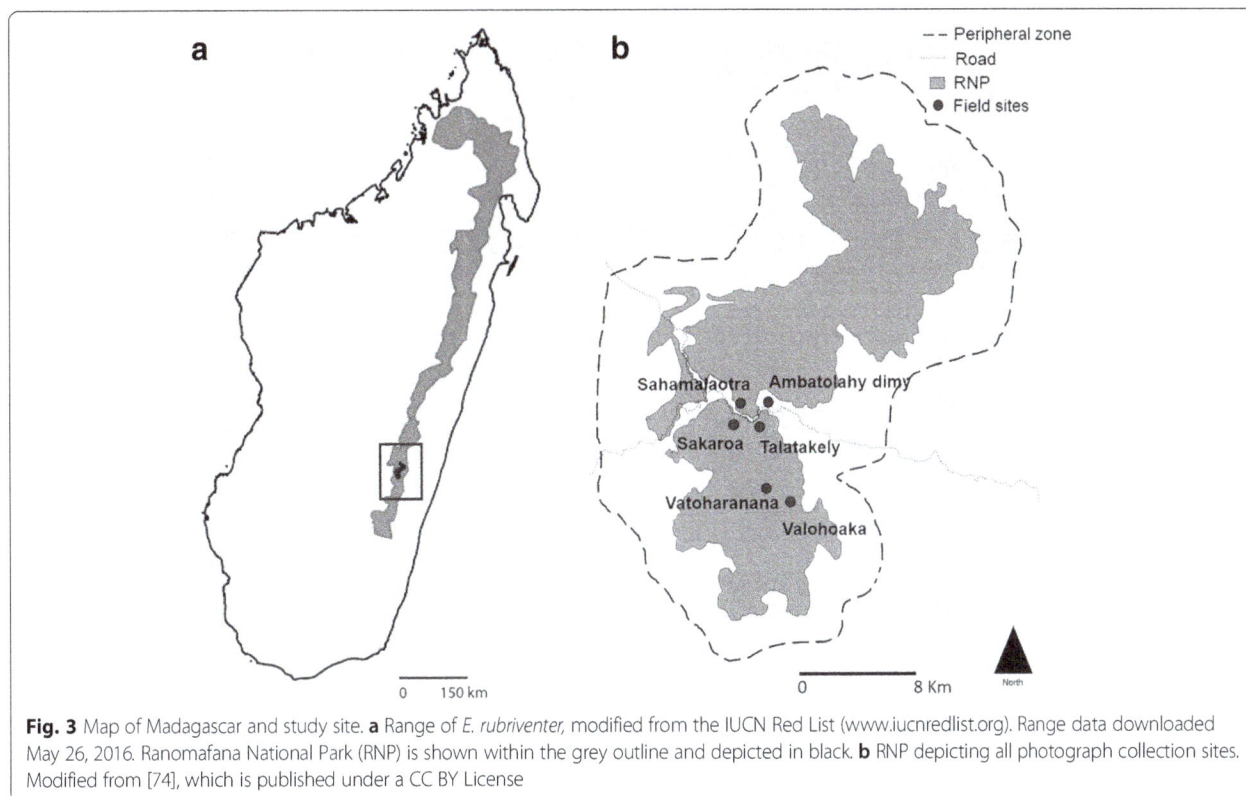

Fig. 3 Map of Madagascar and study site. **a** Range of *E. rubriventer*, modified from the IUCN Red List (www.iucnredlist.org). Range data downloaded May 26, 2016. Ranomafana National Park (RNP) is shown within the grey outline and depicted in black. **b** RNP depicting all photograph collection sites. Modified from [74], which is published under a CC BY License

2014 had been encountered previously. In three cases, there are photographs in the dataset labeled as belonging to two separate individuals that might be of the same individual. These images were treated as belonging to separate individuals when partitioning the dataset for experiments, but if images that might belong to a single individual were matched together, it was counted as a successful match. Figure 5 illustrates the facial similarities and variations present in the dataset. Figure 5a illustrates the similarities and differences between the 80 wild individuals (inter-class similarity), while Fig. 5b shows different images of the same individual (intra-class variability). In addition to the database of red-bellied lemur individuals, a database containing lemurs of other species was assembled. This

database includes 52 images of 31 individuals from Duke Lemur Center and 138 images of lemurs downloaded using an online image search through Google Images. We used only those images with no apparent copyrights. These images were used to expand the size of the gallery for lemur identification experiments.

Recognition system
Figure 6 illustrates the operation of our recognition system (LemurFaceID). This system was implemented using the OpenBR framework (openbiometrics.org; [56]).

Image pre-processing
Eye locations have been found to be critical in human face recognition [40]. The locations of eyes are critical to normalizing the facial image for in-plane rotation. We were unable to design and train a robust eye detector for lemurs because our dataset was not sufficiently large to do so. For this reason, we used manual eye location. Prior to matching, the user marks the locations of the lemur's eyes in the image. Using these two points, with the right eye as the center, a rotation matrix M is calculated to apply an affine transformation to align the eyes horizontally. Let *lex*, *ley*, *rex*, and *rey* represent the x and y coordinates of the left and right eyes, respectively. The affine matrix is defined as:

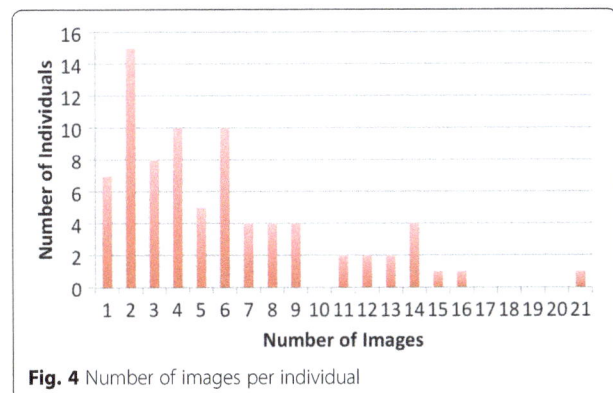

Fig. 4 Number of images per individual

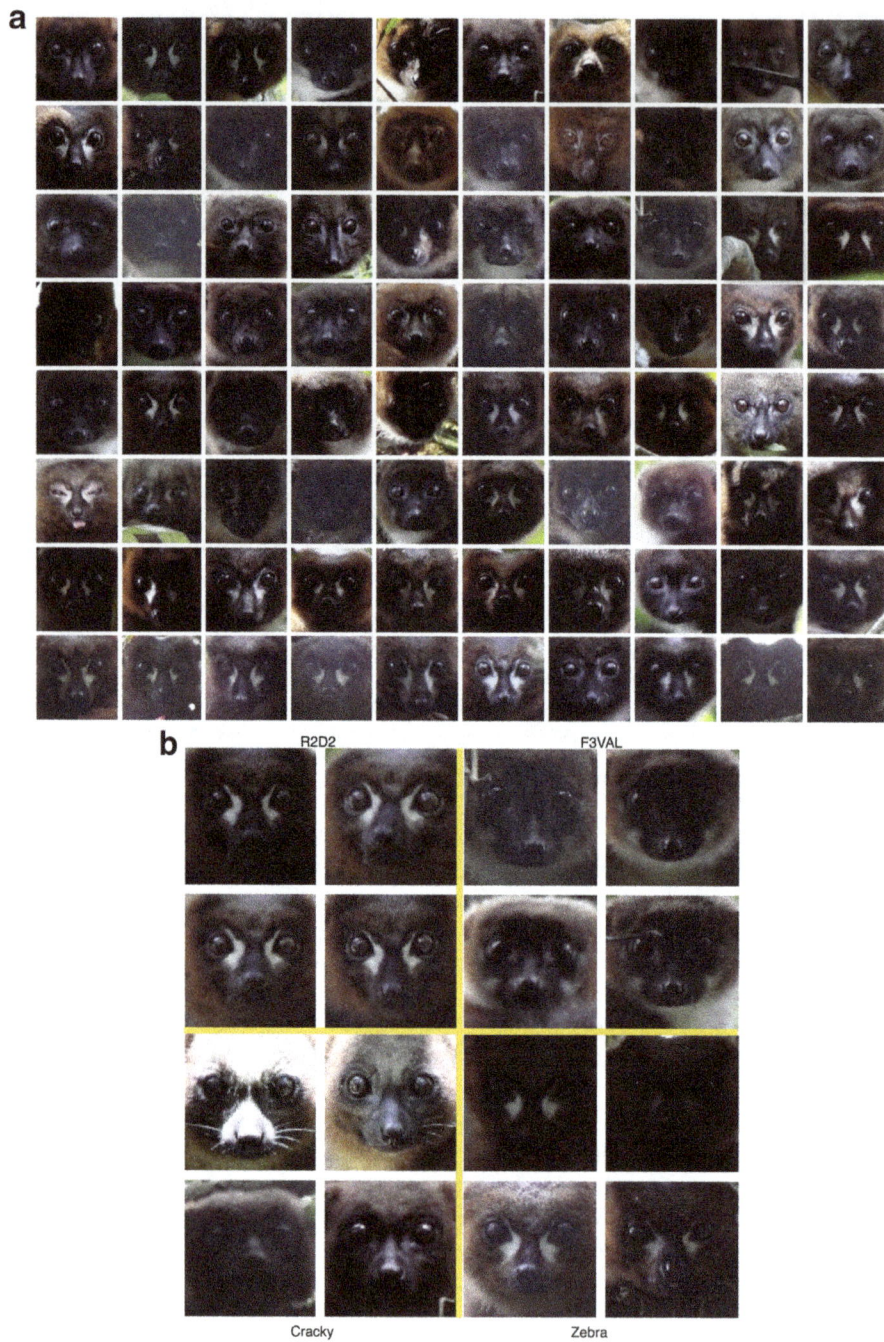

Fig. 5 Variation in lemur face images. **a** Inter-class variation. **b** Intra-class variation. Some images in this figure are modified (i.e., cropped) versions of images that have been previously published in [74] under a CC BY License

$$M = \begin{bmatrix} 0 & 0 & rex \\ 0 & 0 & rey \\ 0 & 0 & 1 \end{bmatrix} \times \begin{bmatrix} cos(\theta) & -sin(\theta) & 0 \\ sin(\theta) & cos(\theta) & 0 \\ 0 & 0 & 1 \end{bmatrix} \times \begin{bmatrix} 0 & 0 & -rex \\ 0 & 0 & -rey \\ 0 & 0 & 1 \end{bmatrix}$$

$$\theta = atan\left(\frac{ley-rey}{lex-rex}\right)$$

The input image is rotated by the matrix M and then cropped based on the eye locations. Rotation is applied prior to cropping so that the area cropped will be as accurate as possible. The Inter-Pupil Distance (IPD) is taken as the Euclidean distance between the eye points. The image is cropped so that the eyes are $\frac{IPD}{2}$ pixels from the nearest edge and $0.7 \times IPD$ pixels from the top edge, with a total dimension of $IPD \times 2$ pixels square. This image is then resized to the final size of 104×104 pixels, which facilitates the patch-wise feature extraction

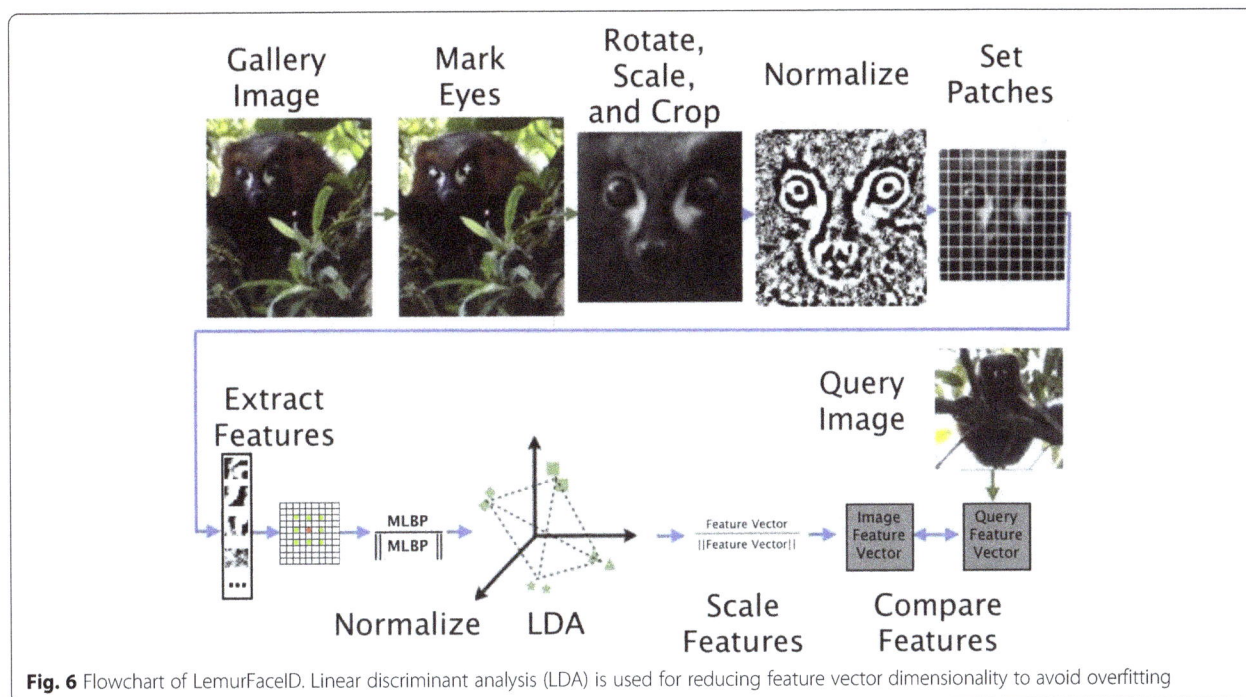

Fig. 6 Flowchart of LemurFaceID. Linear discriminant analysis (LDA) is used for reducing feature vector dimensionality to avoid overfitting

scheme described below. This process is illustrated in Fig. 7. Following rotation and cropping, the image is converted to gray-scale and normalized. Although individual lemurs do show variation in pelage/skin coloration, we disregard color information from the images. In human face recognition studies, skin color is known to be sensitive to illumination conditions and therefore is not considered to be a reliable attribute [57, 58].

Since the primary application of the LemurFaceID system is to identify lemurs from photos taken in the wild, the results must be robust with respect to illumination variations. To reduce the effects of ambient illumination on the matching results, a modified form of the illumination normalization method outlined by Tan and Triggs [48] is applied. The image is first convolved with a Gaussian filter with $\sigma = 1.1$, and is then gamma corrected ($\gamma = 0.2$). A Difference of Gaussians (DoG) operation [48] (with parameters σ_1 and σ_2 corresponding to the standard deviations of the two Gaussians) is subsequently performed on the image. This operation eliminates small-scale texture variations and is traditionally performed with $\sigma_1 = 1$ and $\sigma_2 = 2$. In the case of lemurs, there is an ample amount of hair with a fine texture that varies from image to image within individuals. This fine texture could confuse the face matcher, as changes in hair orientation would result in increased differences between face representations. To reduce this effect in the normalized images, σ_1 is set to 2. The optimal value of σ_2 was empirically determined to be 5. The result of this operation is then contrast equalized using the method outlined in Tan and Triggs [48], producing a face image suitable for feature extraction. Figure 8 illustrates a single lemur image after each pre-processing step.

Feature extraction

Local Binary Pattern (LBP) representation is a method of characterizing local textures in a patch-wise manner [50]. Each pixel in the image is assigned a value based on its relationship to the surrounding pixels, specifically based on whether each surrounding pixel is darker than

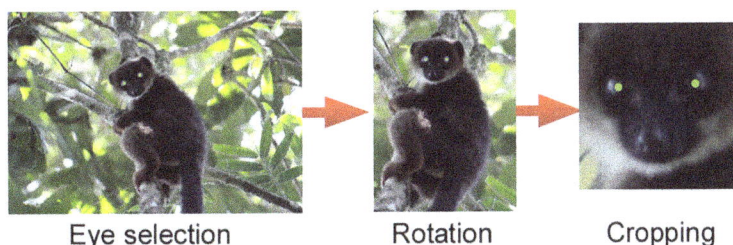

Fig. 7 Eye selection, rotation, and cropping of a lemur image

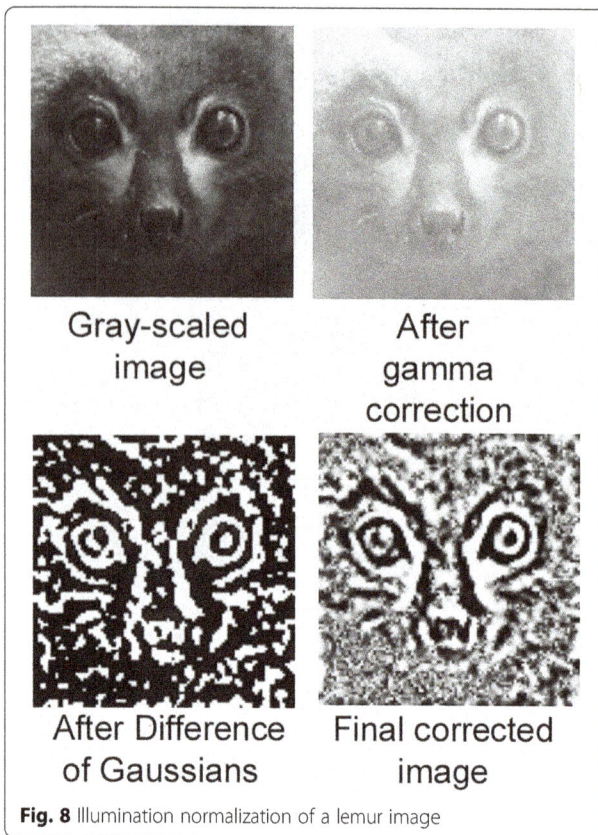

Fig. 8 Illumination normalization of a lemur image

For this application, we used radii of 2, 4, and 8 pixels. Therefore, each patch generates 3 histograms, one per radius, each of which is normalized, and then concatenated and normalized again, both times by L2 norm. This process results in a 177-dimensional feature vector for each 10×10 patch. Figure 10 shows an example of three face images of the same individual with an enlarged grid overlaid. As demonstrated by the highlighted areas, patches from the same area in each image will be compared in matching.

To extract the final feature vector, linear discriminant analysis (LDA) is performed on the 177-dimensional feature vector for each patch. LDA transforms the feature vector into a new, lower-dimensional feature vector such that the new vector still captures 95% of the variation between individuals, while minimizing the amount of variation between images of the same individual. For this transformation to be robust, a large training set of lemur face images is desirable. LDA is trained on a per-patch basis to limit the size of the feature vectors considered. The resulting vectors for all the patches are then concatenated and normalized to produce the final feature vector for the image. Because each patch undergoes its own dimensionality reduction, the final dimensionality of the feature vector will vary from one training set to another. The LemurFaceID system reduces the mean size of the resultant image features from 396,850 dimensions to 7,305 dimensions.

Face matching

In preparation for matching two lemur faces, a gallery (a database of face images and their identities against which a query is searched) is assembled containing feature representations of multiple individual lemurs. The Euclidean distance d between feature vectors of a query image and each image in the gallery is calculated. The final similarity metric is defined as $[1 - log(d + 1)]$; higher values indicate more similar faces. A query can consist of 1 or more images, all of which must be of the same lemur. For each query image, the highest similarity score for each individual represents that individual's match score. The mean of these scores, over multiple query

the central pixel or not. Out of the 256 possible binary patterns in a 3×3 pixel neighborhood, 58 are defined as uniform (having no more than 2 transitions between "darker" and "not darker") [50]. The image is divided into multiple patches (which may or may not overlap), and for each patch a histogram of the patterns is developed. Each of the 58 uniform patterns occupies its own bin, while the non-uniform patterns occupy a 59th bin [50]. This histogram makes up a 59-dimensional feature vector for each patch. In our recognition system, we use 10×10 pixel patches, overlapping by 2 pixels on a side. This results in 144 total patches for the 104×104 face image.

Multi-scale Local Binary Pattern (MLBP) features are a variation on LBP which use surrounding pixels at different radii from the central pixel [50], as shown in Fig. 9.

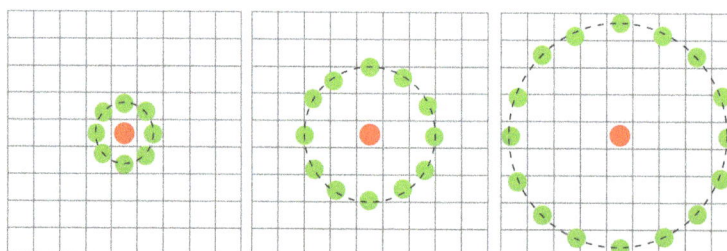

Fig. 9 Local binary patterns of radii 1, 2, and 4. Image from https://upload.wikimedia.org/wikipedia/commons/c/c2/Lbp_neighbors.svg, which is published under the GNU Free Documentation License, Version 1.2 under the Creative Commons

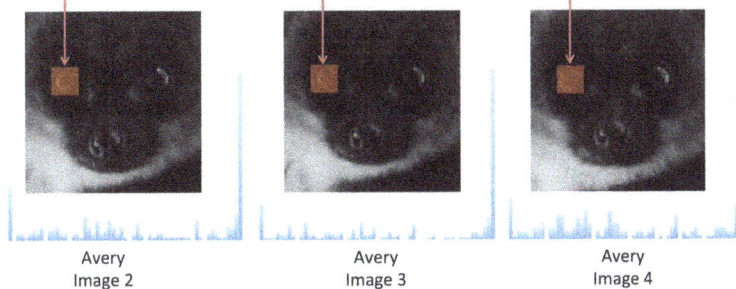

Fig. 10 Patches and corresponding LBP histograms compared across different images of a single lemur (Avery)

images, is calculated to obtain the final individual scores. The top five ranking results (i.e., individuals with the 5 highest scores) are presented in descending order. We evaluated LemurFaceID systems' recognition performance with queries consisting of 1 and 2 images.

Figure 11a shows match score histograms for genuine (comparing 2 instances of the same lemur) vs. impostor (comparing 2 instances of different lemurs) match scores with 1 query image. Figure 11b shows score histograms with fusion of 2 query images. Note that the overlap between genuine and impostor match score histograms is substantially reduced by the addition of a second query image.

Statistical analysis

We evaluated the accuracy of the LemurFaceID system by conducting 100 trials over random splits of the lemur face dataset (462 images of 80 red-bellied lemurs) that we collected. To determine the response of the recognition system to novel individuals, the LDA dimensionality reduction method must be trained on a different set of individuals (i.e., training set) from those used to evaluate matching performance (known as the test set). To satisfy this condition, the dataset was divided into training and testing sets via random split. Two-thirds of the 80 individuals (53 individuals) were designated as the training set, while the remainder (27 individuals) comprised the test set. In the test set, two-thirds of the images for each individual were assigned to the system database (called the 'gallery' in human face recognition literature) and the remaining images were assigned as queries (called the 'probe' in human face recognition literature). Individuals with fewer than 3 images were placed only in the gallery. The gallery was then expanded to include a secondary dataset of other species to increase its size.

Testing was performed in open-set and closed-set identification scenarios. Open-set mode allows for conditions encountered in the wild, where lemurs (query images) may be encountered that have not been seen before (i.e., individuals are not present in the system

database). Queries whose fused match score is lower than a certain threshold are classified as containing a novel individual. Closed-set mode assumes that the query lemur (lemur in need of identification) is represented in the gallery and may be useful for identifying a

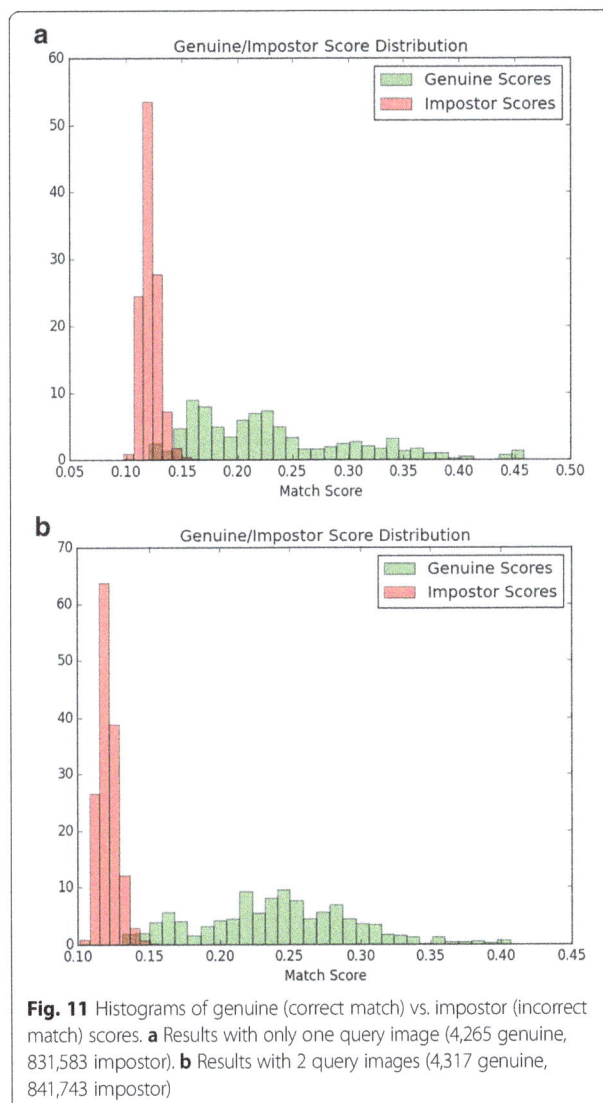

Fig. 11 Histograms of genuine (correct match) vs. impostor (incorrect match) scores. **a** Results with only one query image (4,265 genuine, 831,583 impostor). **b** Results with 2 query images (4,317 genuine, 841,743 impostor)

lemur in situations where the system is guaranteed to know the individual, such as in a captive colony.

For open-set testing, one-third of the red-bellied lemur individuals in the gallery were removed. Their corresponding images in the probe set therefore made up the set of novel individuals. For open-set, the mean gallery size was 266 images, while for closed-set the mean size was 316 images. Across all trials of the LemurFaceID system, the mean probe size was 42 images.

Results

Results of the open-set performance of LemurFaceID are presented in Fig. 12, which illustrates the Detection and Identification Rate (DIR) against the False Accept Rate (FAR). DIR is calculated as the proportion of non-novel individuals that were correctly identified at or below a given rank. FAR is calculated as the number of novel individuals incorrectly matched to a gallery individual at or below a given rank. In general, individuals are correctly identified >95% of the time at rank 5 or higher regardless of FAR, but DIR is lower (<95%) at rank 1, only approaching 95% when FAR is high (0.3).

Rank 1 face matching results for closed-set operation are reported in Table 2, and the Cumulative Match Characteristic (CMC) curves for 1-image query and 2-image fusion (combining matching results for the individual query images) are shown in Fig. 13. This plot shows the proportion of correct identifications at or below a given rank. The mean percentage of correct matches (i.e., Mean True Accept Rate) increases when 2 query images are fused; individuals are correctly identified at Rank 1 98.7% ± 1.81% using 2-image fusion compared to a Rank 1 accuracy of 93.3% ± 3.23% when matching results for a single query image are used.

Discussion

Our initial analyses of LemurFaceID suggest that facial recognition technology may be a useful tool for individual identification of lemurs. This method represents, to our knowledge, the first system for machine identification of lemurs by facial features. LemurFaceID exhibited a relatively high level of recognition accuracy (98.7%; 2-query image fusion) when used in closed-set mode (i.e., all individuals are present in the dataset), which could make this system particularly useful in captive settings, as well as wild populations with low levels of immigration from unknown groups. Given the success of Lemur-FaceID in recognizing individual lemurs, this method could also allow for a robust species recognition system, which would be useful for presence/absence studies.

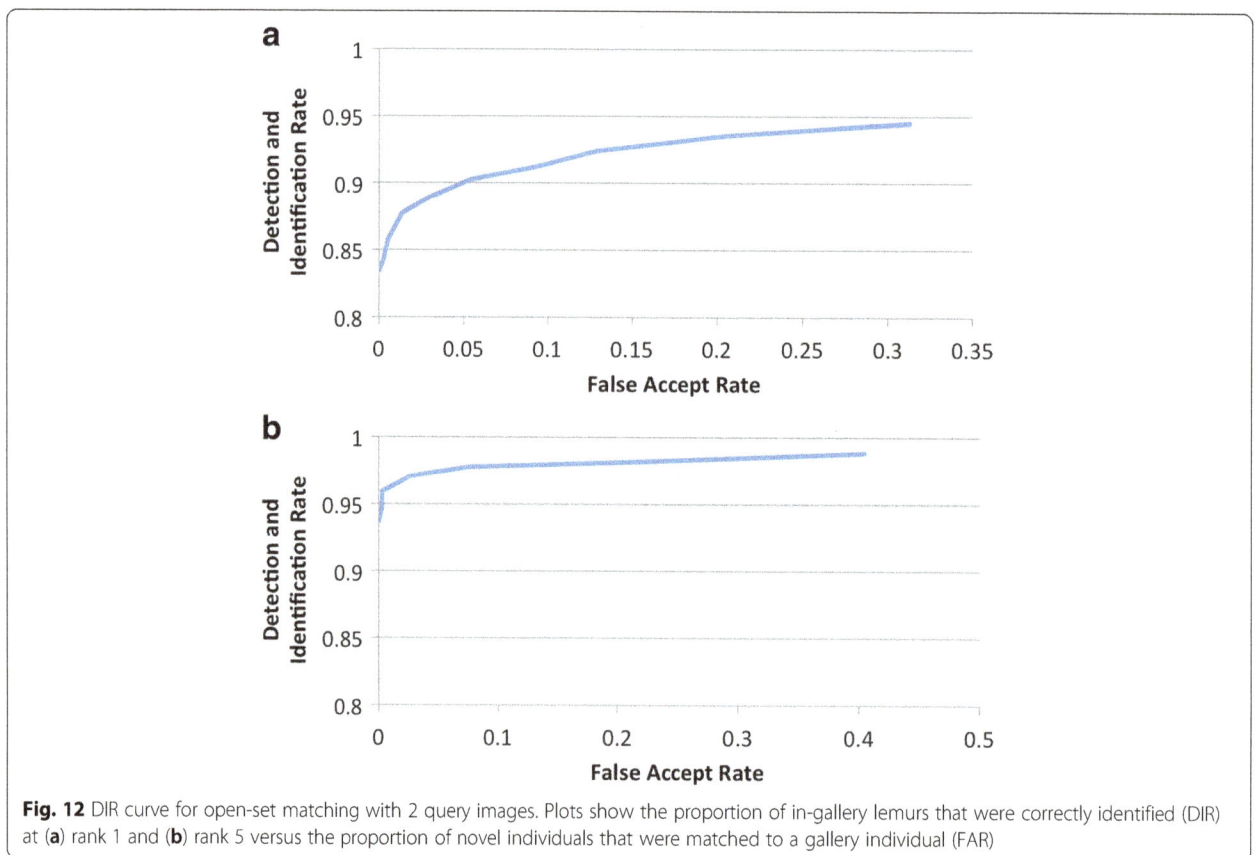

Fig. 12 DIR curve for open-set matching with 2 query images. Plots show the proportion of in-gallery lemurs that were correctly identified (DIR) at (**a**) rank 1 and (**b**) rank 5 versus the proportion of novel individuals that were matched to a gallery individual (FAR)

Table 2 Face matcher evaluation results (Rank 1, closed-set)

Method	Mean (TAR)	SD
Baseline system	81.5%	6.68%
2 query images	98.7%	1.81%
1 query image	93.3%	3.23%

True Accept Rate (TAR) is the percentage of correct matches. Standard deviation (SD) is computed over 100 random splits. The LemurFaceID system is also compared to the earlier (i.e., "Baseline") system (using SIFT) for comparison

The accuracy of our system was lower using open-set mode (i.e., new individuals may be encountered) where, regardless of the False Accept Rate (FAR), non-novel individuals were correctly identified at rank 1 less than 95% of the time and less than 85% of the time given a FAR of 0. These numbers are expected to improve with a larger dataset of photographs and individuals. In our current sample, we also included photographs exhibiting only subtle variation between images. Given that the

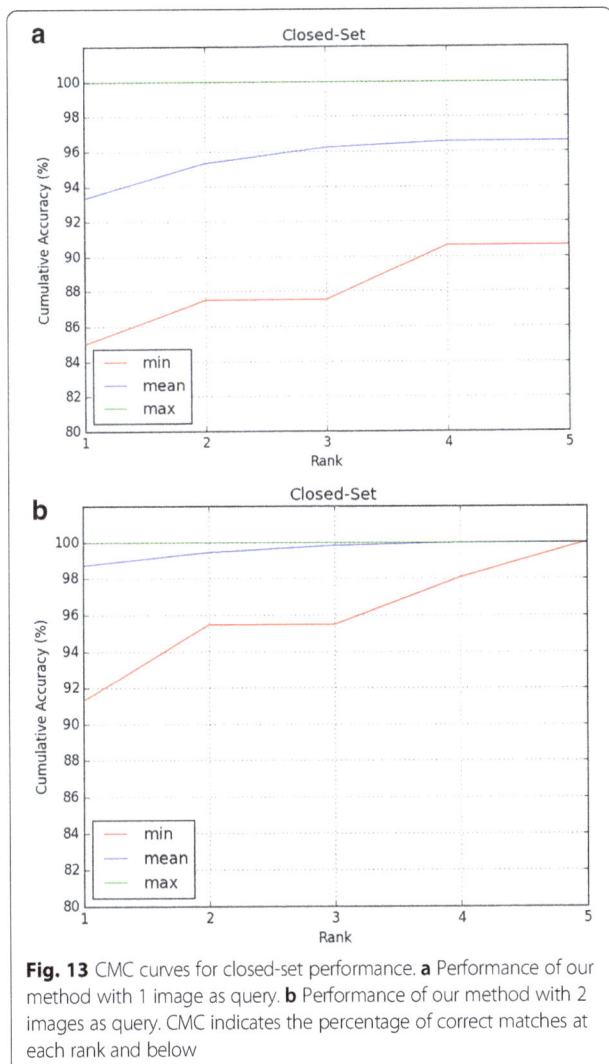

Fig. 13 CMC curves for closed-set performance. **a** Performance of our method with 1 image as query. **b** Performance of our method with 2 images as query. CMC indicates the percentage of correct matches at each rank and below

ultimate goal of LemurFaceID is to provide an alternative, non-invasive identification method for long-term research, it will also be important to test its accuracy using a larger dataset that includes only photographs with large variation (e.g., collected across multiple, longer-term intervals).

We also note that our system focuses specifically on classifying individuals using a dataset of known individuals in a population. Such a tool can be particularly useful for maintaining long-term research on a study population. This approach differs, however, from another potential application of face recognition methods, which would be to identify the number of individuals from a large image dataset containing unknown individuals only (i.e., clustering) [59, 60]. The addition of a clustering technique could allow for more rapid population surveys or facilitate the establishment of new study sites, but such techniques can be challenging as clustering accuracy is expected to be lower than the classification accuracy [59, 60]. That said, in future work, the feature extraction and scoring system of LemurFaceID could potentially be combined with clustering techniques for segmenting datasets of unknown individuals.

Despite some current limitations, LemurFaceID provides the groundwork for incorporating this technology into long-term research of wild lemur populations, particularly of larger-bodied (>2 kg) species. Moving forward, we aim to 1) expand our photographic database, which is necessary to automate the lemur face detector and eye locator, 2) increase open-set performance by improving the feature representation to provide better separation between scores for in-gallery and novel individuals, and 3) field test the system to compare the classification accuracy of LemurFaceID with that of experienced and inexperienced field observers. Once optimized, a non-invasive, computer-assisted program for individual identification in lemurs has the potential to mitigate some of the challenges faced by long-term research using more traditional methods.

For example, facial recognition technology would remove the need to artificially tag individuals, which removes potential risks to animals associated with capturing and collaring; some of these risks, including injury, occur more frequently in arboreal primates [29]. At the same time, many costs incurred using these techniques are removed (e.g., veterinary services, anesthesia), as are potential restrictions on the number of individuals available for study (e.g., local government restrictions on captures). More traditional non-invasive techniques that rely on researchers' knowledge of natural variation can be similarly advantageous, but facial recognition programs can help ensure that data are collected consistently across multiple researchers. That said, we would not recommend researchers become wholly reliant on computer programs for individual identification of study subjects, but training multiple researchers to accurately recognize hundreds of individuals is time-consuming and costly, as well as potentially unrealistic. Facial recognition technology can facilitate long-term monitoring of large populations by removing the need for extensive training, or potentially accelerate training by making phenotypic differences more tangible to researchers and assistants. Moreover, in studies with large sample sizes where immediate recognition of all individuals might be impossible, facial recognition technology can process data more quickly. For example, LemurFaceID takes less than one second to recognize a lemur (using a quad core i7 processor), which will save time identifying individuals when manual comparisons of photographs/descriptions are necessary.

Ultimately then, LemurFaceID can help expand research on lemur populations by providing a method to systematically identify a large number of individuals over extended periods of time. As is the case with other long-term studies of natural populations, this research has the potential to provide substantial contributions to evolutionary biology [1]. More specifically, lemurs are an endemic mammalian lineage that evolved in Madagascar beginning >50 million years ago [61]. Over time, they have greatly diversified with >100 species recognized today [43]. They occupy diverse niches (e.g., small-bodied, nocturnal gummivores; arrhythmic frugivores; large-bodied, diurnal folivores) across Madagascar's varied habitats (e.g., rainforests; spiny, dry forest) [43], and they have recently (in the last ~2,000 years) experienced extensive ecological change owing largely to human impact [62]. Accordingly, this mammalian system provides unique opportunities for studying ecological and evolutionary pressures impacting wild populations.

Data obtained from longitudinal studies of lemurs can also aid in conservation planning and management for this highly endangered group of mammals. Demographic structure and life history parameters documented from long-term research can provide insights into the causes of population change and be used to model extinction risk [63–65]. LemurFaceID also has potential for more direct applications to conservation. One notable threat to lemurs [66, 67], as well as many other animal species [68, 69], is live capture of individuals for the pet trade. LemurFaceID could provide law enforcement, tourists, and researchers with a tool to rapidly report sightings and identify captive lemurs (species and individuals). A database of captive lemurs can help with continued monitoring to determine if individuals remain constant over time.

Importantly, the face recognition methods we developed for LemurFaceID could be useful for individual identification in other primates, as well as other non-

primate species, especially those with similarly variable facial pelage/skin patterns (e.g., bears, red pandas, raccoons, sloths). Furthermore, as camera trapping has become increasingly useful for population monitoring of many cryptic species (e.g., [70, 71]), our facial recognition technology could be potentially incorporated into long-term, individual-based studies conducted remotely. That said, it will be necessary to make unique modifications to methods for different lineages.

To illustrate this point, recent publications also have explored the area of facial recognition for primates. For example, Loos and Ernst's [41] system for recognizing chimpanzees has a similar approach to pre-processing as LemurFaceID, but they use a different illumination normalization method and correct for greater difference in perspective. In feature extraction, their use of speeded-up robust features (SURF), a gradient-based feature similar to SIFT, underscores the difference in lemur and chimpanzee faces, namely the lack of hair/fur in chimpanzees to confound the directionality of the features [41]. Their selection of Gabor features also reflects the relative lack of hair, as such indicators of edgeness would exhibit significantly more noise in lemurs [72]. More recently, Freytag et al. [73] were able to improve upon recognition accuracy of chimpanzees by applying convolutional neural network (CNN) techniques. Their results identify CNNs to be a promising direction of animal face recognition research, but such methods also require datasets that are orders of magnitude larger than our current dataset [73]. Thus, although they are beyond the scope of this study, CNNs could be an interesting avenue for future research in lemur face recognition.

In contrast to these approaches, Allen and Higham [42] use a biologically-based model for identification of guenons. Their feature selection is based on guenon vision models, using the dimensions of facial spots to identify species and individuals [42]. While *E. rubriventer* individuals also possess prominent facial spots, these are not common across different lemur species and therefore unsuitable for use in our system. The wide variety of approaches used underscores that there is no "one size fits all" approach to animal facial recognition, but once developed, this technology has the potential to facilitate long-term research in a host of species, expand the types of research questions that can be addressed, and help create innovative conservation tools.

Conclusions

Our non-invasive, computer-assisted facial recognition program (LemurFaceID) was able to identify individual lemurs based on photographs of wild individuals with a relatively high degree of accuracy. This technology would remove many limitations of traditional methods for individual identification of lemurs. Once optimized, our system can facilitate long-term research of known individuals by providing a rapid, cost-effective, and accurate method for individual identification.

Abbreviations
CMC: Cumulative match characteristic; CNN: Convolutional neural network; DIR: Detection and Identification Rate; FAR: False accept rate; IPD: Inter-pupil distance; LBP: Local binary pattern; LDA: Linear discriminant analysis; MLBP: Multiscale local binary pattern; RNP: Ranomafana National Park; SIFT: Scale invariant feature transform; SURF: Speeded-up robust features; TAR: True accept rate

Acknowledgements
Logistics and permissions for research in Madagascar were facilitated by Benjamin Andriamihaja and MICET, Ministre des Eaux et Forets, Madagascar National Parks, Eileen Larney and the Centre ValBio, and the University of Antananarivo. We would like to thank Samantha Ambler, Caroline Angyal, Alicia S. Arroyo, Bashira Chowdhury, Joseph Falinomenjanahary, Sheila Holmes, Jean Pierre Lahitsara, Avery Lane, Natalee Phelps, Aura Raulo, Soafaniry Razanajatovo, and Jean Baptiste Velontsara for their contribution to collecting face images of lemurs, as well as multiple students, volunteers, and research technicians for assisting with data collection. Finally, we thank the Duke Lemur Center (DLC) staff for logistical support during data collection at the DLC. This is DLC publication number 1336.

Funding
This research was supported by funds from the American Association of Physical Anthropologists to SRT, National Science Foundation (DDIG, BCS 1232535) to RLJ, The Leakey Foundation to RLJ, SRT, and ALB, The Wenner-Gren Foundation to RLJ, Rowe-Wright Primate Fund to RLJ, SRT, and ALB, Stony Brook University to RLJ, IDEAWILD to RLJ and SRT, University of Arizona to SRT, Hunter College-CUNY to ALB, and Michigan State University to AJ, DC, and SK. Funders had no role in the design of the study, data collection, analysis, and interpretation, or preparation of the manuscript.

Authors' contributions
RLJ and SRT conceived of the project. RLJ, SRT, ALB, DC, and ZR acquired data for the project. DC, SK, and AJ conceived and designed the recognition system and experiments. DC, ZR, SK, and AJ performed the experiments and analyzed data. RLJ, SRT, AJ, ALB, and DC drafted the manuscript. All authors read and approved of the final manuscript.

Competing interests
The authors declare that they have no competing interests.

Author details
[1]Department of Computer Science and Engineering, Michigan State University, East Lansing, MI, USA. [2]Department of Anthropology, Center for the Advanced Study of Human Paleobiology, The George Washington University, Washington, DC, USA. [3]Department of Anthropology, Hunter College, City University of New York, New York, NY, USA. [4]The Graduate Center of City University of New York, New York, NY, USA. [5]The New York Consortium in Evolutionary Primatology (NYCEP), New York, NY, USA. [6]School of Anthropology, The University of Arizona, Tucson, AZ, USA. [7]Present address: Samsung Semiconductor, Inc., San Jose, CA, USA. [8]Present address: Noblis, Inc., Annandale, VA, USA.

References
1. Clutton-Brock T, Sheldon BC. Individuals and populations: the role of long-term, individual-based studies of animals in ecology and evolutionary biology. Trends Ecol Evol. 2010;25:562–73.
2. Clutton-Brock T, Pemberton J. Soay sheep: dynamics and selection in an island population. Cambridge: Cambridge University Press; 2004.
3. Clutton-Brock TH. Red deer: the behaviour and ecology of two sexes. Chicago: University of Chicago Press; 1982.
4. Lack D, Gibb J, Owen DF. Survival in relation to brood-size in tits. J Zool. 1957;128:313–26.
5. Grant PR, Grant BR. 40 years of evolution: Darwin's finches on Daphne major island. Princeton: Princeton University Press; 2014.
6. Moss CJ. The demography of an African elephant (Loxodonta africana) population in Amboseli, Kenya. J Zool. 2001;255:145–56.
7. Murray CM, Stanton MA, Wellens KR, Santymire RM, Heintz MR, Lonsdorf EV. Maternal effects on offspring stress physiology in wild chimpanzees. Am J Primatol. 2016, in press. DOI:10.1002/ajp.22525.
8. Wich SA, Utami-Atmoko SS, Setia TM, Rijksen HD, Schürmann C, van Hooff JARAM, van Schaik CP. Life history of wild Sumatran orangutans (Pongo abelii). J Hum Evol. 2004;47:385–98.
9. Alberts SC, Altmann J. The amboseli baboon research project: 40 years of continuity and change. In: Kappeler PM, Watts DP, editors. Long-term field studies of primates. Berlin Heidelberg: Springer; 2012. p. 261–88.
10. Gratten J, Pilkington JG, Brown EA, Clutton-Brock TH, Pemberton JM, Slate J. Selection and microevolution of coat pattern are cryptic in a wild population of sheep. Mol Ecol. 2012;21:2977–90.
11. Albon SD, Coulson TN, Brown D, Guinness FE, Pemberton JM, Clutton-Brock TH. Temporal changes in key factors and key age groups influencing the population dynamics of female red deer. J Anim Ecol. 2000;69:1099–110.
12. Bouwhuis S, Vedder O, Garroway CJ, Sheldon BC. Ecological causes of multilevel covariance between size and first-year survival in a wild bird population. J Anim Ecol. 2015;84:208–18.
13. Grant PR, Grant BR. Unpredictable evolution in a 30-year study of Darwin's finches. Science. 2002;296:707–11.
14. Fernandez-Duque E, Rotundo M. Field methods for capturing and marking Azarai night monkeys. Int J Primatol. 2003;24:1113–20.
15. Van Belle S, Fernandez-Duque E, Di Fiore A. Demography and life history of wild red titi monkeys (Callicebus discolor) and equatorial sakis (Pithecia aequatorialis) in Amazonian Ecuador: A 12-year study. Am J Primatol. 2016;78:204–15.
16. Wasserman MD, Chapman CA, Milton K, Goldberg TL, Ziegler TE. Physiological and behavioral effects of capture darting on red colobus monkeys (Procolobus rufomitratus) with a comparison to chimpanzee (Pan troglodytes) predation. Int J Primatol. 2013;34:1020–31.
17. Wright PC. Demography and life history of free-ranging Propithecus diadema edwardsi in Ranomafana National Park, Madagascar. Int J Primatol. 1995;16:835–54.
18. Richard AF, Dewar RE, Schwartz M, Ratsirarson J. Life in the slow lane? demography and life histories of male and female sifaka (Propithecus verreauxi verreauxi). J Zool. 2002;256:421–36.
19. Irwin MT. Living in forest fragments reduces group cohesion in diademed sifakas (Propithecus diadema) in eastern Madagascar by reducing food patch size. Am J Primatol. 2007;69:434–47.
20. Leimberger KG, Lewis RJ. Patterns of male dispersal in Verreaux's sifaka (Propithecus verreauxi) at Kirindy Mitea National Park. Am J Primatol. 2016. DOI:10.1002/ajp.22455.
21. Fernandez-Duque E. Natal dispersal in monogamous owl monkeys (Aotus azarai) of the Argentinean Chaco. Behaviour. 2009;146:583–606.
22. Wright PC, Erhart EM, Tecot S, Baden AL, Arrigo-Nelson SJ, Herrera J, Morelli TL, Blanco MB, Deppe A, Atsalis S, Johnson S, Ratelolahy F, Tan C, Zohdy S. Long-term research at Centre ValBio, Ranomafana National Park, Madagascar. In: Kappeler PM, Watts DP, editors. Long-term field studies of primates. Berlin Heidelberg: Springer; 2012. p. 67–100.
23. Tecot S, Gerber B, King S, Verdolin J, Wright PC. Risky business: sex ratio, mortality, and group transfer in Propithecus edwardsi in Ranomafana National Park, Madagascar. Behav Ecol. 2013;24:987–96.
24. Zohdy S, Gerber BD, Tecot S, Blanco MB, Winchester JM, Wright PC, Jernvall J. Teeth, sex, and testosterone: aging in the world's smallest primate. PLoS ONE. 2014;9:e109528.
25. Glander KE, Wright PC, Daniels PS, Merenlender AM. Morphometrics and testicle size of rain-forest lemur species from southeastern Madagascar. J Hum Evol. 1992;22:1–17.

26. Sorin AB. Paternity assignment for white-tailed deer (*Odocoileus virginianus*): mating across age classes and multiple paternity. J Mammal. 2004;85:356–62.

27. Loison A, Solberg EJ, Yoccoz NG, Langvatn R. Sex differences in the interplay of cohort and mother quality on body mass of red deer calves. Ecology. 2004;85:1992–2002.

28. Hopkins ME, Milton K. Adverse effects of ball-chain radio-collars on female mantled howlers (*Alouatta palliata*) in Panama. Int J Primatol. 2016;37:213–24.

29. Cunningham EP, Unwin S, Setchell JM. Darting primates in the field: a review of reporting trends and a survey of practices and their effect on the primates involved. Int J Primatol. 2015;36:911–32.

30. Côté SD, Festa-Bianchet M, Fournier F. Life-history effects of chemical immobilization and radiocollars on mountain goats. J Wildlife Manage. 1998; 62:745–52.

31. Moorhouse TP, Macdonald DW. Indirect negative impacts of radio-collaring: sex ratio variation in water voles. J Appl Ecol. 2005;42:91–8.

32. Le Maho Y, Saraux C, Durant JM, Viblanc VA, Gauthier-Clerc M, Yoccoz NG, Stenseth NC, Le Bohec C. An ethical issue in biodiversity science: the monitoring of penguins with flipper bands. C R Biol. 2011;334:378–84.

33. Tecot SR. It's all in the timing: out of season births and infant survival in *Eulemur rubriventer*. Int J Primatol. 2010;31:715–35.

34. Kühl HS, Burghardt T. Animal biometrics: quantifying and detecting phenotypic appearance. Trends Ecol Evol. 2013;28:432–41.

35. Kelly MJ. Computer-aided photograph matching in studies using individual identification: an example from Serengeti cheetahs. J Mammal. 2001;82:440–9.

36. Hiby L, Lovell P, Patil N, Kumar NS, Gopalaswamy AM, Karnath KU. A tiger cannot change its stripes: using a three-dimensional model to match images of living tigers and tiger skins. Biol Lett. 2009;5:383–6.

37. Bolger DT, Morrison TA, Vance B, Lee D, Farid H. A computer-assisted system for photographic mark–recapture analysis. Methods Ecol Evol. 2012;3:813–22.

38. Lahiri M, Tantipathananandh C, Warungu R, Rubenstein DI, Berger-Wolf TY. Biometric animal databases from field photographs: identification of individual zebra in the wild. ICMR. 2011. [http://compbio.cs.uic.edu/~mayank/papers/LahiriEtal_ZebraID11.pdf] Downloaded 1 July 2013.

39. Crouse D, Richardson Z, Jain A, Tecot S, Baden A, Jacobs R. Lemur face recognition: tracking a threatened species and individuals with minimal impact. MSU Technical Report 2015. MSU-CSE-15-8, May 23, 2015.

40. Li SZ, Jain AK. Handbook of face recognition. 2nd ed. London: Springer; 2011.

41. Loos A, Ernst A. An automated chimpanzee identification system using face detection and recognition. EURASIP J Image Video Process. 2013;1:1–17.

42. Allen AL, Higham JP. Assessing the potential information content of multicomponent visual signals: a machine learning approach. Proc R Soc B. 2015;282:20142284.

43. Mittermeier RA, Louis EE, Richardson M, Schwitzer C, Langrand O, Rylands AB, Hawkins F, Rajaobelina S, Ratsimbazafy J, Rasoloarison R, Roos C, Kappeler PM, MacKinnon J. Lemurs of Madagascar. Arlington: Conservation International; 2010.

44. Overdorff DJ. Ecological correlates to social structure in two prosimian primates: *Eulemur fulvus rufous* and *Eulemur rubriventer* in Madagascar. PhD thesis. Duke University, Durham: Department of Biological Anthropology and Anatomy; 1991.

45. Durham DL. Variation in responses to forest disturbance and the risk of local extinction: a comparative study of wild *Eulemurs* at Ranomafana National Park. PhD thesis. University of California, Davis: Department of Animal Behavior; 2003.

46. Tecot SR. Seasonality and predictability: the hormonal and behavioral responses of the red-bellied lemur, *Eulemur rubriventer*, in southeastern Madagascar. PhD thesis. University of Texas at Austin, Austin: Department of Anthropology; 2008.

47. Jacobs RL. The evolution of color vision in red-bellied lemurs (*Eulemur rubriventer*). PhD thesis. Stony Brook University, Stony Brook: Department of Anthropology (Physical Anthropology); 2015.

48. Tan X, Triggs B. Enhanced local texture feature sets for face recognition under difficult lighting conditions. IEEE T Image Process. 2010;19:1635–50.

49. Klum S, Han H, Jain AK, Klare B: Sketch based face recognition: forensic vs. composite sketches. In Biometrics (ICB), 2013 International Conference on Biometrics Compendium, IEEE. 2013:1–8. DOI: 10.1109/ICB.2013.6612993.

50. Ojala T, Pietikainen M, Maenpaa T. Multiresolution gray-scale and rotation invariant texture classification with local binary patterns. IEEE T Pattern Anal. 2002;24:971–87.

51. Lowe DG. Distinctive image features from scale-invariant keypoints. Int J Comput Vision. 2004;60:91–110.

52. Overdorff DJ. Similarities, differences, and seasonal patterns in the diets of *Eulemur rubriventer* and *Eulemur fulvus rufus* in the Ranomafana National Park, Madagascar. Int J Primatol. 1993;14:721–53.

53. Harper GJ, Steininger MK, Tucker CJ, Juhn D, Hawkins F. Fifty years of deforestation and forest fragmentation in Madagascar. Environ Conserv. 2007;34:325–33.

54. Andriaholinirina N, Baden A, Blanco M, Chikhi L, Cooke A, Davies N, Dolch R, Donati G, Ganzhorn J, Golden C, Groeneveld LF, Hapke A, Irwin M, Johnson S, Kappeler P, King T, Lewis R, Louis EE, Markolf M, Mass V, Mittermeier RA, Nichols R, Patel E, Rabarivola CJ, Raharivololona B, Rajaobelina S, Rakotoarisoa G, Rakotomanga B, Rakotonanahary J, Rakotondrainibe H et al.. *Eulemur rubriventer*. The IUCN Red List of Threatened Species 2014, Version 2015.2. Downloaded on 26 May 2016 [www.iucnredlist.org].

55. Wright PC. Primate ecology, rainforest conservation, and economic development: building a national park in Madagascar. Evol Anthropol. 1992;1:25–33.

56. Klontz JC, Klare BF, Klum S, Jain AK, Burge MJ: Open source biometric recognition. In Proceedings of Biometrics: Theory, Applications and Systems (BTAS), 2013 IEEE Sixth International Conference. 2013:1–8. [http://openbiometrics.org/publications/klontz2013open.pdf].

57. Yip AW, Sinha P. Contribution of color to face recognition. Perception. 2002; 31:995–1003.

58. Martinkauppi JB, Hadid A, Pietikainen M. Skin color in face analysis. In: Li SZ SZ, Jain AK, editors. Handbook of face recognition. Secondth ed. London: Springer; 2011. p. 223–49.

59. Jain AK, Dubes RC. Algorithms for clustering data. New Jersey: Prentice Hall; 1988.

60. Otto C, Klare BF, Jain AK. An efficient approach for clustering face images. In Proceedings of IEEE International Conference on Biometrics (ICB). Phuket, Thailand, May 19–22, 2015. (doi: 10.1109/ICB.2015.7139091)

61. Yoder AD, Yang Z. Divergence dates for Malagasy lemurs estimated from multiple gene loci: geological and evolutionary context. Mol Ecol. 2004;13:757–73.

62. Crowley BE, Godfrey LR, Bankoff RJ, Perry GH, Culleton BJ, Kennett DJ, Sutherland MR, Samonds KE, Burney DA. Island-wide aridity did not trigger recent megafaunal extinctions in Madagascar. Ecography. 2016,. DOI:10.1111/ecog.02376.

63. Brook BW, O'Grady JJ, Chapman AP, Burgman MA, Akçakaya HR, Frankham R. Predictive accuracy of population viability analysis in conservation biology. Nature. 2000;404:385–7.

64. Strier KB, Alberts S, Wright PC, Altmann J, Zeitlyn D. Primate life-history databank: setting the agenda. Evol Anthropol. 2006;15:44–6.

65. Strier KB, Altmann J, Brockman DK, Bronikowski AM, Cords M, Fedigan LM, Lapp H, Liu X, Morris WF, Pusey AE, Stoinski TS, Alberts SC. The Primate Life History Database: a unique shared ecological data resource. Methods Ecol Evol. 2010;1:199–211.

66. Reuter KE, Gilles H, Wills AR, Sewall BJ. Live capture and ownership of lemurs in Madagascar: extent and conservation implications. Oryx. 2016;50:344–54.

67. Reuter KE, Schaefer MS. Captive conditions of pet lemurs in Madagascar. Folia Primatol. 2016;2016(87):48–63.

68. Nijman V, Nekaris KAI, Donati G, Bruford M, Fa J. Primate conservation: measuring and mitigating trade in primates. Endang Species Res. 2011;13:159–61.

69. Bush ER, Baker SE, Macdonald DW. Global trade in exotic pets 2006–2012. Conserv Biol. 2014;28:663–76.

70. Jackson RM, Roe JD, Wangchuk R, Hunter DO. Estimating snow leopard population abundance using photography and capture-recapture techniques. Wildlife Soc Bull. 2006;34:772–81.

71. Karanth KU, Nichols JD, Kumar NS, Hines JE. Assessing tiger population dynamics using photographic capture-recapture sampling. Ecology. 2006;87:2925–37.

72. Jain AK, Farrokhnia F. Unsupervised texture segmentation using Gabor filters. Pattern Recogn. 1991;24:1167–86.

73. Freytag A, Rodner E, Simon M, Loos A, Kühl HS, Denzler J. Chimpanzee faces in the wild: Log-Euclidean CNNs for predicting identities and attributes of primates. In: Rosenhahn B, Bjoern A, editors. Pattern recognition 38th German conference, GCPR 2016, Hannover, Germany, September 12–15, 2016, proceedings. Switzerland: Springer International Publishing AG; 2016. p. 51–63.

74. Jacobs RL, Bradley BJ. Considering the influence of nonadaptive evolution on primate colour vision. Plos One. 2016;11:e0149664.

4

Caenorhabditis monodelphis sp. n.: defining the stem morphology and genomics of the genus Caenorhabditis

Dieter Slos[1]* ⓘ, Walter Sudhaus[2], Lewis Stevens[3]*, Wim Bert[1] and Mark Blaxter[3]

Abstract

Background: The genus *Caenorhabditis* has been central to our understanding of metazoan biology. The best-known species, *Caenorhabditis elegans*, is but one member of a genus with around 50 known species, and knowledge of these species will place the singular example of *C. elegans* in a rich phylogenetic context. How did the model come to be as it is today, and what are the dynamics of change in the genus?

Results: As part of this effort to "put *C. elegans* in its place", we here describe the morphology and genome of *Caenorhabditis monodelphis* sp. n., previously known as *Caenorhabditis* sp. 1. Like many other *Caenorhabditis*, *C. monodelphis* sp. n. has a phoretic association with a transport host, in this case with the fungivorous beetle *Cis castaneus*. Using genomic data, we place *C. monodelphis* sp. n. as sister to all other *Caenorhabditis* for which genome data are available. Using this genome phylogeny, we reconstruct the stemspecies morphological pattern of *Caenorhabditis*.

Conclusions: With the morphological and genomic description of *C. monodelphis* sp. n., another key species for evolutionary and developmental studies within *Caenorhabditis* becomes available. The most important characters are its early diverging position, unique morphology for the genus and its similarities with the hypothetical ancestor of *Caenorhabditis*.

Keywords: Taxonomy, Systematics, Evolution, Genome, Phylogeny, Description

Background

The nematode genus *Caenorhabditis* includes the well-known model organism *C. elegans*, which has provided key insights into molecular and developmental biology [1]. Over the past ten years, numerous new *Caenorhabditis* species have been discovered and described [2, 3]. These putative new taxa are generally indistinguishable morphologically, and thus the most recent descriptions of new species within *Caenorhabditis* have been based on DNA sequences and mating tests only [2]. This streamlined species-description methodology has been driven by the need to have names to attach to real biological entities, and the fact that traditional taxonomy

has been unable to keep up with species discovery. The method is relatively simple to implement, and delivers taxa that have a biological reality [2]. However, as the number of species discovered in *Caenorhabditis* grows, traditional, morphological descriptions are still valuable for the understanding of patterns of trait evolution and inference of ecological functions [4, 5]. Although morphology cannot be used to definitively delineate species, it should not be abandoned all together.

M-A Félix, C Braendle and AD Cutter [2] provided new species name designations for 15 biological species, considerably increasing the number of named *Caenorhabditis* species in laboratory culture. However, several key *Caenorhabditis* species remain undescribed. A well-known but undescribed species of *Caenorhabditis*, informally referred to as *Caenorhabditis* sp. 1, has been analysed in several evolutionary and developmental studies [3, 6–8]. *C.* sp. 1 was previously found only once inside a fruiting body of the fungus *Ganoderma applanatum* (Pers.) Pat.

* Correspondence: dieterg.slos@ugent.be; lewis.stevens@ed.ac.uk
[1]Department of Biology, Nematology Research Unit, Ghent University, K.L. Ledeganckstraat 35, 9000 Ghent, Belgium
[3]Institute of Evolutionary Biology, University of Edinburgh, Edinburgh EH9 3FL, UK
Full list of author information is available at the end of the article

(Polyporaceae), growing on the stump of tree in Berlin, Germany. Galleries inside the fungus were frequently visited by beetles of the species *Cis castaneus* (Ciidae), a beetle with a host preference for *Ganoderma* [9]. Associations between nematodes and insects, where the nematode uses the insect as a transport carrier (phoretism), have already been described for several *Caenorhabditis* species, including *Caenorhabditis angaria*, *C. remanei*, and *C. bovis*, and similar phoretic associations could be expected for many or possibly all other *Caenorhabditis* species [10].

Here we use both morphological and molecular analyses to characterise and describe *C.* sp. 1 as a new species, *Caenorhabditis monodelphis* sp. n., and explore its relationship with the beetle *Cis castaneus*. Molecular phylogenetic analysis based on whole genome sequencing of an inbred derivative of the type strain affirms the placement of *C. monodelphis* sp. n. as sister to other analysed *Caenorhabditis*, and we analyse the evolution of phenotypic traits to infer those present in the hypothetical ancestor of *Caenorhabditis*.

Methods
Isolation and culture
Caenorhabditis monodelphis sp. n. (strain SB341) was originally isolated from fruiting bodies of *Ganoderma applanatum* (Pers.) Pat. 1887 collected in Berlin-Grunewald, Germany (April, 2001) and later from four locations in Belgium (strain DSC001 collected from 51°06'24"N, 3°18'13"E, March 2014, strain DSC002 collected from 50°52'7"N, 4°06'54", February 2014, and an uncultured population 51°02'41"N, 3°27'17" June 2014) and from one location in the Botanical Garden in Oslo, Norway (strain JU2884; 59°55'04"N 10°46'01"E, 22 July 2015). These collections were from the same mushroom species. We also found *C. monodelphis* sp. n. in the fruiting body of *Fomes fomentarius* (L.) Fr. 1849 (50°43'02"N, 4°05'06"E, February 2015). Strain SB341 was chosen as type.

Nematodes were extracted from the fruiting bodies of *G. applanatum* using the modified Baermann method [11]. Dauer larvae were isolated from the beetle *Cis castaneus* (Herbst, 1793) that had been extracted from the same mushroom from multiple locations (except the type population and from 51°02'41"N, 3°27'17"). Adults and dauer larvae were picked out and cultured on nutrient agar plates seeded with *E. coli* OP50 at 15 °C.

Morphological characterisation
Cultures of nematodes from two populations (strain SB341 and DSC001) were used for the description. Measurements and drawings were made with an Olympus BX51 equipped with differential interference contrast (DIC). Light microscopic images were taken with a Nikon DS-FI2 camera. For Scanning Electron Microscopy (SEM), two fixation methods were used. For the first fixation method, live animals were fixed in a microwave in Trump's

fixative (2% paraformaldehyde + 2.5% glutaraldehyde in a 0.1 M Sorenson buffer) for a few seconds. Specimens were subsequently washed three times in double-distilled water. For the second method, specimens were put in a refrigerator at 4 °C for 1 h, then Trump's fixative was added and specimens were left overnight at 4 °C. The specimens were then washed with a 0.2 M phosphate buffer followed by 1 h post-fixation in a 1% OsO_4 solution at room temperature and subsequently washed 4 times in double-distilled water. For both methods, the specimens were dehydrated by passing them through a graded ethanol concentration series of 30, 50, 75, 95% (20 min each) and 3x 100% (10 min each). The specimens were critical point-dried with liquid CO_2, mounted on stubs with carbon discs and coated with gold (25 nm) before observation with a JSM-840 EM (JEOL, Tokyo, Japan) at 15 kV. Sperm cells were observed in the female post-uterine sac with Transmission Electron Microscopy (TEM), processing samples as described [12], except for ultramicrotomy with a Leica EM UC7 and 1 h 1% osmium postfixation (Slos et al. unpublished).

Molecular characterisation
For DNA barcoding analyses, temporary slides of individual nematodes were made in tap water and digital light microscope pictures were taken as a morphological voucher. The nematode was then transferred to a PCR tube with a solution containing 10 μl 0.05 M NaOH and 1 μl Tween20, heated for 15 min at 95 °C, and 40 μl of double-distilled water was added. PCR was carried out targeting either the 28S (large subunit) ribosomal RNA gene (nLSU) or the ribosomal internal transcribed spacer 2 (ITS2) locus, and PCR products were cleaned and sequenced directly. Forward and reverse primers for the nLSU were D2Ab (ACAAGTACCGTGAGGGAAAGTTG) and D3b (TCGGAAGGAACCAGCTACTA). For ITS2 we used VRAIN2F (CTTTGTACACACCGCCCGTCGCT) and VRAIN2R (TTTCACTCGCCGTTACTAAGG GAATC). The sequences obtained were 100% identical to published sequences for *Caenorhabditis* sp. 1 [3].

Genome sequencing
Genomic DNA was extracted from an inbred strain, JU1667, of *C. monodelphis* sp. n. (derived from strain SB341), maintained on *E. coli* OP50, using the proteinase K-spin column protocol (detailed in Additional file 1). Total RNA from the same culture was also extracted (methods detailed in Additional file 1). Two paired-end genomic libraries (insert sizes of 300 bp and 600 bp, respectively) and a single paired-end RNA-seq library (insert size 180 bp) were constructed using TruSeq reagents and sequenced on the Illumina HiSeq 2000 by Edinburgh Genomics. We obtained 124.3 million genomic read pairs (100 base, paired end) and 46.2 million pairs of RNA-Seq reads (also 100 base, paired end).

De novo genome assembly and gene prediction

Details of software versions and parameters are available (see Additional file 2). We performed initial quality control of our genomic sequence data using FastQC [13] and used Skewer [14] to remove low quality (Phred score < 30) and adapter sequence. Using blobtools [15], we generated taxon-annotated GC-coverage (TAGC) plots to identify and remove bacterial contamination. Sequence data were assembled with CLC assembler (CLCBio, Copenhagen, Denmark) and reads mapped back to this assembly using CLC mapper. Each assembly contig was compared to the NCBI Nucleotide (nt) database using megablast from the NCBI BLAST+ suite [16]. Genomic read pairs were aligned to genome references from five E. coli (strains: BL21 (DE3), ETEC H10407, K12 substr. DH10B, K-12 substr. MC4100 and B str. REL606) using Bowtie [17], and aligned pairs discarded. We identified laboratory-induced contamination with *Caenorhabditis elegans* in the 600 bp insert library data. To remove this, we aligned read pairs of the uncontaminated 300 bp-insert library to the *C. elegans* N2 reference genome. Regions of similarity between the genomes of *C. monodelphis* sp. n. and *C. elegans* (i.e. those regions of *C. elegans* with aligned *C. monodelphis* sp. n. reads) were masked with Ns using BEDtools [18]. Read pairs of the 600 bp-insert library were subsequently aligned to this masked *C. elegans* reference and any aligned read pairs discarded.

Cleaned sequence data were assembled with ABySS [19] (k = 83) and contigs were scaffolded with transcript evidence using SCUBAT [20]. RepeatModeler [21] was used to identify repetitive regions which were then masked using RepeatMasker [22]. RNA-Seq read pairs were aligned to the assembly using STAR [23], and the resulting BAM file was used to guide the prediction of protein-coding genes by BRAKER [24].

Gene structure comparisons

Genome sequences and annotation GFFs were downloaded from WormBase [25] and imported into a custom Ensembl database (version 84) [26]. Using the Ensembl Perl API, the canonical transcript from each protein-coding gene was identified and exon and intron statistics were calculated. To compare the gene structures of *C. monodelphis* sp. n. with that of *C. elegans,* we identified all orthologous clusters (details below) in which *C. monodelphis* sp. n. and *C. elegans* proteins were present as single-copy. Exon and intron statistics were calculated for each gene pair, as described previously. Plots were generated using the ggplot2 package [27] and GenePainter [28].

Phylogenetic analyses

Pairwise comparisons of protein sequences derived from genomic data for 23 species of *Caenorhabditis* and two outgroup species, *Oscheius tipulae* and *Heterorhabditis bacteriophora,* (see Additional file 3 for details) were performed using NCBI BLAST+ [16] and clustered into orthologous groups using OrthoFinder [29]. The sequences of 303 one-to-one orthologues (allowing for up to two species to have missing data) were extracted and aligned using ClustalOmega [30]. Poorly aligned regions were removed from the alignments using trimAL [31] and trimmed alignments concatenated using FASconCAT [32] to yield a supermatrix. We performed maximum-likelihood (ML) analysis using RAxML [33] (PROTGTR + Γ substitution model) with 1,000 bootstrap replicates. Bayesian analysis was performed using PhyloBayes [34] (CAT-GTR), with two independent Markov chains, and convergence was assessed using Tracer [35].

Nomenclatural acts

This published work and the nomenclatural acts it contains have been registered in Zoobank: http://zoobank.org/urn:lsid:zoobank.org:pub:0E6F137B-9975-4A8E-91F2-D588A57 2076E. The LSID for this publication is: urn:lsid:zoobank.org:pub:0E6F137B-9975-4A8E-91 F2-D588A572076E.

Results

Here we provide a formal description of SB341 as the type strain of *C. monodelphis* sp. n.

Caenorhabditis monodelphis[1] sp. n. Slos & Sudhaus

= *Caenorhabditis* sp. SB341 [7]
= *Caenorhabditis* sp. SB341 and *Caenorhabditis* sp. n. SB341 [36]
= *Caenorhabditis* sp. n. 1 (SB341) and (lapse) *Caenorhabditis* sp. n. 4 (SB341) [10]
= *Caenorhabditis* sp. 1 SB341 [6, 8, 37]
= *Caenorhabditis* sp. 4 SB341 [38]

(Figs. 1, 2, 3 and 4; Table 1)

Adult

Small species (female 0.72 - 1.04 mm, male 0.65 – 0.77 mm); cuticle thin, ca. 1 μm wide and finely annulated, 0.8 μm wide at midbody. Lateral field inconspicuous, about 9% of body width, consisting one ridge that can be traced anteriorly to the level of the median bulb and posteriorly at level of rectum in females and about 1½ spicules length anterior of the cloacal aperture in males. Six lips slightly protruding, each with one apical papilliform labial sensillum and a second circle of four sublateral cephalic sensilla in both sexes; amphids opening on the lateral lips, hardly discernible. Buccal tube long and slender, more than twice the width in lip region, pharyngeal sleeve envelopes nearly half of the stoma, the anterior as well as the posterior end of the tube appear slightly thickened, cheilostom inconspicuous, arcade cells forming the

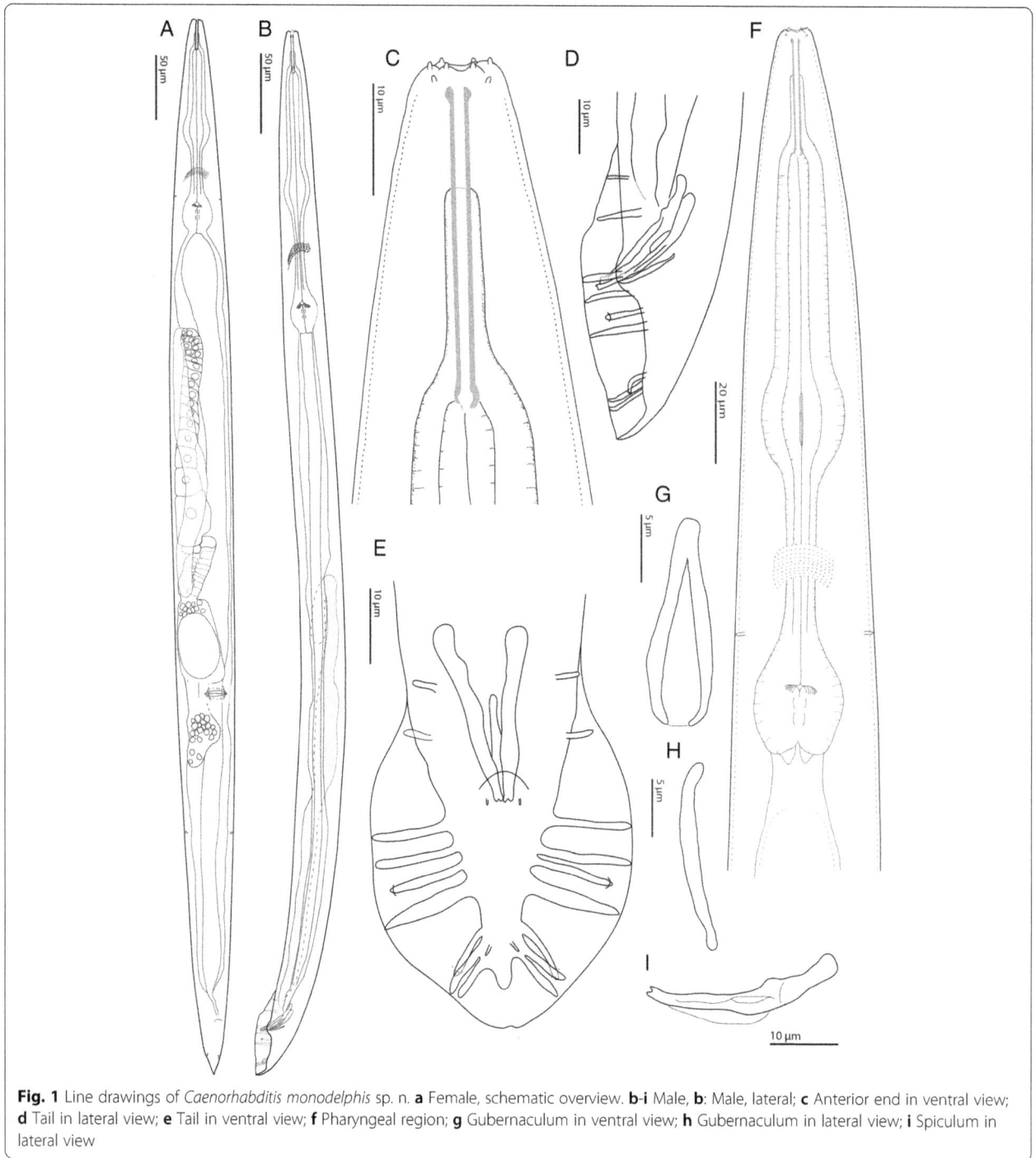

Fig. 1 Line drawings of *Caenorhabditis monodelphis* sp. n. **a** Female, schematic overview. **b-i** Male, **b**: Male, lateral; **c** Anterior end in ventral view; **d** Tail in lateral view; **e** Tail in ventral view; **f** Pharyngeal region; **g** Gubernaculum in ventral view; **h** Gubernaculum in lateral view; **i** Spiculum in lateral view

gymnostom sometimes visible; glottoid apparatus completely absent. Pharynx with a prominent median bulb, diameter more than 90% of diameter of terminal bulb; terminal bulb pyriform, with double chambered haustrulum, the anterior chamber smallish; cardia conspicuous, opens funnel-like in intestine. Nerve ring encircles isthmus in its anterior part in living specimens, more to the middle of the isthmus in heat relaxed or preserved specimens; deirids usually conspicuous in the lateral field at level of beginning of terminal bulb, sometimes not visible in heat relaxed animals; pore of excretory-secretory system hard to discern posterior of deirid level. Two gland cells ventral and slightly posterior of terminal bulb conspicuous in live specimens. Lateral canals visible in live specimens extending anteriorly to two stoma length from the anterior end and ending at rectum level in the female. Postdeirids usually very conspicuous dorsally of the lateral field at about 75% of body length in both

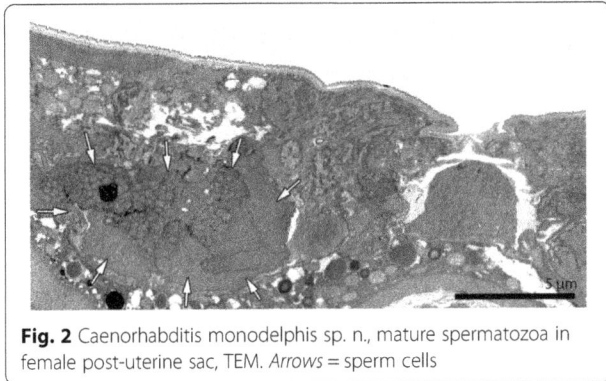

Fig. 2 Caenorhabditis monodelphis sp. n., mature spermatozoa in female post-uterine sac, TEM. *Arrows* = sperm cells

sexes and about half the length between vulva and beginning of rectum (or at level of posterior end of uterus remnant) in females, sometimes not visible in heat relaxed specimens.

Female
Maximum body diameter clearly anterior of the vulva, vulva position 65% body length, a transverse slit, bordered in both ends by cuticular longitudinal flaps, vulva lips moderately protruding, four diagonal vulval muscles conspicuous; one pseudocoelomocyte exists anterior of gonad flexure ventrally. Genital tracts asymmetrical; posterior branch rudimentary, sac like, on the left hand side of intestine, without flexure, almost as long as body diameter at the level of the vulva, containing spermatozoa (Fig. 2); anterior branch right of intestine, reflexed dorsally close to the pharynx, flexure more than half the length of the gonad (measured from vulva to flexure); at the flexure oocytes in several rows, downstream in one row, oocytes predominantly growing in the last position, where granules are stored inside; sphincter between oviduct and uterus, only a few sperm cells in oviduct, most of them in uterus and blind sac; oviparous, one egg at a time in uterus (rarely two), segmentation starts in the uterus. Rectum a little S-shaped, rectal gland cells very small, posterior anal lip slightly protuberant. Tail short, panagrolaimid, dorsally convex, with offset tip tapering, smooth to somewhat telescope-like by cuticle forming a sleeve-like structure; tail tip with tiny hooks, mostly one dorsal, but also subventral (compare with *Poikilolaimus*); opening of phasmids located at 60–65% of tail length, shortly anterior of tip, phasmid glands not reaching anus level.

Male
Testis right of intestine, ventrally reflexed in a certain distance posterior of pharynx; flexure relatively short. One pseudocoelomocyte between pharynx and flexure ventrally. Bursa well developed, peloderan, anteriorly open, with smooth margin and sometimes terminally indented, posterior part of velum transversely striated.

Nine pairs of genital papillae (GP) present, two of them anterior of the cloaca, genital papilla 1 (GP1) and GP2 spaced, GP3 to GP6 and GP7 to GP9 clustered, GP5 and GP7 point to the dorsal side of the velum, GP6 slightly bottle shaped, GP8 and GP9 fused at base, GP2 and GP8 not reaching the margin of velum. Phasmids forming small tubercles to the ventral side posterior of the last GP; formula of GPs: v1,v2/(v3,v4,ad,v5) (pd,v6,v7)ph. Precloacal sensillum small, precloacal lip simple (according to type A of W Sudhaus and K Kiontke [39]), postcloacal sensilla long filamentous. Spicules short and stout, tawny, separate, slightly curved, with prominent head; shaft with a transverse seam, with a prominent longitudinal ridge, a dorsal lamella, and an oval "window", the tip notched. Gubernaculum dorsally projecting, flexible, in the distal part following the contour of the spicules, spoon shaped in ventral view.

Dauer larva
Unsheathed, mouth closed; stoma long, slender. Pharyngeal sleeve covering about half of the stoma; pharynx with well-developed median and terminal bulbs; corpus length ca. 52% of pharynx length. Nerve ring somewhat in the middle between the middle and terminal bulb. Genital primordium at about 60% of body length, elongated oval in shape. Tail conical. Amphids, lateral lines, position excretory pore, deirids and phasmids not observed.

Aberration
In one female a second set of "sensilla" were observed a short distance posterior to postdeirids, possibly a duplication of the postdeirids.

Type carrier and locality
Holotype and paratypes of *Caenorhabditis monodelphis* sp. n. were isolated from the tunnels of *Cis castaneus* (Herbst, 1793) (Ciidae, Coleoptera) in the bracket fungus *Ganoderma applanatum* (Polyporales) on a stump of the common beech (*Fagus sylvatica*) a few centimetres above the ground in Berlin-Grunewald in April 2001. The same sample included individuals of *Diploscapter* sp., *Plectus* sp., *Oscheius dolichura* and one individual dorylaimid and mononchid.

Type material
Holotype male (collection number WT 3684) and five female and four male paratypes (WT 3685, WT 3686) are deposited in the National Plant Protection Organization Wageningen, The Netherlands. In addition, four female and four male paratypes, are deposited in the collection of Museum Voor Dierkunde at Ghent University, Ghent, Belgium, five female and three male paratypes in Museum

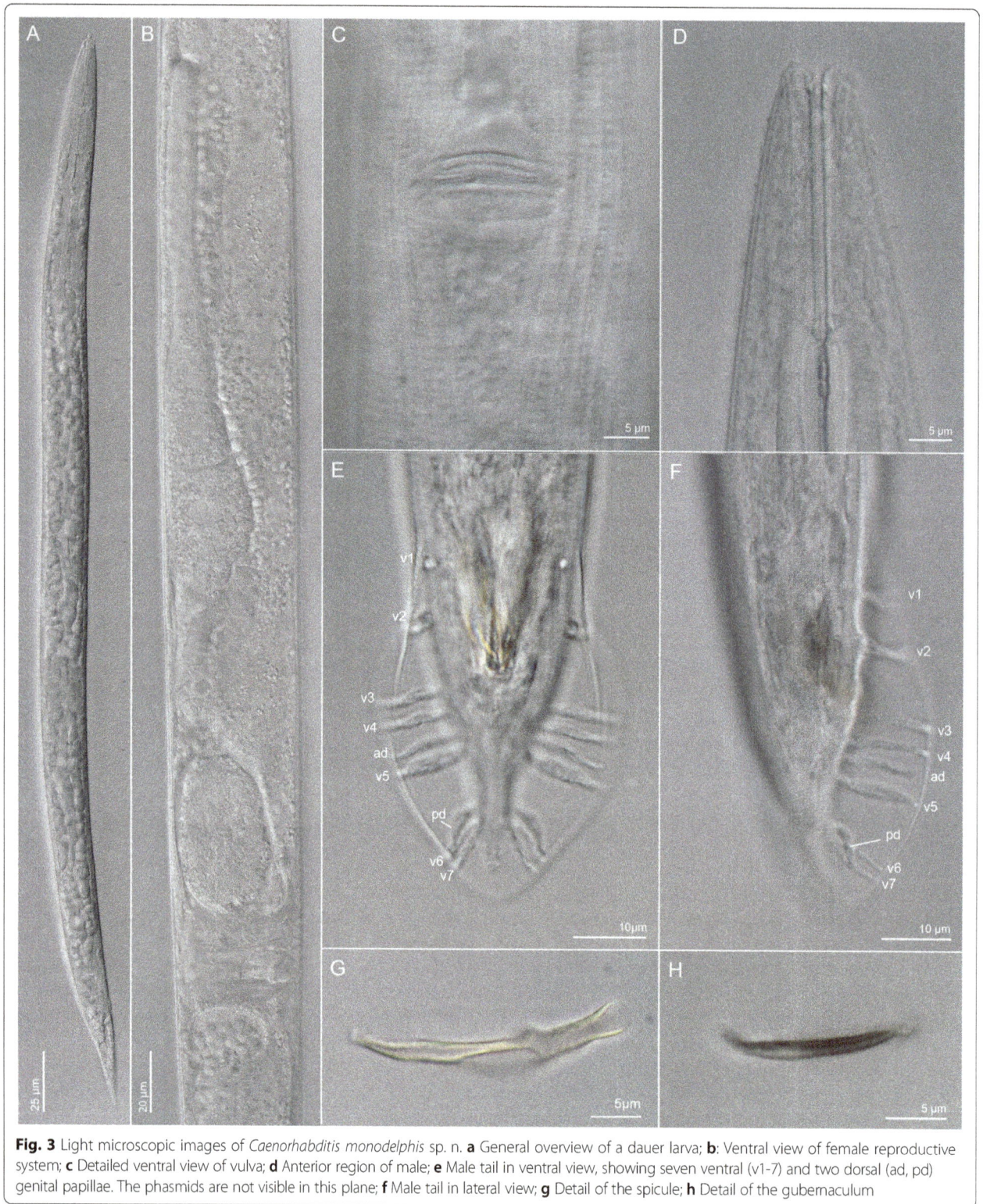

Fig. 3 Light microscopic images of *Caenorhabditis monodelphis* sp. n. **a** General overview of a dauer larva; **b**: Ventral view of female reproductive system; **c** Detailed ventral view of vulva; **d** Anterior region of male; **e** Male tail in ventral view, showing seven ventral (v1-7) and two dorsal (ad, pd) genital papillae. The phasmids are not visible in this plane; **f** Male tail in lateral view; **g** Detail of the spicule; **h** Detail of the gubernaculum

für Naturkunde an der Humboldt-Universität zu Berlin, Berlin, Germany. Additional paratypes are available in the UGent Nematode Collection (slides UGnem158, 159 & 160) of the Nematology Research Unit, Department of Biology, Ghent University, Ghent, Belgium.

Diagnosis and relationship

Caenorhabditis monodelphis sp. n. can be recognised as a *Caenorhabditis* based on the thickened GP6 and the clearly visible postdeirids. *Caenorhabditis monodelphis* sp. n. is distinguished from all other described *Caenorhabditis*

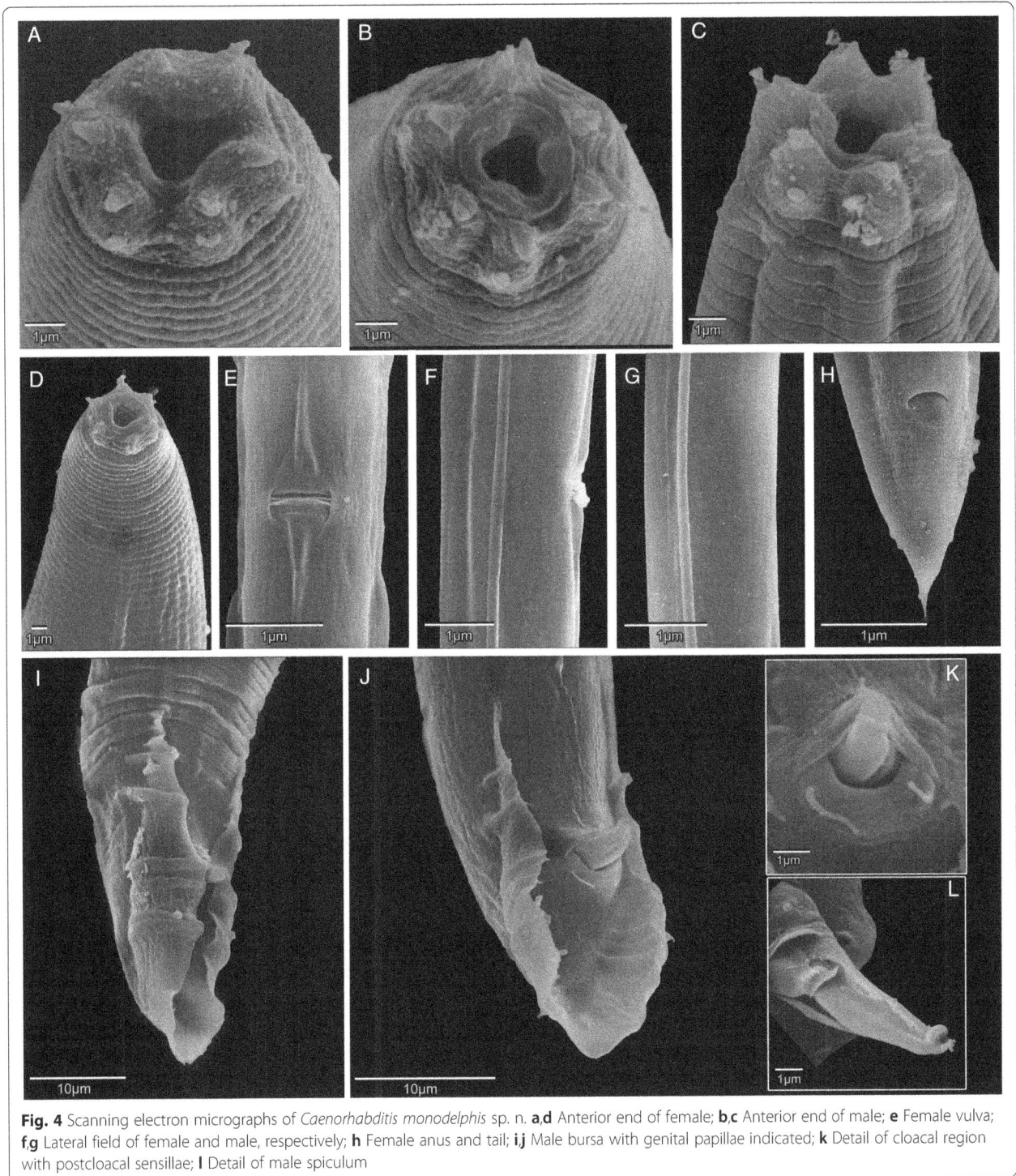

Fig. 4 Scanning electron micrographs of *Caenorhabditis monodelphis* sp. n. **a,d** Anterior end of female; **b,c** Anterior end of male; **e** Female vulva; **f,g** Lateral field of female and male, respectively; **h** Female anus and tail; **i,j** Male bursa with genital papillae indicated; **k** Detail of cloacal region with postcloacal sensillae; **l** Detail of male spiculum

species by the presence of a monodelphic genital tract in the female with a blind sac posterior the vulva, a panagrolaimid female tail shape, adults with only one ridge on the lateral field, a very long and slender stoma without visible glottoid apparatus and male with short, stout spicule with bifurcate tip.

Ecology and biology

Caenorhabditis monodelphis sp. n. is a gonochoristic species with both males and females. Females are oviparous and carry only one egg (rarely two eggs). Development from egg to adult took about 5–6 days in juice prepared from brown algae at room temperature. Development

Table 1 Measurements (in µm) of heat relaxed specimens of *Caenorhabditis monodelphis* sp. n.

Character	Female	Male	Dauer
N	11	10	10
L	870 ± 105	694 ± 36	456 ± 24
A	17.1 ± 0.8	22 ± 1.6	23 ± 1.2
B	4.9 ± 0.5	4.1 ± 0.3	3.6 ± 0.1
C	20.5 ± 2.6	22 ± 2.3	9.8 ± 0.7
c'	1.99 ± 0.17	1.8 ± 0.2	3.9 ± 0.30
V	65 ± 1.8	-	-
Body width	51 ± 6.9	32 ± 3	20 ± 0.6
Stoma length	27 ± 2.3	27 ± 2	21 ± 1.1
Stoma diameter	1.9 ± 0.6	1.2 ± 0.2	0.6 ± 0.1
Cheilostom	2.5 ± 0.2	2.4 ± 0.2	-
Gymnostom	10 ± 0.7	9.7 ± 0.9	-
Stegostom	15 ± 1.5	15 ± 1.5	-
Pharyngeal sleeve	12.42 ± 1.6	13 ± 1.2	-
Pharynx length	150 ± 6.7	141 ± 9.2	107 ± 3.3
Procorpus length	55 ± 3.1	52 ± 3.6	-
Metacorpus length	26 ± 2.1	22.8 ± 1.1	-
Isthmus length	39 ± 3.2	40 ± 4.9	-
Nerve ring to terminal bulb	11 ± 4.9	19 ± 3.3	-
Terminal bulb length	30 ± 1.8	27 ± 1.7	-
Diameter of median bulb	22 ± 2.5	17 ± 1.3	9 ± 0.5
Diameter of terminal bulb	25 ± 2	19 ± 1	11 ± 0.4
Anterior end to deirid	150 ± 8	150 ± 8.3	-
Postdeirid to anus	170 ± 29.8	141 ± 14	-
Length intestine	651 ± 100	494 ± 32	-
Rectum length	25 ± 2.6	24 ± 1.9	-
Anal body width	22 ± 2.1	17 ± 1.1	12 ± 0.6
Tail length	43 ± 4.3	32 ± 3.2	46 ± 2.4
Anus to phasmid distance	26 ± 2.2	-	-
Gonad length[a]	303 ± 68	342 ± 44	-
Gonad flexure length	226 ± 67	46 ± 6.8	-
Postuterine sac	45 ± 6.8	-	-
Sperm diameter	-	9.8 ± 1.3	-
Egg length[b]	53 ± 3.1	-	-
Egg diameter[b]	29 ± 2.9	-	-
Spicule length	-	25 ± 1	-
Gubernaculum length	-	15 ± 0.9	-

[a]from anus to flexure in the female; from cloaca to flexure in the male
[b]n = 7

from dauer larva to adults was completed in less than 3 days at 20 °C on NA seeded with OP50. The lifespan of adults is at minimum 14 days for males and 17 days for females. One pair of adults produced 167 offspring in 8 days and the daily production of fertile eggs was 6–31 (mean 18; n = 14). After the reproductive phase, females lived 9–14 days (n = 3) with males present.

Caenorhabditis monodelphis sp. n. has until now only been found in *Ganoderma* and *Fomes* in Germany and Belgium in relation with the ciid beetle *Cis castaneus*. The *Ganoderma* carrying *C. monodelphis* sp. n. from Oslo was not investigated for the presence of *C. castaneus*. In fungal fruiting bodies lacking the beetle *C. monodelphis* sp. n. was not found. Dauers of *C. monodelphis* sp. n. were found under the elytra of the beetle, but were not found internally when the beetle was further dissected. These findings indicate a phoretic association with the beetle. As only dauer larvae were isolated from beetles, while adults and larvae were present in the fruiting bodies, we infer that *C. monodelphis* sp. n. exit from dauer within the mushroom, develop to adulthood and start to reproduce. The food source of the species in natural conditions is not known, but they survive and reproduce easily on *E. coli* OP50 in culture.

Genome sequence of an inbred strain of Caenorhabditis monodelphis sp. n.

We sequenced the genome of an inbred strain (JU1677) of *C. monodelphis* sp. n. using Illumina sequencing technology to ~110x coverage. The genome was assembled into 6,864 scaffolds, spanning 115.1 Mb with a scaffold N50 of 49.4 kb (Table 2). CEGMA (Core Eukaryotic Gene Mapping Approach) [40] scores suggested the assembly is of high completeness. We predicted 17,180 protein coding gene models using RNA-Seq evidence. These statistics, and the overall gene content and structure of the assembly were largely in keeping with those determined for other *Caenorhabditis* species. The genome was larger than that of *C. elegans* and *C. briggsae*, which are hermaphroditic species, but smaller than that of *C. remanei*, a gonochoristic species.

We carried out preliminary comparisons of the structure and content of the *C. monodelphis* sp. n. genome with those of other sequenced *Caenorhabditis* species. The number of genes identified was lower than estimates for most other *Caenorhabditis* species. To compare the gene structures of *C. monodelphis* sp. n. to that of *C. elegans*, we identified 6,174 orthologous gene pairs and calculated gene structure statistics (Table 3, Fig. 5.). To minimize bias from erroneous gene predictions (such as merged or split genes), orthologous gene pairs which differed in CDS length by 20% were considered outliers. *C. monodelphis* sp. n. genes were typically longer than their orthologues in *C. elegans*. We also found a clear trend toward more coding exons per gene in *C. monodelphis* sp. n. than in *C. elegans* (Fig. 5a). A few examples of *C. monodelphis* sp. n. gene models compared to those of orthologues in *C. elegans* are shown (Fig. 5b). Although introns are, on average, shorter in *C. monodelphis* sp. n. than in *C. elegans*,

Table 2 Genome assembly statistics for *C. monodelphis* sp. n. and other *Caenorhabditis* species

Species	C. monodelphis	C. brenneri	C. briggsae	C. elegans	C. japonica	C. remanei	C. sinica	C. tropicalis
Version	1.0	WS254	WS254	WS254	WS254	WS254	WS254	WS254
Mating type	gonochoristic	gonochoristic	hermaphroditic	hermaphroditic	gonochoristic	gonochoristic	gonochoristic	hermaphroditic
Strain	JU1667	PB2801	AF16	N2	DF5081	PB4641	JU800	JU1373
Span (Mb)	115.12	190.37	108.38	100.29	166.25	118.55	130.76	79.32
Scaffolds (n)[a]	6,864	3,305	367	7	18,808	1,591	11,966	660
N50 (kb)	49.4	381.96	17,485.44	17,493.82	94.15	1,522.09	25,564	20,921.87
Genes (n)	17,180	30,660	21,814	20,362	29,964	26,226	34,696	22,326
GC (%)	43.9	38.6	37.4	35.4	39.2	37.9	39.5	37.7
CEGMA complete/ partial (%)	89.11/ 97.98	98.39/ 99.60	97.98/ 99.19	96.77/ 99.19	78.63/ 97.18	94.35/ 98.79	95.56/ 99.60	97.18/ 98.79

[a]Scaffolds shorter than 500 bp were not considered

C. monodelphis genes typically have a longer total span of introns than *C. elegans* transcripts (Table 3, Fig. 5.).

C. monodelphis sp. n. is sister to other known Caenorhabditis

We clustered a total of 634,564 protein sequences from *C. monodelphis* sp. n., twenty-two other *Caenorhabditis* species, and two rhabditomorph outgroup species (*Oscheius tipulae*; data courtesy of M. A. Félix, and *Heterorhabditis bacteriophora*) to define putative orthologues. We identified 34,425 putatively orthologous groups containing at least two members, 303 of which were either single copy or absent across all 25 species. These single copy orthologues were aligned, and the alignments concatenated and used to perform maximum-likelihood and Bayesian inference analysis using RAxML and PhyloBayes, respectively. Both analysis methods resulted in an identical topology, with the placement of *C. monodelphis* sp. n. arising basally to all other *Caenorhabditis* species (Fig. 6). All branches had maximal support except for three nodes within the *Elegans* super-group. Our analysis included data from several new and currently undescribed putative species of *Caenorhabditis*, including *C.* sp. 21 which is the sister taxon to the *Drosophilae* plus *Elegans* super-groups and *C.* sp. 31 which forms the first branch in the *Elegans* super-group. *C.* sp. 38 is placed within the *Drosophilae* super-group, while *C.* sp.

26, *C.* sp. 32 (sister to *C. afra*) and *C.* sp. 40 (sister to *C. sinica*) are all members of the *Elegans* super-group. From these analyses we conclude that *C. monodelphis* sp. n. is sister to all other known *Caenorhabditis*.

Stemspecies pattern reconstruction

Our phylogenetic analyses were based on species with whole genome data available, and thus did not include the full known diversity of the genus. The stemspecies pattern was reconstructed based on ingroup and outgroup comparison. Previous molecular phylogenetic analyses of *Caenorhabditis* species using a small number of marker genes [10] placed *C. monodelphis* sp. n. and *C. sonorae* [41] as sister species, again arising at the base of the genus.

The following morphological synapomorphies can be hypothesised to support a *C. monodelphis* sp. n. – *C. sonorae* clade: mouth opening triangular (Fig. 4b), spicule having a complicated tip (notched or dentated) and a longish thin walled "window" in the blade (Figs. 1i, 4l), postcloacal sensilla being filiform (Fig. 4k), and the female tail shortened to less than three times anal body width. Other similarities between both these species are plesiomorphic.

Caenorhabditis and its sister group constitute the monophylum Anarhabditis within the Rhabditina. For convenience, we will call the sister clade of *Caenorhabditis* Protoscapter (Fig. 7): it comprises "*Protorhabditis*", *Prodontorhabditis*, *Diploscapter* and *Sclerorhabditis* [42]. To reconstruct the characters of the stemspecies of *Caenorhabditis* it is necessary to consider the morphologies of all these taxa, and not only the taxa for which we have molecular data. "*Protorhabditis*" is paraphyletic. The *Oxyuroides* group is sister taxon of *Prodontorhabditis* [43, 44], and the *Xylocola* group may be sister taxon of *Diploscapter/Sclerorhabditis*. However, the two species *Protorhabditis elaphri* (Hirschmann in Osche, 1952) and *P. tristis* [45] appear to represent basal branches in Protoscapter (compare [43]). These last two species, despite

Table 3 Gene structure comparison of orthologous gene pairs from *C. monodelphis* sp. N. and *C. elegans*

	C. monodelphis sp. n.	C. elegans
Gene length (bp)	3359	2854
Coding exon length (bp)	109	144
Coding exon count (n)	10	6
CDS span (bp)[a]	1167	1182
Intron length (bp)	69	76
Total intron span per gene (bp)	1918	1187

All values are medians
[a]orthologous gene pairs which differed in CDS length by 20% were not included

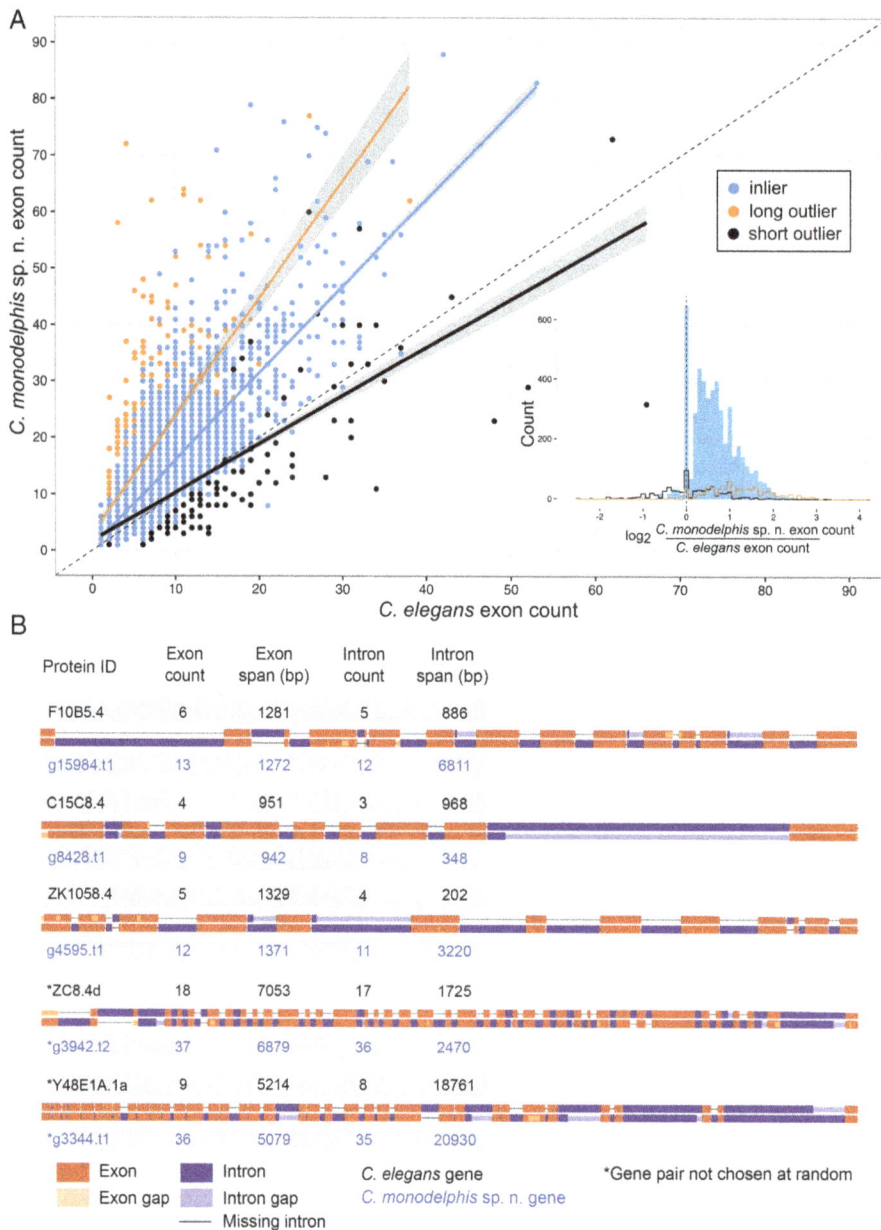

Fig. 5 Comparison of exon counts in single-copy orthologues between *C. monodelphis* sp. n. and *C. elegans*. **a** Exon counts in 6,174 single-copy orthologous gene pairs. *C. monodelphis* sp. n. genes which had transcripts with CDS lengths 20% longer (*orange*) or shorter (*black*) than *C. elegans* were defined as outliers. Linear regression lines are shown. Inset: Frequency histogram of log2 ratio of *C. monodelphis* sp. n. exon counts to *C. elegans* exon counts. **b** Comparison of gene structures of five orthologous gene pairs. Three gene pairs were selected at random and a further two were selected because they showed a large divergence in exon count

the paucity of information available for them, are crucial for comparisons that will illuminate the stemspecies patterns of Anarhabditis, Protoscapter and *Caenorhabditis*.

By ingroup comparison we reconstruct the following characters of the stemspecies of Anarhabditis without differentiating them into apo- or plesiomorphies (on apomorphies see the legend of Fig. 7):

– adults of small size (less than 1 mm); – lips not offset from anterior end; – four cephalic sensilla present in male and female; – stoma with pharyngeal sleeve (stegostom length nearly that of gymnostom); – median bulb of pharynx strongly developed, corpus intima with transverse ridging, terminal bulb with double haustrulum; – gonochoristic; – female tail elongate conoid; – gonads amphidelphic, the anterior branch right and the posterior left of intestine; – vulva at midbody, a transverse slit; – oviparous, usually only one egg at a time in the uteri; – male gonad on the right side, reflexed to the ventral; –

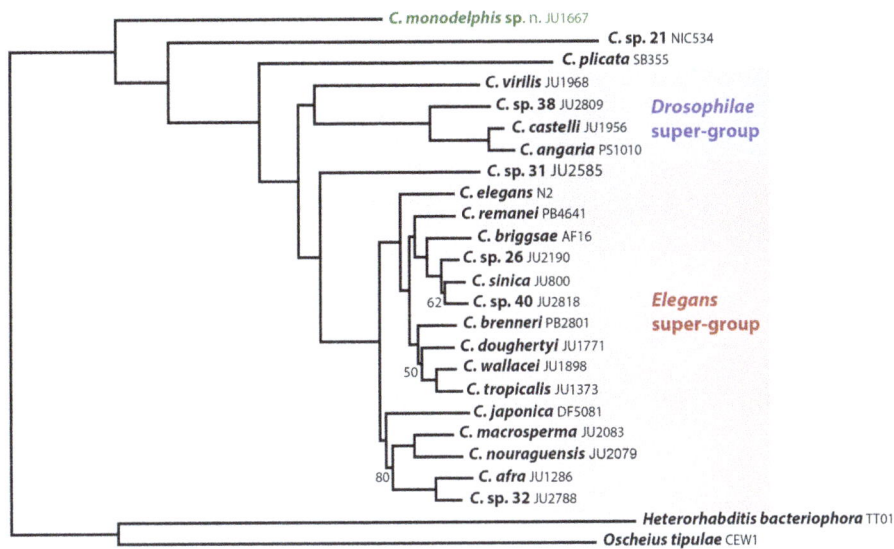

Fig. 6 Phylogenetic relationships of *Caenorhabditis* species and two outgroup species. Maximum-likelihood based tree from RAxML. Bootstrap support values (1000 replicates) were 100 all branches, unless noted otherwise as numbers on branches. Bayesian Posterior Probabilities were 1 for all branches and are not shown. Coloured boxes indicate supergroups, as defined in [3]. Strain names from which protein sequence were derived are noted

bursa peloderan and anteriorly open, oval-shaped in ventral view, with smooth margin, terminally not notched; – 9 pairs of even genital papillae, two precloacal largely spaced, GP3–6 evenly spaced, the last three GPs forming a tight cluster; GP1, GP5 and GP7 terminate on the dorsal surface of the bursa velum; – phasmids open behind GP9, inconspicuous; – bursa formula thus v1,v2/v3,v4,ad,v5 (pd,v6,v7)ph; – male tail tip present; – 1 + 2 circumcloacal sensilla inconspicuous, precloacal lip simple; – spicules separate, stout, head not rounded, behind the shaft a slight ventral projection, dorsal part of blade weakly cuticularised (velum), its tip possibly not even (argued below); – gubernaculum simple spatulate; – dauerlarvae with double cuticle (ensheathed), not waving.

Discussion
Taxonomy of *Caenorhabditis monodelphis* sp. n.
Caenorhabditis monodelphis sp. n. is a new species of *Caenorhabditis* supported by its phylogenetic position as inferred from 303 molecular markers, morphology, habitat and specific association with *Cis castaneus* (Coleoptera). Morphologically, it could be confused with "*Protorhabditis*" species because of the absence of a clear glottoid apparatus. A glottoid apparatus has been lost 5–6 times independently within "Rhabditidae" [46] and, as illustrated here, also in *C. monodelphis* sp. n. This species resembles species from "*Protorhabditis*" with a very long stoma without glottoid apparatus, but differs from the *Oxyuroides*-group within "*Protorhabditis*" in having an open bursa and GP1 not anterior of the bursa. It is differentiated from the *Xylocola*-group within "*Protorhabditis*" in having nine genital papillae.

Previously, *Caenorhabditis* has been characterised as having the following apomorphic characteristics: the presence of a dorsal velum on the spicule, a lateral field with three ridges, an unsheathed dauer juvenile and a slightly thickened GP6 [42]. With the discovery and description of *C. monodelphis* sp. n. the number of lateral ridges is no longer an apomorphic character of *Caenorhabditis*, since *C. monodelphis* sp. n. only has one lateral ridge.

Association with fungivorous beetles
Species of *Caenorhabditis* are known to occur in soil, compost, cadavers of insects, some plant material and the intestine of birds [10], and can most easily be isolated from rotting fruits, flowers and stems [3]. *Caenorhabditis elegans* has also been found infesting cultures of the mushroom *Agaricus bisporus* [47]. Wild mushrooms are an under-explored habitat for this genus, but our limited geographical sampling indicates that they could be an important habitat. *Caenorhabditis monodelphis* sp. n. was present in galleries made by *Cis castaneus* inside *Ganoderma applanatum* in Belgium, Norway, Germany and in an old fruiting body of *Fomes fomentarius* in Belgium. Although the true distribution of *C. monodelphis* sp. n. is not yet known, it is expected that this species will be found throughout Europe where *Ganoderma* (or in lesser extent *Fomes*) co-occurs with the mycophagous beetle *Cis castaneus*.

That *Caenorhabditis* species have phoretic relationships with insects and other invertebrates is well known [10]. For *C. monodelphis* sp. n., all records are from

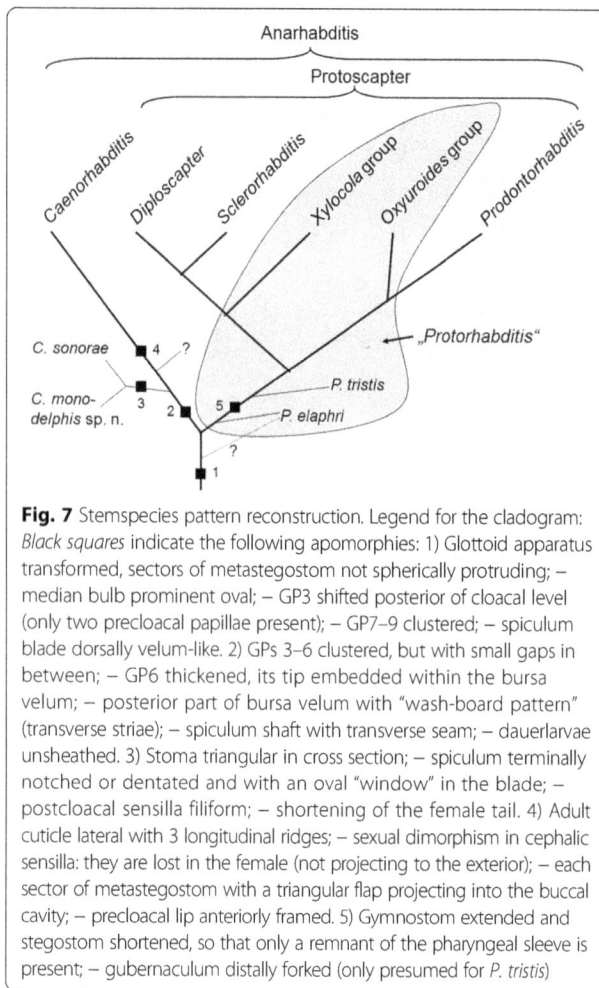

Fig. 7 Stemspecies pattern reconstruction. Legend for the cladogram: *Black squares* indicate the following apomorphies: 1) Glottoid apparatus transformed, sectors of metastegostom not spherically protruding; – median bulb prominent oval; – GP3 shifted posterior of cloacal level (only two precloacal papillae present); – GP7–9 clustered; – spiculum blade dorsally velum-like. 2) GPs 3–6 clustered, but with small gaps in between; – GP6 thickened, its tip embedded within the bursa velum; – posterior part of bursa velum with "wash-board pattern" (transverse striae); – spiculum shaft with transverse seam; – dauerlarvae unsheathed. 3) Stoma triangular in cross section; – spiculum terminally notched or dentated and with an oval "window" in the blade; – postcloacal sensilla filiform; – shortening of the female tail. 4) Adult cuticle lateral with 3 longitudinal ridges; – sexual dimorphism in cephalic sensilla: they are lost in the female (not projecting to the exterior); – each sector of metastegostom with a triangular flap projecting into the buccal cavity; – precloacal lip anteriorly framed. 5) Gymnostom extended and stegostom shortened, so that only a remnant of the pharyngeal sleeve is present; – gubernaculum distally forked (only presumed for *P. tristis*)

mushroom fruiting bodies that are also inhabited by different insect groups, and dauer larvae were found under the elytra of *Cis castaneus*. Based on this evidence, we propose that *C. monodelphis* sp. n. propagates in galleries generated by Cisidae and the dauer larvae are transported by these beetles to uninfested mushrooms. Records of *C. monodelphis* sp. n. in both *Ganoderma* and *Fomes*, respectively the preferred [9] and the known [48] host indicate a beetle-specific rather than a mushroom-specific relationship. The only other known *Caenorhabditis* species which appears to be phoretically associated with fungivorous organisms, most likely insects, is *C. auriculariae* Tsuda & Futai, [49] of the *Elegans* super-group. This species was found only once in the fruit bodies of *Auricularia polytricha* (Agaricomycetes) in Japan, but the vector needed to infest the mushroom is unknown [49]. *C. elegans* was also found to infest cultures of the champignon mushroom *Agaricus bisporus* [47], but most likely originated from mushroom compost where it can be frequently found. Several samples of different mushrooms on wood in Europe, USA and Japan did not yield other *Caenorhabditis* spp. However, given that many more insect species are known

to feed and reproduce on mushrooms [50] and Rhabditida are known to use insects as a phoretic transport carrier [51], it is possible that mushroom species are habitats for many other rhabditid species, including new species of *Caenorhabditis*.

Genome sequence and gene structures of *C. monodelphis* sp. n.

Using next generation sequencing technologies and advanced bioinformatics toolkits, we have generated a good first-draft genome sequence for an inbred line derived from the type strain of *C. monodelphis* sp. n.. Although assembly metrics and CEGMA scores indicate the assembly is relatively contiguous and complete, it is likely that a proportion of *C. monodelphis* sp. n. genes are assembled only partially. This may have affected gene prediction, with the number of predicted gene models (17,180) being lower than estimates from most other *Caenorhabditis* species with available sequence data [25]. Comparisons of orthologous gene pairs revealed a significant divergence in gene structure between *C. monodelphis* sp. n. and *C. elegans*. *C. monodelphis* sp. n. genes are typically longer, contain more coding exons and a longer span of introns than *C. elegans* genes (Table 3). This increase in gene length may, in part, account for the difference in genome span between *C. monodelphis* sp. n. and *C. elegans*. The clear trend towards more coding-exons in *C. monodelphis* sp. n. relative to *C. elegans* (Fig. 5) could be explained by extensive intron loss or gain in either species. Previous studies using a small number of genes have shown that intron losses have been far more common in *Caenorhabditis* evolution than intron gains [7, 52, 53]. Thus, it is possible that the gene structures seen in *C. monodelphis* sp. n. reflect an intron-rich ancestral state, and intron loss has predominated during the evolution of *C. elegans*. In *Pristionchus pacificus*, which is distantly related to *Caenorhabditis*, genes typically have roughly twice as many introns as their orthologues in *C. elegans* [54]. Further analysis using genomes from more closely related outgroup species and other *Caenorhabditis* species is necessary before we can infer the dynamics of intron evolution in the genus.

Phylogenetic analyses

Phylogenetic analysis of 303 clusters of putatively orthologous protein sequences derived from whole genome sequence data of 23 species of *Caenorhabditis* and two outgroup species resulted in a well resolved phylogenetic diagram and confirmation of *C. monodelphis* sp. n. as basal to all other analysed *Caenorhabditis* species (Fig. 6). The topology is largely congruent with previously published analyses performed using a smaller number of molecular loci [1]. However, in contrast to the analyses of Kiontke *et. al.* [1] and Felix *et. al.* [2] which show *C. brenneri* and *C. doughertyi* as sister species, our phylogenetic hypothesis places *C. doughertyi* as more closely related to *C. wallacei* and *C. tropicalis*. This node, however, has low bootstrap support. Genome sequencing projects for several *Caenorhabditis* species, including those from the currently under-sampled *Drosophilae* super-group, are currently underway. These data will be essential to resolving the phylogenetic relationships of this important genus where morphology can be misinformative and/or misleading.

Reconstruction of stemspecies pattern
Details of our inference of ASR depends on the placement of *P. elaphri* in "*Protorhabditis*" *versus* as sister taxon of Anarhabditis (because of its distinct pharynx morphology) (Fig. 7). Molecular data resolving this issue are urgently required. *Caenorhabditis monodelphis* sp. n. and *P. elaphri* share a conspicuously long and narrow stoma due to an extended stegostom (long pharyngeal sleeve) without a glottoid apparatus (bulging of the three sectors of metastegostom) and one ridge in the lateral field. Based on current evidence, we interpret these peculiar similarities as homologous and thus as further characters of the Anarhabditis stemspecies as well as the *Caenorhabditis* stemspecies. The narrowing of the buccal cavity could have restricted the formation of sectoral swellings of the metastegostom, so that the typical glottoid apparatus disappeared. This happened in parallel in the rhabditid *Matthesonema eremitum* [55]. The hypothesis of a reduction of the glottoid apparatus and its denticles in the stemspecies of Anarhabditis is in conflict with the structure of the metastegostom in most species of *Caenorhabditis*, where it looks like a transformation of a glottoid apparatus [39], and in *C. sonorae* is credibly described as a glottoid apparatus [41]. To solve this conflict we must assume a partial reversion both in *C. sonorae* and in the sister-lineage of *C. sonorae/C. monodelphis* sp. n. However, instead of proposing two independent reversions, the possibility of an independent reduction of the glottoid apparatus in Protoscapter and *C. monodelphis* sp. n. remains an equally parsimonious alternative. A reinvestigation of *P. elaphri* could resolve this question.

In the stemspecies pattern of Anarhabditis the morphology of the tip of the spicules remains unclear. In the description of *P. elaphri* some drawings show the tip to be nearly pointed [45], but in other drawings (Figure twelve m of [45]) it is terminally notched. In *P. tristis*, I Andrássy [56] depicted a small terminal hook. These characters were not mentioned in the text in either species' description. Nevertheless, the dentation of the spicule tips in the first branching *Caenorhabditis sonorae/C. monodelphis* sp. n. is different and distinct enough to judge this character as synapomorphic for these sister species (Fig. 7). Starting from the characters of the last common stemspecies of both these species, in *C. sonorae* the lateral ridge must have been reduced, so that its lateral field is smooth, and

the male tail tip was retracted, so that the tail ends obtusely between the last GPs. In *C. monodelphis* sp. n. both these characters remain plesiomorphic, but the female posterior gonad branch is in the process of reduction. The ecological requirements of *C. sonorae* (inhabitant of cactus rot) and *C. monodelphis* sp. n. (living in the tunnels of Ciidae beetles in bracket fungi) are so different, that no statement on the ecology of their last common stemspecies is possible. However, as *P. elaphri* and *C. monodelphis* sp. n. exhibit a phoretic relationship with beetles and their dauer larvae seek a place under the elytra, we cautiously suggest that this behaviour could be found in the stemspecies of Anarhabditis and that of *Caenorhabditis*, respectively.

Transformations from the stemspecies pattern of Anarhabditis to *Caenorhabditis* can be traced in the cladogram (Fig. 7). With respect to the hypothesis of the stemspecies pattern of *Caenorhabditis* formulated by W Sudhaus and K Kiontke [39] only the character of the lateral field must be revised: a single ridge in the lateral field of adults must be assumed in the stemspecies pattern of Anarhabditis and of *Caenorhabditis*, respectively. Therefore, the evolution of three lateral cuticular ridges must have occurred first within *Caenorhabditis* (Fig. 7).

Degenerative evolution towards monodelphy
Uniquely for *Caenorhabditis* species, in *C. monodelphis* sp. n. the posterior female gonad branch has been reduced to a blind sac without gamete forming function. This vestigial branch serves mainly in storing sperm. In contrast to most mono-prodelphic rhabditids, the vulva is not shifted posteriorly in *C. monodelphis* sp. n.. A relict posterior gonad together with a nearly median vulva also occurs in *Oscheius guentheri* (Sudhaus & Hooper, 1994) [57] and an undescribed *Diplogastrellus* species from India (Sudhaus, unpublished data). Remarkable, in all these cases the anterior branch does not extend into the body posterior to the vulva, in contrast to monodelphic cephalobids, panagrolaimids and the rhabditid *Rhabpanus*. In *Rhabpanus ossiculus* Massey, [58] and *R. uniquus* Tahseen, Sultana, Khan & Hussain, [59] the prodelphic reflexed gonad reaches almost to the rectum while the vulva is located at 65–69% of body length and a short post-uterine sac filled with sperm is present [58, 59]. In contrast to species of *Acrobeloides*, *Cephalobus*, *Mesorhabditis* and *Panagrolaimus*, the posterior branch of the gonad of *O. guentheri* is not reduced by apoptosis of the distal tip cell [60], and the vestigial branch is very variable within this species [57]. These patterns argue for a relatively recent reduction in *O. guentheri*. Based on the similarities (in the female gonad and nearly median vulva) between *C. monodelphis* sp. n. and *O. guentheri*, the gonadal system of female *C. monodelphis* sp. n. may also represent a relatively recent evolutionary shift.

Conclusions

The basal position and the unique characters of *C. monodelphis* sp. n. in the genus *Caenorhabditis* and its similarity with the hypothetical ancestor of *Caenorhabditis* makes *C. monodelphis* sp. n. a key species for future evolutionary and developmental studies within *Caenorhabditis*. Importantly we present here, alongside traditional morpological diagnosis of this new species a complete genome draft, which we believe is the first time this has been done for a metazoan species description. Release of the draft genome sequence of *C. monodelphis* sp. n., along with its formal description will, we hope, promote forward- and reverse-genetic analyses of its biology. In particular, CRISPR-Cas9 gene editing technologies, which require sequence knowledge for design of targeting oligonucleotides, are immediately facilitated.

While publication of marker sequence alongside species description is becoming commonplace [61], formal publication of whole genome data alongside species descriptions has historically been limited to prokaryotic taxa. In Eukaryota, this practice is just gaining traction, with the recent publication of the description of a fungal taxon with genome data (*Epichloë inebrians*, an ergot fungus [62]). Additionally, novel arthropod taxa used in phylogenomic analyses have had species descriptions published independently, but near-concurrently, with their genome data (*Mengenilla moldrzyki*, a strepsipteran insect [63, 64] or transcriptome data (the centipede *Eupolybothrus cavernicolus* [65]). For *Caenorhabditis* species, where morphology can be misinformative and/or misleading, phylogenomic analyses – and thus determination of genome sequence – will be essential for resolution of relationships. We suggest that genome scale data allied to species description should become commonplace.

Endnote

[1]Named after the monodelphic reproductive system in the female.

Acknowledgements

We are grateful to Dr. Karin Kiontke for providing a culture of *Caenorhabditis monodelphis* sp. n. We thank Dr. Marie-Anne Félix for sharing new data about *Caenorhabditis monodelphis* sp. n. found in Norway and for the isolation of DNA and RNA samples from her inbred strain JU1667. We are thankful for Marjolein Couvreur for providing SEM images and Myriam Claeys for providing the TEM image. Sequencing was performed by Edinburgh Genomics, The University of Edinburgh. We also thank Roderic Page, Chrstopher Schardl, Christoph Bleidorn and Jason Stajich for responses on twitter concerning genome data allied to Eukaryotic species descriptions.

Funding

This work was supported by by a special research fund UGent 01 N02312 and the Foundation for Scientific Research, Flanders grant FWOKAN2013001201. Edinburgh Genomics is partly supported through core grants from NERC (R8/H10/56), MRC (MR/K001744/1) and BBSRC (BB/J004243/1). L.S. is funded by a Baillie Gifford Studentship, University of Edinburgh.

Authors' contributions

DS, WS, LS and MB conceived the study. DS, WS, LS collected and analysed data and wrote the manuscript. All authors provided comments on early drafts of the manuscript. WB, LS and MB funded this study. All authors read, revised, and approved the final manuscript.

Competing interests

The authors declare that they have no competing interests.

Author details

[1]Department of Biology, Nematology Research Unit, Ghent University, K.L. Ledeganckstraat 35, 9000 Ghent, Belgium. [2]Institut für Biologie/Zoologie, Freie Universität Berlin, Königin-Luise-Str. 1-3, 14195 Berlin, Germany. [3]Institute of Evolutionary Biology, University of Edinburgh, Edinburgh EH9 3FL, UK.

References

1. Girard LR, Fiedler TJ, Harris TW, Carvalho F, Antoshechkin I, Han M, et al. WormBook: the online review of *Caenorhabditis elegans* biology. Nucleic Acids Res. 2007;35:D472–5.
2. Félix M-A, Braendle C, Cutter AD. A streamlined system for species diagnosis in *Caenorhabditis* (Nematoda: Rhabditidae) with name designations for 15 distinct biological species. PLoS ONE. 2014;9:e94723.
3. Kiontke K, Félix M-A, Ailion M, Rockman M, Braendle C, Penigault J-B, et al. A phylogeny and molecular barcodes for *Caenorhabditis*, with numerous new species from rotting fruits. BMC Evol Biol. 2011;11:339.
4. Abebe E, Mekete T, Thomas WK. A critique of current methods in nematode taxonomy. Afr J Biotechnol. 2013;10:312–23.
5. Huys R, Llewellyn-Hughes J, Olson PD, Nagasawa K. Small subunit rDNA and Bayesian inference reveal *Pectenophilus ornatus* (Copepoda incertae sedis) as highly transformed Mytilicolidae, and support assignment of Chondracanthidae and Xarifiidae to Lichomolgoidea (Cyclopoida). Biol J Linn Soc. 2006;87:403–25.
6. Kiontke K, Barrière A, Kolotuev I, Podbilewicz B, Sommer R, Fitch DHA, et al. Trends, stasis, and drift in the evolution of nematode vulva development. Curr Biol. 2007;17:1925–37.
7. Kiontke K, Gavin NP, Raynes Y, Roehrig C, Piano F, Fitch DHA. *Caenorhabditis* phylogeny predicts convergence of hermaphroditism and extensive intron loss. Proc Natl Acad Sci U S A. 2004;101:9003–8.
8. Nuez I, Félix M-A. Evolution of susceptibility to ingested double-stranded RNAs in *Caenorhabditis* nematodes. PLoS ONE. 2012;7:e29811.
9. Guevara R, Rayner ADM, Reynolds SE. Orientation of specialist and generalist fungivorous ciid beetles to host and non-host odours. Physiol Entomol. 2000;25:288–95.
10. Kiontke K, Sudhaus W. Ecology of *Caenorhabditis* species. In: Community TCeR: WormBook, editor. WormBook. 2006.
11. Hooper DJ. Extraction of free-living stages from soil. In: Laboratory Methods for Work with Plant and Soil Nematodes. Edited by Southey JF. London: Her Majesty's Stationery Office; 1986: 5–30.
12. Yushin VV, Claeys M, Bert W. Ultrastructural immunogold localization of major sperm protein (MSP) in spermatogenic cells of the nematode *Acrobeles complexus* (Nematoda, Rhabditida). Micron. 2016;89:43–55.
13. Andrews S. FastQC: a quality control tool for high throughput sequence data. 2010. http://www.bioinformatics.babraham.ac.uk/projects/fastqc/.
14. Jiang HS, Lei R, Ding SW, Zhu SF. Skewer: a fast and accurate adapter trimmer for next-generation sequencing paired-end reads. BMC Bioinformatics. 2014;15:182.
15. Laetsch D. Blobtools. https://github.com/DRL/blobtools.
16. Camacho C, Coulouris G, Avagyan V, Ma N, Papadopoulos J, Bealer K, et al. BLAST plus: architecture and applications. BMC Bioinformatics. 2009;10:421.
17. Langmead B, Salzberg SL. Fast gapped-read alignment with bowtie 2. Nat Meth. 2012;9:357–U354.
18. Quinlan AR, Hall IM. BEDTools: a flexible suite of utilities for comparing genomic features. Bioinformatics. 2010;26:841–2.

19. Simpson JT, Wong K, Jackman SD, Schein JE, Jones SJM, Birol I. ABySS: a parallel assembler for short read sequence data. Genome Res. 2009;19:1117–23.

20. Koutsovoulos G. SCUBAT2. https://github.com/GDKO/SCUBAT2.

21. Smit AF, Hubley R. RepeatModeler Open-1.0. 2008–2015. http://www.repeatmasker.org.

22. Smit AF, Hubley R, Green P. RepeatMasker Open-4.0. 1996–2010. http://www.repeatmasker.org.

23. Dobin A, Davis CA, Schlesinger F, Drenkow J, Zaleski C, Jha S, et al. STAR: ultrafast universal RNA-seq aligner. Bioinformatics. 2013;29:15–21.

24. Hoff KJ, Lange S, Lomsadze A, Borodovsky M, Stanke M. BRAKER1: unsupervised RNA-Seq-based genome annotation with GeneMark-ET and AUGUSTUS. Bioinformatics. 2016;32:767–9.

25. Howe KL, Bolt BJ, Cain S, Chan J, Chen WJ, Davis P, et al. WormBase 2016: expanding to enable helminth genomic research. Nucleic Acids Res. 2016; 44:D774–80.

26. Aken BL, Ayling S, Barrell D, Clarke L, Curwen V, Fairley S, et al. The Ensembl gene annotation system. Database. 2016;2016:baw093.

27. Wickham H. ggplot2: Elegant Graphics for Data Analysis: Springer-Verlag New York: Springer; 2009.

28. Mühlhausen S, Hellkamp M, Kollmar M. GenePainter v. 2.0 resolves the taxonomic distribution of intron positions. Bioinformatics. 2015;31:1302–4.

29. Emms DM, Kelly S. OrthoFinder: solving fundamental biases in whole genome comparisons dramatically improves orthogroup inference accuracy. Genome Biol. 2015;16:157.

30. Sievers F, Wilm A, Dineen D, Gibson TJ, Karplus K, Li WZ, et al. Fast, scalable generation of high-quality protein multiple sequence alignments using clustal omega. Mol Syst Biol. 2011;7:539.

31. Capella-Gutierrez S, Silla-Martinez JM, Gabaldon T. trimAl: a tool for automated alignment trimming in large-scale phylogenetic analyses. Bioinformatics. 2009;25:1972–3.

32. Kueck P, Longo GC. FASconCAT-G: extensive functions for multiple sequence alignment preparations concerning phylogenetic studies. Front Zool. 2014;11:81.

33. Stamatakis A. RAxML version 8: a tool for phylogenetic analysis and post-analysis of large phylogenies. Bioinformatics. 2014;30:1312–3.

34. Lartillot N, Lepage T, Blanquart S. PhyloBayes 3: a Bayesian software package for phylogenetic reconstruction and molecular dating. Bioinformatics. 2009;25:2286–8.

35. Rambaut A, Suchard M, Xie D, Drummond A. Tracer v1. 6. 2014. http://tree.bio.ed.ac.uk/software/tracer/.

36. Kiontke K, Fitch DHA. The phylogenetic relationships of Caenorhabditis and other rhabditids. In: Community TCeR: WormBook, editor. WormBook. 2005.

37. Félix M-A. Cryptic quantitative evolution of the vulva intercellular signaling network in Caenorhabditis. Curr Biol. 2007;17:103–14.

38. Akimkina T, Yook K, Curnock S, Hodgkin J. Genome characterization, analysis of virulence and transformation of Microbacterium nematophilum, a coryneform pathogen of the nematode Caenorhabditis elegans. FEMS Microbiol Lett. 2006;264:145–51.

39. Sudhaus W, Kiontke K. Phylogeny of Rhabditis subgenus Caenorhabditis (Rhabditidae, Nematoda). J Zool Syst Evol Res. 1996;34:217–33.

40. Parra G, Bradnam K, Korf I. CEGMA: a pipeline to accurately annotate core genes in eukaryotic genomes. Bioinformatics. 2007;23:1061–7.

41. Kiontke K. Description of Rhabditis (Caenorhabditis) drosophilae n. sp. and R. (C.) sonorae n. sp. (Nematoda: Rhabditida) from saguaro cactus rot in Arizona. Fundam Appl Nematol. 1997;20:305–15.

42. Sudhaus W. Phylogenetic systematisation and catalogue of paraphyletic "Rhabditidae" (Secernentea, Nematoda). J Nematode Morphol Syst. 2011;14:113–78.

43. Sudhaus W. Vergleichende Untersuchungen zur Phylogenie, Systematik, Ökologie, Biologie und Ethologie der Rhabditidae (Nematoda). Zoologica. 1976;43:1–229.

44. Sudhaus W, Fitch D. Comparative studies on the phylogeny and systematics of the Rhabditidae (Nematoda). J Nematol. 2001;33:1–70.

45. Hirschmann H. Die Nematoden der Wassergrenze mittelfränkischer Gewässer. Zoologische Jahrbücher (Systematik). 1952;81:313–407.

46. Sudhaus W. Order Rhabditina: "Rhabditidae". In: Schmidt-Rhaesa A, editor. Handbook of zoology gastrotricha, cycloneuralia and gnathifera, vol. 2. Nematodath ed. Berlin, Boston: Walter De Gruyter; 2014. p. 537–55.

47. Grewal PS, Richardson PN. Effects of Caenorhabditis elegans (Nematoda: Rhabditidae) on yield and quality of the cultivated mushroom Agaricus bisporus. Ann Appl Biol. 1991;118:381–94.

48. Økland B. Insect fauna compared between six polypore species in a southern Norwegian spruce forest. Fauna Norv Ser B. 1995;42:21–6.

49. Tsuda K, Futai K. Description of Caenorhabditis auriculariae n. sp. (Nematoda: Rhabditida) from fruiting bodies of Auricularia polytricha. Jpn J Nematol. 1999;29:18–23.

50. Hammond PM, Lawrence JF. Appendix - Mycophagy in Insects: a Summary. In: Insect-fungus Interactions. Wilding N, Collins NM, Hammond PM, Webber JF, editors, vol. 14. London: Academic Press; 1989. p. 275–324.

51. Timper P, Davies KG. Biotic interactions. In: Nematode Behaviour. Gaugler R, Bilgrami AL, editors. Wallingford, UK: CABI; 2004. p. 277–308.

52. Cho S, Jin S-W, Cohen A, Ellis RE. A phylogeny of Caenorhabditis reveals frequent loss of introns during nematode evolution. Genome Res. 2004;14:1207–20.

53. Hoogewijs D, De Henau S, Dewilde S, Moens L, Couvreur M, Borgonie G, et al. The Caenorhabditis globin gene family reveals extensive nematode-specific radiation and diversification. BMC Evol Biol. 2008;8:1.

54. Dieterich C, Clifton SW, Schuster LN, Chinwalla A, Delehaunty K, Dinkelacker I, et al. The Pistionchus pacificus genome provides a unique perspective on nematode lifestyle and parasitism. Nat Genet. 2008;40:1193–8.

55. Sudhaus W. Matthesonema eremitum n. sp. (Nematoda, Rhabditida) associated with hermit crabs (Coenobita) from the Philippines and its phylogenetic implications. Nematologica. 1986;32:247–55.

56. Andrássy I. Erd- und Süßwassernematoden aus Bulgarien. Acta Zool Acad Sci Hung. 1958;4:1–88.

57. Sudhaus W, Hooper DJ. Rhabditis (Oscheius) guentheri sp. n., an unusual species with reduced posterior ovary, with observations on the Dolichura and Insectivora groups (Nematoda: Rhabditidae). Nematologica. 1994;40:508–33.

58. Massey CL. Two new genera of nematodes parasitic in the eastern subterranean termite, Reticulitermes flavipes. J Invertebr Pathol. 1971;17:238–42.

59. Tahseen Q, Sultana R, Khan R, Hussain A. Description of two new and one known species of the closely related genera Artigas, 1927 and Massey, 1971 (Nematoda: Rhabditidae) with a discussion on their relationships. Nematology. 2012;14:555–70.

60. Félix MA, Sternberg PW. Symmetry breakage in the development of one-armed gonads in nematodes. Development. 1996;122:2129–42.

61. Sommer R, Carta LK, Kim S-y, Sternberg PW. Morphological, genetic and molecular description of Pristionchus pacificus sp. n. (Nematoda: Neodiplogasteridae). Fundam Appl Nematol. 1996;19:511–22.

62. Chen L, Li XZ, Li CJ, Swoboda GA, Young CA, Sugawara K, et al. Two distinct Epichloë species symbiotic with Achnatherum inebrians, drunken horse grass. Mycologia. 2015;107:863–73.

63. Niehuis O, Hartig G, Grath S, Pohl H, Lehmann J, Tafer H, et al. Genomic and morphological evidence converge to resolve the enigma of Strepsiptera. Curr Biol. 2012;22:1309–13.

64. Pohl H, Niehuis O, Gloyna K, Misof B, Beutel R. A new species of Mengenilla (Insecta, Strepsiptera) from Tunisia. ZooKeys. 2012;198:79–102.

65. Edmunds SC, Hunter CI, Smith V, Stoev P, Penev L. Biodiversity research in the "big data" era: GigaScience and pensoft work together to publish the most data-rich species description. GigaScience. 2013;2:14.

5

Tracking the migration of a nocturnal aerial insectivore in the Americas

Philina A. English[1*], Alexander M. Mills[2], Michael D. Cadman[3], Audrey E. Heagy[4], Greg J. Rand[5], David J. Green[1] and Joseph J. Nocera[6]

Abstract

Background: Populations of Eastern Whip-poor-will (*Antrostomus vociferous*) appear to be declining range-wide. While this could be associated with habitat loss, declines in populations of many other species of migratory aerial insectivores suggest that changes in insect availability and/or an increase in the costs of migration could also be important factors. Due to their quiet, nocturnal habits during the non-breeding season, little is known about whip-poor-will migration and wintering locations, or the extent to which different breeding populations share risks related to non-breeding conditions.

Results: We tracked 20 males and 2 females breeding in four regions of Canada using geolocators. Wintering locations ranged from the gulf coast of central Mexico to Costa Rica. Individuals from the northern-most breeding site and females tended to winter furthest south, although east-west connectivity was low. Four individuals appeared to cross the Gulf of Mexico either in spring or autumn. On southward migration, most individuals interrupted migration for periods of up to 15 days north of the Gulf, regardless of their subsequent route. Fewer individuals showed signs of a stopover in spring.

Conclusions: Use of the southeastern United States for migratory stopover and a concentration of wintering locations in Guatemala and neighbouring Mexican provinces suggest that both of these regions should be considered potentially important for Canadian whip-poor-wills. This species shows some evidence of both "leapfrog" and sex-differential migration, suggesting that individuals in more northern parts of their breeding range could have higher migratory costs.

Keywords: Geolocator, Nightjar, Whip-poor-will, *Antrostomus vociferous*, Migration, Stopover, Leapfrog, Sex-differential migration, Recapture rate, Trans-Gulf

Background

At high latitudes, over 80% of bird species are migratory [1]. Migration increases exposure to novel challenges, including pathogens, predators, and anthropogenic threats at geographically disparate locations [1, 2]. Cumulatively the energetic, time, and fatality costs associated with these long journeys can account for most annual mortality for some species [3–5] and can influence survival and productivity in subsequent seasons [6–9]. Depending on the relative costs and benefits [5, 10, 11], individual strategies relating to the timing and speed of migration, migratory routes, and winter destinations vary widely both within

[12–14] and between species [15–17]. Some birds build up large reserves of fat to fuel long flights across inhospitable habitats or barriers [18, 19], while others employ fly-and-forage strategies that allow lower weight burdens and reduced time spent at stopover locations [20]. Crossing barriers, such as large bodies of water, likely increases the time required to build up fuel reserves and increases risks associated with abrupt changes in weather, but may help migrants to avoid predation and reduce transit time associated with longer over-land detours [21].

Migratory strategies that allow individuals to track seasonal variation in resources may be particularly important for temperate breeding aerial insectivores (i.e., birds that specialize in catching and eating flying insects while they themselves are also in flight). In temperate climates, insect flight periods are ultimately limited by seasonal

* Correspondence: paenglis@lakeheadu.ca
[1]Department of Biological Sciences, Simon Fraser University, Burnaby, BC V5A 1S6, Canada
Full list of author information is available at the end of the article

changes in temperature [22]. While some insectivorous birds that pursue dormant prey in sheltered hiding places can overwinter in temperate regions (e.g., woodpeckers), most aerial insectivores must migrate to ensure an adequate supply of flying insects. Even when prey is abundant, this foraging strategy is sensitive to the high energetic costs of flight for both predator and prey during inclement weather. Unseasonably cold, or extreme, weather can kill or make prey less accessible to predators [23, 24]. This sensitivity to weather could increase selective pressure on the timing, migration routes, and choice of winter habitat [11, 25].

Population declines among many temperate breeding aerially insectivore birds may be partially due to recent increases in the frequency of extreme weather events [26], interacting with existing costs of migration and a reliance on weather-sensitive prey [27–30]. For example, long-term decreases in body mass found in a declining swallow population, which could not be explained by changes in breeding habitat quality, suggest a carry-over of change in migration or wintering conditions [31]. In addition, the degree of connectivity between populations on the breeding and wintering grounds can buffer or exacerbate a loss of habitat at other locations used throughout the annual cycle [32–34]. Therefore, to understand and mitigate threats to aerial insectivores, it is important to identify the year-round geographic and habitat requirements, migratory routes, and temporal constraints of individuals belonging to threatened populations [35, 36].

Nightjars may be especially sensitive to inclement weather, because they are limited to foraging on flying insects only at dawn and dusk, or on moonlight nights, when there is adequate light to see their prey [37, 38]. The only two species of Neotropical migrant nightjars that occur at high latitudes in North America differ in foraging strategy, migratory distance, and breeding site fidelity, and still both are listed as threatened. The Eastern Whip-poor-will (*Antrostomus vociferous*) is a sally-foraging, medium-distance migrant, with high breeding site fidelity, whose populations appear to be declining range-wide. Due to their quiet, nocturnal habits during the non-breeding season, little is known about when and where changes in food availability could influence this population. We seek to fill this knowledge gap by identifying wintering locations, migratory routes and stopovers, and variation in timing of movements, with respect to breeding origin and sex. This is not only the first examination of these parameters for Eastern Whip-poor-wills, but the first for any Neotropical migrant nightjar.

Methods
Study locations/sites
We deployed light-logging geolocation tags (Fig. 1), hereafter "geolocators", in four regions spanning a 1000 km

Fig. 1 Female Eastern Whip-poor-will wearing a geolocator tag. The light stalk on this recently deployed tag is clearly visible, but feathers soon covered the light stalk much of the time. Photo credit: PA English

stretch of the species' range in Ontario, Canada: Rainy River District, Norfolk County, Muskoka District Municipality, and Frontenac County (Fig. 2). The Rainy River site (48° 49–59'N 94° 0–21'W) consisted of a 40000-hectare mosaic of agriculture, poplar (*Populus* sp.), coniferous forests, logged areas, and wetlands. The Norfolk County site (42° 42'N 80° 21–28'W) was St. Williams Conservation Reserve, which consists of two forest patches totaling 1035 hectares of pine-oak sand barrens and pine reforestation in a zone of intensive agriculture. The Muskoka district sites (including portions of neighbouring Parry Sound District and Simcoe County; 44° 22–56'N 79° 08–47'W) contained extensive pine-oak rock barrens. The Frontenac County site (44° 28–34'N 76° 20–25'W) was Queen's University Biological Station, which consists of over 3200 hectares of deciduous forest and abandoned farmland in various stages of succession, both with scattered small rock barrens.

Field methods/geolocator deployment
We captured and banded adult/after hatch-year whip-poor-wills between 5 May and 25 July in 2011–2013. We captured male whip-poor-wills at night using mist nets and song playback at all sites. We only targeted females in Frontenac, where we captured them on nests by placing a soft mesh fishnet over them while they were incubating. All birds received a numeric aluminum leg-band issued by the Canadian Wildlife Service.

Geolocator tags record and store time and light-level data that can be used to estimate latitude and longitude based on sunrise and sunset timing. Birds must be recaptured to retrieve the tags and download the data. We deployed 65 LightBug geolocator tags (Lotek, Newmarket, Ontario, Canada; Fig. 1) during the 2011 and 2012 breeding seasons. We fitted tags to individual birds using a leg loop harness [39] made of 2.5 mm Teflon ribbon and secured with a cyanoacrylate-glued square

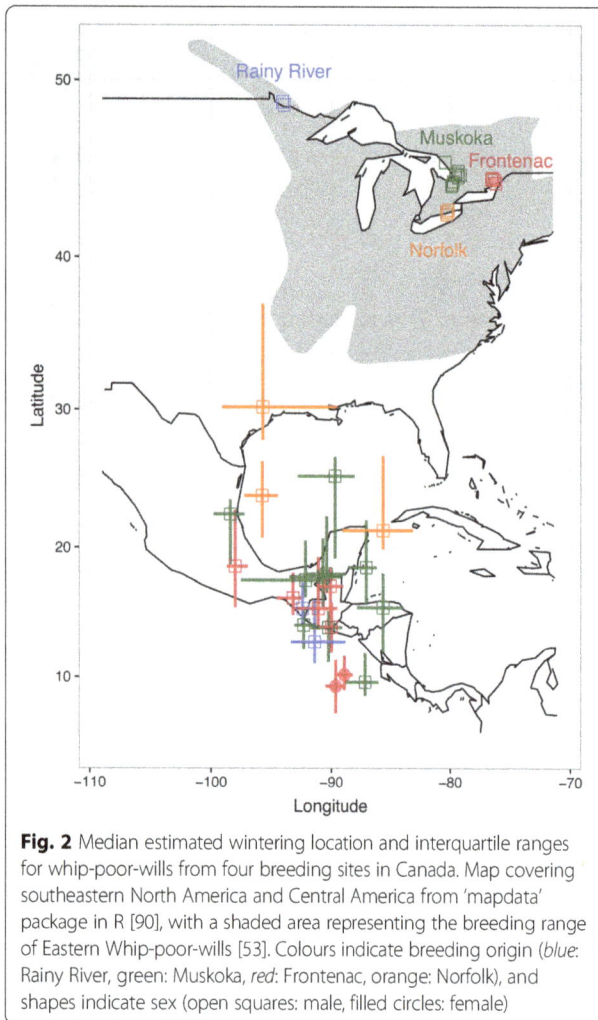

Fig. 2 Median estimated wintering location and interquartile ranges for whip-poor-wills from four breeding sites in Canada. Map covering southeastern North America and Central America from 'mapdata' package in R [90], with a shaded area representing the breeding range of Eastern Whip-poor-wills [53]. Colours indicate breeding origin (*blue*: Rainy River, green: Muskoka, *red*: Frontenac, orange: Norfolk), and shapes indicate sex (open squares: male, filled circles: female)

knot. Total weight of the tag and gear was approximately 2.7 g. We deployed 59 tags on males (Rainy River: 5, Norfolk: 14, Muskoka: 24, Frontenac: 16) and 6 on females. Six returning males received tags in two consecutive years. Whip-poor-wills captured in this study weighed 46.7–67.5 g (mean = 57.8 g), but no birds weighing < 54 g were fitted with geolocators. Geolocators with harnesses amounted to 4–5% of body mass [40, 41]. An additional 36 birds weighing > 54 g were banded, but did not receive geolocators (Rainy River: 10, Norfolk: 2, Frontenac: 24).

Recapture/return rates
We compared the combined effect of survival and site fidelity for banded birds with and without geolocators. We attempted to recapture birds at all sites and banding locations where birds had been fitted with geolocator tags the previous year, but effort varied in duration, date, moon phase, and weather between sites and years. To retrieve tags from females, we searched for nests on all territories on which females were tagged in the previous

year. We also attempted to capture birds in territories adjacent to those where geolocators had been deployed the previous year. We were unable to recapture all birds occupying sites where geolocators had been deployed the previous year; therefore, we estimated return rates only for territories on which individuals of the same sex were successfully captured in two consecutive years.

Geolocator analysis/data processing
LightBug geolocators were programmed to record the intensity of blue light every 8 min for up to one year. Horizon clutter and clouds affect blue light less than other wavelengths [42]. Using Lotek's LAT Viewer Studio Software, these recorded light values were compared with a template of how blue light levels should change at twilight and location estimates were produced along with an error estimate based on the fit of the data [43]. The template fit method is less sensitive to daily variation in cloud cover and ambient light intensity than the threshold method [41, 43–45]. This method also allows for the possibility of estimating latitude during the equinox, although with greater error than at other times of the year. The template fit method is still sensitive to short term fluctuations in light conditions, including those resulting from the behaviour of crepuscular animals like whip-poor-wills [44]. Because our tags used a proprietary data format and our light-level data was extremely noisy (making it necessary to manually select which peaks qualified as true sunrises or sunsets for most analysis packages), we could not easily apply recent advances in movement modeling, such as FlightR, to our data [45]. Our template fit method instead provided an objective way of assessing reliability of individual light curves by incorporating deviations from a smooth curve into error estimates [46].

We used a series of criteria to filter the daily latitude and longitude estimates to exclude points with limited precision or that were biologically impossible. Latitude and longitude were analyzed independently because they respond to noise in the light signals differently [43, 47]. First, we included only location estimates within the species' plausible geographic range (between 0° and 58° latitude and −60° and −110° longitude). This resulted in average exclusion of 26% of latitude estimates and 4% of longitude estimates. Second, we excluded points with error estimates (provided by LAT Viewer) of > 15° and > 5° for latitude and longitude respectively. We used different thresholds because estimates of latitude have more error than estimates of longitude [43, 47]. These thresholds excluded another 20% percent of plausible estimates. Third, we removed estimates that required birds to travel > 800 km in a day (similar to Fraser et al. 2012). We chose 800 km as a cut-off distance because it allows for some error beyond the maximum average migration rate recorded for small birds of 500–600 km day^{-1} [48, 49]. Finally, we excluded estimates that required a redundant movement of 800 km (i.e., movements of 800 km away from and back to average weekly longitude or latitude) even when daily movements were < 800 km. The resulting proportion of missing days per bird per year averaged 31% (range: 4–72%) for longitude and 62% (range: 37–93%) for latitude. To evaluate the accuracy of these location estimates we compared capture locations with average longitude and latitude values obtained for the breeding season (15 May to 31 Aug). The average difference between median of breeding season estimates and the actual capture location was -0.20° (-2.96°–0.93°) for longitude and −0.43° (−7.90°–2.17°) for latitude [see Additional file 1: Figure S1].

Wintering range

A qualitative examination of latitude and longitude estimates plotted independently against time (see [Rakhimberdiev et al. 2016] for example plots) provided no evidence that whip-poor-will used multiple wintering sites (i.e. no shifts away from the median value that consistently exceeded the variance in our estimates). Therefore, we defined wintering location of each bird as the median latitude and longitude estimates obtained between 15 Dec (the latest date individuals arrive on their wintering grounds, see Results) and 28 Feb (day before the earliest estimated start of spring migration, see Results). We illustrate the uncertainty in this estimate using interquartile ranges.

Migratory behaviour

For Ontario whip-poor-wills, departure from both breeding and wintering grounds occurs near the equinoxes, so longitude data were used to estimate the start of migratory behaviour for birds from the more eastern study sites (Norfolk, Muskoka, and Frontenac). The two Rainy River birds were not included in this analysis because their tags did not detect any longitudinal movement at the start of autumn migration and both tags stopped collecting data prior to spring migration. Latitude data were used in estimating the end dates of migration only when the wintering/breeding latitude was reached after reaching the wintering/breeding longitude.

Due to variation in the number of retained location estimates, and the variance in the precision of these estimates, we used a range of dates to estimate migratory transitions. We defined the start of migration as the mid-point between the last day in a series of 2 consecutive samples (< one week apart) that are within 1 standard deviation of the mean breeding ground longitude (68% probability that the bird is still at the breeding ground longitude) and the day prior to first 2 consecutive samples that are in the direction of subsequent movement and outside 1 standard deviation (68%

probability that the bird is no longer at breeding ground longitude). Similarly, arrival on the wintering grounds was defined as the midpoint between the last day in a series of 2 consecutive samples that are in the direction of previous movement and outside 1 standard deviation of the wintering longitude and latitude (68% probability that the bird is not yet at the wintering longitude) and the first 2 consecutive samples that are within 1 standard deviation of mean wintering longitude (68% probability that the bird has reached wintering longitude). This threshold produces a larger and more conservative range of estimates for departure/arrival dates than a 95% probability threshold. The degree of uncertainty in each of these estimates was defined as the number of days between the two dates used to calculate each midpoint. For statistical analysis, we excluded those estimates with > 7 days uncertainty.

While the lack of latitude estimates for many days during this migratory period makes identification of precise migratory routes impossible, broad patterns in the duration of migration, stopover use, and route around the Gulf of Mexico were identifiable for some individuals. We defined duration of migration as the time between the estimated start and end of autumn and spring migrations, including any time spent at stopover locations. Duration estimates derived from start and end dates with total combined uncertainty of > 14 days were excluded from further analysis. Stopovers were identified by visual inspection of temporal changes in longitude to identify periods of at least 4 days without any consecutive days with forward progress of > 2° longitude.

At least 3 days would be required for a bird flying at a maximum of 500 km/day (487 km/day was maximum rate estimated for another nightjar *Caprimulgus europaeus* [50]) to travel the 1500 km of the gulf shoreline that lies furthest west, between 95° and 98°W. Therefore, flight over some portion of the Gulf of Mexico was assumed to have occurred where mean wintering latitude was south of 25°N and east of 95°W, and when < 3 consecutive samples during the migratory period were west of 95° and any periods of missing data during this stage of migration were also < 3 days.

In total, we were able to estimate: timing of departure from breeding longitude for 11 individual annual cycles (Norfolk: 1, Frontenac: 7, Muskoka: 3), arrival at winter longitude for 15 (Frontenac: 8, Muskoka: 7), duration of autumn migration for 11 (Frontenac: 7, Muskoka: 4), departure from wintering longitude for 7 (Norfolk: 1, Frontenac: 2, Muskoka: 4), duration of spring migration for 11 (Norfolk: 1, Frontenac: 5, Muskoka: 6), and arrival at breeding longitude for 15 (Norfolk: 1, Frontenac: 7, Muskoka: 7). We were able to estimate the location and time spent at stopover sites for 12 autumn (Frontenac: 6, Muskoka: 6) and 4 spring migrations (Norfolk: 1,

Frontenac: 2, Muskoka: 1), and to determine whether individuals crossed or travelled around the Gulf of Mexico for 12 autumn (Frontenac: 5, Muskoka: 7) and 10 spring migrations (Frontenac: 3, Muskoka: 7).

Statistical analysis

We examined i) the correlation between the wintering latitudes and longitudes with the breeding origins for all males (if data were obtained for two years we used the year with more winter locations), ii) sex differences in winter latitude and longitude using data from birds captured in central Ontario (Muskoka and Frontenac sites that share similar latitude and habitat types), and iii) interannual variation in wintering latitude for males tracked twice. For all tracks with sufficient daily resolution (see Migratory Behaviour), we classified migratory routes qualitatively and estimated duration of stopovers. Finally, we estimated i) the variation among individual males in the departure, arrival, and duration of autumn and spring migration (when timing did not differ significantly between years, we pooled inter-individual variation in migratory timing across years), and ii) interannual variation in the timing of migration for males tracked twice. We report raw differences in timing between the sexes, but do not apply statistical tests due to the small sample sizes. We used non-parametric statistical tests (Kendall rank correlation or Wilcox rank sum tests) in R [51].

Results

Return rates

We captured territorial males at 45 of the 59 sites where a geolocator was deployed in the previous year, and in 23 of those 45 cases we recaptured the same individual. We retrieved two additional tags: one two years after it was deployed, and one from a male that had moved to an adjacent territory. We captured a territorial male at 19 of the 36 sites where males weighing > 54 g were banded but did not receive geolocators; 12 of these were returning males. A combination of annual survival and territory fidelity resulted in territory specific return rates of 51% (23 of 45) for males with geolocators and 63% (12 of 19) for banded males without geolocators (chi-square = 0.68, df = 1, p = 0.41). We only captured females on 3 territories in which females were tagged the previous year and 2 (67%) were returning females.

Wintering range

Light data were recorded on 24 of 25 geolocator tags retrieved from males, and both tags retrieved from females. Four males were tracked successfully for two consecutive years. Therefore, we determined the winter locations for 22 individual birds [see Additional file 2: Table S1]. These locations ranged from the gulf coast of

central Mexico to Costa Rica (Fig. 2). Median wintering latitudes for 20 males ranged from 10° to 30°N (10° to 24°N for 15 males with > 20 estimates). Males from more northern breeding sites wintered further south (Kendall's rank correlation tau = -0.33 z = -2, p = 0.04, N = 20). Female median wintering latitudes were 9° and 10°N, which are both farther south than all but one male from similar breeding latitude (Wilcox rank sum W = 3, p = 0.03, N = 15, 2). Median wintering longitudes ranged from −86° to −98°, were not related to breeding longitude for males (Kendall's rank correlation tau =−0.005, z =−0.03, p = 1, N = 20), and did not differ between the sexes (Wilcox rank sum W = 24, p = 0.2, N = 15, 2). Three of the four males tracked in two years appeared to overwinter in the same location; interquantile ranges for both latitude and longitude estimates overlap between years (Fig. 3). The winter site fidelity of the remaining male was uncertain because we only obtained estimates of its winter location for 7 days in 2012/13.

Migratory route and stopovers

In autumn, ten individual males (including both years for one bird) and both females, all from central Ontario populations, appeared to stop migrating for between 4 and 15 days along the north coast of the Gulf of Mexico between 83° and 96°W (pooled median = 30°N, 89°W; Fig. 4a). After these stopovers, one male crossed the gulf, another continued west to winter on the gulf coast of Mexico (19°N, 98°W; QU907), 8 individuals travelled

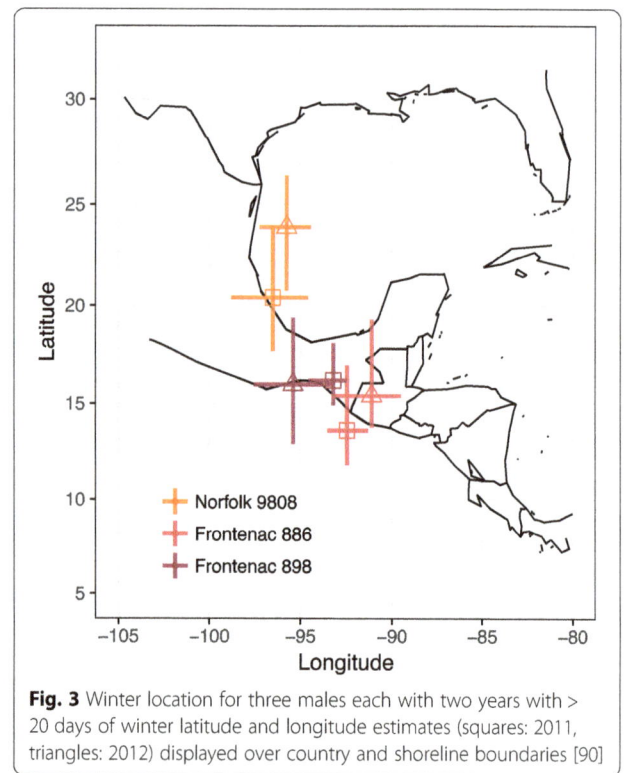

Fig. 3 Winter location for three males each with two years with > 20 days of winter latitude and longitude estimates (squares: 2011, triangles: 2012) displayed over country and shoreline boundaries [90]

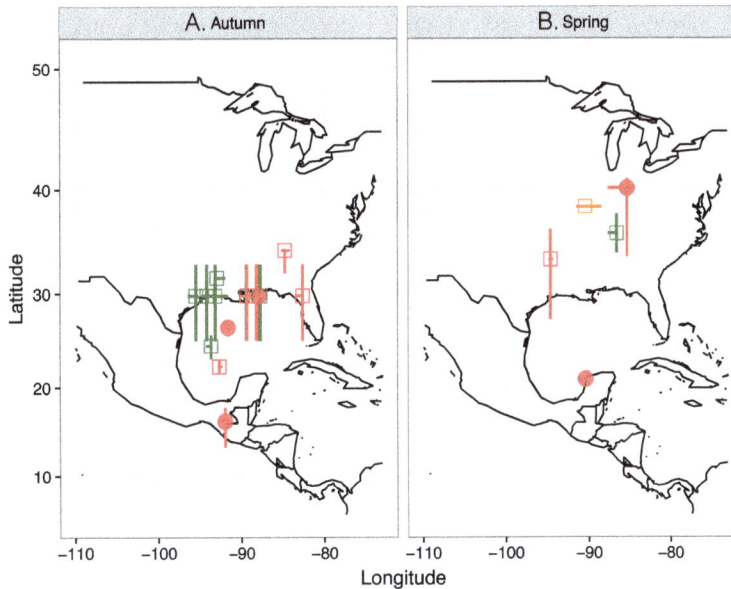

Fig. 4 Stopover locations (median with interquartile ranges) for all birds that halted longitudinal progress for ≥ 4 days during either migratory period. We estimate latitude for individuals where possible, or based on pooled estimates for all individuals showing signs of stopover during the same time period. Map outlines [90], colours indicating breeding origin (green: Muskoka, red: Frontenac, orange: Norfolk), and shapes indicating sex (open squares: male, filled circle: female) are the same as in Figure 2

southwest around the gulf and then east into southwestern Mexico or Central America, and for two the path was uncertain. Another male crossed the Gulf of Mexico during southward migration without stopping for a detectable length of time (4 days). One female and 1 male appeared to stopover a second time south of the Gulf of Mexico (26–13°N, 91–94°W) before continuing south another ~5° latitude.

In spring, one female appeared to stop on the Yucatan Peninsula for 10 days (1–11 Mar) and showed evidence of a 6-day stopover north of the Gulf (1–7 May), but it was unclear whether she crossed or circumnavigated the Gulf. Three males, at least one of which circumnavigated the Gulf, showed evidence of a stopover north of the Gulf for between 7 and 12 days at a median of 30°N and 90°W (start: 2–22 Apr 2012; end: 8–29 Apr 2012; Fig. 4b). Two males crossed the Gulf without evidence of any stopovers.

Variation in timing

Across all males, variation in timing of migratory behaviour was much less (<18 days) for both departure from and arrival at breeding longitudes than for arrival at and departure from wintering longitudes (>38 days; Fig. 5), but overall duration of spring and autumn migration was not different (autumn: median = 42.5, range = 26.5–68; spring: median = 37, range = 23–58; W = 31, p = 0.4). Arrival dates on both wintering and breeding grounds were not correlated with timing of departure, or wintering latitude or longitude (all p > 0.2).

Mean male dates of departure from the breeding grounds, departure from the wintering grounds, and return to the breeding grounds did not differ between years (N ≥ 6 and 2, p ≥ 0.1). However, males arrived at their wintering longitude later in 2012 than 2011 (mean = 1 Dec and 9 Nov respectively, N = 10 and 4, p = 0.02). Only one individual (Frontenac 898 in Fig. 3) had reliable timing estimates for both years; he left the breeding grounds earlier (3–9 days), but arrived on (7–15 days) and left from (1–33 days) the wintering grounds later in the 2012/2013 non-breeding season than in the previous year. In contrast, this male appeared to arrive on the breeding grounds on the same day in both years (range: 5 days later to 7 days earlier).

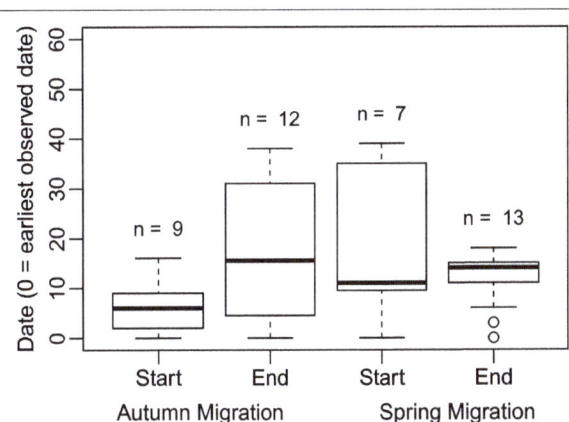

Fig. 5 Variation in the timing of migration for males from all breeding sites (except Rainy River)

In autumn, male whip-poor-wills departed from breeding longitudes in Ontario between 25 Sept and 11 Oct (mean = 2 Oct, N = 9). They arrived on wintering longitudes between 2 Nov and 3 Dec (mean = 16 Nov, N = 12). The minimum duration of travel was 27 ($\pm$ 4) days for a male from Frontenac, which wintered in central Mexico (19.11°N, 97.91°W), covering an estimated minimum distance of 2260 km at a rate of 135 km/day. The next shortest duration of 32 ($\pm$ 1) days belonged to a male from Muskoka, which wintered furthest south of all males with reliable duration estimates at (13.9°N, 90.25°W), requiring a minimum travel distance of 3637 km, and yet also travelled an average of 135 km/day. We could not assess whether males that crossed the gulf spent less time on migration, because no birds with $\leq$ 14 days uncertainty in duration appeared to cross the Gulf.

In spring, male whip-poor-wills departed from wintering longitudes between 1 Mar and 9 Apr (mean = 21 Mar, N = 7). Arrival at breeding longitude ranged from 19 Apr to 7 May (mean = 1 May, N = 13). The shortest migration time was 23 ($\pm$ 2) days for a Muskoka male that wintered in Campeche, Mexico and travelled west around the Gulf covering an estimated 4160 km at a mean rate of 180 km/day. Of two males that crossed the Gulf, the only male with accurate timing estimates took only 24 ($\pm$ 4) days to cover a minimum of 3650 km for an average travel rate of 152 km/day.

Sex differences
The two females from which we retrieved geolocators appear to have departed later than males (1 Oct and 13 Oct), spent more time on autumn migration (53 and 58 days versus mean of 45 days for males), arrived later on wintering grounds (28 Nov and 5 Dec), and departed earlier from winter longitude (27 Feb and 17 Mar). The two females also took on average 30 days longer (56 and 75 days vs. mean = 36.1, N = 8, SD = 12.6 for males) and arrived at breeding longitudes after 10 May in contrast to a mean arrival of 30 Apr for males (N = 12, SD = 5.65). Neither female appeared to cross the Gulf in either season.

Discussion
Winter location and connectivity
Our results suggest that whip-poor-wills breeding in the more northern parts of their breeding range may experience different wintering conditions and have higher migratory costs, in terms of energy expenditure, novel threats, and ability to adjust arrival time to track breeding ground conditions [52], than more southern breeding populations. Whip-poor-wills from sites across their Ontario breeding range showed some evidence of "leapfrog", and perhaps sex-differential, migration. More northerly breeding individuals and females wintered to

the south of more southerly breeding individuals, and the vast majority of males. While most individuals wintered within the well-established winter range for this species [53], 3 birds (including both females) wintered south of the Honduras-Nicaragua border, a latitude where whip-poor-wills are described as only "a casual to very rare winter resident" [54]. Given that Ontario is on the northern edge of the breeding range, this pattern is reinforced by these 3 birds appearing to winter south of the usual winter range, and not finding any birds overwintering within the most northern portions of the known winter range (with the possible exception of a single bird with only 9 days of winter latitude data). In contrast, both eastern and western-most breeding individuals wintered together, concentrated in Guatemala and neighbouring provinces of Mexico, suggesting low connectivity between breeding longitude and wintering location [33, 55]. Although population data for whip-poor-wills lacks the precision to effectively compare regional population trajectories, given that there are regional differences in population trends for other aerial insectivores [56], leapfrog migration patterns may help explain regional differences in breeding ground population trends that are not obviously linked to local changes in habitat.

Both inter-population leapfrog migration patterns and differential migration between sexes have been attributed to differences in the importance of arrival timing, asymmetric competition, or differences in cold tolerance due to body size differences [52]. Males often experience higher net benefits of early arrival on the breeding grounds [57–60] and may therefore accept higher costs of wintering further north [52]. Likewise, populations breeding further south may benefit more from being able to track spring phenology more closely [61]. The earlier spring arrival and shorter migration times we found for male whip-poor-wills suggest that early arrival on breeding grounds is more beneficial for males, potentially allowing occupation of higher quality territories. Females could be forced to migrate further by lower competitive abilities, or to exploit more abundant resources at lower latitudes [52]. However, more information on winter territoriality and resource use by whip-poor-wills is required before these hypotheses could be fully developed and tested.

For a few individuals, our geolocator data suggest biologically impossible wintering locations that are over the open ocean. Wintering locations estimated using the timing of dusk and dawn could be biased for two reasons: i) Steep mountain slopes could consistently skew sunrise or sunset by shading from the terrain [47, 62]. This could explain the aberrant points if the three southern-most birds were wintering on a west-facing slope of the continental divide in Central America, where sunrise was skewed later, causing them to appear further west and north than the actual wintering location. ii) Abrupt changes in light levels can cause smoothed light curves to appear steeper than if shading remained constant. However, this cannot explain our southern-most points, because this would cause the true winter latitude would be closer to the equator (by up to 1.4°) than the estimated location, placing the actually winter location even further into the ocean [44].

Variation in migratory route and stopover

It is generally assumed that most whip-poor-wills travel overland through Mexico and Central America [53]. Our data, however, suggest that flights across some portion of the Gulf of Mexico were undertaken by at least two individuals in autumn and two different individuals in the spring. That at least some whip-poor-wills attempt Gulf crossings is supported by vagrant records for Cuba and the Caribbean islands [53, 63] and by one e-bird (http://ebird.org/) record from off-shore in the Gulf from 12 Oct 2011.

Similar numbers of Gulf crossings in both seasons are somewhat surprising given that loop migrations in which spring migration routes are west of autumn routes seem to be most common in both Neotropical [17, 21, 64, 65] and Afro-Palaearctic migrants [14, 36, 66–69], although the reverse is seen as well [17, 70–72]). It has been suggested that these patterns are a response to prevailing winds and/or availability of resources along the different routes. The choice to cross the Gulf of Mexico is likely less risky in autumn when passing cold fronts provide tailwinds, while in spring such a cold front would be a substantial obstacle and cannot be easily anticipated when setting out from the Yucatán [73]. As a result, the dominant pattern for species migrating between eastern North America and South and Central America seems to involve more frequent over-ocean flights in autumn and more individuals taking longer over-land routes around the western side of the Gulf of Mexico in spring, with an increasing tendency to circumnavigate with more westerly breeding longitudes [49, 65, 74, 75].

Species often show within population variation in migration patterns with respect to large bodies of water [14, 21, 36, 69, 74]. Individuals tracked over multiple years, often show considerable variation in route choice [21, 49, 67, 69]. What causes individuals to make different choices in different seasons remains unclear, but could relate to individual differences in physiological condition, age, resource availability at stopover sites, or local weather patterns [76–79].

Migratory stopovers appeared to be more frequent and were of longer duration in autumn than in spring. Due to low resolution for both migration timing and route, we cannot link stopover behaviour with timing or Gulf-crossing behaviour [80]. But evidence from swallows in Europe suggest that even diurnal aerial insectivores, which employ a fly-and-forage migration strategy, use stopovers before crossing major ecological barriers [81]. In autumn, more than half of whip-poor-wills appeared to stop for up to 15 days somewhere near the north coast of the Gulf of Mexico (median = 30°N). Stopovers of similar length by northbound *Catharus* thrushes in Columbia have been shown to allow for sufficient fat storage to fuel direct flights across both the Caribbean and the Gulf of Mexico [19]. In spring, fewer individual whip-poor-wills showed evidence of stopovers that were of sufficient length to be detected, and those that did appeared to stop further north (~37°N). In fact, all evidence of spring stopovers by males occurred in 2012, which was a much earlier spring (by the end of March, e-bird records reach 39°N in 2012 and 35°N in 2013), suggesting that whip-poor-wills may track spring phenology and adjust timing of arrival by adding or lengthening stopovers depending on the conditions they find en route. Whether these stopovers were used to accumulate fat to fuel rapid travel through inhospitable habitats (e.g., Gulf crossings), or to wait for better weather conditions, the temporal and energetic demands associated with migration may make populations exceptionally sensitive to even minor alterations in habitat quality or food abundance at these sites.

Temporal variability in the annual cycle

Across individuals, similarity in duration and variability between autumn and spring migratory timing contrasts with the expectation of greater time-constraint in pre-breeding movements [67, 82, 83]. The much larger variability in timing of departure from the wintering grounds than in arrival on the breeding grounds could largely be the result of differences in geographic spread between breeding and wintering sites (< 3° versus > 15° latitude respectively) rather than evidence of an increase in time pressure with proximity to breeding and a selective advantage to early or synchronous arrival [84, 85]. Likewise, although timing of migratory transitions have been found to be related to timing of previous events within the annual cycle for many species of migratory birds [14, 15, 21, 49], we found no evidence of any relationship suggesting either a unique lack of population-level time-limitation, or that conditions vary between individual migration routes and at different wintering sites [86, 87].

Most studies that track individuals over multiple years have found much less variation in timing than in route choice [21, 49, 67]. While we have little data to assess intra-individual differences in timing of migration, we did find that for a single individual arrival date on breeding grounds was the same in both years despite differences between years in the timing of all other transitions. Also, consistent with increasing time pressure in spring, the fastest migration rate we observed was 180 km/day in spring by a male that circumnavigated the Gulf. Still, given our expectation that migratory aerial insectivores would experience time constraints in their annual cycle, high variability in timing of migration could represent evidence of either phenotypic plasticity or genetic variation, either of which could be beneficial under a changing climate [88].

Conclusion

With increases in activity during the critical dusk and dawn periods, light-based geolocation might appear an unlikely tool for tracking movements of a crepuscular bird [44]. However, we were able to identify wintering areas, migratory routes and stopovers, and to document the variability in timing of migratory movements for a threatened nightjar population. Migratory stopovers in the southeastern and central United States and wintering locations in southern Mexico and Central America both appear important for Eastern Whip-poor-will's at the northern edge of their range, such as those we studied in Canada. Determining the precise location of these sites, and how they are used by whip-poor-wills, will soon be possible using new technologies like archival GPS tags [89]. Ultimately, we hope protection of habitat and insect populations throughout the whip-poor-will's range, including at migratory stopover locations, may help a higher proportion of individuals survive the pressures of long migrations and a changing climate. Regardless, our results will help to better target both research and conservation efforts for this enigmatic species.

Additional files

> **Additional file 1: Figure S1.** Variation in accuracy of geolocation estimates on breeding grounds as illustrated by median and interquartile ranges in latitude and longitude estimates between 15 May and 31 Jul. Black dots: locations where geolocator tags were deployed. The absolute error averaged across all sites was 1.3° for latitude and 0.56° for longitude. (PDF 55 kb)
>
> **Additional file 2: Table S1.** Non-breeding location estimates for 22 eastern whip-poor-wills breeding in Ontario, Canada. M and F in the bird ID indicates males and females. (PDF 45 kb)

Acknowledgements
We thank the numerous field assistants and volunteers who supported the deployment and retrieval of geolocator tags including: M Conboy, M Timpf, E Suenaga, C Freshwater, E Dobson, A Zunder, E Purves, T Willis, Z Southcott, D Okines, and M Falconer. We also appreciate the comments and suggestions of three reviewers that helped to improve the final version of this manuscript.

Funding
Funding and material support was provided by the Canadian Wildlife Service and Ontario Ministry of Natural Resources and Forestry (OMNRF) Species at Risk Research and Stewardship grants (SARRFO6-10-SFU and 114-11-QUEENSU), NSERC Postgraduate Doctoral Fellowship (PAE), NSERC Discovery Grants (JJN and DJG), NSERC Early Career Researcher Supplement (JJN), the OMNRF Science and Research Branch, Environment and Climate Change Canada, and Simon Fraser University.

Authors' contributions
PAE, MDC and AMM conceived the idea, with contributions to study design from all authors. AMM, PAE, AEH and GJR supervised deployment of geolocators. PAE, DJG and JJN analyzed data and wrote paper. JJN and MDC provided majority of funding for equipment. All authors read and approved the final manuscript.

Competing interests
The authors declare that they have no competing interests.

Author details
[1]Department of Biological Sciences, Simon Fraser University, Burnaby, BC V5A 1S6, Canada. [2]Department of Biology, York University, Toronto, ON M3J 1P3, Canada. [3]Canadian Wildlife Service, Environment and Climate Change Canada, Burlington, ON L7S 1A1, Canada. [4]Bird Studies Canada, P.O. Box 160 115 Front Street, Port Rowan, ON N0E 1M0, Canada. [5]Canadian Museum of Nature, 1740 Pink Road, Gatineau, QC J9J 3N7, Canada. [6]Faculty of Forestry and Environmental Management, University of New Brunswick, Fredericton, NB E3B 5A3, Canada.

References
1. Newton I. Bird migration. London: Collins; 2010.
2. Palacín C, Alonso JC, Martín CA, Alonso JA. Changes in bird migration patterns associated with human-induced mortality. Conserv Biol. 2017;31:106–15.
3. Sillett TS, Holmes RT. Variation in survivorship of a migratory songbird throughout its annual cycle. J Anim Ecol. 2002;71:296–308.
4. Klaassen RHG, Hake M, Strandberg R, Koks BJ, Trierweiler C, Exo K-M, et al. When and where does mortality occur in migratory birds? Direct evidence from long-term satellite tracking of raptors. J Anim Ecol. 2014;83:176–84.
5. Lok T, Overdijk O, Piersma T. The cost of migration: spoonbills suffer higher mortality during trans-Saharan spring migrations only. Biol Lett. 2015;11:20140944.
6. Smith RJ, Moore FR. Arrival fat and reproductive performance in a long-distance passerine migrant. Oecologia. 2003;134:325–31.
7. Newton I. Can conditions experienced during migration limit the population levels of birds? J Ornithol. 2006;147:146–66.
8. Drake A, Rock CA, Quinlan SP, Martin M, Green DJ. Wind speed during migration influences the survival, timing of breeding, and productivity of a Neotropical migrant, Setophaga petechia. PLoS One. 2014;9:e97152.
9. Latta SC, Cabezas S, Mejia DA, Paulino MM, Almonte H, Miller-Butterworth CM, et al. Carry-over effects provide linkages across the annual cycle of a Neotropical migratory bird, the Louisiana Waterthrush Parkesia motacilla. Ibis. 2016;158:395–406.
10. Bell CP. Seasonality and time allocation as causes of leap-frog migration in the Yellow Wagtail Motacilla flava. J Avian Biol. 1996;27:334.
11. Buehler DM, Piersma T. Travelling on a budget: predictions and ecological evidence for bottlenecks in the annual cycle of long-distance migrants. Philos Trans R Soc B Biol Sci. 2008;363:247–66.
12. Bächler E, Hahn S, Schaub M, Arlettaz R, Jenni L, Fox JW, et al. Year-round tracking of small trans-Saharan migrants using light-level geolocators. PLoS One. 2010;5:e9566.
13. Contina A, Bridge ES, Seavy NE, Duckles JM, Kelly JF. Using geologgers to investigate bimodal isotope patterns in Painted Buntings (Passerina ciris). Auk. 2013;130:265–72.
14. Lemke HW, Tarka M, Klaassen RHG, Åkesson M, Bensch S, Hasselquist D, et al. Annual cycle and migration strategies of a trans-Saharan migratory songbird: a geolocator study in the great reed warbler. PLoS One. 2013;8:e79209.
15. Callo PA, Morton ES, Stutchbury BJM. Prolonged spring migration in the Red-eyed Vireo (Vireo olivaceus). Auk. 2013;130:240–6.

16. Jahn AE, Cueto VR, Fox JW, Husak MS, Kim DH, Landoll DV, et al. Migration timing and wintering areas of three species of flycatchers (tyrannus) breeding in the great plains of north America. Auk. 2013;130:247–57.

17. La Sorte FA, Fink D, Hochachka WM, Kelling S. Convergence of broad-scale migration strategies in terrestrial birds. Proc R Soc B Biol Sci. 2016;283:20152588.

18. Bayly NJ, Gómez C, Hobson KA, González AM, Rosenberg KV. Fall migration of the Veery (Catharus fuscescens) in northern Colombia: determining the energetic importance of a stopover site. Auk. 2012;129:449–59.

19. Bayly NJ, Gómez C, Hobson KA. Energy reserves stored by migrating Gray-cheeked Thrushes Catharus minimus at a spring stopover site in northern Colombia are sufficient for a long-distance flight to North America. Ibis. 2013;155:271–83.

20. Alerstam T. Optimal bird migration revisited. J Ornithol. 2011;152:5–23.

21. Stanley CQ, MacPherson M, Fraser KC, McKinnon EA, Stutchbury BJM. Repeat tracking of individual songbirds reveals consistent migration timing but flexibility in route. PLoS One. 2012;7:e40688.

22. Sparks TH, Yates TJ. The effect of spring temperature on the appearance dates of British butterflies 1883–1993. Ecography. 1997;20:368–74.

23. Brown CR, Brown MB. Intense natural selection on body size and wing and tail asymmetry in Cliff Swallows during severe weather. Evolution. 1998;52:1461–75.

24. Newton I. Weather-related mass-mortality events in migrants. Ibis. 2007;149:453–67.

25. Both C, Van Turnhout CAM, Bijlsma RG, Siepel H, Van Strien AJ, Foppen RPB. Avian population consequences of climate change are most severe for long-distance migrants in seasonal habitats. Proc R Soc B Biol Sci. 2010;277:1259–66.

26. Peterson TC, Zhang X, Brunet-India M, Vázquez-Aguirre JL. Changes in North American extremes derived from daily weather data. J Geophys Res Atmospheres. 2008;113:D07113.

27. Stokke BG, Møller AP, Sæther B-E, Rheinwald G, Gutscher H. Weather in the breeding area and during migration affects the demography of a small long-distance passerine migrant. Auk. 2005;122:637–47.

28. Nebel S, Mills A, McCracken JD, Taylor PD. Declines of aerial insectivores in North America follow a geographic gradient. Avian Conserv Ecol. 2010;5:1.

29. Smith AC, Hudson M-AR, Downes CM, Francis CM. Change points in the population trends of aerial-insectivorous birds in North America: synchronized in time across species and regions. PLoS One. 2015;10:e0130768.

30. Michel NL, Smith AC, Clark RG, Morrissey CA, Hobson KA. Differences in spatial synchrony and interspecific concordance inform guild-level population trends for aerial insectivorous birds. Ecography. 2016;39:774–86.

31. Paquette SR, Pelletier F, Garant D, Bélisle M. Severe recent decrease of adult body mass in a declining insectivorous bird population. Proc R Soc Lond B Biol Sci. 2014;281:20140649.

32. Esler D. Applying metapopulation theory to conservation of migratory birds. Conserv Biol. 2000;14:366–72.

33. Webster MS, Marra PP. The importance of understanding migratory connectivity and seasonal interactions. In: Greenberg R, Marra PP, editors. Birds Two Worlds Ecol Evol Migr Baltimore. Maryland: Johns Hopkins University Press; 2005. p. 199–209.

34. Taylor CM, Norris DR. Population dynamics in migratory networks. Theor Ecol. 2010;3:65–73.

35. Martin TG, Chadès I, Arcese P, Marra PP, Possingham HP, Norris DR. Optimal conservation of migratory species. PLoS One. 2007;2:e751.

36. Hewson CM, Thorup K, Pearce-Higgins JW, Atkinson PW. Population decline is linked to migration route in the common cuckoo. Nat Commun. 2016;7:12296.

37. Mills AM. The influence of moonlight on the behavior of goatsuckers (Caprimulgidae). Auk. 1986;103:370–8.

38. Jetz W, Steffen J, Linsenmair KE. Effects of light and prey availability on nocturnal, lunar and seasonal activity of tropical nightjars. Oikos. 2003;103:627–39.

39. Naef-Daenzer B. An allometric function to fit leg-loop harnesses to terrestrial birds. J Avian Biol. 2007;38:404–7.

40. Gaunt AS, Oring LW, Able KP, Anderson DW, Baptista LF, Barlow JC, et al. Guidelines to the use of wild birds in research. Washington: The Ornithological Council; 1997.

41. Bridge ES, Kelly JF, Contina A, Gabrielson RM, MacCurdy RB, Winkler DW. Advances in tracking small migratory birds: a technical review of light-level geolocation. J Field Ornithol. 2013;84:121–37.

42. Ekstrom PA. Blue twilight in a simple atmosphere. 2002. p. 73–81.

43. Ekstrom PA. An advance in geolocation by light. Mem Natl Inst Polar. 2004;58:210–26.

44. Cresswell B, Edwards D. Geolocators reveal wintering areas of European Nightjar (Caprimulgus europaeus). Bird Study. 2013;60:77–86.

45. Rakhimberdiev E, Winkler DW, Bridge E, Seavy NE, Sheldon D, Piersma T, et al. A hidden Markov model for reconstructing animal paths from solar geolocation loggers using templates for light intensity. Mov Ecol. 2015;3:1–15.

46. Ekstrom P. Error measures for template-fit geolocation based on light. Deep Sea Res Part II Top Stud Oceanogr. 2007;54:392–403.

47. Lisovski S, Hewson CM, Klaassen RHG, Korner-Nievergelt F, Kristensen MW, Hahn S. Geolocation by light: accuracy and precision affected by environmental factors. Methods Ecol Evol. 2012;3:603–12.

48. McKinnon EA, Fraser KC, Stutchbury BJM. New discoveries in landbird migration using geolocators, and a flight plan for the future. Auk. 2013;130:211–22.

49. Fraser KC, Stutchbury BJM, Kramer P, Silverio C, Barrow J, Newstead D, et al. Consistent range-wide pattern in fall migration strategy of purple martin (progne subis), despite different migration routes at the gulf of Mexico. Auk. 2013;130:291–6.

50. Norevik G, Åkesson S, Hedenström A. Migration strategies and annual space-use in an Afro-Palaearctic aerial insectivore – the European Nightjar Caprimulgus europaeus. J. Avian Biol. 2017;in press.

51. R Core Team. R: A language and environment for statistical computing. Vienna: R Foundation for Statistical Computing; 2015.

52. Bell CP. Inter- and intrapopulation migration patterns. In: Greenberg R, Marra PP, editors. Birds Two worlds. Baltimore: JHU Press; 2005. p. 41–52.

53. Cink CL. Eastern Whip-poor-will (Antrostomus vociferus). In: Poole A, Gill F, editors. Birds N Am Online Ithaca. New York: Cornell Lab of Ornithology; 2002.

54. Stiles FG, Skutch AF. Guide to the birds of Costa Rica. New York: Comstock Publishing Associates, Ithaca; 1989.

55. Webster MS, Marra PP, Haig SM, Bensch S, Holmes RT. Links between worlds: unraveling migratory connectivity. Trends Ecol Evol. 2002;17:76–83.

56. Shutler D, Hussell DJT, Norris DR, Winkler DW, Robertson RJ, Bonier F, et al. Spatiotemporal patterns in nest box occupancy by tree swallows across north America. Avian Conserv Ecol. 2012;7:3.

57. Tøttrup AP, Thorup K. Sex-differentiated migration patterns, protandry and phenology in North European songbird populations. J Ornithol. 2007;149:161–7.

58. Canal D, Jovani R, Potti J. Multiple mating opportunities boost protandry in a Pied Flycatcher population. Behav Ecol Sociobiol. 2011;66:67–76.

59. Morbey YE, Coppack T, Pulido F. Adaptive hypotheses for protandry in arrival to breeding areas: a review of models and empirical tests. J Ornithol. 2012;153:207–15.

60. McKellar AE, Marra PP, Ratcliffe LM. Starting over: experimental effects of breeding delay on reproductive success in early-arriving male American redstarts. J Avian Biol. 2013;44:495–503.

61. Lundberg S, Alerstam T. Bird migration patterns: conditions for stable geographical population segregation. J Theor Biol. 1986;123:403–14.

62. McKinnon EA, Stanley CQ, Fraser KC, MacPherson MM, Casbourn G, Marra PP, et al. Estimating geolocator accuracy for a migratory songbird using live ground-truthing in tropical forest. Anim Migr. 2013;1:31–8.

63. Garrido OH, Kirkconnell A. Field guide to the birds of Cuba. Ithaca: Comstock Pub; 2000.

64. Ross JD, Bridge ES, Rozmarynowycz MJ, Bingman VP. Individual variation in migratory path and behavior among Eastern Lark Sparrows. Anim Migr. 2014;29–33.

65. Hobson KA, Kardynal KJ, Wilgenburg SLV, Albrecht G, Salvadori A, Cadman MD, et al. A continent-wide migratory divide in North American breeding Barn Swallows (Hirundo rustica). PLoS One. 2015;10:e0129340.

66. Klaassen RHG, Strandberg R, Hake M, Olofsson P, Tøttrup AP, Alerstam T. Loop migration in adult Marsh Harriers Circus aeruginosus, as revealed by satellite telemetry. J Avian Biol. 2010;41:200–7.

67. Vardanis Y, Klaassen RHG, Strandberg R, Alerstam T. Individuality in bird migration: routes and timing. Biol Lett. 2011;7:502–5.

68. Willemoes M, Strandberg R, Klaassen RHG, Tøttrup AP, Vardanis Y, Howey PW, et al. Narrow-front loop migration in a population of the common cuckoo *cuculus canorus*, as revealed by satellite telemetry. PLoS One. 2014;9:e83515.

69. Trierweiler C, Klaassen RHG, Drent RH, Exo K-M, Komdeur J, Bairlein F, et al. Migratory connectivity and population-specific migration routes in a long-distance migratory bird. Proc R Soc B Biol Sci. 2014;281:20132897.

70. Tøttrup AP, Klaassen RHG, Strandberg R, Thorup K, Kristensen MW, Jørgensen PS, et al. The annual cycle of a trans-equatorial Eurasian–African passerine migrant: different spatio-temporal strategies for autumn and spring migration. Proc. R. Soc. Lond. B Biol. Sci. 2011;279:1008–16.

71. Schmaljohann H, Buchmann M, Fox JW, Bairlein F. Tracking migration routes and the annual cycle of a trans-Sahara songbird migrant. Behav Ecol Sociobiol. 2012;66:915–22.

72. Mellone U, López-López P, Limiñana R, Piasevoli G, Urios V. The trans-equatorial loop migration system of Eleonora's falcon: differences in migration patterns between age classes, regions and seasons. J Avian Biol. 2013;44:417–26.

73. Rappole JH, Ramos MA. Factors affecting migratory bird routes over the Gulf of Mexico. Bird Conserv Int. 1994;4:251–62.

74. Fraser KC, Silverio C, Kramer P, Mickle N, Aeppli R, Stutchbury BJM. A trans-hemispheric migratory songbird does not advance spring schedules or increase migration rate in response to record-setting temperatures at breeding sites. PLoS One. 2013;8:e64587.

75. Stanley CQ, McKinnon EA, Fraser KC, Macpherson MP, Casbourn G, Friesen L, et al. Connectivity of Wood Thrush breeding, wintering, and migration sites based on range-wide tracking. Conserv Biol. 2015;29:164–74.

76. Gauthreaux SA. A radar and direct visual study of passerine spring migration in southern Louisiana. Auk. 1971;88:343–65.

77. Schmaljohann H, Naef-Daenzer B. Body condition and wind support initiate the shift of migratory direction and timing of nocturnal departure in a songbird. J Anim Ecol. 2011;80:1115–22.

78. Woodworth BK, Mitchell GW, Norris DR, Francis CM, Taylor PD. Patterns and correlates of songbird movements at an ecological barrier during autumn migration assessed using landscape- and regional-scale automated radiotelemetry. Ibis. 2015;157:326–39.

79. Deppe JL, Ward MP, Bolus RT, Diehl RH, Celis-Murillo A, Zenzal TJ, et al. Fat, weather, and date affect migratory songbirds' departure decisions, routes, and time it takes to cross the Gulf of Mexico. Proc Natl Acad Sci. 2015;112:E6331–8.

80. Alerstam T. Detours in bird migration. J Theor Biol. 2001;209:319–31.

81. Rubolini D, Gardiazabal Pastor A, Pilastro A, Spina F. Ecological barriers shaping fuel stores in Barn Swallows *Hirundo rustica* following the central and western Mediterranean flyways. J Avian Biol. 2002;33:15–22.

82. McNamara JM, Welham RK, Houston AI. The timing of migration within the context of an annual routine. J Avian Biol. 1998;29:416.

83. Conklin JR, Battley PF, Potter MA. Absolute consistency: individual versus population variation in annual-cycle schedules of a long-distance migrant bird. PLoS One. 2013;8:e54535.

84. Kokko H. Competition for early arrival in migratory birds. J Anim Ecol. 1999;68:940–50.

85. Gunnarsson TG, Gill JA, Sigurbjörnsson T, Sutherland WJ. Pair bonds: arrival synchrony in migratory birds. Nature. 2004;431:646.

86. Tøttrup AP, Thorup K, Rainio K, Yosef R, Lehikoinen E, Rahbek C. Avian migrants adjust migration in response to environmental conditions en route. Biol Lett. 2008;4:685–8.

87. Conklin JR, Battley PF, Potter MA, Fox JW. Breeding latitude drives individual schedules in a trans-hemispheric migrant bird. Nat Commun. 2010;1:67.

88. Nussey DH, Postma E, Gienapp P, Visser ME. Selection on heritable phenotypic plasticity in a wild bird population. Science. 2005;310:304–6.

89. Hallworth MT, Marra PP. Miniaturized gps tags identify non-breeding territories of a small breeding migratory songbird. Sci Rep. 2015;5:11069.

90. Brownrigg R, Becker RA, Wilks AR. Mapdata: Extra map databases. R package version 2.2-6. 2016.

6

Rediscovering a forgotten canid species

Suvi Viranta[1*†] (iD), Anagaw Atickem[2,3,4†], Lars Werdelin[5] and Nils Chr. Stenseth[2,4*]

Abstract

Background: The African wolf, for which we herein recognise *Canis lupaster* Hemprich and Ehrenberg, 1832 (Symbolae Physicae quae ex Itinere Africam Borealem er Asoam Occidentalem Decas Secunda. Berlin, 1833) as the valid species name (we consider the older name *Canis anthus* Cuvier, 1820 [Le Chacal de Sénégal, Femelle. In: Geoffroy St.-Hilaire E, Cuvier F, editors. Histoire Naturelle des Mammifères Paris, A. Belin, 1820] a *nomen dubium*), is a medium-sized canid with wolf-like characters. Because of phenotypic similarity, specimens of African wolf have long been assigned to golden jackal (*Canis aureus* Linnaeus, 1758 [Systema Naturae per Regna Tria Naturae, Secundum Classes, Ordines, Genera, Species, cum Characteribus, Differentiis, Synonymis, Locis. Tomus I. Editio decima, reformata, 1758]).

Results: Here we provide, through rigorous morphological analysis, a species description for this taxonomically overlooked species. Through molecular sequencing we assess its distribution in Africa, which remains uncertain due to confusion regarding possible co-occurrence with the Eurasian golden jackal. *Canis lupaster* differs from all other *Canis* spp. including the golden jackal in its cranial morphology, while phylogenetically it shows close affinity to the Holarctic grey wolf (*Canis lupus* Linnaeus, 1758 [Systema Naturae per Regna Tria Naturae, Secundum Classes, Ordines, Genera, Species, cum Characteribus, Differentiis, Synonymis, Locis. Tomus I. Editio decima, reformata, 1758]). All sequences generated during this study clustered with African wolf specimens, consistent with previous data for the species.

Conclusions: We suggest that the estimated current geographic range of golden jackal in Africa represents the African wolf range. Further research is needed in eastern Egypt, where a hybrid zone between Eurasian golden jackal and African wolf may exist. Our results highlight the need for improved studies of geographic range and population surveys for the taxon, which is classified as 'least concern' by the IUCN due to its erroneous identification as golden jackal. As a species exclusively distributed in Africa, investigations of the biology and threats to African wolf are needed.

Keywords: African wolf, Canidae, *Canis lupaster*, *Canis aureus*, Taxonomy, Conservation

Background

Most canids (Family Canidae) are easy to recognize by their characteristic long muzzle, long limbs and bushy tails. They have a conservative body plan retaining traits of early mammals, including a primitive dental formula (I 3/3, C 1/1, P 4/4, M 2/3 in the majority of Canidae) [1]. Morphological variation within the family is relatively slight [1, 2], which creates problems of species recognition and classification. Wolves are the largest members of the Canidae. They are charismatic species with a long special relationship with people. They are also the ancestors of the first domesticate, the dog [3, 4]. During historic times and into the present wolves have been persecuted due to fear of predation on domestic animals and attacks on people. Once widespread across the Holarctic, wolves are now absent in many areas of North America and Eurasia [5]. Wolves have been thought to be absent from Africa. Instead the large and medium sized canids in Africa are the African wild dog (*Lycaon pictus* Temminck, 1820 [6]) and the two jackals: side-striped jackal (*Lupulella adusta* (Sundevall, 1847) [7]) and black-backed jackal (*Lupulella mesomelas* (Schreber, 1775) [8]). The fourth medium sized canid species, the African wolf (*Canis lupaster*), was until recently equated with the Eurasian golden jackal (*Canis aureus*). Recent papers, including this one,

* Correspondence: suvi.viranta-kovanen@helsinki.fi; n.c.stenseth@ibv.uio.no
†Equal contributors
¹Department of Anatomy, Faculty of Medicine, University of Helsinki, PO Box 63, 00014 Helsinki, Finland
²Department of Biosciences, Centre for Ecological and Evolutionary Synthesis (CEES), University of Oslo, P.O. Box 1066 Blindern, N-0316 Oslo, Norway
Full list of author information is available at the end of the article

show that it is a separate species, *Canis lupaster*. In the phylogenetic tree the African wolf groups with other *Canis* species, whereas *Lupulella* and *Lycaon* fall outside this clade, resulting in identification of separate genera (Additional file 1).

The presence of a wolf relative in North and West Africa was indicated in the early literature [9–12], but until recently [13–15] largely ignored in the modern literature. Here we demonstrate the presence of a species closely related to the Holarctic wolf in Africa and discuss its taxonomic status and morphology. We provide the first formal taxonomic description of the African wolf.

A medium-sized canid with a wide distribution in North, West, and East Africa has been described under various names, but is today mistakenly equated with the golden jackal, *Canis aureus* Linnaeus, 1758 [16, 17]. Recent publications [13–15] have identified this animal as a separate species, more closely related to the Holarctic grey wolf than to the golden jackal. Gaubert et al. [13] suggested the existence of both the golden jackal and African wolf in North and West Africa. Their mtDNA analysis revealed a close relationship between specimens morphologically assigned as golden jackals and those assigned as the African wolf, differentiating them from Indian golden jackal. Morphological features characteristic of the African wolf are heavy build and wider head, as well as some traits of the pelage. Koepfli et al. [15], using both mtDNA and autosomal loci, found evidence for African and Eurasian golden jackals as distinct species and found no evidence for the existence of both the golden jackal and the African wolf in Africa. They also estimated the divergence times and found an estimate of 1.9 Ma for the golden jackal and the African wolf and 1.3. Ma for the African wolf and the grey wolf. They also identified some morphological traits and provided evidence for apparent convergent evolution having resulted in the similarity of the golden jackal and African wolf. Rueness et al. [14] concluded, based on yet another sample of mtDNA, that the African wolf is a separate species, more closely related to the grey wolf than to the golden jackal.

This species, which we here call the African wolf, has, however, only cursorily been described morphologically, and a detailed investigation of its taxonomic status has not previously been undertaken. Furthermore, the putative presence of Eurasian golden jackal in Africa remains unclear and has led to confusion among researchers. With a formal taxonomic description and the demonstrated distinct evolutionary history of the African wolf, the need for a reassessment of the geographic distribution and population abundance of this species is evident.

The fact that the phylogenetic uniqueness of the African wolf has escaped the attention of science for over a century serves as a cautionary example of reliance on outdated authority and a lack of proper taxonomic research. Biodiversity research, as well as conservation studies, is only valuable when built on solid taxonomic work [18, 19]. The erroneous merging of two distinct species (the African wolf and the golden jackal) into one as 'golden jackal' has resulted in confusing phylogenetic trees and false interpretations of intraspecific biological variation and evolutionary history.

Methods

We studied crania of canids labelled by earlier scholars or museum curators as *Canis aureus*, *Canis lupaster* or *Canis anthus* in the collections of Swedish Museum of Natural History, Stockholm, Sweden (NRM); Museum für Naturkunde, Berlin, Germany (ZMB); Natural History Museum of Denmark, Copenhagen, Denmark (ZMUC), and Finnish Museum of Natural History, Helsinki, Finland (FMNH). We also studied specimens of the closely related Old World canids *Lupulella mesomelas*, *L. adusta*, *C. simensis* Rüppell, 1840 [20], and *C. lupus* in the same institutes. Moreover, we studied crania collected from road kills for this project in Ethiopia. In the case of the type specimens, housed in the Museum für Naturkunde, Berlin, the skins were also studied. For skulls with a skin with the same specimen number (presumed to be from the same individual), the skin was sampled for DNA data ($n = 20$). We sampled scats ($n = 31$) and blood samples ($n = 14$) from different African countries. Eleven skin samples also were obtained from museum collections (Additional file 2: Table S1).

A total of 31 dental and 22 cranial measurements were taken on skulls using dial calipers. Additional measurements were obtained from the data files of Björn Kurtén (curated by LW). Measurement data are provided in Additional file 3. The skins were photographed and the head and body length were measured using a tape measure. By convention lower case letters are used for lower teeth and upper case letters for upper teeth.

The DNA extraction from scat samples was carried out using Dynabeads MyOneTM SILANE as given in detail in [21] and the Phenol chloroform method was used for museum and blood samples [22, 23]. Polymerase chain reaction (PCR) was carried out at two fragments of mtDNA (12S ribosomal RNA and Cytb region) for samples from blood and scat. The 12S rRNA was amplified using primers 12S3 and 12S2 [24]. The DNA extracts from museum samples were amplified using internal primers developed to sequence short sequences (Additional file 2: Table S2). Sequences were aligned using MEGA 5.2-clustal parameters [25]. The mtDNA amplification was performed in 15 µl reactions containing 2.5 µl HotStar PCR buffer (QIAGEN GmbH Hamburg, Germany), 5 nmol dNTP, 0.01 mg BSA

(New England Biolabs), 50 nmol Mgcl2, 1.25 units Hot-Star Taq polymerase, 8 pmol of each primer, 50–100 ng template DNA and mqH20. The program for the PCR consisted of initial denaturation at 95 °C for 15 min followed by 45 cycles of 94 °C for 1 min, 55 °C for 1 min and a final extension at 72 °C for 10 min for Cytb1 and 12S rRNA. The PCR cycle parameters for DNA extracts from museum samples were similar except for a higher annealing temperature of 58 °C and 60 °C (Additional file 2: Table S2). Additional nucleotide sequences of canids were obtained from GenBank (Additional file 2: Table S3). Phylogenetic relationships were analysed using Bayesian approach in BEAST 1.8 [26]. Site model and clock model were set as unlinked between the two partitions. A HKY + G (4 classes) + I substitution model with empirical base frequency and a strict clock-rate were set for both partitions. The Yule Process was used as a tree prior model. Three replicates were run for 10 000 000 generations and convergence of parameters was checked on Tracer 1.5 ([27, 28]. The phylogenetic tree was then drawn in FigTree 1.4 [28, 29]. Median-joining network analysis was carried out using PopART Network analysis [27]. Regional genetic variation was estimated using the DnaSP software [29].

The statistical analyses of the morphological data were carried out using the PAST software (version 2) [30].

Nomenclatural acts
This published work and the nomenclatural acts it contains have been registered in Zoobank: http://zoobank.org/NomenclaturalActs/2D51EA46-45D3-4F31-BCC5-7AA122 1F66DB. The LSID for this publication is: lsid:zoobank.org:act:2D51EA46-45D3-4F31-BCC5-7AA1221F66DB.

Results
Systematics
Canis lupaster Hemprich and Ehrenberg, 1832 [9].

Synonymy (selected, for an expanded list see Additional file 4)
Canis anthus (Cretzschmar, 1826 [31] non *C. anthus* Cuvier, 1820 [10])
Dieba anthus (Gray, 1869) [32]
Canis anthus (De Winton, 1899) [12]
Canis lupaster (Hilzheimer, 1906) [33]
Canis aureus lupaster (Schwarz, 1926) [34]
Thos aureus lupaster (Allen, 1939) [17]

Original description (Hemprich and Ehrenberg, 1832) [9]
CANIS Lupaster H. et E. Dib, Sib
Vulpe maior, Lupo affinior, inferior, longius pilosus cineracente flavidus, fusco nigroque obsolete varius, capite incrassato, ore subacuto, vertice auribus naso pedibusque flavis, cauda brevi laxius pilosa, apicibus pilorum et macula prope basin nigricantibus aut rufis. C. Anthus Cretzschmar nec Frid. Cuvier. In Fayum vulgaris. Lupus Aegypti.

"Large fox, similar to wolf but smaller; long hair, ash-yellow to dark black pelage; head thickened, ears pointed, mouth, ears, nose and feet yellow; short tail sparsely furred, tips of hairs reddish and blackish spot near the base. *C. anthus* of Cretschmar, not F. Cuvier; Common in Fayum; Egyptian wolf." (our translation)

Holotype
Three specimens, all from the governate of Fayum (Fayium, Fayoum), Egypt, are marked as types in the collections of the Museum für Naturkunde, Berlin: ZMB_mam_833, a skull with worn teeth and damaged occipital region, sex unknown; ZMB_mam_834, a skull and skin of an adult female; ZMB_mam_835, a skull of a young female individual with deciduous dentition and erupting permanent teeth. Of these, ZMB_mam_834 is considered the holotype of *C. lupaster* [34]. Of the other two specimens, ZMB_mam_833 becomes a paratype as it is part of the type series [34]. Specimen, ZMB_mam_835, on the other hand, is the type specimen of *Canis sacer* Hemprich and Ehrenberg, 1832 [9], a putative synonym of *C. lupaster* [34].

Description of ZMB_Mam_834
A female individual collected by Friedrich Wilhelm Hemprich and Christian Gottfried Ehrenberg from Fayum, Egypt in the early 19th century. It consists of a complete skull and a skin (Figs. 1 and 2). Measurement data for this specimen are given in Additional file 2: Tables S4–S6.

Skull and dentition
The skull (Fig. 1) is that of a medium-sized canid. The upper and lower postcanine teeth are slightly crowded, with diastemata between the upper canine and the third incisor and between the lower canine and the first premolar.

The mandible (Fig. 1) is robust with well-developed masseteric fossa and elevated coronoid. The condyloid process has a short neck. The angular process is long and convex with a pointed tip. Two mental foramina are located below p3 and just mesial to p2. The hemimandibles have been separated at the symphysis and are now glued together, so the natural angle between the two is lost. A small and round m3 is present bilaterally. The m2 is elongated and has four distinct cusps that, in accordance with other Canidae, are protoconid, metaconid, entoconid, and hypoconid. In the m1 both the trigonid and talonid are well developed. The metaconid is distinct from the protoconid and located

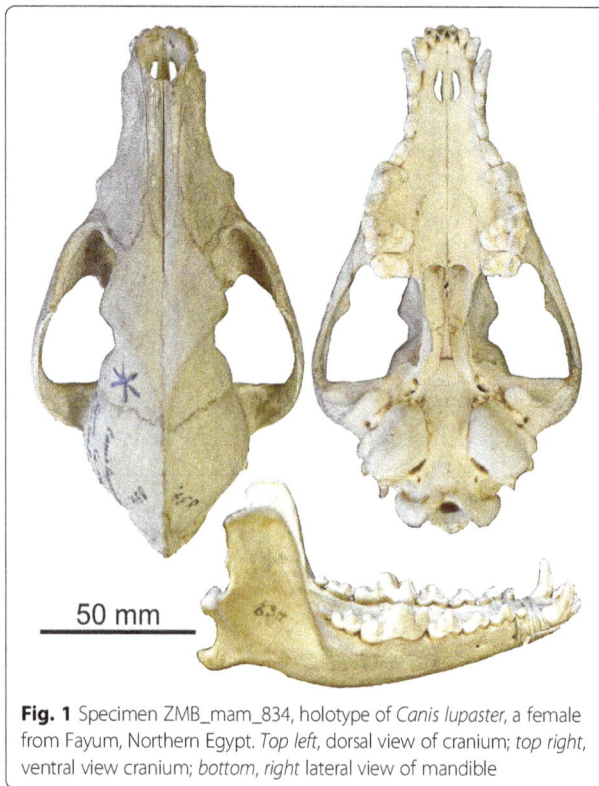

Fig. 1 Specimen ZMB_mam_834, holotype of *Canis lupaster*, a female from Fayum, Northern Egypt. *Top left*, dorsal view of cranium; *top right*, ventral view cranium; *bottom*, *right* lateral view of mandible

distolingually to it. The talonid has three cusps, entoconid, hypoconid and hypoconulid. The p4 is >50% of the length of the m1 and has three cusps and a lingual cingulum. The mesial cusp has a mesial crest. The p3 and p2 are of about equal length. They both have a main cusp, a distal accessory cusp, and a cingulum with a distal elevation. The p1 is round and has a sharp anterior cusp. The lower canine is mediolaterally flattened. The incisors are crowded. The i2 and i3 have two cusps.

The cranium is dome-shaped with a ca. 20° angle between the rostrum and the braincase (forehead). Sutures between bones are clearly visible and the skull has a moderate sagittal crest. The widest part of the rostrum is at the posterior end of the P4. The premolar

Fig. 2 Specimen ZMB_mam_834, holotype of *Canis lupaster*, a female from Fayum Northern Egypt. Skin; head to right, tail to left

and molar rows are angled at about 30° to each other. The incisive foramina are long, extending from the anterior end of the canines to the level of P1. There are three palatine foramina on the right side and two on the left. They are convex in shape. The infraorbital foramen is well developed and placed above the P3. The postorbital process is large, but blunt. The auditory bullae are inflated, oval and placed at 45° to the sagittal line.

The upper incisors are crowded and have lingual cingula. The upper canines are convex. The left canine has wear that appears to be ante mortem. The reason for this is not known. The P1 is small and pointed. The P2 has two cusps and the P3 three cusps. The P4 has a protocone that is clearly separate and placed lingual to the paracone. It lies at about 100° to a line drawn through the metacone and paracone. The M1 is distally convex and has a cingulum and four cusps, paracone, protocone, metacone, and hypocone. The M2 is smaller, but displays the same cingulum and cusps.

Skin

The skin of ZMB_Mam_834 is incomplete, with the distal parts of the limbs and tail missing (Fig. 2). There is a median dorsal ruff extending from the neck to the tail, composed of hairs with black tips and ginger and white bases. The head is ginger with agouti on the forehead and ears. The hair on the limbs and ventral side is short and yellow.

Differential diagnosis

We compared the cranial and dental measurements of 69 African wolves to the measurements taken on *Canis* species and the jackals. Based on skull size *Canis lupaster* is smaller than the smallest grey wolves (*Canis lupus arabs* Pocock, 1934 [35], *C. l. pallipes* Sykes, 1831 [36], *C. l. chanco*, Gray, 1863 [37]) (Additional file 2: Table S2; Additional file 5: Figure S1).

Canis lupaster differs from grey wolves in having a lower coronoid process of the mandible. The palatine bone is relatively longer and the distance between the upper canines smaller in *C. lupaster*. The molar row is relatively longer as compared to the premolar row (Additional file 5: Figure S1).

Canis lupaster is larger than the two African jackals (*Lupulella adusta* and *Lupulella mesomelas*) and differs from them by its relatively shorter palatine and larger skull.

The Eurasian golden jackal (*C. aureus*) has a wider and shorter palate and also relatively greater interorbital breadth than *C. lupaster*. The upper canine is mediolaterally flatter in *C. lupaster* than in *C. aureus* (Additional file 5: Figure S2).

The Ethiopian wolf (*C. simensis*) is a larger species and has a longer rostrum than *C. lupaster*. It also has a very distinct pelage with white markings, while *C. lupaster* is tawny or rufous with black and grey on the dorsum.

Canis lupaster shows considerable variation in size, but sexual dimorphism has not been detected in our data (Additional file 5: Figure S3).

Separation from *Canis aureus*

We ran a discriminant analysis on the 52 morphological characters obtained for the study. Using $\log_{10}$-transformed data for 65 individuals we obtained a correct classification of 68.3% (jackknifed) for the comparison *C. aureus* – *C. lupaster*. When only characters we considered most likely to be diagnostic were included, 89.7% correct classification was obtained (Additional file 5: Figure S4).

A total of 64 nucleotide sequences from Ethiopia, South Sudan, Egypt and Western Sahara newly generated for this study, as well as the 39 additional sequences of *C. lupaster* from GenBank (Additional file 2: Tables S3, S7), clustered to the African wolf lineage (Fig. 3; Additional file 5: Figure S5). A single *Canis aureus* haplotype has been reported from Egypt (Fig. 4) [15]. This specimen is from the Sinai Peninsula, close to the border between Egypt and Israel.

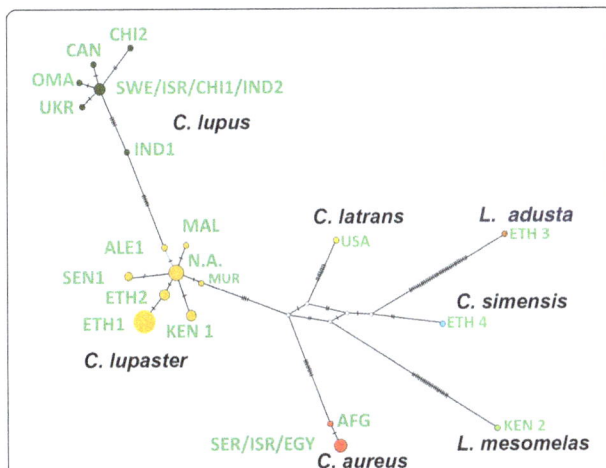

Fig. 3 Median-joining network of canid haplotypes based on cytochrome b (380 bp). Circle size and branch lengths are proportional to haplotype frequency and number of mutational steps among haplotypes, respectively. Nucleotide sequences of *C. lupus*, *C. lupaster* and *C. aureus* are represented based on their geographic sources as follows: SWE = Sweden, ISR = Israel, CHI = China, CAN = Canada, IND = India, UKR = Ukraine, OMA = Oman, N.A. = North Africa (Algeria Egypt Mali and Morocco), MUR = Mauritania, KEN = Kenya, ETH = Ethiopia, SEN = Senegal, ALE = Algeria), MAL = Mali, AFG = Afghanistan; SER = Serbia ISR = Israel and EGY = Egypt, USA = United States of America. Details of nucleotide sequences used are presented in the Supplementary Information (Additional file 2: Table S3)

Geographic and intrapopulation variation

Several authors have noted the existence of two morphotypes of African wolf (see, e.g. [13]). Our data show that there are significant differences in size between populations of *C. lupaster*, with East African individuals being smaller than North and West African ones. This is not manifest in a bimodal distribution, however. On the other hand, our metric data do show a higher coefficient of variation (CV) in *C. lupaster* than in our *C. aureus* sample, which comes from specimens with a broad geographic distribution across Eurasia. This may be a signal of some morphotype differences within *C. lupaster* that are unmatched in *C. aureus*. Further subdividing the *C. lupaster* material into North, West and East African samples shows that all three have higher CV that the entire *C. aureus* sample. Among the three sub-samples of *C. lupaster*, the North African one has the highest CV (Additional file 2: Table S9). The *C. lupaster* population in Ethiopia has higher genetic diversity compared to the population in the northern African countries (Egypt, Algeria, Morocco; Additional file 2: Table S8).

Taxonomy and nomenclature

Accepting the African wolf as a distinct species leads to the question of the appropriate species name. Previous authors have alternated between *Canis lupaster* (e.g., [13, 14, 38]) and *Canis anthus* [13]. Of these, *C. anthus* F. Cuvier, 1820 [10] has priority. It is based on the description of a female individual from Senegal. In a later publication, Cuvier described a male individual he ascribed to *C. anthus* [11]. However, the two specimens are markedly different and are unlikely to belong to a single species. This, and the fact that the holotype is missing (a search in the Muséum National d'Histoire Naturelle, Paris was unsuccessful; G. Veron, pers. comm. to LW) render the status of *C. anthus* very unsatisfactory. It is, in fact, possible that the holotype is a specimen of *Lupulella adusta* (side striped jackal), which was not formally described until 1847 [7]. The description and illustration in Cuvier's work are not adequate to distinguish between the two. Thus, we consider *C. anthus* a *nomen dubium* and use *C. lupaster* as the name for the African wolf. A longer discussion of the taxonomic history of these names is provided in Appendix 2 (Additional file 6). It should also be noted that the publication of the Symbolae Physicae of Hemprich and Ehrenberg as a whole is dated 1833, but the section on *Canis lupaster* is dated November, 1832, which is the date of publication of the name.

Phylogenetic position within the Canidae

The fact that the majority of recent phylogenetic studies have considered the African wolf and Eurasian golden jackal to be conspecific makes them useless when

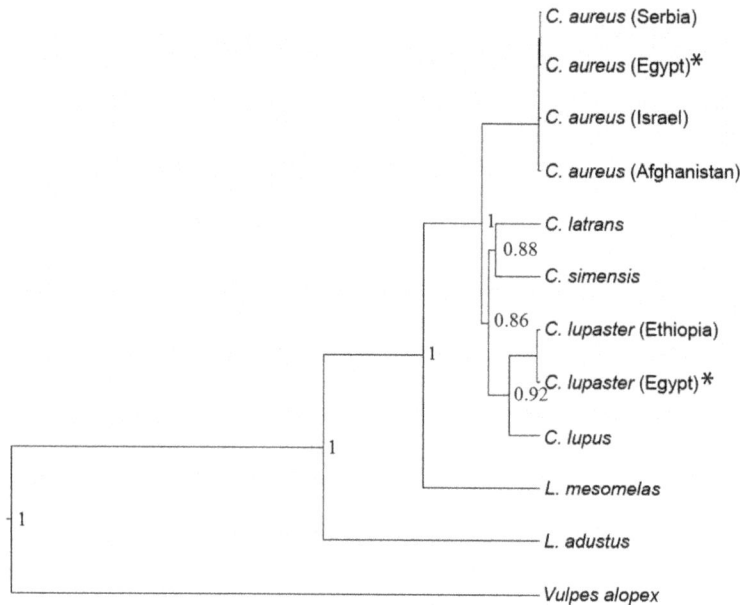

Fig. 4 Bayesian phylogenetic analysis of canids with posterior probabilities of nodal support based on cytochrome b (1140 bp). Branches marked with asterisks: C. aureus (Egypt) was obtained from GenBank (KT447732), while C. lupaster (Egypt) was generated from this study. Details of the sequences used in this analysis are given in Supplementary Information (Additional file 2: Table S3)

tracing the phylogenetic position of the African wolf. Early studies that used mitochondrial DNA sequences in phylogenetic analyses of canids, including an exclusively African 'C. aureus', resulted in a position outside a crown clade *Canis* including Holarctic grey wolf, coyote, and Ethiopian wolf [39]. Separating 'C. aureus' samples into a Eurasian and an African component and including a wide range of molecular markers shows the African sample to be closer phylogenetically to Holarctic grey wolf and coyote than are either Ethiopian wolf or Eurasian *C. aureus* [15].

Discussion

History of the African wolf

We have provided evidence for and described the African wolf as a distinct taxonomic entity clearly separate from the Eurasian golden jackal and as a species closely related to the Holarctic grey wolf. It should be noted, however, that the Holarctic grey wolf might not be a single species. Some wolf lineages, e.g., in India and North America may deserve species status as well [40–42].

From the first descriptions of African wolf [9–11, 31] until the 1920s the majority of authors maintained a distinction between the African wolf and the Eurasian golden jackal. Some also maintained a distinction between *C. anthus* and *C. lupaster* based on the original descriptions [43, 44].

The African wolf and golden jackal were synonymized by Schwarz [34] for reasons that are not clear in that

publication, and Allen accepted this synonymy in his highly influential checklist of African mammals [17]. Allen's viewpoint was rapidly accepted in both the ecological and evolutionary research communities and since that time few authors have considered the African wolf a distinct species, despite there being a few notable exceptions: Keimer mentions *C. lupaster* in his faunal work on Egypt [45]; Kurtén lists the wolf jackal (*C. lupaster*) for a fossil collection from the Levant and suggests the presence of extant *C. lupaster* in North Africa [46]; Ferguson studied *C. lupaster* crania from Israel and concluded that *C. aureus lupaster* differs from *C. aureus* and represents a small desert race of *C. lupus* [38]. Most recently an m1 from Middle Pleistocene deposits in the Nefud Desert, Saudi Arabia, has been identified as *C. anthus* [47].

It is only recently that new data from molecular genetic studies have resurrected the African wolf [13–15]. Thus far, presence of African wolf has been confirmed from southern Ethiopia to Egypt and West African countries, covering the former range of golden jackal delineated by IUCN [48] (Fig. 5). The golden jackal haplotype from Egypt alluded to above needs further study to determine whether it is from an isolated golden jackal in Egypt or from golden jackal-African wolf hybrids in the region. In Egypt, in particular in the Sinai Peninsula, which serves as a land bridge between Asia and Africa, hybrid canids could be a possibility. Eurasian golden jackals from Israel are reported to show signals of hybridization with grey wolves, dogs, and the African wolf [15].

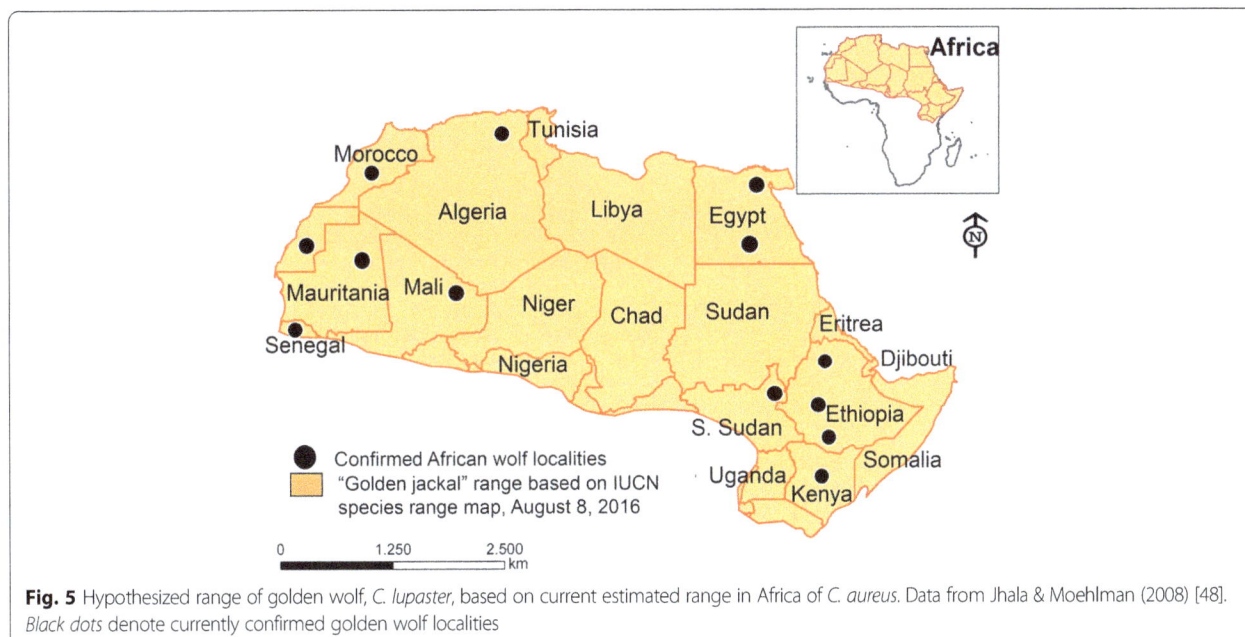

Fig. 5 Hypothesized range of golden wolf, *C. lupaster*, based on current estimated range in Africa of *C. aureus*. Data from Jhala & Moehlman (2008) [48]. *Black dots* denote currently confirmed golden wolf localities

Population status of the African wolf

The basic biology and population status of the African wolf are insufficiently known. Our biological knowledge of the African wolf is further complicated by the fact that many ecological and behavioural conclusions are made based on observations of golden jackals and assuming taxonomic identity between the two. The African wolf is likely to face threats from the growing human population, although it seems to habituate to human propinquity relatively well [49].

There are no data on distribution patterns for the African wolf in recent times and African wolf is still cited as golden jackal in recent publications [50]. However, the geographic range of golden jackal in Africa given by IUCN [48] may be considered as the potential range of the African wolf (Fig. 5). This shows African wolf documented from the Ethiopian highlands to the Rift Valley, across North Africa and the Sahara desert, to the west coast of the continent (but not to the coast of the Bay of Benin). It is thus distributed across a wide range of ecological zones.

Persecution by pastoral communities as a result of livestock predation is probably the greatest challenge for the African wolf populations. Several studies document African wolf as one of the most important livestock predators [51–55].

All wolves living near human occupation risk interbreeding with domestic dogs. All *Canis* spp. share the same chromosome number (2n = 78) [56] and occasionally interbreed in the wild [57, 58]. The domestic dog, as a descendant of the wolf, mates with wild canids [59, 60], including the Ethiopian wolf [61]. To our knowledge no record of hybridization with the African wolf exists, although Rueness et al. [14] found evidence of introgression in one of their samples.

There are five species of large and medium sized canids in Africa (side-striped and black-backed jackals, (*Lupulella* spp.) African wolf (*Canis lupaster*), Ethiopian wolf (*Canis simensis*) and African wild dog [*Lycaon pictus*]). The jackal and African wild dog lineages have long fossil records in Africa [62, 63] and can be considered endemic taxa following initial entry of Canidae into Africa in the latest Miocene. The two species of *Canis* are likely to be relatively recent immigrants from lineages originating in Eurasia. Neither lineage has a definitive fossil record in Africa or elsewhere, so their evolutionary history remains to be discovered, including why they were able to successfully colonize Africa in the face of the presence of the endemic lineages already there.

Conclusions

The erroneous inclusion of the African wolf (*Canis lupaster*) in the taxonomic envelope of Eurasian golden jackal (*Canis aureus*) has obscured the unique evolutionary history of the species. For a century, the African wolf was considered as a part of a widely distributed species with a recent history of immigration into Africa [13]. New research is now needed to assess the evolutionary history and population status of *C. lupaster* and to understand the biology of this species. While there is little evidence for the presence of Eurasian golden jackal in Africa, further study is needed to confirm whether it may be present in eastern Egypt.

Additional files

Additional file 1: Comment on use of *Lupulella*. (DOCX 115 kb)

Additional file 2: Additional information: tables. The file contains tables from S1 to S9. (DOC 358 kb)

Additional file 3: Metric data. Contains all the morphometric measurements taken for the study. (XLS 127 kb)

Additional file 4: Expanded synonymy of *Canis lupaster*. (DOCX 85 kb)

Additional file 5: Additional information: figures. The file contains figures from S1 S5. (PDF 5588 kb)

Additional file 6: Taxonomy and nomenclatural history of the African wolf. (DOCX 3905 kb)

Acknowledgments
We thank the Ethiopian Wildlife Conservation Authority for giving us permission to conduct this research. We also thank curators Daniel Klingberg Johansson (Copenhagen), Christiane Funk (Berlin) and Ilpo Hanski (Helsinki) for their help to access the specimens in their care. Dr. Jakob Kiepenheuer is thanked for providing material from West Sahara. Two reviewers provided very good comments that helped to improve the manuscript.

Funding
The Rufford Small Grants for Nature Conservation to AA, core funding from the Norwegian Research Council (RCN) to the Centre for Ecological and Evolutionary synthesis (CEES), University of Oslo, and Grants from the Swedish Research Council to LW.

Authors' contributions
SV and AA did most of the research and writing; LW assembled the historic literature used, helped acquire data and assisted in the writing; NCS supervised the study and helped interpret the data as well as assisted in the writing. All authors gave final approval for publication.

Competing interests
The authors declare that they have no competing interests.

Author details
[1]Department of Anatomy, Faculty of Medicine, University of Helsinki, PO Box 63, 00014 Helsinki, Finland. [2]Department of Biosciences, Centre for Ecological and Evolutionary Synthesis (CEES), University of Oslo, P.O. Box 1066 Blindern, N-0316 Oslo, Norway. [3]Cognitive Ethology Laboratory, German Primate Center, Kellnerweg 4, 37077 Göttingen, Germany. [4]Department of Zoological Sciences, Addis Ababa University, P. O. Box 1176, Addis Ababa, Ethiopia. [5]Department of Palaeobiology, Swedish Museum of Natural History, Box 50007, S-10405 Stockholm, Sweden.

References
1. Clutton-Brock J, Corbet GB, Hills M. Review of the family Canidae, with a classification by numerical methods. Bull Br Mus Nat Hist Zool. 1976;29:117–99.
2. Werdelin L, Wesley-Hunt GD. The biogeography of carnivore ecomorphology. In: Goswami A, Friscia A, editors. Carnivoran evolution: new views on phylogeny, form, and function. Cambridge: Cambridge University Press; 2010. p. 225–45.
3. Clutton-Brock J. Man-made dogs. Science. 1977;197:1340–2.
4. Wayne RK, VonHoldt BM. Evolutionary genomics of dog domestication. Mamm Genome. 2012;23:3–18.
5. Mech LD, Boitani L. (IUCN SSC Wolf Specialist Group). Canis lupus. The IUCN Red List of Threatened Species 2010: e.T3746A10049204 http://dx.doi.org/10.2305/IUCN.UK.2010-4.RLTS.T3746A10049204.en. Downloaded on 14 November 2016.
6. Temminck CJ. Sur le genre Hyène, et description d'une espèce nouvelle, découverte en Afrique. Annales Générales des Sciences Physiques. 1820;3: 46–57.
7. Sundevall CJ. Nya mammalia från Sydafrika. Öfversigt af Kongl. Vetenskapsakademiens Förhandlingar. 1847;3:118–21.
8. Schreber JCD. Die Säugthiere in Abbildungen nach der Natur mit Beschreibungen (Erster Theil): Der Mensch. Der Affe. Der Maki. Die Fledermaus. Erlangen: Verlag Wolfgang Walther; 1775.
9. Hemprich FG, Ehrenberg CG. Symbolae Physicae quae ex Itinere Africam Borealem er Asoam Occidentalem Decas Secunda. Berlin: Ex Officina Academica; 1833.
10. Cuvier F. Le Chacal de Sénégal, Femelle. In: Geoffroy St. -Hilaire E, Cuvier F, editors. Histoire Naturelle des Mammifères Paris, A. Belin. 1820. p. 1–3.
11. Cuvier F. Chacal du Sénégal, Male. In: Geoffroy St. -Hilaire E, Cuvier F, editors. Histoire Naturelle des Mammifères Paris, A. Belin. 1830. p. 1–2.
12. De Winton WE. On the species of Canidae found on the continent of Africa. In: Proceedings of the Zoological Society of London. 1899. p. 533–52.
13. Gaubert P, Bloch C, Benyacoub S, Abdelhamid A, Pagani P, Djagoun CA, Couloux A, Dufour S. Reviving the African wolf Canis lupus lupaster in North and West Africa: a mitochondrial lineage ranging more than 6,000 km wide. PLoS One. 2012;7:e42740. doi:10.1371/journal.pone.0042740.
14. Rueness EK, Asmyhr MG, Sillero-Zubiri C, Macdonald DW, Bekele A, Atickem A, Stenseth NC. The cryptic African wolf: Canis aureus lupaster is not a golden jackal and is not endemic to Egypt. PLoS One. 2011;6: e16385. doi:10.1371/journal.pone.0016385.g001.
15. Koepfli KP, Pollinger J, Godinho R, Robinson J, Lea A, Hendricks S, Schweizer RM, Thalmann O, Silva P, Fan Z, et al. Genome-wide evidence reveals that African and Eurasian golden jackals are distinct species. Curr Biol. 2015;25: 2158–65. doi:10.1016/j.cub.2015.06.060.
16. Linnaeus C. Systema Naturae per Regna Tria Naturae, Secundum Classes, Ordines, Genera, Species, cum Characteribus, Differentiis, Synonymis, Locis, vol. I. Stockholm: Laurentii Salvii; 1758. p. 824.
17. Allen GA. A checklist of African mammals. Bull Mus Comp Zool. 1939;83:3–763.
18. Watson MF, Lyal CHC, Pendry C. Descriptive taxonomy: the foundation of biodiversity research. Cambridge: Cambridge University Press; 2015.
19. Costello MJ, Vanhoorne B, Appeltans W. Conservation of biodiversity through taxonomy, data publication, and collaborative infrastructures. Conserv Biol. 2015;29:1094–9.
20. Rüppell E. Neue Wirbelthiere zu der Fauna von Abyssinien gehörig. Säugethiere. 1840;1:27–35.
21. Atickem A, Loe LE, Langangen Ø, Rueness EK, Bekele A, Stenseth NC. Population genetic structure and connectivity in the endangered Ethiopian mountain Nyala (Tragelaphus buxtoni): recommending dispersal corridors for future conservation. Conserv Genet. 2013;14:427–38.
22. Sambrook J, Fritsch EF, Manlatis T. Molecular cloning: a laboratory manual. 2nd ed. Cold Spring Harbor: Cold Spring Harbor Laboratory Press; 1989.
23. Sambrook J, Russell DW. Purification of nucleic acids by extraction with phenol:chloroform. CSH Protoc. 2006. http://dx.doi.org/10.1101/pdb.prot4455.
24. Janczewski DN, Modi WS, Stephens JC, O'Brien SJ. Molecular evolution of mitochondrial 12S RNA and cytochrome b sequences in the pantherine lineage of Felidae. Mol Biol Evol. 1995;12:690–707.
25. Tamura K, Stecher G, Peterson D, Filipski A, Kumar S. MEGA6: molecular evolutionary genetics analysis version 6.0. Mol Biol Evol. 2013;30:2725–9.
26. Drummond AJ, Rambaut A. BEAST: Bayesian evolutionary analysis by sampling trees. BMC Evol Biol. 2007;7:214.
27. Bandelt H, Forster P, Röhl A. Median-joining networks for inferring intraspecific phylogenies. Mol Biol Evol. 1999;16:37–48.
28. Rambaut A, Suchard MA, Xie D, Drummond AJ. Tracer v. 1.5. 2013. http://tree.bio.ed.ac.uk/software/tracer/.
29. Librado P, Rozas J. DnaSP v5: a software for comprehensive analysis of DNA polymorphism data. Bioinformatics. 2009;25:1451–2.

30. Hammer Ø, Harper DAT, Ryan PD. PAST: Paleontological statistics software package for education and data analysis. Palaeontol Electron. 2001;4:1. http://palaeo-electronica.org/2001_1/past/issue1_01.htm. (2.10 ed.).

31. Cretzschmar JC. Atlas zu der Reise im nördlichen Afrika von Eduard Rüppell, Säugethiere. Frankfurt am Main: Senckenbergischen naturforschenden Gesellschaft; 1826. p. 78.

32. Gray JE. Catalogue of carnivorous, pachydermatous, and edentate Mammalia in the British Museum. London: British Museum (Natural History); 1869. p. 398.

33. Hilzheimer M. Die geographische Verbreitung der afrikanischen Grauschakale. Zoologischer Beobachter. 1906;47:363–73.

34. Schwarz E. Über Typenexemplare von Schakalen. Senckenbergiana. 1926; 8:39–47.

35. Pocock RI. Preliminary diagnoses of some new races of south Arabian mammals. Ann Mag Nat Hist. 1934;14(10):635–6.

36. Sykes WH. Canis pallipes. In: Proceedings of the Committee of Science and Correspondence of the Zoological Society of London 1830–1831, vol. 1. 1831. p. 101.

37. Gray JE. Notice of the chanco or golden wolf (Canis chanco) from Chinsese Tartary. Proc Zool Soc London. 1863;31:94.

38. Ferguson WW. The systematic position of Canis aureus lupaster (Carnivora: Canidae) and the occurrence of Canis lupus in North Africa, Egypt and Sinai. Mammalia. 1981;45:460–5.

39. Wayne RK, Geffen E, Girman DJ, Koepfli K-P, Lau LM, Marshall CR. Molecular systematics of the Canidae. Syst Biol. 1997;46:622–53.

40. Wilson PJ, Grewal S, Lawford ID, Heal JN, Granacki AG, Pennock D, Theberge JB, Theberge MT, Voigt DR, Waddell W, Chambers RE, Paquet PC, Goulet G, Cluff D, White BN. DNA profiles of the eastern Canadian wolf and the red wolf provide evidence for a common evolutionary history independent of the gray wolf. Can J Zool. 2000;78:2156–66.

41. Sharma DK, Maldonado JE, Jhala YV, Fleischer RC. Ancient wolf lineages in India. Proc R Soc B Biol Sci. 2004;271 Suppl 3:S1–4. doi:10.1098/rsbl.2003.0071.

42. Aggarwal RK, Kivisild T, Ramadevi J, Singh L. Mitochondrial DNA coding region sequences support the phylogenetic distinction of two Indian wolf species. J Zoolog Syst Evol Res. 2007;45:163–72.

43. Anderson J, De Winton WE. Zoology of Egypt. Mammalia. London: Hugh Rees Limited; 1901.

44. Cabrera A. Algunos carnívoros africanos nuevos. Bol R Soc Esp Hist Nat. 1921;21:261–4.

45. Keimer L. Jardins zoologiques d'Égypte. Cahiers D'histoire Égyptienne. 1954; 6:81–159.

46. Kurtén B. The Carnivora of the Palestine caves. Acta Zool Fenn. 1965;107:1–74.

47. Stimpson CM, Lister A, Parton A, Clark-Balzan L, Breeze PS, Drake NA, Groucutt HS, Jennings R, Scerri EML, White TS, et al. Middle Pleistocene vertebrate fossils from the Nefud Desert, Saudi Arabia: implications for biogeography and palaeoecology. Quat Sci Rev. 2016;143:13–36. doi:10.1016/j.quascirev.2016.05.016.

48. Jhala YV, Moehlman PD. Canis aureus. IUCN red list of threatened species. Gland: IUCN; 2008. Version 2011.1.

49. Admasu E, Thirgood SJ, Bekele A, Laurenson KM. Spatial ecology of golden jackal in farmland in the Ethiopian Highlands. Afr J Ecol. 2004;42:144–52.

50. Eshete G, Tesfay G, Bauer H, Ashenafi ZT, de Iongh HH, Marino J. Community resource uses and Ethiopian wolf conservation in Mount Abune Yosef. Environ Manag. 2015;56:684–94.

51. Marino J. Threatened Ethiopian wolves persist in small isolated Afroalpine enclaves. Oryx. 2003;37:62–71.

52. Simeneh G. Habitat use and diet of golden jackal (Canis aureus) and human - carnivore conflict in Guassa community conservation area, Menz. M. Sc. Thesis. Addis Ababa: Addis Ababa University; 2010.

53. Yihune M, Bekele A, Ashenafi ZT. Human-Ethiopian wolf conflict in and around the Simien Mountains National Park, Ethiopia. Int J Ecol Environ Sci. 2008;34:149–55.

54. McShane TO, Grettenberger JF. Food of the golden jackal (Canis aureus) in central Niger. Afr J Ecol. 1984;22:49–53.

55. Atickem A, Williams S, Bekele A, Thirgood S. Livestock predation in the Bale Mountains, Ethiopia. Afr J Ecol. 2010;48:1076–82. doi:10.1111/j.1365-2028.2010.01214.x.

56. Wayne RK, Nash WG, O'Brien SJ. Chromosomal evolution of the Canidae. Cytogenet Cell Genet. 1987;44:134–41.

57. Rutledge LY, Devillard S, Boone JQ, Hohenlohe PA, White, BN. RAD sequencing and genomic simulations resolve hybrid origins within North American Canis. Biol Lett. 2015; doi:10.1098/rsbl.2015.0303.

58. vonHoldt BM, Pollinger JP, Earl DA, Knowles JC, Boyko AR, Parker H, Geffen E, Pilot M, Jedrzejewski W, Jedrzejewska B, et al. A genome-wide perspective on the evolutionary history of enigmatic wolf-like canids. Genome Res. 2011;21:1294–305.

59. Wronski T, Macasero W. Evidence for the persistence of Arabian Wolf (Canis lupus pallipes) in the Ibex Reserve, Saudi Arabia and its preferred prey species. Zool Middle East. 2008;45:11–8.

60. Koshravi R, Rezaei HR, Kaboli M. Detecting hybridization between Iranian wild wolf (Canis lupus pallipes) and free-ranging domestic dog (Canis familiaris) by analysis of microsatellite markers. Zoolog Sci. 2013;30:27–34.

61. Gotelli D, Sillero-Zubiri C, Applebaum GD, Girman D, Roy M, García-Moreno J, Ostrander E, Wayne RK. Molecular genetics of the most endangered canid: the Ethiopian wolf, Canis simensis. Mol Ecol. 1994;3:301–12.

62. Werdelin L, Lewis ME. Plio-Pleistocene Carnivora of eastern Africa: species richness and turnover patterns. Zool J Linn Soc. 2005;144:121–44.

63. Hartstone-Rose A, Werdelin L, de Ruiter DJ, Berger LR, Churchill SE. The Plio-Pleistocene ancestor of wild dogs, Lycaon sekowei n. sp. J Paleo. 2010; 84:299–308.

7

Survey of avifauna of the Gharana wetland reserve: implications for conservation in a semi-arid agricultural setting on the Indo-Pakistan border

Pushpinder S. Jamwal[1], Pankaj Chandan[1], Rohit Rattan[1], Anupam Anand[2], Prameek M. Kannan[3] and Michael H. Parsons[4,5*]

Abstract

Background: The Gharana wetland conservation reserve (GWCR) is a semi-arid wetland adjacent to agricultural areas on the Indo-Pakistani border. Despite being declared an Important Bird Area (IBA) by Birdlife International, the occurrence and distribution of birds has not been well-documented in this area. Our aims were to systematically document the composition, relative abundance and feeding guilds of all avian fauna in order to form a baseline to monitor changes from—and to underwrite—future conservation actions.

Results: From 24 surveys over 1 year, we recorded 151 species from 45 families and 15 orders. 41% of species were listed as 'rare' and only 22% were 'very common'. The largest number of families belonged to the order Passeriformes (40%), followed by Charadriiformes (14%) and Coraciiformes (11%). The most species (12%), were found in the family Anatidae (Anseriformes—widely recognized as bio-indicators), followed by Accipitridae (Falconiformes;12%) and Muscicapidae (Passeriformes; 6%). Carnivores and insectivores were the feeding guilds most frequently observed. Indeed, more than 50% of all species fed on the abundant fish, mollusks and insects and larvae. Bark-feeders and nectarivores were the least common.

Conclusions: Winter visitors were frequently found, while summer visitors were rare, reinforcing the importance of GWCR as a wintering site for high-altitude species. The conservation of this wetland is especially crucial for nine globally-threatened species. We have provided baseline documentation to help future monitoring efforts for this region, and a template to initiate the implementation of conservation plans for other remote IBAs.

Keywords: Biodiversity, Biological indicators, Feeding guilds, Relative abundance, Residential status, Wetland conservation

Background

Global avian diversity has been reviewed intermittently over the last 75 years [1–4], and is not complete, especially in Asia. This lack of documentation is especially prominent in India, which has one of the highest biodiversity indices in the world and includes 12% of the world's avifauna fauna. However, almost 25% of the bird species found in India (1224 species belonging to 78 families and 17 orders) are dependent on wetlands [5] at a time when wetland loss is considered the prime threat to waterfowl across the globe [6]. Eighty percent of the population decline in Asian flyways near wetlands are a result of human encroachment, increased agriculture and climate change, and militarization near borders [7, 8].

The Gharana wetland conservation reserve (GWCR) is recognized as an Important Bird Area (IBA) by Birdlife International [9]. IBAs ensue from a global network that identifies focal areas for conservation implementation [10].

* Correspondence: Parsons.HMichael@gmail.com
[4]Department of Biology, Hofstra University, Hempstead, NY 11549, USA
[5]Department of Biological Sciences, Fordham University, Bronx, NY 10458, USA
Full list of author information is available at the end of the article

Criteria for inclusion into an IBA are based on the abundance of avian species, the presence of globally-threatened or restricted-range species, and/or their vulnerability to climate change [9] GWCR is especially important because it consists of a semi-arid wetland on the international border between the Indian states and the four provinces of Pakistan, and provides a unique habitat not only for birds, but also for many meso-predators and small carnivores, herbivores, primates and reptiles. The primary threats to this wetland are human encroachment and its corollaries such as cattle grazing, bathing, stray dogs and military shelling across the Indo-Pakistan border.

In order to draft conservation plans for the remaining avifauna in accordance with the IBA designation, it is essential that a number of criteria are documented: including the presence and abundance of bird species across all seasons, and their feeding guilds which relate to food abundance, quality, and availability of perching, roosting and nesting sites. These factors are important, not only because they influence the abundance and diversity of birds, but may have indirect effects on other animal and plant taxa throughout the ecosystem. For instance, granivorous birds can reduce seed survival of plant/crop species [11, 12], while insectivores can decrease the abundance of herbivorous arthropods [13, 14]. Frugivorous birds influence seed dispersal [15, 16] and the survival and reproduction of herbaceous and woody plants. They influence these processes directly through seed predation, and indirectly, by reducing the abundance of herbivorous insects and seed dispersal [17].

The avifauna has been minimally documented in Gharana. Sharma and Saini [18] recorded 21 waterfowl species in the region, while Pandotra and Sahi [19] reported the presence of 57 species of waterfowl and terrestrial birds. No complete documentation has been available, however, and no study has reported feeding guilds for either the resident or visiting species. Thus, it is unclear what resources from the wetland are attracting migrants.

Objectives

Our objectives were to comprehensively document the species composition, relative abundance and feeding guilds of all avian fauna over 1 year in GWCR, inclusive of the surrounding agricultural fields.

Results

The maximum number of families (Table 1) belonged to the order Passeriformes, 18 (40% of total) followed by Charadriiformes, 6 (14%). Most identified species belonged to Anatidae 19 (12%), followed by Accipitridae 18 (12%) and Muscicapidae 9 (6%). After ranking avifauna into three categories based on their cumulative abundance (Fig. 1), we learned that 62 (41% of total) species were rare, 56 species (37% of total) were

common, and 33 (22% of total) species were very common. Nine globally-threatened species were identified: Painted Stork *Mycteria leucocephala*, Wooly-necked Stork *Ciconia episcopus*, Black-necked Stork *Ephippiorhynchus asiaticus*, Black-headed (White) Ibis *Threskiomis melanocephalus*, Ferruginous Duck *Aythya nyroca*, Greater Spotted Eagle *Aquila clanga*, Egyptian Vulture *Neophron percnopterus*, Pallid Harrier *Circus macrourus* and Indian River Tern *Sterna aurantia*. Among 151 total species (Table 1), 74 (49%) were winter visitors, 54 (36%) were resident, 11 (7%) were vagrant and 12 (8%) were summer visitors (Fig. 1).

Birds of GWCR primarily utilized eight feeding guilds: herbivores, bark feeders, carnivores, frugivores, granivores, insectivores, nectarivores and omnivores. Among these families, 19 (13%) were herbivores, bark feeders 2 (1%), carnivores 46 (36%), frugivores 6 (4%), graminivores 7 (5%), insectivores 40 (26%), nectarivore 1 (1%) and omnivores 30 (20%).

Discussion

We have provided baseline data for an under-reported, but vulnerable, wetland near a border in remote Asia. We recorded 151 species including 62 waterfowl and 89 terrestrial species. This provides a substantial update to the 21 and 57 species already documented [18, 19]. Most of the high-altitude bird species are known to migrate towards lower altitude sites such as GWCR during winter [20], and this was also observed in our study. In particular, the high number of winter visitors likely suggests that Gharana and its adjoining agricultural fields provide appropriate habitat for thousands of winter migratory birds as well as important wintering and stopover site for several other migratory species.

The high prevalence of the Anatidae affirms notions that this region provides particularly suitable habitat and abundant food for ducks, geese and swans. The Accipitridae are ideal indicators of ecosystem health because they are near the top of local trophic levels. As top-order predators, the Accipitridae are key bio-indicators to understanding the dynamics of local ecosystems. In GWCR, their presence likely reflects the greater availability of small mammals, birds, reptiles, amphibians and insects. Indeed, over 70% of the total feeding guilds were carnivorous (36%), insectivorous (26%) or omnivorous (20%).

The regional diversity of birds commonly varies with factors such as climate of the area (temperature, humidity and rainfall), altitude, food availability [21]. While some of these factors were beyond the remit of our study, and will be updated in furture reports, we were able to note the presence of a large number of species of fish, mollusks, amphibians and aquatic insects and their larvae, that these birds fed upon. These resources are important to document as thoroghly as possible because

Table 1 Comprehensive list of bird species recorded utilizing Gharana wetland conservation reserve and associated agricultural fields

Species (no.)	Order	Family	Common name	Scientific name	Residental status	Abundance	Feeding	IUCN status
1	Podicipediformes	Podicipedidae	Little Grebe	*Tachybaptus ruficollis*	R	VC	C	LC
2	Pelecaniformes	Phalacrocoracidae	Great Cormorant	*Phalacrocorax carbo*	WV	VC	C	LC
3			Little Cormorant	*Phalacrocorax niger*	WV	VC	C	LC
4	Ciconiiformes	Ardidae	Yellow Bittern	*Ixobrychus sinensis*	WV	R	C	LC
5			Black-crowned Night Heron	*Nycticorax nycticorax*	WV	C	C	LC
6			Indian Pond Heron	*Ardeola grayii*	R	VC	C	LC
7			Cattle Egret	*Bubulcus ibis*	R	VC	C	LC
8			Little Egret	*Egretta garzetta*	R	VC	C	LC
9			Intermediate Egret	*Mesophoyx intermedia*	R	C	C	LC
10			Great Egret	*Casmerodius albus*	WV	C	C	LC
11			Purple Heron	*Ardea purpurea*	R	VC	C	LC
12			Grey Heron	*Ardea cinerea*	R	VC	C	LC
13		Ciconiidae	Painted Stork	*Mycteria leucocephala*	WV	R	C	NT
14			Black Stork	*Ciconia nigra*	WV	R	C	LC
15			Wooly-necked Stork	*Ciconia episcopus*	WV	R	C	VU
16			Black-necked Stork	*Ephippiorhynchus asiaticus*	WV	R	C	NT
17		Threskiornithidae	Black-headed (White) Ibis	*Threskiomis melanocephalus*	WV	R	C	NT
18			Red-naped Ibis	*Pseudibis papillosa*	WV	R	C	LC
19			Glossy ibis	*Plegadis falcinellus*	WV	R	C	LC
20			Eurasian Spoonbill	*Platalea leucorodia*	WV	R	C	LC
21	Anseriformes	Anatidae	Lesser Whistling Duck	*Dendrocygna javanica*	WV	VC	H	LC
22			Greylag Goose	*Anser anser*	WV	R	H	LC
23			Greater White-fronted Goose	*Anser albifrons*	WV	R	H	LC
24			Indian Cotton Teal	*Nettapus coromandelianus*	WV	VC	H	LC
25			Bar-headed Goose	*Anser indicus*	WV	C	H	LC
26			Ruddy Shelduck	*Tadorna ferruginea*	WV	R	H	LC
27			Comb Duck	*Sarkidiornis melanotos*	WV	R	H	LC
28			Eurasian Wigeon	*Anas penelope*	WV	C	H	LC
29			Gadwall	*Anas strepera*	WV	VC	H	LC
30			Eurasian Teal	*Anas crecca*	WV	VC	H	LC
31			Mallard	*Anas platyrhynchos*	WV	R	H	LC
32			Indian Spot-billed Duck	*Anas poecilorhyncha*	WV	R	H	LC
33			Northern Pintail	*Anas acuta*	WV	C	H	LC
34			Garganey	*Anas querquedula*	SV	R	H	LC
35			Northern Shoveler	*Anas clypeata*	WV	VC	H	LC
36			Red-crested Pochard	*Netta rufina*	WV	R	H	LC
37			Common Pochard	*Aythya ferina*	WV	C	H	LC
38			Ferruginous Duck	*Aythya nyroca*	WV	R	H	NT
39			Tufted Duck	*Aythya fuligula*	WV	R	H	LC
40	Falconiformes	Accipitridae	Black-shouldered Kite	*Elanus caeruleus*	R	C	C	LC
41			Black Kite	*Milvus migrans*	R	C	C	LC
42			Steppe Eagle	*Aquila nipalensis*	WV	C	C	LC
43			Greater Spotted Eagle	*Aquila clanga*	WV	R	C	VU

Table 1 Comprehensive list of bird species recorded utilizing Gharana wetland conservation reserve and associated agricultural fields
(Continued)

44			Eurasian Marsh-Harrier	Circus aeruginosus	WV	R	C	LC
45			Eurasian Sparrowhawk	Accipiter nisus	V	R	C	LC
46			Himalyan buzzard	Buteo buteo	WV	R	C	LC
47			Long-legged Buzzard	Buteo rufinus	WV	R	C	LC
48			Besra	Accipiter virgatus	WV	R	C	LC
49			Northern Goshawk	Accipiter gentilis	WV	R	C	LC
50			Booted Eagle	Hieraaetus pennatus	WV	R	C	LC
51			Egyptian Vulture	Neophron percnopterus	SV	C	C	NT
52			Shikra	Accipiter badius	R	C	C	LC
53			Hen Harrier	Circus cyaneus	WV	C	C	LC
54			Eurasian Marsh-Harrier	Circus aeruginosus	WV	VC	C	LC
55			Pallid Harrier	Circus macrourus	WV	R	C	NT
56			Short-toed snake Eagle	Circaetus gallicus	WV	C	C	LC
57		Falconidae	Eurasian Hobby	Falco subbuteo	WV	R	C	LC
58	Galliformes	Phasianidae	Gray Francolin	Francolinus pondicerianus	R	VC	O	LC
59	Gruiformes	Rallidae	Water Rail	Rallus aquaticus	WV	C	O	LC
60			White-breasted Waterhen	Amaurornis phoenicurus	R	VC	O	LC
61			Common Moorhen	Gallinula chloropus	R	VC	O	LC
62			Purple Swamphen	Porphyrio porphyrio	R	VC	O	LC
63			Common Coot	Fulica atra	WV	C	O	LC
64	Charadriiformes	Jacanidae	Pheasant-tailed Jacana	Hydrophasianus chirurgus	SV	C	O	LC
65		Charadriidae	Red-wattled Lapwing	Vanellus indicus	R	VC	O	LC
66			Little Ringed Plover	Charadrius dubius	R	R	O	LC
67			White-tailed Plover	Vanellus leucurus	WV	R	O	LC
68		Scolopacidae	Greenshank	Tringa nebularia	WV	C	I	LC
69			Common Snipe	Gallinago gallinago	WV	R	I	LC
70			Common Redshank	Tringa totanus	V	R	I	LC
71			Common Sandpiper	Actitis hypoleucos	WV	C	I	LC
72			Green sandpiper	Tringa ochropus	WV	R	I	LC
73			Curlew Sandpiper	Calidris ferruginea	V	R	I	LC
74			Little Stint	Calidris minuta	V	R	I	LC
75			Ruff	Philomachus pugnax	WV	VC	I	LC
76		Recurvirostridae	Black-winged Stilt	Himantopus himantopus	WV	C	I	LC
77		Glareolidae	Oriental Pratincole	Glareola maldivarum	V	R	I	LC
78			Little Pratincole	Glareola lactea	R	C	I	LC
79		Laridae	Indian River Tern	Sterna aurantia	SV	C	C	NT
80			Common Tern	Sterna hirundo	V	R	C	LC
81			White-winged Black Tern	Chlidonias leucopterus	V	R	C	LC
82	Columbiformes	Columbidae	Eurasian Collared-Dove	Streptopelia decaocto	R	VC	O	LC
83			Spotted Dove	Streptopelia chinensis	WV	R	O	LC
84			Rock Pigeon	Columba livia	R	VC	O	LC
85	Psittaciformes	Psittacidae	Rose-ringed Parakeet	Psittacula krameri	R	C	F	LC
86			Plum-headed Parakeet	Psittacula cyanocephala	WV	R	F	LC
87	Cuculiformes	Cuculidae	Greater Coucal	Centropus sinensis	R	C	C	LC

Table 1 Comprehensive list of bird species recorded utilizing Gharana wetland conservation reserve and associated agricultural fields *(Continued)*

88			Asian Koel	*Eudynamys scolopaceus*	SV	C	O	LC
89			Pied Cuckoo	*Clamator jacobinus*	SV	R	O	LC
90			Eurasian Cuckoo	*Cuculus canorus*	SV	R	O	LC
91	Strigiformes	Strigidae	Spotted Owlet	*Athene brama*	R	C	C	LC
92	Coraciiformes	Alcedinidae	White throated Kingfisher	*Halcyon smyrnensis*	R	VC	C	LC
93			Common Kingfisher	*Alcedo atthis*	WV	C	C	LC
94			Crested Kingfisher	*Megaceryle lugubris*	R	VC	C	LC
95		Meropidae	Green Bee-eater	*Merops orientalis*	R	VC	I	LC
96			Blue-tailed Bee-eater	*Merops philippinus*	SV	C	I	LC
97		Coraciidae	Indian Roller	*Coracias benghalensis*	R	C	I	LC
98		Upupidae	Eurasian Hoopoe	*Upupa epops*	R	C	I	LC
99		Bucerotidae	Indian Grey Hornbill	*Ocyceros birostris*	R	R	F	LC
100	Piciformes	Picidae	Lesser goldenback	*Dinopium benghalense*	R	C	BF	LC
101			Yellow-crowned Woodpecker	*Dendrocopos mahrattensis*	R	R	BF	LC
102		Capitonidae	Coppersmith Barbet	*Megalaima haemacephala*	R	R	F	LC
103	Passeriformes	Alaudidae	Crested Lark	*Galerida cristata*	R	C	O	LC
104		Hirundinidae	Wire-tailed Swallow	*Hirundo smithii*	SV	C	I	LC
105			Barn Swallow	*Hirundo rustica*	WV	R	I	LC
106			Plain Martin	*Riparia paludicola*	R	R	I	LC
107		Motacillidae	Gray Wagtail	*Motacilla cinerea*	WV	C	I	LC
108			Paddyfield Pipit	*Anthus pratensis*	R	C	I	LC
109			Tree Pipit	*Anthus trivialis*	V	R	I	LC
110			Rosy Pipit	*Anthus roseatus*	WV	C	I	LC
111			White Wagtail	*Motacilla alba*	WV	C	I	LC
112			Citrine Wagtail	*Motacilla citreola*	WV	R	I	LC
113			White-browed Wagtail	*Motacilla madaraspatensis*	R	VC	I	LC
114		Campephagidae	Small Minivet	*Pericrocotus cinnamomeus*	R	R	I	LC
115		Pycnonotidae	Red-vented Bulbul	*Pycnonotus cafer*	R	VC	F	LC
116		Laniidae	Bay-backed Shrike	*Lanius vittatus*	R	C	O	LC
117			Long-tailed Shrike	*Lanius schach*	SV	C	O	LC
118		Muscicapidae	Pied Bushchat	*Saxicola caprata*	R	C	G	LC
119			Variable Wheatear	*Oenanthe picata*	WV	R	G	LC
120			Isabelline Wheatear	*Oenanthe isabellina*	V	R	G	LC
121			Black Redstart	*Phoenicurus ochruros*	WV	R	I	LC
122			Oriental Magpie-Robin	*Copsychus saularis*	R	VC	I	LC
123			Gray Bushchat	*Saxicola ferreus*	WV	C	I	LC
124			Indian Robin	*Copsychus fulicatus*	R	VC	I	LC
125			Bluethroat	*Luscinia svecica*	R	R	I	LC
126			White-tailed Stonechat	*Saxicola leucurus*	V	R	G	LC
127		Paridae	Great Tit	*Parus major*	WV	C	F	LC
128		Nectariniidae	Purple Sunbird	*Nectarinia asiatica*	SV	C	N	LC
129		Zosteropidae	Oriental White-eye	*Zosterops palpebrosus*	R	C	I	LC
130		Estrildidae	Scaly breasted munia	*Lonchura punctulata*	WV	VC	G	LC
131		Passeridae	House Sparrow	*Passer domesticus*	R	VC	G	LC

Table 1 Comprehensive list of bird species recorded utilizing Gharana wetland conservation reserve and associated agricultural fields *(Continued)*

132		Sind Sparrow	*Passer pyrrhonotus*	WV	R	G	LC
133	Ploceidae	Baya Weaver	*Ploceus philippinus*	WV	C	O	LC
134		Black-breasted weaver	*Ploceus benghalensis*	WV	R	O	LC
135	Sturnidae	Brahminy Starling	*Temenuchus pagodarum*	WV	R	O	LC
136		Common Starling	*Sturnus vulgaris*	WV	C	O	LC
137		Bank Myna	*Acridotheres ginginianus*	R	VC	O	LC
138		Asian Pied Starling	*Gracupica contra*	V	C	O	LC
139		Common Myna	*Acridotheres tristis*	R	C	O	LC
140	Oriolidae	Eurasian Golden Oriole	*Oriolus oriolu*	WV	R	O	LC
141	Dicruridae	Black Drongo	*Dicrurus macrocercus*	R	C	I	LC
142		Ashy Drongo	*Dicrurus leucophaeus*	SV	C	I	LC
143	Corvidae	House Crow	*Corvus splendens*	R	VC	O	LC
144		Rufous Treepie	*Dendrocitta vagabunda*	R	C	O	LC
145		Large-billed Crow	*Corvus macrorhynchos*	WV	R	O	LC
146	Cisticolidae	Ashy Prinia	*Prinia socialis*	R	C	I	LC
147		Striated Prinia	*Prinia crinigera*	R	C	I	LC
148		Common Tailorbird	*Orthotomus sutorius*	R	C	I	LC
149		Plain Prinia	*Prinia inornata*	R	C	I	LC
150		Common Chiffchaff	*Phylloscopus collybita*	WV	C	I	LC
151		Zitting Cisticola	*Cisticola juncidis*	R	R	I	LC

Residential status: *WV* winter visitors, *R* resident, *V* vagrant and *SV* summer visitors. Abundance: *C* common, *VC* very common, *R* rare. Feeding: *BF* bark feeder, *C* carnivorous, *F* frugivorous, *G* granivorous, *H* herbivorous, *I* insectivorous, *N* nectarivorous, *O* omnivorous. IUCN Status (as of the time of manuscript preperation): *LC* least concern, *NT* near threatened, *VU* = vulnerable

they serve as attractive food sources for resident and migrants. In particular, wader species were found to regularly visits the agricultural fields surrounding GWCR, likely owing to the shallow water and presence of high numbers of aquatic insects.

Importantly, we have documented nine globally threatened species (5% of the total species). These species epitomize the need for further monitoring and conservation actions related to GWCR and its associated agricultural fields. The exceptional arthropod diversity provides abundant food for these guilds, and included a substantial number of unknown arachnids whose description warrants detailed scientific studies. Hence, the Gharana wetland is not only an ideal

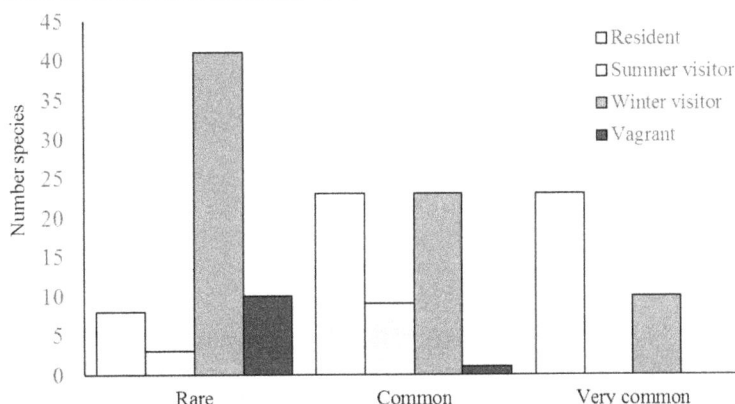

Fig. 1 Residential status and abundance of bird species observed in Gharana wetland conservation reserve and associated agricultural fields in Jammu and Kasmir, India from July 2012 to June 2013

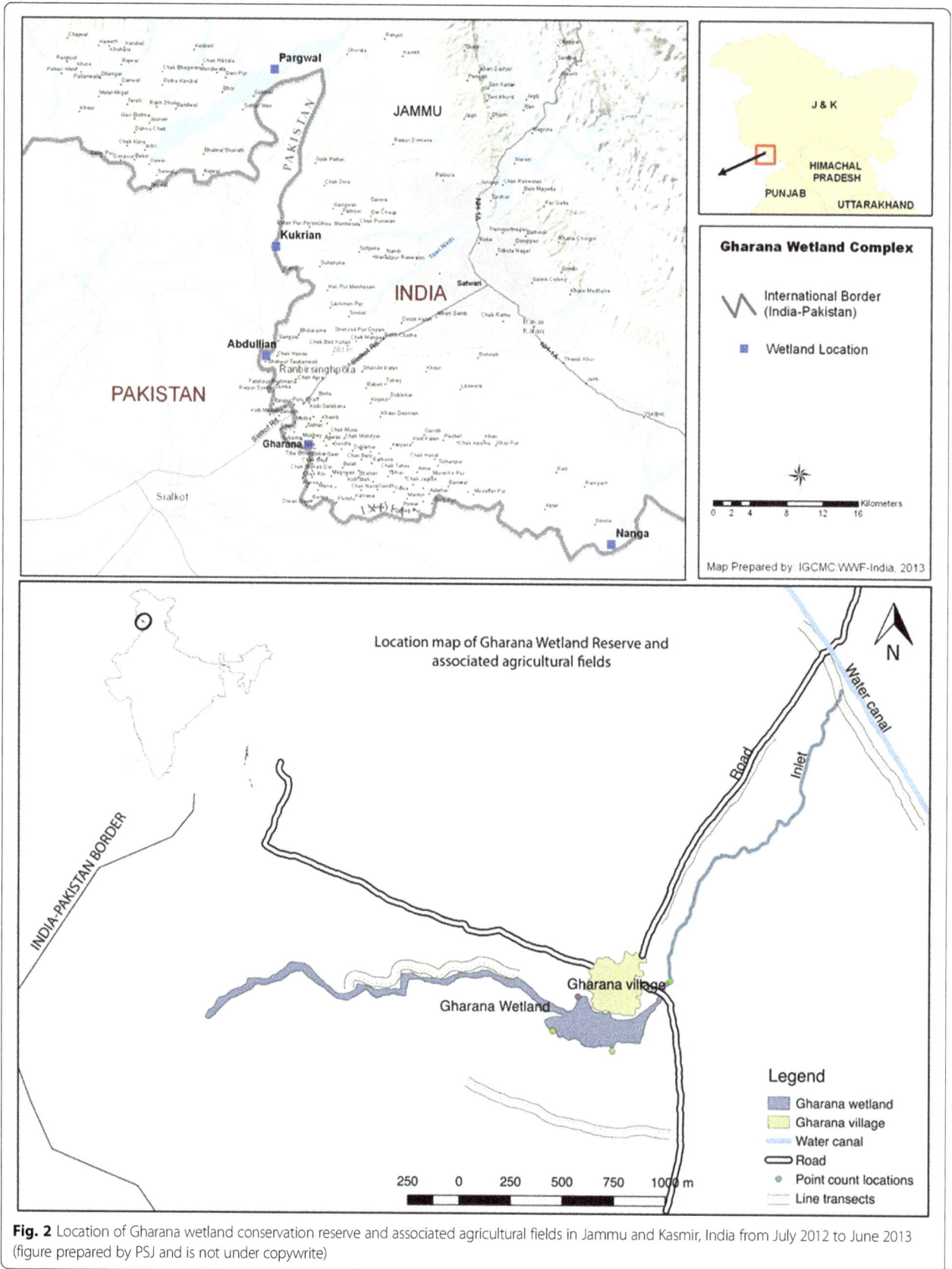

Fig. 2 Location of Gharana wetland conservation reserve and associated agricultural fields in Jammu and Kasmir, India from July 2012 to June 2013 (figure prepared by PSJ and is not under copywrite)

Survey of avifauna of the Gharana wetland reserve: implications for conservation in a semi-arid...

79

place for the conservation of endemic and globally threatened birds, but also for a complex array of flora and fauna that attract such a broad range of bird species.

Conclusions

Winter visitors were frequently found in GWCR, while summer visitors were rare, reinforcing the importance of this region as a wintering site for high-altitude species. The conservation of this wetland is especially crucial for nine globally-threatened species. We have provided baseline documentation to help future monitoring efforts for this region, and a template to initiate the implementation of conservation plans for other remote IBAs.

Methods
Study site
Gharana 32°32'28" N; 74°41'27" E; 281 m asl (Fig. 2) is located on the international India-Pakistan border in the south-western part of Jammu province in the Indian state of Jammu and Kashmir. It is a naturally maintained, rain-fed swamp with a bottom surface of loamy clay with decaying vegetation. Surrounding plants include macrophytes such as *Eicchornia spp.* and *Hydrilla spp.* [22] and the Common reed (*Typha spp.*). Additional sources of water are spillover from a nearby canal (the Ranbir Canal) and surface runoff from agricultural areas [19].

This wetland and its adjacent agricultural fields are in the subtropical climatic zone where summer temperatures may reach 46 °C maximum and winter minima decrease to as low as 2 °C. Annual rainfall is around 1331 mm, with most precipitation occurring when the south-western monsoon winds arrive from July-September. The agricultural fields adjacent to Gharana village also provides both suitable habitat, and concomitant threats, for a diverse group of bird taxa. Owing to the wide diversity of avifauna, and also being a wintering ground for many threatened and migratory waterfowl, GWCR was also declared as Important Bird Area (IBA) by the Bombay Natural History Society and BirdLife International [23].

Data collection
We conducted twenty-four surveys from July 2012 to June 2013, covering all seasons; summer (April–June), monsoon (July–Sept), autumn (Oct–Novem) and winter (Dec–March). Our surveys (Fig. 2) followed well established methods including line transects and point count methods, as per [24]. Bird counts were direct visual sightings only. Counts were performed twice per month at all sites by a team of ten individuals in the early morning (07:00–10:00) during the time of highest bird activity [25] and lowest human disturbance. Experts with over 200 h of wetland bird identification and post-doctoral training were consulted throughout the period.

We classified all species as common/rare, resident/migratory status of the birds as per [26] For instance, VC = very common species encountered during 80% of all surveys); C = common species encountered frequently (50–70%) and R = rare species which are encountered less frequently (10–20%). Likewise, if we only documented a particular species between December and March, then we considered it as a winter visitor. Whereas, presence between April and June was documented as a summer visitation. If we documented a bird throughout a year in and around GWCR, then it was considered as a resident. Feeding guilds were identified from the literature, rather than what birds were seen feeding on at the time. Nikon Monarch 10 × 42 binoculars were used during surveys for taking observations and on-the-spot identification. We used photographs and/or video to validate any unidentified species. The checklist was prepared using the standardized common and scientific names assigned in [27]. All data collected were observational and did not involve any manipulation or alteration of any animals, plants or humans.

Limitations
The limitations of our study are due to the lack of hypotheses testing, and is purely descriptive. Post-hoc analyses may be performed using our data set which has been submitted to a public repository (details in the declarative statement).

Abbreviations
GWCR: Gharana wetland conservation reserve; IBA: Important Bird Area

Acknowledgements
We thank the Department of Wildlife Protection, Jammu and Kashmir State for granting permission and providing the necessary logistic support and cooperation for this extensive study. We are particularly appreciative of the support from Mr. Ravi Singh, Mr. A. K. Singh, Dr. Sejal Worah, Dr. Dipankar Ghose Mr. Asif M. Sagar, Mr. Tahir Shawl, Mr. Raja Sayeed, Mr. Shakeel Ahmed and Mr. Ram Saroop.

Funding
No external funding was received and thus the authors are not declaring any funding sources.

Authors' contributions
PSJ, PC, RR and AA designed the study and collected all data. PSJ and MHP analyzed and presented the data and drafted the manuscript. PMK assisted the analysis and all drafts of the manuscript. All authors read and approved the final manuscript.

Competing interests

The authors declare that they have no competing interests.

Author details

¹Western Himalayas Landscape, WWF-India, New Delhi 110003, India. ²Global Land Cover Facility, University of Maryland, College Park, MD 20742, USA. ³Department of Biology, Pace University, 861 Bedford Road, Pleasantville, NY 10570, USA. ⁴Department of Biology, Hofstra University, Hempstead, NY 11549, USA. ⁵Department of Biological Sciences, Fordham University, Bronx, NY 10458, USA.

References

1. Ali S. The book of Indian birds: Bombay natural history society Bombay. 1979.
2. Parkes KC, Stiles FG, Skutch AF, Ridgely RS. Special review: a guide to the birds of Costa Rica. Wilson Bull. 1991;103(2):316–20.
3. Monroe BL, Sibley CG. A world checklist of birds. Connecticut: Yale University Press; 1997.
4. Dickinson EC, Bahr N, Dowsett R, Pearson D, Remsen V, Roselaar C, Schodde D. The Howard and Moore complete checklist of birds of the world. London: Christopher Helm; 2004.
5. Kumar A. Handbook on Indian wetland birds and their conservation. 2005.
6. O'Connell M. Threats to waterbirds and wetlands: implications for conservation, inventory and research. Wildfowl. 2000;51(51):1–16.
7. Piersma T, Lok T, Chen Y, Hassell CJ, Yang HY, Boyle A, Slaymaker M, Chan YC, Melville DS, Zhang ZW. Simultaneous declines in summer survival of three shorebird species signals a flyway at risk. J Appl Ecol. 2016;53(2):479–90.
8. Studds CE, Kendall BE, Murray NJ, Wilson HB, Rogers DI, Clemens RS, Gosbell K, Hassell CJ, Jessop R, Melville DS, et al. Rapid population decline in migratory shorebirds relying on Yellow Sea tidal mudflats as stopover sites. Nat Commun. 2017;8:14895.
9. Butchart SH, Scharlemann JP, Evans MI, Quader S, Arico S, Arinaitwe J, Balman M, Bennun LA, Bertzky B, Besancon C. Protecting important sites for biodiversity contributes to meeting global conservation targets. PLoS One. 2012;7(3):e32529.
10. Heath MF, Evans MI, Hoccom D, Payne A, Peet N. Important bird areas in Europe priority sites for conservation. v. 1: Northern Europa. v. 2: Southern Europe. Cambridge: Birdlife International; 2000.
11. Marone L, Rossi B, Casenave LD. Granivore impact on soil-seed reserves in the central Monte desert, Argentina. Funct Ecol. 1998;12(4):640–5.
12. Kelt DA, Meserve PL, Gutiérrez JR. Seed removal by small mammals, birds and ants in semi-arid Chile, and comparison with other systems. J Biogeogr. 2004;31(6):931–42.
13. Hooks CR, Pandey RR, Johnson MW. Impact of avian and arthropod predation on lepidopteran caterpillar densities and plant productivity in an ephemeral agroecosystem. Ecol Entomol. 2003;28(5):522–32.
14. Van Bael SA, Brawn JD, Robinson SK. Birds defend trees from herbivores in a neotropical forest canopy. Proc Natl Acad Sci. 2003;100(14):8304–7.
15. Herrera CM, Jordano P, Lopez-Soria L, Amat JA. Recruitment of a mast-fruiting, bird-dispersed tree: bridging frugivore activity and seedling establishment. Ecol Monogr. 1994;64(3):315–44.
16. Ingle NR. Seed dispersal by wind, birds, and bats between Philippine montane rainforest and successional vegetation. Oecologia. 2003;134(2):251–61.
17. Nogales M, Delgado J, Medina FM. Shrikes, lizards and Lycium intricatum (Solanaceae) fruits: a case of indirect seed dispersal on an oceanic island (Alegranza, Canary Islands). J Ecol. 1998;86(5):866–71.
18. Sharma K, Saini M. Impact of anthropogenic pressure on habitat utilization by the waterbirds in Gharana Wetland (reserve), Jammu (J&K, India). Int J Environ Sci. 2012;2(4):2050–62.
19. Pandotra A, Sahi D. Avifaunal assemblages in suburban habitat of Jammu, J&K, India. Res J Environ Sci. 2014;3(6):17–24.
20. Chopra G, Sharma SK. Avian biodiversity in and around major wetlands of "Lower Shivalik Foothills" (India). Nat Sci. 2012;10(7):86–93.
21. Laiolo P. Diversity and structure of the bird community overwintering in the Himalayan subalpine zone: is conservation compatible with tourism? Biol Conserv. 2004;115(2):251–62.
22. Tara J, Kour R, SHharma S. A record of aquatic hemiptera of Gharan wetland, Jammu. Bioscan. 2011;6(4):649–55.
23. Zafar-ul-Islam M, Rahmani AR. Important bird areas in India: priority sites for conservation. India: Bombay Natural History Society; 2004.
24. Bibby CJ. Bird census techniques. Amsterdam: Elsevier; 2000.
25. Buckland ST, Anderson DR, Burnham KP, Laake JL, Borchers DL, Thomas L. Introduction to distance sampling estimating abundance of biological populations. Oxford: Oxford University Press; 2001.
26. Saikia P, Saikia M. Diversity of bird fauna in NE India. J Assam Sci Soc. 2000; 41(2):379–96.
27. Kazmierczak K, Perlo Bv. Field guide to the birds of the Indian Subcontinent. Connecticut: Yale University Press; 2000.

Shoaling promotes place over response learning but does not facilitate individual learning of that strategy in zebrafish (*Danio rerio*)

Claire L. McAroe[1,2], Cathy M. Craig[2] and Richard A. Holland[1,3*]

Abstract

Background: Flexible spatial memory, such as "place" learning, is an important adaptation to assist successful foraging and to avoid predation and is thought to be more adaptive than response learning which requires a consistent start point. Place learning has been found in many taxonomic groups, including a number of species of fish. Surprisingly, a recent study has shown that zebrafish (*Danio rerio*), a common species used in cognitive research, demonstrated no significant preference for the adoption of either a place or a response strategy during a plus maze task. That being said, a growing body of research has been looking at how group living influences navigational decisions in animals. This study aims to see how zebrafish, a shoaling species, differ in their ability to perform a maze task when learning in a shoal and as an individual.

Results: Results suggest that shoals of zebrafish are able to learn to perform the spatial memory task in a significantly shorter time than individual fish and appear to show place learning when tested from a novel start point. Interestingly, zebrafish who were trained first in a shoal but were then tested as individuals, did not show the same level of consistency in their choice of navigation strategy.

Conclusions: These findings suggest that shoaling influences navigation behaviour, resulting in faster group learning and convergence on one spatial memory strategy, but does not facilitate the transfer of the strategy learned to individuals within the shoal.

Keywords: Zebrafish, *Danio rerio*, Spatial memory, Spatial cognition, Navigation, Shoal

Background

Being able to navigate through a familiar environment is critically important in the lives of all animals. Successful spatial memory helps an animal find food and avoid predation [1–3]. Previous studies have suggested that animals can use a variety of methods to find their way through an environment [4–6], using two predominant mechanisms for encoding spatial locations to memory, namely egocentric and allocentric [7, 8]. Allocentric encoding, so called "place learning", is thought to be more complex and

cognitively demanding as it involves building up relationships between multiple features within the environment and is often synonymous with a cognitive map [6, 8], although relationships between multiple features can arise through associative mechanisms [9–13]. On the other hand, egocentric encoding is based more on learning a particular set of responses to reach the goal, for example, learning a route and is thus often referred to as response learning [8, 14]. By creating complex mental representations of the landmarks within the environment not solely reliant on associative processes formed with single cues, allocentric spatial memory is thought to be a more flexible strategy as it allows the animal to locate a goal from a novel start point without the need to recapitulate a previously learned route [15]. Previous experiments have

* Correspondence: r.holland@bangor.ac.uk
[1]School of Biological Sciences, Queen's University Belfast, Medical Biology Centre, 97 Lisburn Road, Belfast BT9 7BL, Northern Ireland
[3]Current address, School of Biological Sciences, Bangor University, Deiniol Road, Bangor LL57 2UW, UK
Full list of author information is available at the end of the article

typically used a maze task to assess which strategy animals prefer when encoding location in spatial memory [7, 16, 17]. Such experiments suggest that a number of taxonomic groups are capable of using a predominantly allocentric strategy but are also capable of switching between different strategies in order to accommodate changes within the environment [4]. Other factors such as age, or experience may also influence whether animals use an allo- or egocentric strategy [14, 18].

Individuals within a group may prefer different strategies for facilitating spatial learning, and it is possible that group-moving animals are presented with conflicting directional preferences among group members. How the individuals, and the group as a whole, deal with this potential difference in spatial memory mechanisms within the group poses an interesting question. Although the role of group living in navigational decision-making has been relatively understudied, it has started to receive more attention in the last 10 years. The effect of group living on navigational accuracy [19] and/or collective decisions [20] has been explored theoretically. Empirical studies have looked at how information affects group cohesion [21–24] and how differential experience influences group decisions [25, 26]. However, the question of whether being a member of a group influences overall choice of navigational strategy remains unanswered. Related to this is how group membership influences the spatial learning of individuals within the group. Some evidence suggests that individual guppies can learn foraging routes from others [27], but in other cases, the so called "passenger effect" occurs (e.g. in pigeons), in which following another animal to a goal does not facilitate individual learning [28] (although see [29]). In general however, the extent to which behaviours formed as a group are retained by individuals, is poorly understood [30].

Fish are often selected as preferred animals of study as they are relatively easy to keep compared to other vertebrates (e.g. mammals) and also have comparative cognitive ability [31]. Indeed, the general use of fish in learning and memory experiments has significantly increased in recent years [32, 33]. Shoalling is also a common occurrence in fish species which makes them ideal models for studying group behaviour [31–34]. A recent study by the authors used a plus maze task to explore the individual spatial memory of four different species of fish. The results showed that three species of fish (killifish, goldfish and Siamese fighting fish) demonstrated a preference for the use of the more complex, allocentric place strategy, whereas the fourth species, zebrafish, showed no significant preference for either a place (allocentric) or a response (egocentric) strategy. Furthermore, the zebrafish were found to take significantly longer to learn the task

than any of the other three species [16]. The fact that zebrafish are a naturally shoaling species [34–36] presents an opportunity to investigate how different individual preferences may influence the overall strategy of the shoal and indeed the general cohesiveness of the group.

With this in mind, our study aims to address two questions: [1] does shoaling result in a more consistent navigation strategy, and; [2] does it facilitate or impede transfer of learning to individuals within the shoal? To answer these questions, we will compare the predominant navigation strategies (allocentric (place learning) vs egocentric (response learning)) adopted by individual zebrafish, shoals of zebrafish, and individual zebrafish trained in shoals using the classic plus maze paradigm.

Methods
Subjects
Twenty individual zebrafish (*Danio rerio*) were trained and tested as part of a previous experiment [16] and form part of the analysis of the results presented here. Forty shoals consisting of five fish (200 individual zebrafish in total) were tested during the course of the current experiment. Two groups of 20 shoals were tested under two different experimental procedures on the probe trial (see below). The fish were all adults although exact age was unknown. The sex of the fish was not known. All animals were experimentally naïve and were commercially sourced from two local suppliers, due to availability of stock (the 20 trained and tested as individuals from *Exotic Aquatics, Belfast, N. Ireland* and fish in shoals from *Grosvenor Tropicals, Lisburn, N. Ireland*). All fish were introduced to the laboratory a minimum of 1 week before any experiments began. This was to allow the animals time to acclimatise to laboratory conditions and also to allow natural shoaling to occur in the relevant fish.

Housing conditions
All apparatus was commercially sourced (from sources listed in the *Subjects* section and also from *Maidenhead Aquatics, Newtownabbey, N. Ireland*). Individual zebrafish were each housed separately in 2 L glass jars) during the course of experimentation for identification purposes. A maximum of 10 jars were placed together on a heat mat at any one time so that individuals could see other conspecifics. Shoals were kept in 25 L tanks with a density of five individual fish in each. Water was maintained at an average temperature of 25 °C. When not completing experiments, all fish were fed commercial flaked food. pH and waste levels in all tanks were monitored regularly using *API Freshwater Master Test Kit* and water changes were carried out on a regular basis.

Waste levels were kept within safe ranges (0 ppm ammonia & nitrite; <40 ppm nitrate). pH range was maintained at a range of 7.7 ± 0.3. All fish were maintained in a 13: 11 h light: dark cycle at all times during the laboratory.

Experimental design
Apparatus
The exact apparatus used in a previous related study [16] was also used during the course of these experiments and consisted of a plus maze made from acrylic Perspex panels glued to the inside of a square tank measuring 63 cm × 63 cm × 43 cm (Fig. 1). pH, waste levels and temperature were maintained at the same levels as the housing conditions and water changes were also carried out regularly in the experimental tank. All trials (both training and probe) were recorded using a *Sony HDR-X190E Handycam* video camera mounted above the tank. Trials were timed using a standard stopwatch.

Experimental design
Training-individuals
Training was conducted between 19/03/2013 and 14/03/2014. A training block consist of a total of 10 trials. Each fish would complete a maximum of one training block per day. Training began at 9 am and carried on until each fish had completed one block. Fish were randomly assigned to receive a bloodworm reward at the arm either to the left or to the right of the start arm (*n* = 10 for each side). A trial was considered complete when the tail fin of the fish had passed fully into either arm of the maze. If the fish swam to its assigned rewarded arm, it would receive bloodworm immediately administered by the experimenter using tweezers and the fish would then be moved back to the start arm for the next trial. If the

fish turned to their correct side in 8 out of 10 of these trials, this training block was considered 'successful'. If the fish had 3 consecutive 'successful' training blocks, then the fish was considered to have reached training criterion. The probability of at least 24/30 trials correct occurring by chance is <0.0001 and this is consistent with other studies using this method as a criterion e.g. [6]. If the fish swam to the unrewarded arm, the exit from that arm would be blocked using a removable piece of Perspex and the fish would receive a two-minute "time out" (no reward given), to mimic the amount of time the fish spent feeding before being moved back to the start arm for the next trial.

The water in the tank was disturbed between each trial to help minimise the risk of the fish using olfactory cues to navigate. The tank would also be fully filtered for a minimum of 20 min between each individual training block. Potential intramaze visual cues were reduced or eliminated where possible, e.g. the heater was removed from the tank during experiments, and the tubing of the external filter was mirrored in the maze layout using additional pieces of tubing. Outside the maze, there was a wall at the end of the left arm, while there was no wall at the end of the right arm. Potential extramaze cues included housing tanks and pieces of paper and plastic on the wall. No attempt was made to control access to these global cues. The location of the experimenter varied across individual trials, moving to different locations relative to the arms of the maze. However, due to, the nature of the setup, namely goal arms being perpendicular to the edge of the bench, the experimenter was constrained to the right side (with respect to the training start box) of the maze.

Shoals
Twenty shoals of five fish were used and received all their training and testing as shoals. Training was completed 17/03/2014 and 17/04/2014 In a similar fashion to that of individual training. Again training would start at approximately 9 am and continue until all shoals had completed a full block. During training trials, half of the shoals in each group would receive a food reward in the left hand arm of the maze, and the other half in the right. In these instances, a trial would be considered complete when all five fish were in either the left or the right arm of the maze at the same time. This was the only difference in the training of individual fish and shoals. As with individuals, if the shoal swam to their allocated rewarded arm, the shoal would immediately receive a bloodworm reward administered with tweezers. If, however, the shoal swam to the other arm, they would receive the same two-minute "time out". As with individual fish, a training block was considered successful if 8 out of 10 trials were correct and the shoal was

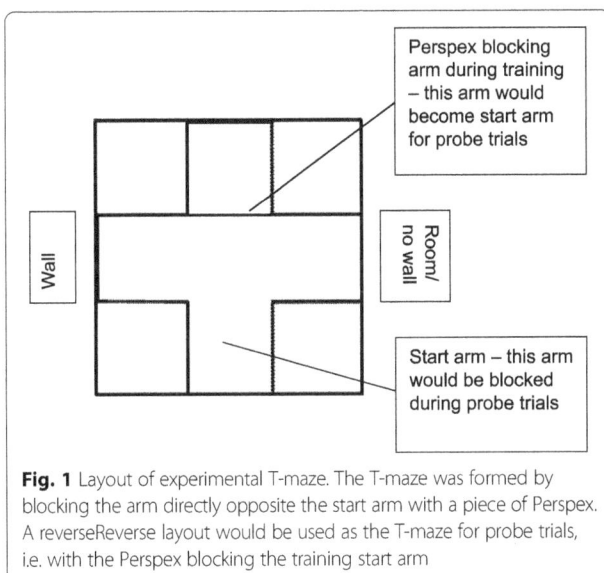

Fig. 1 Layout of experimental T-maze. The T-maze was formed by blocking the arm directly opposite the start arm with a piece of Perspex. A reverseReverse layout would be used as the T-maze for probe trials, i.e. with the Perspex blocking the training start arm

Labels in figure:
- Perspex blocking arm during training – this arm would become start arm for probe trials
- Room/ no wall
- Start arm – this arm would be blocked during probe trials
- Wall

considered to have reached criterion when 3 consecutive training blocks were achieved.

Shoals tested as individuals

To assess whether individuals trained as a shoal displayed a different distribution of navigational choices compared to individuals trained individually, a second group of 20 shoals of five fish were also trained following the same procedure used in the training of the first batch of shoals. These experiments were completed between 19/08/2014 and 26/11/2014. As above, training started at 9 am and continued until all shoals had completed a full block of trials. The only difference was that upon reaching criterion and being tested with probe trials, these shoals were tested as the five individual members rather than as a shoal (see below).

Probe trial- individuals

On reaching criterion, an individual would immediately receive a probe trial. This trial would begin in the opposite arm from training (Fig. 1), with the original start arm now blocked. Again a trial would be considered complete when the tail of the fish had passed into either the left or right hand arm. If the individual moved to the previous rewarded arm, this was recorded as a place strategy. If it swam using the same turning direction as on training trials (i.e. the opposite location to where it was rewarded), it was recorded as a response strategy. No reward was administered during probe trials. After the probe trial, the animal would be returned to their housing tanks and experimentation for that animal would be complete.

Probe trial – Shoals

As for individuals, for the first 20 shoals, on reaching criterion, a shoal would immediately receive a probe trial, in which they would start in the opposite arm from training with the original start arm blocked. Again, a trial would be considered complete on the first occasion that all five fish were either in the left or right hand arm at the same time. If the shoal moved to the previous rewarded arm, this was recorded as a place strategy. If the shoals swam using the same turning direction as on training trials (i.e. the opposite location to where it was rewarded), it was recorded as a response strategy. Again no reward was administered for probe trials. After the probe trial, the animals would be returned to their housing tanks and experimentation for those animals would be complete.

Probe trial – Individuals trained as shoals

A second set of 20 shoals was involved in the experiment to assess the navigational choice used by individual fish following training as members of shoals. Upon reaching training criterion, the shoal was immediately removed from the experimental tank and was placed in a beaker containing water from the experimental tank. A single fish would be placed into the probe start arm and would complete the probe trial alone with its choice recorded before it was removed and returned to its housing tank. As per the previous two groups, no reward was administered during probe trials. The water would be disturbed and allowed to settle before another fish would be placed into the start arm to complete the probe trial, and so on until all five fish had completed a probe trial.

Shoal cohesion

To assess whether all individuals within shoals made the same decision we noted whether a shoal was cohesive or non cohesive. Cohesive was defined as all fish entering the same goal arm on their first choice, whereas non cohesive was defined as the shoal splitting in their first choice of goal arm.

Due to a technical issue, only 18 of 20 probe trial videos were available for this analysis.

Statistical analyses

Data were analysed using SPSS statistical package (v20.0). A generalized linear model with an underlying poisson loglinear distribution was used to assess the effect of experimental group on the number of blocks required to reach training criterion. This performs better than a square root transformation and analysis assuming a Gaussian distribution [37]. Individual binomial tests were used to assess whether there was a significant preference for either a place or a response strategy on navigation choices made in each experimental group, and to assess the prevalence of shoal cohesion on the probe trial.

Results

Acquisition time

Acquisition time was the number of blocks required to reach training criterion by each shoal or individual fish (Fig. 2). There was a significant main effect of experimental group on task acquisition time: $Wald\ \chi^2$ (df = 2) = 54.15; $P < .001$. Posthoc analyses showed that individual zebrafish took significantly more blocks to learn the training task than either set of shoals ($P < .001$). There was no significant difference between the two sets of shoals on acquisition time ($P = .290$). (See Fig. 3).

Navigational strategy

Figure 4 shows the percentage number of times in each experimental group that the fish adopted a place or response strategy during the probe trial. Individual

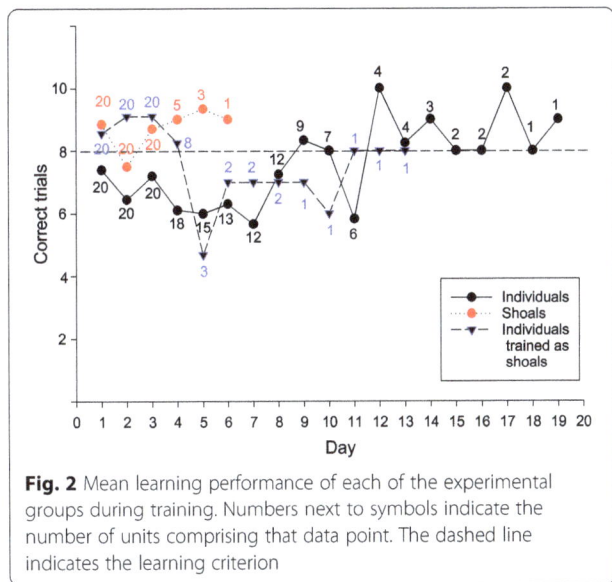

Fig. 2 Mean learning performance of each of the experimental groups during training. Numbers next to symbols indicate the number of units comprising that data point. The dashed line indicates the learning criterion

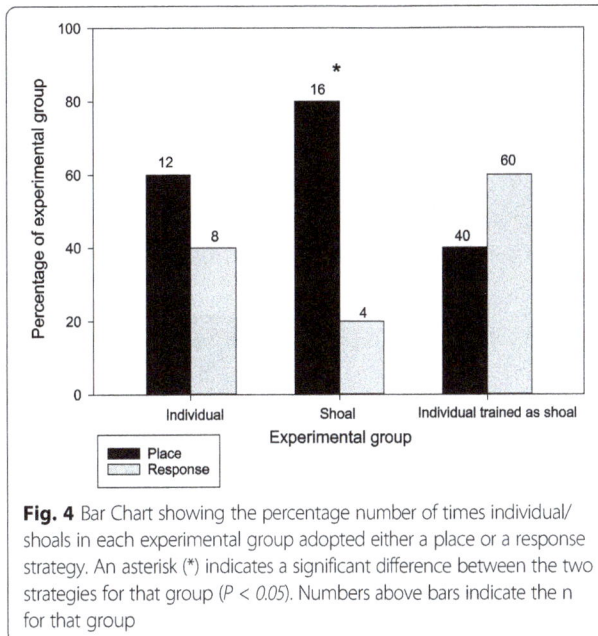

Fig. 4 Bar Chart showing the percentage number of times individual/shoals in each experimental group adopted either a place or a response strategy. An asterisk (*) indicates a significant difference between the two strategies for that group (P < 0.05). Numbers above bars indicate the n for that group

binomial tests showed that the experimental group that was trained and tested on the probe trial in shoals had a significant preference for choosing a place strategy (binomial test: $N = 20$, $P = .012$) whereas the group of fish who completed training and probe trials as individuals, and the group that completed training trials as shoals but the probe trial as individuals showed no significant preference for either a place or response strategy (binomial test: $N = 20$, $P = .507$ and $N = 100$, $P = .271$ respectively).

There was no significant difference in the time taken to reach a decision on the probe trial for each group. As the variance differed between groups (Levines test: $W_{2,137} = 3.57$, $p = 0.031$), the Welch test was used (ANOVA, Welch test, $F_{2,39.56} = 0.94$, $p = 0.399$).

Shoal cohesion
Individuals within shoals were significantly more likely to choose the same side as all shoal mates, i.e. remain cohesive than to choose different sides i.e. be non cohesive during their first choice of side on the probe trial; (Fig. 5, Binomial test: $N = 18$, $P = .008$). There was no relationship between cohesion and strategy choice (Fig. 6), with a place strategy dominating in both cases.

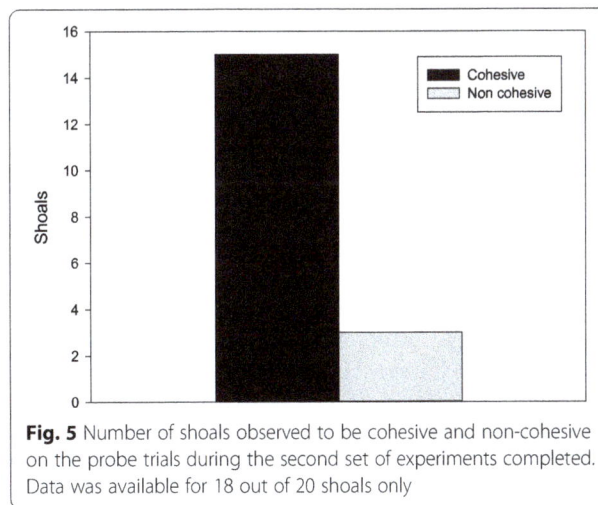

Fig. 3 Boxplot showing task acquisition time for each experimental group where "*Individual*" represents the group of 20 individual zebrafish that formed part of a previous study. "*Shoal*" was the group that were trained and received probe trials as shoals, and "*Individual Probe/Shoal Training*" indicates the group that received all training trails as shoals but completed probe trials as individuals. *Rectangular boxes* display 25th & 75th quartiles and the median. In both the "Shoal" and the "Individual Probe/Shoal Training" groups the median was equal to the minimum value of 3. Whiskers represent the 10th and 90th percentile of the data.Outliers outside this range are marked with *circles*

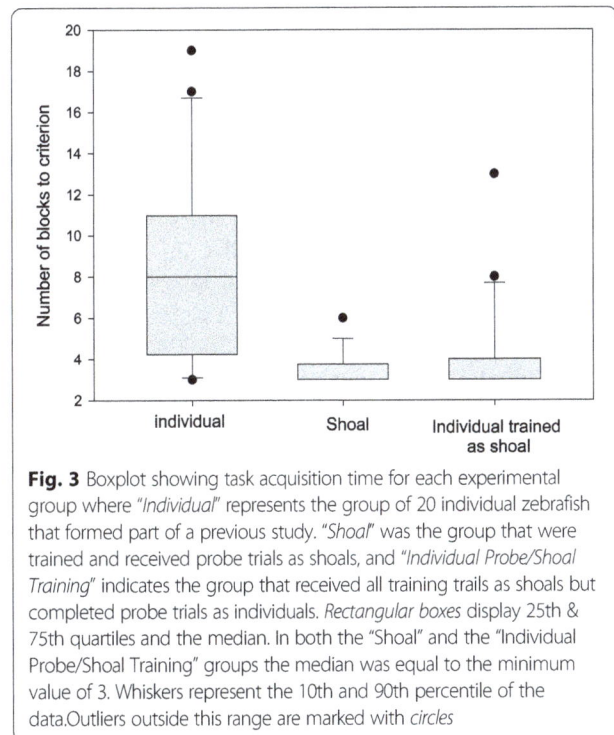

Fig. 5 Number of shoals observed to be cohesive and non-cohesive on the probe trials during the second set of experiments completed. Data was available for 18 out of 20 shoals only

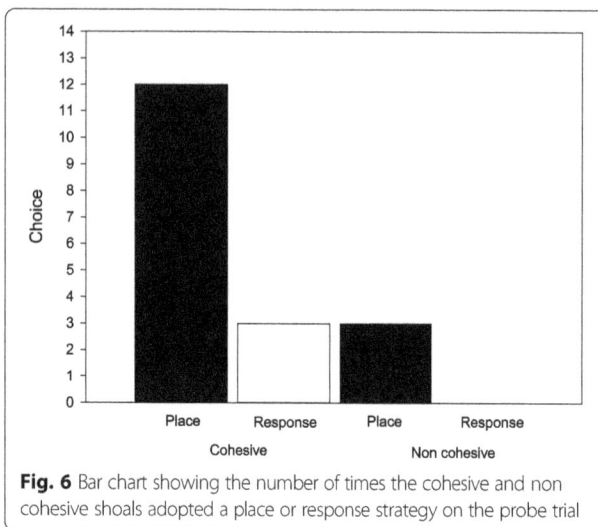

Fig. 6 Bar chart showing the number of times the cohesive and non cohesive shoals adopted a place or response strategy on the probe trial

Discussion

This study investigated the effect of shoaling on the adoption of a navigation strategy in zebrafish (*Danio rerio*). Results showed that fish who completed the task in shoals took a shorter period of time to learn than individual fish. Unlike individuals, when tested on the crucial probe trial, shoals adopted a more consistent navigational strategy across the population with a significant preference for the allocentric "place" strategy being shown. As a result, shoals were more likely than chance to adopt a navigational strategy that took them to the location of the food reward in training when tested from a novel start point. These findings suggest that whereas individual zebrafish do not show a consistent navigation strategy, to learn the location of a food reward as a shoal, the most flexible navigation strategy is adopted from a novel start point (i.e. the one that takes them to the location of the food reward that was learned in training). Some caution is warranted in the fact that the fish from the individual experiments came from a different commercial supplier than those used in the shoaling experiments. A number of factors such as age, sex, rearing environment and nutritional status prior to housing were unknown. However, the consistency between individual responses on probe trials whether trained as an individual or as a shoal would suggest that the responses seen here here are robust.

Many experiments, including this study, use a binary choice which places navigational decisions in conflict with group cohesion and thus may confound our understanding of the factors involved in collective decision making [38]. However, in the case of our experiment there was no a priori reason to expect that individuals that had learned a place strategy would be more likely to influence the group decision than those that had learned a response strategy. How this mechanism occurs

requires further investigation. Place learning is a common phenomenon across a number of vertebrate taxonomic groups and appears to be the preferred strategy [39], but the preference for a place strategy is also influenced by a number of factors including age and quantity of training [14, 18]. One possibility is that if there are, on average, slightly more "place" learners than "response" learners within each group, then the majority may win, with the response learners sacrificing individual learning to maintain cohesion within the group. This is suggested by the fact that in this current experiment, more shoals than predicted by chance stayed cohesive, i.e. all chose the same goal arm as their first choice. However, further experimentation is possibly required, in which the number of place learners and response learners is controlled for, to explore this in more depth. Another possibility is that more dominant individuals use a place strategy, and thus those leading the shoal are more likely to choose place. Given that age is a factor in place learing, this remains an alternative explanation that also needs further investigation.

Zebrafish trained as shoals did not show a significant preference for either a place or a response strategy when tested from a novel start point as individuals. Whether this is because of the passenger effect, where following others overshadowed learning in some individuals, or because individuals within the group learn and maintain individual strategies in the training task, cannot be disentangled in the current study. A recent study on homing pigeons navigating as a flock has, however, suggested that leader-follower relationships may not only occur naturally, but may be inevitable in moving groups [23]. The study also suggests that those individual animals that take on the leadership role show more consistent and effective spatial learning when travelling alone, suggesting that, in some animals, individual learning may have an impact on the overall movement of the group. In spite of this, the results found in the study presented here do suggest that the more flexible information that determines the strategy of the shoal is not consistently transferred to individual zebrafish. This is perhaps not suprising as the training task could be solved equally successfully by either a place or a response strategy and so there appears no conceivable mechanism by which those using a place strategy could demonstrate this to others in the group.

As a shoaling species zebrafish are more likely to move around their environment in a group [34–36] and this has been argued to be beneficial for navigation (e.g. through the "many wrongs principle") [40]. The results presented here confirm that living socially has an impact on the cognitive performance of a group of fish. First, it took individual fish significantly more time to reach training criterion, suggesting that moving in isolation has an effect on the learning abilities of this species. It is possible that having the animals from this social species complete the task alone could have caused stress. No visible signs of stress were observed, however, and all fish ate the food reward administered on correct trials – with avoidance of food being a common sign of stress in fish [41].

Another issue to take into consideration is the wall at the end of one arm of the maze. This wall could have provided a significant salient cue meaning that the "place" learning in this experiment may have been due to beaconing rather than the use of allocentric processes. If this was the case, it might have been expected that individuals would also show a significant preference for learning location, however neither individuals trained alone, nor those trained as shoals but tested alone, showed a significant preference for either strategy. This suggests that completing trials alone did affect spatial memory. A similar argument could be made for the restriction on which side of the maze the experimenter had to stand. Again, further investigation is required to fully assess what cues could have been used by the animals in this experiment. Previous studies have tried to do this by moving or altering the position of particular landmarks, to investigate which cues animals adhere to most when moving through a maze, or by totally eliminating landmarks [42–47]. Furthermore, the telencephalon has been identified as the area of the brain in fish responsible for encoding allocentric spatial information [48]. The same experimental design could be used again with ablated fish to assess whether such subjects would then show a preference for the response strategy. This was, however, beyond the scope of the current experiment.

Conclusions

This study demonstrates that moving in groups may cause individuals with different navigation strategies to converge on using just one strategy. In this case, it was the more flexible allocentric strategy that allows location of a goal from a novel start point that was used more often by shoals of fish. Whether this was due to do a majority rule, or a consequence of leader follower relationships through the choice of dominant individuals in the group, remains to be seen. Many species of fish show that shoal membership is fluid and is subject to fusion-fission on a regular basis [49]. Some fish species known to shoal have been shown to spend more than half of their time moving in isolation from conspecifics [49]. Because individual zebrafish do not appear to learn and transfer all relevant knowledge gained as a shoal member when navigating alone, this may indicate that individuals are at much greater risk alone than when in shoals as spatial memory is crucial for foraging and also for avoiding predators [50, 51].

Acknowledgements
We thank Kyriacos Kareklas, Kelly McCullogh and Gill Riddell for help with husbandry.

Funding
CLM and the project were funded through a Northern Ireland Department of Employment and Learning PhD scholarship. The funding body played no role in the design of the study, collection, analysis, and interpretation of data or in writing the manuscript.

Authors' contributions
CLM, CMC and RAH designed the experiments. CLM carried out the experiments. CLM and RH analysed the data. CLM, CMC and RAH wrote the paper. All authors read and approved the final manuscript.

Competing interests
Richard Holland is an Associate Editor of BMC Zoology.

Author details
[1]School of Biological Sciences, Queen's University Belfast, Medical Biology Centre, 97 Lisburn Road, Belfast BT9 7BL, Northern Ireland. [2]School of Psychology, Queen's University Belfast, University Road, Belfast BT7 1NN, Northern Ireland. [3]Current address, School of Biological Sciences, Bangor University, Deiniol Road, Bangor LL57 2UW, UK.

References
1. White GE, Brown C. A comparison of spatial learning and memory capabilities in intertidal gobies. Behav Ecol Sociobiol. 2014;68(9):1393–401.
2. Schluessel V, Bleckmann H. Spatial learning and memory retention in the grey bamboo shark (*Chiloscyllium griseum*). Zool. 2012;115(6):346–53.
3. Wolbers T, Hegarty M. What determines our navigational abilities? Trends Cogn Sci. 2010;14(3):138–46.
4. Iglói K, Zaoui M, Berthoz A, Rondi-Reig L. Sequential egocentric strategy is acquired as early as allocentric strategy: parallel acquisition of these two navigation strategies. Hippocampus. 2009;19(12):1199–211.
5. Burgess N. Spatial cognition and the brain. Ann N Y Acad Sci. 2008;1124:77–97.
6. Salas C, Rodriguez F, Vargas JP, Duran E, Torres B. Spatial learning and memory deficits after Telencephalic ablation in goldfish trained in place and turn maze procedures. Behav Neurosci. 1996;110(5):965–80.
7. van Gerven DJH, Schneider AN, Wuitchik DM, Skelton RW. Direct measurement of spontaneous strategy selection in a virtual Morris water maze shows females choose an allocentric strategy at least as often as males do. Behav Neurosci. 2012;126(3):465–78.
8. O'Keefe J, Nadel L. The hippocampus as a cognitive map. Oxford: Clarendon Press; 1978.
9. Pearce JM. Evaluation and development of a connectionist theory of configural learning. Anim Learn Behav. 2002;30:73–95.
10. Miller NY, Shettleworth SJ. Learning about environmental geometry: an associative model. J Exp Psychol Anim Behav Proc. 2007;33:191–212.
11. Farina FR, Burke T, Coyle D, Jeter K, McGee M, O'Connell J, Taheny D, Commins S. Learning efficiency: the influence of cue salience during spatial navigation. Behav Proc. 2015;116:17–27.
12. Rodrigo T, Gimeno E, Ayguasanosa M, Chamizo VD. Navigation with two landmarks in rats (Rattus Norvegicus): the role of landmark salience. J Comp Psychol. 2014;128:378–86.

13. Sánchez-Moreno J, Rodrigo T, Chamizo VD, Mackintosh NJ. Overshadowing in the spatial domain. Anim Lear Behav. 1999;27:391–8.

14. Packard MG, McGaugh JL. Inactivation of hippocampus or caudate nucleus with lidocaine differentially affects expression of place and response learning. Neurobiol Learn Mem. 1996;65(1):65–72.

15. Bridgeman B, Peery S, Anand S. Interaction of cognitive and sensorimotor maps of visual space. Percept Psychophys. 1997;59(3):456–69.

16. McAroe CL, Craig CM, Holland RA. Place versus response learning in fish: a comparison between species. Anim Cogn. 2016;19(1):153–61.

17. Hamilton DA, Johnson TE, Redhead ES, Verney SP. Control of rodent and human spatial navigation by room and apparatus cues. Behav Process. 2009;81(2):154–69.

18. Barnes CA. Memory deficits associated with senescence: a neurophysiological and behavioral study in the rat. J Comp Physiol Psychol. 1979 Feb;93(1):74–104.

19. Nesterova AP, Flack A, van Loon EE, Marescot Y, Bonadonna F, Biro D. Resolution of navigational conflict in king penguin chicks. Anim Behav. 2014;93:221–8.

20. Miller N, Garnier S, Hartnett AT, Couzin ID. Both information and social cohesion determine collective decisions in animal groups. Proc Natl Acad Sci U S A. 2013;110(13):5263–8.

21. Bode NWF, Franks DW, Wood AJ, Piercy JJB, Croft DP, Codling EA. Distinguishing Social from Nonsocial Navigation in Moving Animal Groups. Am Nat. 2012;179(5):621–32.

22. Biro D, Sumpter DJT, Meade J, Guilford T. From Compromise to Leadership in Pigeon Homing. Curr Biol. 2006;16(21):2123–8.

23. Pettit B, Zsuzsa A, Vicsek T, Biro D. Speed determines leadership and leadership determines learning during pigeon flocking. Curr Biol. 2015;25:1–6.

24. Flack A, Pettit B, Freeman R, Guilford T, Biro D. What are leaders made of? The role of individual experience in determining leader–follower relations in homing pigeons. Anim Behav. 2012;83(3):703–9.

25. Freeman R, Mann R, Guilford T, Biro D. Group decisions and individual differences: route fidelity predicts flight leadership in homing pigeons (Columba livia). Biol Lett. 2011;7(1):63–6.

26. Couzin ID, Krause J, Franks NR, Levin S a. Effective leadership and decision-making in animal groups on the move. Nature. 2005 Feb 3;433(7025):513–6.

27. Laland KN, Williams K. Shoaling generates social learning of foraging information in guppies. Anim Behav. 1997;53:1161–9.

28. Burt de Perera T, Guilford T. The social transmission of spatial information in homing pigeons. Anim Behav. 1999;57:715–9.

29. Pettit B, Flack A, Freeman R, Guilford T, Biro D. Not just passengers: pigeons, Columba livia, can learn homing routes while flying with a more experienced conspecific. Proc R Soc B Biol Sci. 2013;280(1750):20122160

30. Biro D, Sasaki T, Portugal S. Bringing a Time-Depth Perspective to Collective Animal Behaviour. Trends Ecol Evol. 2016;31(7):550–62.

31. Borski RJ, Hodson RG. Fish Research and the Institutional Animal Care and Use Committee. ILAR J. 2003;44(4):286–94.

32. Brown C. Fish intelligence, sentience and ethics. Anim Cogn. 2014;18:1–17.

33. Brown C, Laland KN. Social learning in Fishes : a review. Fish Fish. 2003;4: 280–88.

34. Spence R, Gerlach G, Lawrence C, Smith C. The behaviour and ecology of the zebrafish, Danio rerio. Biol Rev Camb Philos Soc. 2008;83(1):13–34.

35. Miller N, Gerlai R. Quantification of shoaling behaviour in zebrafish (Danio rerio). Behav Brain Res. 2007;184(2):157–66.

36. Wright D, Rimmer LB, Pritchard VL, Krause J, Butlin RK. Inter and intra-population variation in shoaling and boldness in the zebrafish (Danio rerio). Naturwissenschaften. 2003;90(8):374–7.

37. O'Hara RB, Kotze DJ. Do not log-transform count data. Methods Ecol Evol. 2010;1(2):118–22.

38. Miller N, Garnier S, Hartnett AT, Couzin ID. Both information and social cohesion determine collective decisions in animal groups. Proc Natl Acad Sci. 2013;110(13):5263–8.

39. Salas C, Broglio C, Rodriguez F. Evolution of forebrain and spatial cognition in vertebrates: conservation across diversity. Brain, Behav Evol. 2003;62(2):72–82.

40. Simons AM. Many wrongs: the advantage of group navigation. Trends Ecol Evol. 2004 Sep;19(9):453–5.

41. Carr JA. Stress, neuropeptides, and feeding behavior: a comparative perspective. Integr Comp Biol. 2002;42(3):582–90.

42. Durán E, Ocaña FM, Martín-Monzón I, Rodríguez F, Salas C. Cerebellum and spatial cognition in goldfish. Behav Brain Res. 2014;259:1–8.

43. Saito K, Watanabe S. Experimental analysis of spatial learning in goldfish. Psychol Rec. 2005;55:647–62.

44. Tommasi L, Gagliardo A, Andrew RJ, Vallortigara G. Separate processing mechanisms for encoding of geometric and landmark information in the avian hippocampus. Eur J Neurosci. 2003;17(8):1695–702.

45. Kamil AC, Jones JE. The seed storing corvid Clark's nutcracker learns geometric relationships among landmarks. Nature. 1997;390:276–9.

46. Collett TS, Cartwright BA, Smith BA. Landmark learning and visuo-spatial memories in gerbils. Journal of Comparative Physiology A. 1986;158:835–51.

47. Diviney M, Fey D, Commins S. Hippocampal contribution to vector model hypothesis during cue-dependent navigation. Learn Mem. 2013;20(7):367–78.

48. Rodriguez F, Duran E, Vargas J, Torres B, Salas C. Performance of goldfish trained in allocentric and egocentric maze procedures suggests the presence of a cognitive mapping system in fishes. Anim Learn Behav. 1994; 22(4):409–20.

49. Croft DP, Arrowsmith BJ, Bielby J, Skinner K, White E, Couzin ID, et al. Mechanisms underlying shoal composition in the Trinidadian guppy, Poecilia reticulata. 2009;3:429–38.

50. Griffiths NW, Magurran AE. Familiarity in schooling fish: how long does it take to acquire? Anim Behav. 1997;(1994):945–9.

51. Day RL, MacDonald T, Brown C, Laland KN, Reader SM. Interactions between shoal size and conformity in guppy social foraging. Anim Behav. 2001;62(5): 917–25.

52. Guidelines for the treatment of animals in behavioural research and teaching. Anim Behav. 2012;83(1):301–9.

9

Whole mitochondrial genomes provide increased resolution and indicate paraphyly in deer mice

Kevin A. M. Sullivan[1], Roy N. Platt II[1], Robert D. Bradley[1,2] and David A. Ray[1*]

Abstract

Background: Recent phylogenies of deer mice, genus *Peromyscus*, have relied heavily on mitochondrial markers. These markers provided resolution at and below the level of species groups, but relationships among species groups and *Peromyscus* affiliated genera have received little support. Here, we present the mitochondrial genomes of 14 rodents and infer the phylogeny of *Peromyscus* and related taxa.

Results: Our analyses support results from previous molecular phylogenies, but also yield support for several previously unsupported nodes throughout the *Peromyscus* tree. Our results also confirm several instances of paraphyly within the clade and suggest additional taxonomic work will be required to clarify some relationships.

Conclusions: Our findings greatly enhance our understanding of the evolution of *Peromyscus* providing support for previously unsupported relationships. However, the results also highlight the need to address paraphyly that may exist in this clade.

Keywords: Mitochondrial genome, Paraphyletic, *Peromyscus*, Phylogenetics

Background

Mitochondrial loci have been the most popular phylogenetic markers in animals for over three decades. Their ease of amplification [1, 2], uniparental inheritance, lack of recombination in mammals [2–4], differential selection among genes [2], and general synteny [5] combine to make mitochondrial phylogenetic markers an excellent choice for many study systems. Recent increases in DNA sequencing capabilities have provided an opportunity to sequence full mitochondrial genomes (mitogenomes) rather than partial, one, or a few mitochondrial genes. As a result, previous phylogenies that relied on a portion of the mitochondrion are now undergoing re-analysis in an effort to expand support for or re-affirm previous conclusions [6].

The genus *Peromyscus*, commonly referred to as North American deer mice, encompasses more than 70 species that diverged within the last 6–10 million years [7]. Species including *P. maniculatus* [8] and *P. leucopus* [9] are among the most common mammals in North

America and have been studied extensively for over 100 years [10]. Despite extensive study, new species [11] and subspecies [12] are being described on a regular basis. The large number of species, both described and undescribed, as well as substantial cryptic variation, has yielded numerous distinct phylogenetic hypotheses.

Previous classifications of *Peromyscus* recognized seven morphologically distinct subgenera (*Habromys, Haplomylomys, Isthmomys, Megadontomys, Osgoodomys, Peromyscus,* and *Podomys*) [13, 14]. Many of those subgenera (*Habromys, Isthmomys, Megadontomys, Osgoodomys,* and *Podomys*) have at times been elevated to generic status [15]. Some recent molecular phylogenies, however, show paraphyly [7]. *Isthmomys* is sister to the genus *Peromyscus*, representing a distinct genus, as is currently accepted, but *Habromys, Megadontomys, Osgoodomys, Neotomodon,* and *Podomys*, distinct genera, were found within *Peromyscus*, rendering the genus paraphyletic [7, 16, 17], though *Neotomodon alstoni* [18] previously had been *Peromyscus alstoni* [14, 19].

Recent attempts to resolve the history of this clade by Bradley et al. [16], Miller and Engstrom [17] and Platt et al. [7] were based on various combinations of nuclear

* Correspondence: david.4.ray@gmail.com
[1]Department of Biological Sciences, Texas Tech University, Lubbock, TX 79409, USA
Full list of author information is available at the end of the article

and mitochondrial genes. All three studies arrived at similar topologies, but varied in their levels of support at some nodes. Species-level relationships were well-supported but the relationships among some genera and species groups were not. Given that approaches that use cytochrome b (*cytb*) alone were somewhat successful in resolving *Peromycus*, we reasoned that the use of whole mitogenomes, a more data-rich approach, may be an avenue to more robust results at the generic and species group level, thus providing additional clarity. Here, we analyze whole mitogenomes from *Habromys, Isthmomys, Neotomodon, Peromyscus, Podomys*, and three outgroups, *Sigmodon, Neotoma*, and *Reithrodontomys*, as identified from previous molecular phylogenies [7, 16, 17].

Methods
Sampling, DNA preparation and sequencing
Taxa selected for this study were identified based on the findings of Platt et al. [7]. Our objective was to reduce taxon sampling for some groups to isolate specific clades. To do so, we chose representatives from selected subclades to serve as proxies. Specifically, one individual here is a proxy for a larger clade (generally represented as a species group). Members of *Neotoma, Reithrodontomys*, and *Isthmomys*, which have determined to be closely related clades by several morphological and molecular analyses [7, 15, 16] were included. Species identifications, museum catalog numbers and are included in Table 1.

Whole genomic DNA was isolated using a Qiagen DNeasy Blood and Tissue kit (Qiagen, Valencia, CA) via the manufacturer's protocol and fragmented to ~ 400 bp. Sizes were verified on an Agilient 2100 Bioanalyzer.

Illumina compatible sequencing libraries were prepared from the fragmented DNA using the NEB/KAPA library preparation. Each sample was tagged with a unique index and pooled in equal proportions, after which, the pooled libraries were sequenced on single run of an Illumina MiSeq (2 × 250 bp). Raw data are available in GenBank under BioProject ID PRJNA308567.

Mitochondrial genome assembly and annotation
Raw sequence reads were filtered and processed using Trimmomatic version 0.35 [20]. Specifically, we clipped Illumina adapters, disregarded read ends whose Phred scores fell below 20, and utilized a four-base sliding window to trim reads once the average quality fell below 25. We assembled mitogenomes through a custom Bash script (https://github.com/KevinAMSullivan/Mitochondrial_Genomes/tree/MIRA) The script utilized two major programs, MITObim [21] and MIRA [22] to map the filtered reads to a reference genome before assembling the mitogenomes. We selected *Akodon montensis* [23], a Sigmodontine rodent and the most closely related organism with a fully sequenced mitogenome, as the reference (GenBank accession number KF769456).

MITOS [24] was used to annotate the mitochondrial genes for each genome. Putative genes were submitted to BLASTn to confirm sequence length. When gaps were noted, we manually checked for frameshifts but none were observed. Acceptable results were those whose top hits were to a *Peromyscus* or closely related species and that covered nearly 100% of the putative gene. The putative gene was shortened or lengthened to match the length of the BLAST hit if a different size. All

Table 1 Basic read, coverage and size data for each taxon. Museum ID refers to a special identification or catalogue number for speciments at the Natural Science Research Laboratory, Museum of Texas Tech University

Taxon	Museum ID	Reads	Avg. coverage	Genome size
Habromys ixtlani (Goodwin, 1964) [35]	TK 93158	1,912,125	22	16,515
Isthmomys pirrensis (Goldman, 1912) [31]	TK 22583	1,689,780	106	16,628
Neotomadon alstoni (Merriam, 1898) [16]	TK 148849	1,401,025	106	16,776
Neotoma mexicana (Baird, 1855) [30]	TK 93257	2,967,246	102	16,697
Peromyscus attwateri (Allen, 1895) [37]	TK 116711	1,082,677	122	16,679
Peromyscus aztecus (Saussure, 1860) [32]	TK 93385	2,568,999	50	16,692
Peromyscus crinitus (Merriam, 1891) [27]	TK 119629	1,389,297	65	16,703
Peromyscus megalops (Merriam, 1898) [48]	TK 93381	1,444,002	33	16,925
Peromyscus mexicanus (Sassure, 1860) [32]	TK 93144	1,524,641	243	16,860
Peromyscus pectoralis (Osgood, 1904) [36]	TK 148816	1,771,386	238	16,833
Peromyscus polionotus (Wagner, 1984) [33]	TK 24228	1,059,501	30	16,413
Podomys floridanus (Chapman, 1889) [49]	TK 92501	1,738,547	62	16,517
Reithrodontomys mexicanus (Sassure, 1860) [32]	TK 104488	1,576,095	190	16,606
Sigmodon hispidus (Say and Ord, 1825) [26]	160,320	1,055,119	27	16,394
Mean		1,655,746	114	16,664

sequences were deposited in GenBank under the accession numbers KY707299–707312. Genome coverage was estimated using bedtools genomecov [25] in combination with mapping of the processed reads to each MITObim assembly with BWA via a custom script (https://github.com/KevinAMSullivan/Mitochondrial_Genomes/tree/MIRA).

Assembly validation
To test the reliability of mitogenome assemblies, we used two *cytb* genes per taxon from GenBank as controls (Additional file 1: Table S1), and worked under the assumption that our reference assembled *cytb* gene and those from GenBank should form monophyletic clades. We chose full length, non-identical entries from the selected species, aligned them with *cytb* from our assemblies, and used RAxML [26] to estimate the best tree from 1000 maximum likelihood (ML) searches. Support for each node was generated with bipartition frequencies from 10,000 bootstrap replicates. The GTR+GAMMA+I model of nucleotide substitution was used for both ML analyses.

Phylogenetic analysis of whole mitochondrial genomes
We concatenated the mitochondrial protein coding sequences and aligned them with Muscle [27]. That alignment is available as Additional file 2. Bayesian and ML phylogenetic analyses were accomplished using MrBayes [28] and RAxML [26], respectively. For the Bayesian analysis, four independent runs on five chains were implemented for 3,000,000 generations. We evaluated stability of the final tree by continuing to an average standard deviation of split frequency less than 0.01. The data were also partitioned by codon. No substitution model was specified as a prior to let the program search across all possible models to determine the most appropriate model of evolution. We then used the selected model, GTR+GAMMA, for our ML analysis, for which we implemented 1000 bootstrap replicates. In both analyses, the *A. montensis* and *S. hispidus* [29] mitogenomes were specified as outgroups when drawing the trees as Neotominae and Sigmodontinae are sister subfamilies whose relationships have been supported by previous analyses [30, 31]. Our unrooted tree supports this relationship. The implementation scripts for both analyses are available on github (https://github.com/KevinAMSullivan/MIRA-MitoBim/tree/Phylogenies).

Results
We sequenced and assembled whole mitogenomes of 14 rodents in the subfamily Neotominae and one member of Sigmodontinae, *Sigmodon hispidus*. Taxon sampling included seven genera and six of the 13 species groups within *Peromyscus* [16]. Genome assembly data, including accession numbers, read coverage, and mean read totals across all taxa, are listed in Table 1.

Each assembled mitogenome comprises 22 tRNA, 2 rRNA, and 13 protein coding regions, totaling 37 genes. Although most genes are transcribed from the positive (heavy) strand, several genes, mainly tRNAs but also including *nad6*, are transcribed from the negative (light strand). Additionally, the 3′ end of *atp6* overlaps with the 5′ end of *atp8*, a common characteristic in mitogenomes. Noncoding regions typical of mitogenomes are present, including a control region downstream of *cytb*. As these mitogenomes were not closed assemblies, such variation in mitogenome size is thought to be due to the highly variable and often heteroplasmic control region.

Validation of MITObim assembled and MITOS-annotated *cytb* took three forms. First, reconstructed genes clustered with loci amplified and sequenced using more traditional methods (Additional file 3: Figure S1), suggesting that the remaining protein coding sequences were accurately assembled and annotated for each taxon. Second, differences that were present are within acceptable ranges for such rapidly evolving loci. Despite forming monophyletic clades for each species, some of the *cytb* genes differed from those in GenBank. For example, *cytb* from our *P. crinitus* [32] mitogenome exhibited 43 and 44 differences, respectively, when compared to their GenBank counterparts, i.e. ~ 96% identity. Divergences in the range may be expected for such a rapidly evolving locus within a species [33]. Third, we have high confidence in our base calls. For example, average coverage of our *cytb* loci is 142× (data not shown), and the average percentage of those reads indicating any given nucleotide is 99.3%. Given such high support, as well as the fact that *P. crinitus* is the top BLAST hit to the gene (96% identify, 0.0 E value), we have confidence in the veracity of our *cytb* sequences over alternate hypotheses for such sequence dissimilarity. Similar results were observed for *cytb* sequences in all taxa (Additional file 4: Table S2). The single exception was for *Neotoma mexicana*. The best match (100%) for our reconstructed *cytb* sequence was to *N. isthmica*. However, *N. isthmica* is closely related to *N. mexicana* [34], and the second best hit was to *N. mexicana* (99%), confirming its validity as an outgroup [35].

ML and Bayesian phylogenies inferred from mitochondrial protein coding genes recovered identical topologies (Fig. 1). This topology is similar to recent molecular phylogenies. First, *S. hispidus*, *N. mexicana*, *I. pirrensis* [36], and *R. mexicanus* [37] are positioned outside of *Peromyscus*. Unlike previous phylogenies however, high posterior probabilities are found at these nodes. In fact, every node, save that linking *P. crinitus* and *P. polionotus* [8], has high support. Second, our analyses indicated a paraphyletic *Peromyscus*. Finally, *Isthmomys* and *Reithrodontomys* comprise a clade sister to what is currently considered to *Peromyscus* and its

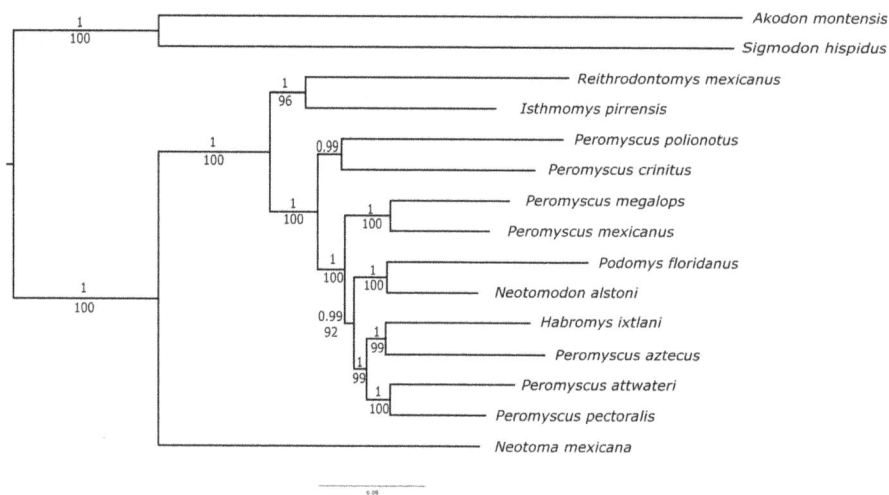

Fig. 1 Phylogeny inferred by the partitioned maximum likelihood analysis. The topology of the Bayesian tree was identical. *A. montensis* was included as a previously sequenced outgroup. Bootstrap values (*above*) and posterior probabilities (*below*) are provided at all internodes

affiliated genera. This pairing is of note, as both genera have shown conflicting relationships depending on the marker used [7, 16, 17, 38].

Our phylogeny using whole mitogenomes largely mirrors that of Platt et al. [7], save the quartet of *H. ixtlani* [39], *P. aztecus* [37], *P. pectoralis* [40], and *P. attwateri* [41]. Platt et al. found *P. attwateri* and *P. aztecus* to form a clade, with *P. pectoralis* as sister to this grouping and *H. ixtlani* being the most closely related species to a clade containing all three. The whole mitogenome tree is in partial agreement with Bradley et al. [16] with regard to these same four species. Our close pairing of *P. attwateri* and *P. pectoralis* matches, but their phylogeny places *P. aztecus* as a sister group to a clade that encompasses *H. ixtlani*, *P. attwateri*, and *P. pectoralis*.

Discussion

Regardless of the differences or similarities, our tree (Fig. 1) provides support for previously unsupported nodes within the *Peromyscus* phylogeny. Previous trees were largely resolved, but lacked significant support at key nodes, mainly those in middle regions of the tree that identify relationships among species groups [7, 16, 17]. Platt et al. [7] provided substantial support (posterior probabilities ≥ 95% and bootstrap values ≥ 85) at nodes grouping the *crinitus*, *eremicus*, and *californicus* species groups; the *mexicanus*, *megalops*, and *melanophrys* species groups; and the *aztecus* and *boylii* species groups. However, support in the Bayesian mitogenome phylogeny has posterior probabilities (> 0.95) at every node and bootstrap support (> 90) in the ML phylogeny at all nodes save one.

Two branches of our inferred phylogeny had not yet been reinforced by any molecular analysis. The first concerns *H. ixtlani*, *P. aztecus*, *P. attwateri*, and *P.*

pectoralis. Although some analyses had suggested a close relationship among these taxa, no node in this clade had received substantial support in any previous molecular phylogeny. Our analysis is the first to provide such support for these relationships.

The second branch of interest suggests a close relationship between *Isthmomys* and *Reithrodontomys*. Although no phylogenetic analysis ever suggested the genus *Reithrodontomys* should be nested within *Peromyscus*, *Isthmomys* had previously been subsumed as a subgenus [13]. The pairing is counterintuitive given their morphological differences. Although both occupy the same geographic region, there are obvious incongruities in size. *Isthmomys pirrensis* can reach well over 100 g (averaging 140 g in one study) and 300 mm [42, 43], whereas rodents in *Reithrodontomys* are much smaller. *R. mexicanus* ranges from 167 to 190 mm in length [19], and a collection of different surveys gave a range from 7.9–9.5 g [42]. Although their pairing has been previously suggested [7, 16, 17], our analysis is the first to provide high support values.

One major weakness of our study is its lack of complete taxon sampling. This makes comprehensive phylogenetic reconstruction uncertain due to missing critical (i.e. representing additional ingroups) taxa. The addition of excluded taxa, especially *Osgoodomys* and *Megadontomys*, should provide a more comprehensive view of phylogenetic relationships within the genus. That being said, the high support values we recover suggest that the relationships at the deeper nodes are valid.

The *Peromyscus* phylogeny has been studied repeatedly and revised often, using data from several sources [42]. This latest attempt, using next-generation sequencing, continues to suggest paraphyly. Indeed, every molecular phylogeny has been unequivocal in suggesting paraphyly.

Given this observation, addressing the status of genera subsumed within *Peromyscus* should be seen as a priority within rodent taxonomy.

Aside from acknowledging paraphyly, future work on *Peromyscus* phylogenetics should focus on further clarifying the true phylogeny via more taxa and markers. The inclusion of additional mitogenomes will elucidate relationships within the genus to provide increased detail. However, despite their advantages, mitochondrial loci are imperfect markers. Mitochondria genomes are susceptible to incomplete lineage sorting (ILS) and introgression [44]. Hybrids, especially asymmetric hybrids, exemplify a similar problem in that they may all have the same mitochondrial DNA and species boundaries in those cases would not be reflected [45–47]. Supplementing the mitochondrial phylogeny with large numbers of nuclear markers such as ultraconserved elements (UCEs) [48, 49] or retrotransposons [50, 51] is an important next step in understanding phylogenetic relationships within *Peromyscus*.

Conclusions

Here, we present analyses of whole mitochondrial genomes, for 14 species, ten of which are *Peromyscus* or close relatives. The data yield phylogenies with significant support at previously unsupported nodes, particularly the pairing of *Isthmomys* and *Reithrodontomys*, but also suggest paraphyly within the genus that could be resolved by elevating monophyletic groups to genera or subsuming currently recognized genera as subgenera. Our analyses provide evidence that additional data will help clarify the evolutionary history of this genus.

Additional files

Additional file 1: Table S1. Title of data: Cytochrome b GenBank information. Taxa names and accession numbers for cytochrome b genes downloaded from GenBank in Additional file 3: Figure S1. (XLS 28 kb)

Additional file 2: Whole mitochondrial exomes. Description: Alignment of the protein coding genes of each assembled mitogenome along with *Akodon montensis*, downloaded from GenBank. (TXT 257 kb)

Additional file 3: Figure S1. Cytochrome b phylogenetic validation. ML tree of the assembled cytochrome b sequences compared with cytochrome b sequences downloaded from GenBank. These results showed if the sequences from our data correlated with known *cytb* sequences. (PNG 60 kb)

Additional file 4: Table S2. v2. Cytochrome b blast validation. Best hits to GenBank entries for cytochrome b genes from assembled mitochondrial genomes. For *N. mexicana*, the best hit was to a closely related species, with the second best hit to *N. mexicana*. (XLS 38 kb)

Abbreviations

Cytb: Cytochrome-b; Mitogenome: Mitochondrial genome; ML: Maximum likelihood

Acknowledgements

The Natural Sciences Research Laboratory at the Museum of Texas Tech University kindly provided tissues via the Genetic Resources Collection. Additional thanks goes to RTL Genomics for sequencing efforts.

Funding

This work was supported by the National Science Foundation, DEB-1355176. Additional support was provided by College of Arts and Sciences at Texas Tech University.

Authors' contributions

DAR conceived the project, aided in mitochondrial genome assembly and assisted in writing. RDB provided essential information regarding *Peromyscus* and related taxa, devised the taxon sampling scheme and assisted writing the manuscript. RNP aided in the phylogenetic analyses. KAMS analyzed the data and led writing of the manuscript. All authors read, edited, and approved the final manuscript.

Competing interests

The authors declare they have no competing interests.

Author details

[1]Department of Biological Sciences, Texas Tech University, Lubbock, TX 79409, USA. [2]Museum of Texas Tech University, Lubbock, TX 79409, USA.

References

1. Falkenberg M, Larsson NG, Gustafsson CM. DNA replication and transcription in mammalian mitochondria. Annu Rev Biochem. 2007;76:679–99.
2. Galtier N, Nabholz B, Glemin S, Hurst GD. Mitochondrial DNA as a marker of molecular diversity: a reappraisal. Mol Ecol. 2009;18(22):4541–50.
3. Birky CW Jr. Uniparental inheritance of mitochondrial and chloroplast genes: mechanisms and evolution. Proc Natl Acad Sci U S A. 1995;92(25):11331–8.
4. Ballard JW, Whitlock MC. The incomplete natural history of mitochondria. Mol Ecol. 2004;13(4):729–44.
5. Boore JL. Animal mitochondrial genomes. Nucleic Acids Res. 1999;27(8):1767–80.
6. Williams ST, Foster PG, Littlewood DTJ. The complete mitochondrial genome of a turbinid vetigastropod from MiSeq Illumina sequencing of genomic DNA and steps towards a resolved gastropod phylogeny. Gene. 2014;533(1):38–47.
7. Platt RN, Amman AM, Keith MS, Thompson CW, Bradley RD. What is *Peromyscus*? Evidence from nuclear and mitochondrial DNA sequences suggests the need for a new classification. J Mammal. 2015;96(4):708–19.
8. Wagner A. Diagnosen neuer Arten Brasiliscer Handfluger. Arch Naturgesch. 1843;9:366.
9. Rafinesque CS. Further discoveries in natural history made during a journey through the western region of the United States. American Monthly Magazine and Critical Review. 1818;3:445–7.
10. Osgood WH. Revision of the mice of the American genus *Peromyscus*, vol. 28 Washington DC: North American Fauna; 1909.
11. Bradley RD, Carroll DS, Haynie ML, Martinez RM, Hamilton MJ, Kilpatrick CW. A new species of *Peromyscus* from western Mexico. J Mammal. 2004;85(6):1184–93.

12. Bradley RD, Schmidly DJ, Amman BR, Platt RN 2nd, Neumann KM, Huynh HM, Muniz-Martinez R, Lopez-Gonzalez C, Ordonez-Garza N. Molecular and morphologic data reveal multiple species in *Peromyscus pectoralis*. J Mammal. 2015;96(2):446–59.

13. Hooper ETMG. Notes on classification of the rodent genus *Peromyscus*. Occasional Papers of the Museum of Zoology, University of Michigan. 1964;635:1–13.

14. Wilson D, Reeder DM. Mammal species of the world, vol. vol. 2. 3rd ed. Baltimore: Johns Hopkins University Press; 2005.

15. Carleton MD. Phylogenetic relationships of neotomine–peromyscine rodents (Muroidea) and a reappraisal of the dichotomy within new world Cricetinae. Miscellaneous Publications of the Museum of Zoology, University of Michigan. 1980;157:1–146.

16. Bradley RD, Durish ND, Rogers DS, Miller JR, Engstrom MD, Kilpatrick CW. Toward a molecular phylogeny for *Peromyscus*: evidence from mitochondrial cytochrome-b sequences. J Mammal. 2007;88(5):1146–59.

17. Miller JR, Engstrom MD. The relationships of major lineages within peromyscine rodents: a molecular phylogenetic hypothesis and systematic reappraisal. J Mammal. 2008;89(5):1279–95.

18. Merriam CH. A new genus (*Neotomodon*) and three new species of murine rodents from the mountains of the southern Mexico. P Biol Soc Wash. 1898;12:127–9.

19. Williams SL, Ramirez-Palido J, Baker RJ. Peromyscus alstoni. Mammalian Species. 1985;242:1–4.

20. Bolger AM, Lohse, M., & Usadel, B. Trimmomatic: a flexible trimmer for Illumina sequence data. Bioinformatics: btu170; 2014.

21. Hahn C, Bachmann L, Chevreux B. Reconstructing mitochondrial genomes directly from genomic next-generation sequencing reads-a baiting and iterative mapping approach. Nucleic Acids Res. 2013;41(13):1–9.

22. Chevreux B, Wetter T, Suhai S. Genome sequence assembly using trace signals and additional sequence information. Hannover: Computer Science and Biology: Proceedings of the German Conference on Bioinformatics; 1999. p. 45–6.

23. Thomas O. New forms of *Akodon* and *Phyllotis*, and a new genus for "*Akodon*" teguina. Ann Mag Nat Hist. 1913;11:404–9.

24. Bernt M, Donath A, Juhling F, Externbrink F, Florentz C, Fritzsch G, Putz J, Middendorf M, Stadler PF. MITOS: improved de novo metazoan mitochondrial genome annotation. Mol Phylogenet Evol. 2013;69(2):313–9.

25. A.R Q. BEDTools: the Swiss-Army tool for genome feature analysis. Curr Protoc Bioinformatics. 2014;47(11):1–11.

26. Stamatakis A. RAxML version 8: a tool for phylogenetic analysis and post-analysis of large phylogenies. Bioinformatics. 2014;30(9):1312–3.

27. Edgar RC. MUSCLE: a multiple sequence alignment method with reduced time and space complexity. BMC Bioinformatics. 2004;5:113.

28. Huelsenbeck JP, Ronquist F. MRBAYES: Bayesian inference of phylogenetic trees. Bioinformatics. 2001;17(8):754–5.

29. Say T, Ord G. Description of a new species of Mammalia, whereon a genus is proposed to be founded. J Acad Natl Sci Phila. 1825;4:352–6.

30. D'Elia G. Phylogenetics of sigmodontine (Rodentia, Muroidea, Cricetidae), with special reference to the akodont group, and with additional comments on historical biogeography. Cladistics. 2003;19(4):307–23.

31. Teta P, Canon C, Patterson BD, Pardinas UFJ. Phylogeny of the tribe Abrotrichini (Cricetidae, Sigmodontinae): integrating morphological and molecular evidence into a new classification. Cladistics. 2017;33(2):153–82.

32. Merriam CH. Mammals of Idaho. Washington DC: North American fauna; 1891. p. 31–88.

33. Baker RJ, Bradley RD. Speciation in mammals and the genetic species concept. J Mammal. 2006;87(4):643–62.

34. Baird SF. Characteristics of some new species of north American Mammalia. Proc Acad Natl Sci Phila. 1855;7:333–7.

35. Ordonez-Garza N, Thompson CW, Unkefer MK, Edwards CW, Owen JG, Bradley RD. Systematics of the *Neotoma mexicana* species group (Mammalia: Rodentia: Cricetidae) in Mesoamerica: new molecular evidence on the status and relationships of N. Ferruginea tomes, 1862. P Biol Soc Wash. 2014;127(3):518–32.

36. Goldman E. New mammals from eastern Panama. Smithsonian Miscellaneous Collections. 1912;60(2):1–18.

37. Saussure MH. Note Sur Quelques Mamiferes de Mexique. Revue et Magasin de Zoologie. 1860;2(13):103–9.

38. FB Stangl RB. Evolutionary relationships in *Peromyscus*: congruence in chromosomal, genic, and classical data sets. J Mammal. 1984;65(4):643–54.

39. Goodwin GG. A new species and a new subspecies of *Peromyscus* from Oaxaca. Mexico American Museum Novitates. 1964;2183:1–8.

40. Osgood WH. Thirty new mice of the genus *Peromyscus* from Mexico and Guatemala. P Biol Soc Wash. 1904;17:55–77.

41. Allen JA. Descriptions of new north American mammals. Bull Am Mus Nat Hist. 1895;7:327–40.

42. Carleton MD, G. L. Kirkland, and J. N. Layne. Advances in the study of *Peromyscus* (Rodentia); 1989.

43. Hill RW. Metabolism, thermal conductance, and body-temperature in one of largest species of *Peromyscus, Peromyscus pirrensis*. J Therm Biol. 1976;1(2):109–12.

44. Tang QY, Liu SQ, Yu D, Liu HZ, Danley PD. Mitochondrial capture and incomplete lineage sorting in the diversification of balitorine loaches (Cypriniformes, Balitoridae) revealed by mitochondrial and nuclear genes. Zool Scr. 2012;41(3):233–47.

45. Sarver BA, Demboski JR, Good JM, Forshee N, Hunter SS, Sullivan J. Comparative phylogenomic assessment of mitochondrial introgression among several species of chipmunks (Tamias). Genome Biol Evol. 2017;9(1):7–19.

46. Good JM, Hird S, Reid N, Demboski JR, Steppan SJ, Martin-Nims TR, Sullivan J. Ancient hybridization and mitochondrial capture between two species of chipmunks. Mol Ecol. 2008;17(5):1313–27.

47. Li G, Davis BW, Eizirik E, Murphy WJ. Phylogenomic evidence for ancient hybridization in the genomes of living cats (Felidae). Genome Res. 2016;26(1):1–11.

48. Faircloth BC, McCormack JE, Crawford NG, Harvey MG, Brumfield RT, Glenn TC. Ultraconserved elements anchor thousands of genetic markers spanning multiple evolutionary timescales. Syst Biol. 2012;61(5):717–26.

49. Crawford NG, Faircloth BC, McCormack JE, Brumfield RT, Winker K, Glenn TC. More than 1000 ultraconserved elements provide evidence that turtles are the sister group of archosaurs. Biol Lett. 2012;8(5):783–6.

50. Ray DA, Xing J, Salem AH, Batzer MA. SINEs of a nearly perfect character. Syst Biol. 2006;55(6):928–35.

51. Platt RN 2nd, Zhang Y, Witherspoon DJ, Xing J, Suh A, Keith MS, Jorde LB, Stevens RD, Ray DA. Targeted capture of phylogenetically informative Ves SINE insertions in genus *Myotis*. Genome Biol Evol. 2015;7(6):1664–75.

10

White-nose syndrome fungus, *Pseudogymnoascus destructans,* on bats captured emerging from caves during winter in the southeastern United States

Riley F. Bernard[1,2*], Emma V. Willcox[3†], Katy L. Parise[4,5†], Jeffrey T. Foster[4,5†] and Gary F. McCracken[1†]

Abstract

Background: Emerging infectious diseases in wildlife are an increasing threat to global biodiversity. White-nose syndrome (WNS) in bats is one of the most recently emerged infectious diseases in North America, causing massive declines in eastern bat populations. In the Northeast, winter behavior of bats during the hibernation period, such as flying during the day or in cold weather, has been attributed to WNS. However, winter emergence of bats in the southeastern United States, where winters are warmer, has received little attention. The goals of this study were to determine if winter emergence results from infection by *Pseudogymnoascus destructans,* the causative pathogen of WNS, and to investigate how pathogen load and prevalence vary by species, site, and over time.

Results: We collected epidermal swab samples from 871 active bats of 10 species captured outside of hibernacula in Tennessee during winters 2012–2013 and 2013–2014. Deoxyribonucleic acid (DNA) from *P. destructans* was not detected on 54% of these bats, suggesting that winter emergence occurs regardless of fungal infection. Among infected bats, *Perimyotis subflavus* (tri-colored bats) had the highest mean fungal load, whereas *Myotis lucifugus* (little brown bats) had the highest infection prevalence of all individuals captured. Less than 18% ($n = 59$ of 345 individuals sampled) of all *M. grisescens* (gray bats) captured had detectible *P. destructans* DNA on their forearms and muzzle. Hibernacula with large *M. grisescens* populations had lower fungal loads than sites used by other species; however, mean load per species did not significantly differ between *M. grisescens* and non-*M. grisescens* sites.

Conclusions: We found that pathogen load and prevalence were higher on bats captured during winter 2012–2013 than in the following winter, indicating that fungal loads on bats did not increase the longer a site was presumably contaminated. Repeated low-dose exposure, mild temperatures, and availability of prey during winter in the Southeast may provide a regional refuge for surviving bat populations.

Keywords: Bats, White-nose syndrome, Hibernation, Tennessee, Body condition

* Correspondence: rbernar3@vols.utk.edu
†Equal contributors
[1]Department of Ecology and Evolutionary Biology, University of Tennessee, Knoxville, TN, USA
[2]Department of Ecosystem Science and Management, Pennsylvania State University, University Park, Pennsylvania, USA
Full list of author information is available at the end of the article

Background

Emerging infectious diseases in wildlife pose an increasing threat to global biodiversity and conservation [1, 2]. A significant proportion of these diseases are the result of "pathogen pollution": the introduction by humans or livestock of novel pathogens into naïve wildlife populations [2, 3]. Prominent examples of pathogen pollution causing mass mortality are African rinderpest and amphibian chytridiomycosis. In the 1880's rinderpest killed 90% of Kenya's buffalo population, resulting in downstream effects on predator populations and ecosystem health [2]. Chytridiomycosis has infected over 50% of all amphibian species and can kill 80% of a population within 4–5 months of its introduction [4]. Such emerging infectious diseases are devastating to native species, with deleterious effects that pervade ecosystems [2, 5].

White-nose syndrome (WNS) is a recently emerged infectious disease that has rapidly spread through eastern populations of cave hibernating bats in North America. It is caused by the psychrophilic fungus *Pseudogymnoascus destructans*, and was first documented in North America in February 2006 at a cave in upstate New York [6, 7]. This invasive pathogen, which is hypothesized to have originated in Eurasia [8–10], has since spread to more than half of the United States (U.S.) and five Canadian provinces and has killed over 5.7 million bats [11]. Currently, at least six bat species are experiencing detectable population losses due to WNS, wherein once abundant species are now threatened with regional extinction [11–14]. Population declines and the loss of bat species due to WNS are likely to have major ecological and economic consequences, with expected increases in crop and forest pest populations [15, 16].

Pseudogymnoascus destructans colonizes the cutaneous membranes of the muzzle, ears, wings and tail of bats, eroding the epidermis and invading the underlying skin and connective tissue [17]. Once invasion occurs, *P. destructans* disrupts critical physiological functions such as cutaneous respiration, blood circulation, and water balance [18–21]. These physiological changes result in more frequent arousals from torpor and increased depletion of energy reserves needed for hibernation [21, 22]. Recent studies suggest infected individuals can elicit the initial stages of an immune response (e.g. transcription of cytokines); however, a protective response does not occur due to hibernation [23–25]. Bats with WNS also exhibit aberrant behavior in winter, including movement from thermally stable cave environments to locations near the cave entrance, daytime emergence, and flying in cold winter temperatures [7, 12, 26].

Species-specific behaviors during hibernation, such as microclimate preference, may also play a role in disease susceptibility and survival [14, 27, 28]. In North America, small bodied bats have been known to hibernate at microclimate temperatures ranging from 0 to 10 °C [20] and relative humidity as high as 90–100%, which fall within the optimal growing conditions for *P. destructans*. Whereas larger bodied species, like *E. fuscus* (mean = 12 g) and *M. grisescens* (mean = 10 g), often roost in colder, drier sites in a hibernacula [29]. European bats, such as *M. myotis*, a 30 g species, have been found to hibernate at microclimate temperatures ranging from –4 to 12 °C [30], suggesting that there is no optimal microclimate temperature for hibernating bat species, with individual-specific microclimate preferences within a species ranging widely [31, 32]. *Myotis myotis* is the most frequent bat in Europe documented with *P. destructans* and ulcerations leading to the manifestation of WNS [27, 33–35]. Naturally occurring *P. destructans* infections on *M. myotis* have been found to be quite extensive, yet have not lead to widespread mortality of the species [33]. Overall, the most affected bat species in Europe are larger bodied species, whereas in North America, small bodied individuals have experienced the largest population declines due to WNS [13, 14, 36].

In northeastern North America, where winters are severe and prey is limited, bats flying outside during the hibernation period are likely suffering the effects of WNS. However, bats in the southeastern U.S. are known to leave hibernacula to feed on warm winter nights [Bernard et al. unpublished], suggesting that winter activity in the South may not be a consequence of disease [37]. As an example, minimum night time temperatures throughout January in Tennessee over the last four years ranged from –17 °C to –6 °C, whereas external cave temperatures in Vermont ranged from –27 °C to –17 °C, consistently 10 °C colder [38]. As WNS has now spread throughout much of the southeastern U.S. [39] the possible effects of winter activity on the epizootiology of WNS remain unknown. Winter foraging on insects may provide bats hibernating in southern latitudes with energy not available to bats in the North. Further, arousing from torpor to engage in episodic feeding during winter will raise body temperature, which should activate the immune system and possibly bolster immunological defenses against *P. destructans*. Evidence from rabies in bats [40], as well as other host-pathogen systems [41, 42], demonstrates that host immunity can result from repeated low-level exposure to pathogens. Behaviorally and physiologically, bats in the South may be different from northern bats in ways that enable them to survive WNS. To investigate possible effects of winter activity on *P. destructans* infections on bats in southern latitudes, we examined fungal load and prevalence on bats captured outside of hibernacula during winter.

In this study, we assessed prevalence and fungal load of *P. destructans* and identified lesions and ulcerations caused by penetration of *P. destructans* into wing and tail membranes for ten species of bats captured while active outside of hibernacula during two winters in Tennessee. Our goals were to determine if emergence during winter is caused by the presence of *P. destructans* and to examine if there are relationships between winter activity, fungal load and prevalence, and bat species. To address these goals, we tested the following hypotheses: 1) active bats leaving caves during winter in the Tennessee will show signs of WNS as demonstrated by fungal load or ultraviolet fluorescence; 2) fungal load and prevalence will be higher on small-bodied cave hibernating species, such as *M. lucifugus* (little brown bats), *M. septentrionalis* (northern-long eared bats) and *Perimyotis subflavus* (tricolored bats), than larger bodied species, such as *Eptesicus fuscus* (big brown bats) and *M. grisescens* (gray bats).

Methods

We conducted our study at the entrances of five hibernacula in Tennessee from October to April 2012–2013 and 2013–2014 (Fig. 1). Prior to the emergence of WNS, Blount Cave was the largest known endangered *M. sodalis* (Indiana bat) hibernaculum in the state of Tennessee, with an estimated population of 9500 individuals in February 2013 [43]. Small numbers of *M. lucifugus* and *P. subflavus* also occur in the cave. Hawkins and Warren Caves are two of the largest hibernacula for endangered *M. grisescens* in the state, with estimated populations of 150,000 and 400,000 *M.*

grisescens, respectively. Both caves also contain a small population of *M. sodalis* during winter [44]. Campbell and White Caves contain populations of *M. leibii* (eastern small-footed bats), *M. lucifugus*, *M. septentrionalis*, and *M. sodalis*, with fewer than 1000 individuals in each cave [45]. Bats in Blount and Hawkins Caves were confirmed positive for *P. destructans* in the winters of 2009–2010 and 2010–2011, respectively, with all other sites confirmed by winter 2012–2013 [44, 46].

We captured bats at each site once a month using mist-nets (Avinet Inc., Dryden, NY; mesh diameter: 75/ 2, 2.6 m high, 4 shelves, 6–12 m wide). Site-specific single-, double- and triple-high nets were deployed 30 min before civil sunset at cave entrances and along corridors within 100 m of the cave. We kept the nets open for 5 h or until we captured 30 bats and closed them when temperatures dropped below 0 °C. After capture, individual bats were placed in paper bags and held for 30 to 60 min in an insulated box with four hand-warmers (HotHands®, Dalton, GA). *Myotis grisescens* and *M. sodalis* were held for a maximum of 30 min. We recorded species, reproductive condition, forearm length (mm), weight (g), mite load [47] and wing-damage index (WDI, [48]), and collected epidermal swab samples from each bat following established protocols (see below). During the winter of 2013–2014, we examined bats for the presence of WNS-related fluorescence by transilluminating the wings with ultraviolet

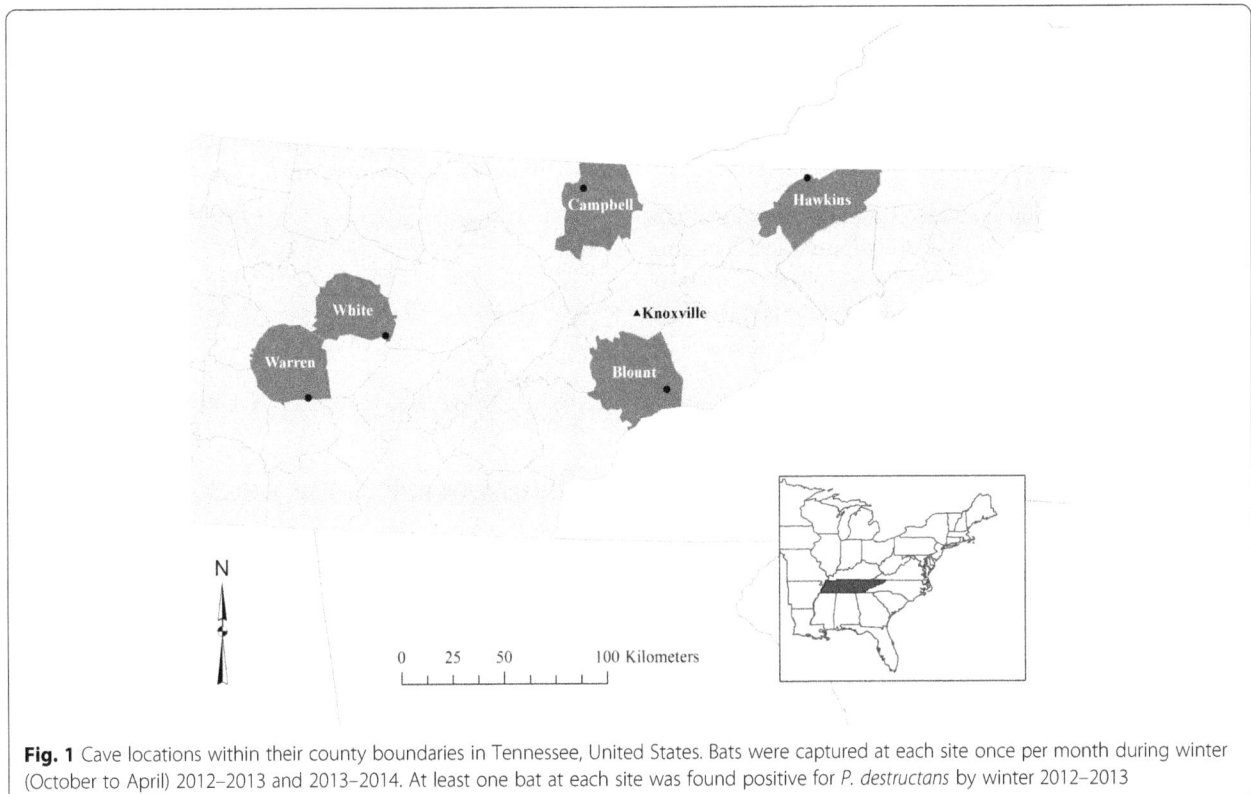

Fig. 1 Cave locations within their county boundaries in Tennessee, United States. Bats were captured at each site once per month during winter (October to April) 2012–2013 and 2013–2014. At least one bat at each site was found positive for *P. destructans* by winter 2012–2013

(UV) light (wavelengths 385–390 nm, [49, 50]). If *P. destructans* has penetrated the skin, lesions fluoresce yellow-orange under UV illumination [49]. Fungal samples for each individual were collected using a sterile epidermal swab dipped in sterile deionized water and rubbed on the bat's forearm and muzzle five times each [51]. Swabs were placed in RNAlater® Tissue Stabilization Solution (Life Technologies, Grand Island, NY) and stored at 4 °C. All cave-roosting species were banded with either 2.4 mm or 2.9 mm numbered, lipped alloy forearm bands (Porzana, Ltd., Icklesham, East Sussex, UK) and released at the site of capture. Due to the distance between sites and the lack of evidence to suggest movement occurs between caves during winter in the region, we assumed each cave was a closed population.

We extracted fungal DNA from each swab sample using DNeasy 96 Blood & Tissue kits (Qiagen Inc., Valencia, CA; [52]). All samples, as well as negative control wells distributed across each polymerase chain reaction (PCR) plate, were tested for the presence of *P. destructans* DNA using a Real-Time PCR assay targeting the intergenic spacer (IGS) region of the ribosomal ribonucleic acid (rRNA) gene complex [53]. All plates were run in duplicate with a quantified standard of isolate *P. destructans* 20,631-21. Any reaction that crossed the threshold baseline in fewer than 40 cycles on either plate was considered positive for *P. destructans* DNA and, when relevant, the average *P. destructans* load, hereafter referred to as fungal load, in nanograms (ng) was calculated in each sample based on the cycle threshold (C^t) value and a generated standard curve based on serial dilutions ([34]; nanograms *P. destructans* = $10^{-3.348 \times Ct + 22.049}$). Fungal load values of *P. destructans* were averaged across both runs.

We followed field decontamination protocols in accordance with the United States Fish and Wildlife Service WNS Decontamination Guidelines and recommendations by the state of Tennessee [54]. All capture and handling techniques were approved by the University of Tennessee Institute of Animal Care and Use Committee (IACUC 2026–0514) and were consistent with the guidelines issued by the American Society of Mammalogists [55]. We obtained both federal (USFWS TE-71613A; GRSM-2013-SCI-1053; GRSM-2014-SCI-1053) and state (TWRA 3716; TDEC 2011–031) permits to capture and handle bats at winter hibernacula for this study.

Statistical analysis
Fungal load data were log transformed prior to analyses to meet assumptions of normality and homogeneity of variance. We used separate generalized linear models (function glm in package lme4 [56] in Program R v 3.1.2 [57]) to compare changes in load and prevalence of *P. destructans* for each species over time. Models were run as either binomial (prevalence) or Gaussian

(fungal loads) distributions, and were tested for significance using likelihood ratio tests. To determine the change in *P. destructans* load and prevalence over time within each model, we used a modified time axis similar to Langwig et al. 2015 [36] where time-0 represented the start of hibernation (October 1). Infection prevalence was calculated by dividing the total number of infected individuals by the total number of individuals captured during the same time period. All means are reported ± standard error. The results presented herein represent bats captured outside of each site, not of the hibernating population as a whole.

Results
We captured and swabbed 871 bats of 10 species (593 males, 276 females, 2 unknowns due to escape; Table 1). Of these, 408 individuals were positive for the presence of *P. destructans* DNA (Pd+) by Real-Time PCR analysis. At least one individual from all species captured was Pd+, including two *Corynorhinus rafinesquii* (Rafinesque's big-eared bat), two *Lasiurus borealis* (eastern red bat) and one *Lasionycteris noctivagans* (silver-haired bat) ([58]; Table 1). However, these three species were excluded from the comparative analyses due to small sample sizes. Capture rates of *M. septentrionalis*, *M. sodalis*, and *P. subflavus* dramatically declined during winter 2013–2014, with *M. septentrionalis* rarely captured after December 2013.

Fifty-one percent of the bats captured (n = 245/480) during winter 2012–2013 were Pd+, whereas only 41.6% of the bats (n = 163/391) were Pd + in winter 2013–2014. When pooled by season, mean fungal loads were significantly higher during the first year of sampling (likelihood ratio test: X^2 = 17.978, p < 0.0001). Excluding species with low sample size, there were significant differences in load

Table 1 Total bats captured and swabbed at five caves in Tennessee during winters 2012–2013 and 2013–2014

Species	Winter 2012–2013			Winter 2013–2014		
	Pd+	Pd-	Total	Pd+	Pd-	Total
Corynorhinus rafinesquii [a]	1	2	3	1	4	5
Eptesicus fuscus	8	11	19	8	11	19
Lasiurus borealis [a]	2	0	2	0	3	3
Lasionycteris noctivagans [a]	0	0	0	1	2	3
Myotis grisescens	24	139	163	35	147	182
Myotis leibii	21	23	44	24	27	51
Myotis lucifugus	11	1	12	14	5	19
Myotis septentrionalis	84	31	115	46	9	55
Myotis sodalis	38	11	49	14	8	22
Perimyotis subflavus	55	18	73	21	11	32

Tallies are provided for the total number of individuals determined positive (Pd+) or negative (Pd-) for *P. destructans* through real-time PCR analysis
[a]Data for these species are explained further in Bernard et al. [58]

of *P. destructans* per species over time (likelihood ratio test: X^2 = 278.06, p < 0.0001, Fig. 2). Fungal loads were lowest when bats entered hibernation in October (-4.99 ± 0.328 log$_{10}$ ng) and peaked for most species during mid-hibernation (December–February; -2.80 ± 0.095 log$_{10}$ ng; Fig. 2 and Additional file 1). Thereafter, mean fungal loads on six of the seven species remained stable through the end of the hibernation period in April. However, mean fungal load on *P. subflavus*, the seventh species, continued to increase through the end of hibernation, reaching levels twice as high as those recorded in December (Fig. 2). *Perimyotis subflavus* had the highest mean *P. destructans* load (-2.34 ± 0.091 log$_{10}$ ng), whereas *M. grisescens* had the lowest mean *P. destructans* load of all species sampled (-4.89 ± 0.075 log$_{10}$ ng). Infection prevalence varied among species, with large-bodied species, such as *E. fuscus* and *M. grisescens*, experiencing the lowest prevalence on average (Fig. 3). Fungal loads on active bats in the Southeast were also lower than on torpid bats sampled in a separate study in the northeastern U.S. (Table 2).

During the second season of sampling, we examined 481 bats for WNS-related fluorescence. Ultraviolet fluorescence revealed varying degrees of damage due to the fungus, from small pin-sized lesions to large coalescing regions of fluorescence and infiltration corresponding with increased pathogen loads. Only 15 bats that fluoresced showed some signs of wing damage, varying from slight depigmentation to pin holes (WDI = 1). All bats captured during early hibernation (October and November) were negative for UV fluorescence. The highest percent of UV-infected bats were captured during mid-hibernation (December 35.9%, January 29.5%). A total of 66 bats were positive by both PCR and UV, with only two UV positives not detected as Pd + by PCR. A total of 181 individuals were Pd + from PCR but UV negative, whereas 232 bats were negative for both *P. destructans* and UV. Bats that were Pd + by both PCR and UV had higher fungal loads than individuals that were determined *P. destructans* positive only by PCR ($t_{200.6}$ = 8.83, p < 0.0001). As noted in Zukal et al. 2016 [8], we did not observe a threshold with which the presence of UV fluorescence corresponded to a minimum fungal load (UV positive: range – 4.74 – −0.22 log$_{10}$ ng; UV negative: range – 5.73 – 0.36 log$_{10}$ ng).

Discussion

Our study demonstrates that the emergence of bats during winter in Tennessee is not indicative of WNS, as less than half of all bats captured outside of the caves sampled during the hibernation period were positive for *P. destructans*. Although we can only make inference directly to bats captured outside of hibernacula in Tennessee, we find a wide range of *P. destructans* loads on captured bats. As we are not only sampling high-load individuals leaving the hibernacula we, therefore, can assume that our results are representative of the entire population. By capturing bats active during winter and coupling winter activity with measures of prevalence and load of *P. destructans* on bats, we highlight the regional differences in the responses of WNS affected species within their greater geographic range. This study demonstrates that as WNS continues to spread throughout North America, it

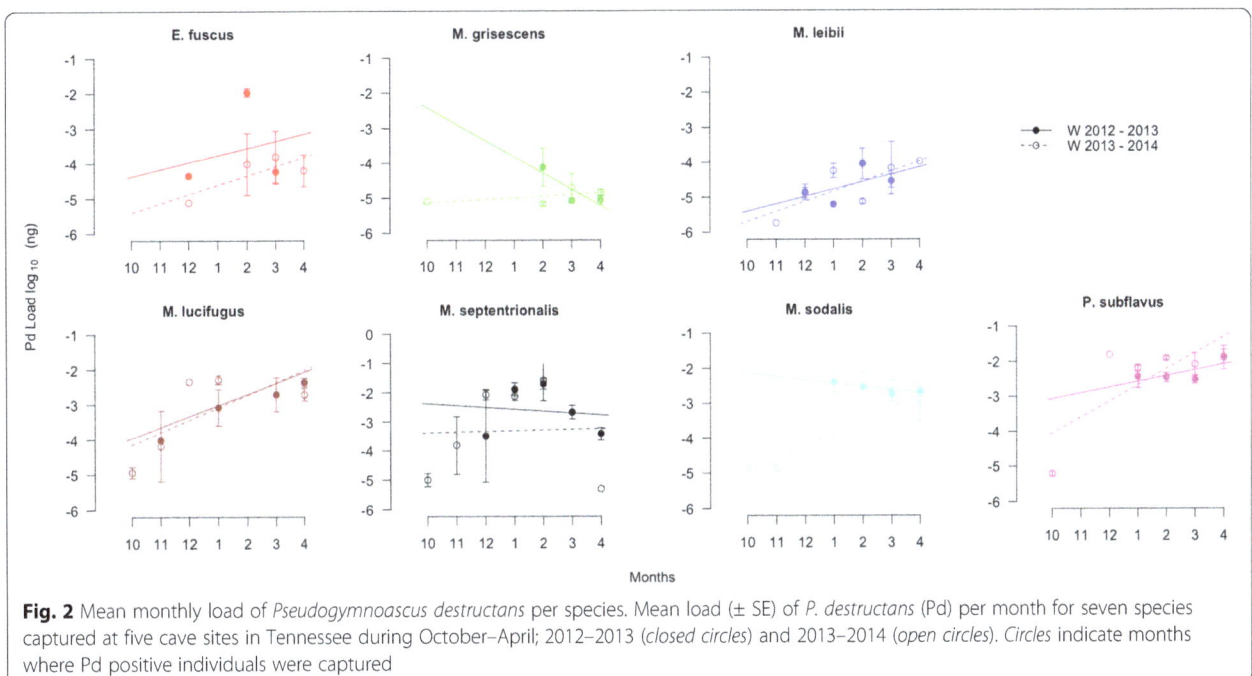

Fig. 2 Mean monthly load of *Pseudogymnoascus destructans* per species. Mean load (± SE) of *P. destructans* (Pd) per month for seven species captured at five cave sites in Tennessee during October–April; 2012–2013 (*closed circles*) and 2013–2014 (*open circles*). *Circles* indicate months where Pd positive individuals were captured

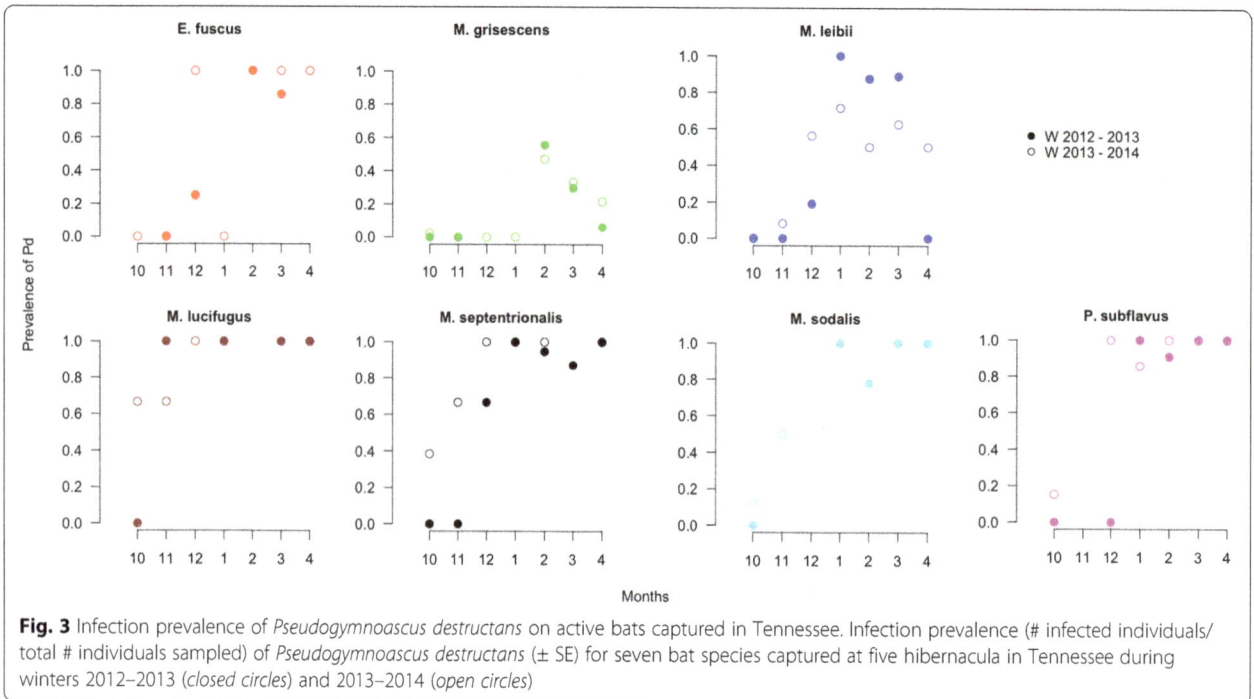

Fig. 3 Infection prevalence of *Pseudogymnoascus destructans* on active bats captured in Tennessee. Infection prevalence (# infected individuals/ total # individuals sampled) of *Pseudogymnoascus destructans* (± SE) for seven bat species captured at five hibernacula in Tennessee during winters 2012–2013 (*closed circles*) and 2013–2014 (*open circles*)

should not be assumed that all individuals within a species will react similarly to the disease. In the Northeast, a secondary symptom of the disease is activity during winter, specifically aberrant behavior, or bats flying during cold weather or during daylight hours [12, 13, 26, 59]. Although, cases of unusual winter behavior have been reported in the Southeast [37, 46], our data highlight that nighttime emergences from hibernacula in Tennessee are not always associated with *P. destructans* infection. Therefore, the effects of WNS on bats in the southeastern U.S. are not directly comparable to those in the North, as regional and species-specific differences, like degree of winter activity, body condition [60] and susceptibility to disease [14, 36], likely vary significantly from those in the Northeast.

Both infection prevalence and loads varied considerably among species, with all small-bodied bats, except *M. leibii*, having higher fungal loads and prevalence than *E. fuscus* and *M. grisescens*. Species, such as *M. septentrionalis* and *P. subflavus*, with high rates of fungal loads and prevalence, were consistently found with the largest regions of fluorescence and wing damage, indicating high rates of tissue invasion by *P. destructans* [48]. Whereas, *M. grisescens*, which have low fungal loads and prevalence, were often found with substantial discoloration, wing damage, and tissue loss unrelated to WNS based on negative UV and PCR results, as well as WNS WDI scoring [48, 61]. According to survey records in Tennessee, *M. grisescens* was often observed with discolored wing membranes and significant scarring and tissue loss prior to *P. destructans* North

American introduction (John Lamb and Troy Best, personal communication).

Although transmission of *P. destructans* among individuals could be associated with the accumulation of the fungus within a hibernaculum based on the time since initial introduction, number of bats, and internal cave conditions, it is important to also consider species specific biology and behaviors. When the Northeast was the leading edge of WNS infection, small, solitary bats, such as *M. septentrionalis* and *P. subflavus*, had significantly higher fungal loads than similar sized colonial species [11, 28, 41], suggesting *P. destructans* was spreading via density-dependent transmission in these two species as cluster size increased [14]. In contrast, our findings suggest that *M. sodalis* and *M. lucifugus*, species known to cluster in tight aggregations during hibernation, had fungal loads and prevalence similar to those of solitary species. Therefore, transmission of *P. destructans* among more colonial species in southern hibernacula may be a function of the frequency, or rate, of infection among individuals within the cluster, rather than the cluster size [14, 34, 62]. Interestingly, *M. grisescens*, the largest bat that hibernates exclusively in caves in the Midwest and southeastern U.S., had the lowest fungal loads and prevalence of *P. destructans* among all species sampled (Figs. 2 and 3). This finding contrasts with patterns observed in Europe, where *M. myotis* and other large bodied hibernating bats have the highest incidence of *P. destructans* [8, 35, 63]. In the Northeast, disease impacts on *M. lucifugus*, *M. septentrionalis* and *P. subflavus* increased with higher humidity and temperature within

Table 2 P. destructans loads (log$_{10}$ ng) on bats captured in the Southeast compared to values collected from torpid bats in the Northeast

Species												
	EPFU		MYLE		MYLU		MYSE		MYSO		PESU	
Month	Southeast	Northeast	Southeast	Northeast	Southeast	Northeast	Southeast	Northeast	Southeast	Northeast	Southeast	Northeast
Oct.					**-4.94 ± 0.16**	-3.96 ± 0.84	-5.00 ± 0.23	-5.18	-4.83	-4.83 ± 0.21	**-5.20 ± 0.07**	-4.20 ± 0.95
Nov.		-1.99	-5.73		**-4.12 ± 0.59**	-2.75 ± 0.35	-3.81 ± 0.99	**-2.03 ± 0.98**	-4.84	-4.43 ± 0.36		-3.12 ± 0.53
Dec.	**-4.72 ± 0.38**	-2.23	-4.86 ± 0.12		-2.34		-2.24 ± 0.23				-1.80	
Jan.		-3.30	-4.40 ± 0.23	-4.17	-2.78 ± 0.34	**-3.11**	-2.03 ± 0.12		-2.38 ± 0.29		-2.28 ± 0.13	
Feb.	-2.98 ± 0.69		-4.28 ± 0.38				-2.52 ± 0.17		-2.52 ± 0.44		-2.40 ± 0.12	
Mar.	**-4.12 ± 0.29**	-2.74 ± 0.25	**-4.39 ± 0.29**	-2.48 ± 0.49	**-2.70 ± 0.48**	-1.19 ± 0.18	**-2.68 ± 0.24**	**-1.35 ± 0.50**	-2.70 ± 0.11	-2.43 ± 0.21	**-2.40 ± 0.12**	-0.82 ± 0.12
Apr.	**-4.19 ± 0.45**	-3.80	-3.96		**-2.57 ± 0.14**	-1.64 ± 0.22	-3.50 ± 0.22		-2.86 ± 0.26	-2.89 ± 0.60	-1.86 ± 0.17	-1.46 ± 0.45

Values for fungal loads on northeastern bats are based on the published reports of samples collected from 30 hibernacula colonies in New York, Vermont, Massachusetts, Virginia, New Hampshire and Illinois where *P. destructans* has been present for at least one year [36, 82]. All values are presented with ± standard error, unless only one individual was sampled. Empty cells indicate no *P. destructans* was detected on the species during the sample month. Bold values signify which load is lowest in regional comparison per species per month, with similar loads in both regions italicized. Only species that were captured in both regions are included in the table. Samples from both studies were analyzed at the same lab by two of the authors (KLP and JTF)

roosts, such that individuals sampled in the coldest and driest roosts had significantly lower fungal loads [14, 27]. *Myotis grisescens*, however, hibernate in aggregations of 100,000 to 1,500,000 individuals [64] in cold air traps varying from 1 to 9 °C [65], which are the lowest temperatures at which *P. destructans* growth occurs [66]. As of spring 2017, *M. grisescens* have yet to experience any WNS-related declines and their populations appear to have remained stable within Tennessee. Although some *M. grisescens* that we captured have been identified with secondary fungal infections, skin discoloration, and/or substantial tissue loss [Bernard unpublished data, John Lamb and Troy Best personal communication], we have yet to identify how the species is surviving WNS. Several behavioral traits, such as preferred microclimates within hibernacula, sustained activity and foraging throughout winter [37] and year-round cave use [67, 68], may enable this species to prevent or minimize the colonization of *P. destructans* during torpor.

When all seven species with samples sizes ≥12 individuals were combined, mean fungal load was highest during mid-hibernation, December through February, the coldest period of the year in the Southeast. *Perimyotis subflavus*, however, continued to experience an increase in fungal load through the end of hibernation, which could be attributed to the microclimate (11 to 23 °C; ≥ 80% Relative Humidity) used by the species during hibernation [69–71]. Alternatively, in vitro growth curves suggest that *P. destructans* may reproduce more quickly in cave environments that maintain more moderate temperatures of 10 to 15 °C in winter [66], which could result in increased growth rates of the fungus in southern hibernacula, and therefore lead to a peak in fungal load within species hibernating within that temperature range.

Contrary to our prediction and the findings of studies from northern hibernacula [69], both pathogen load and prevalence were lower in the second year of the study for seven of the ten bat species captured [58]. By the second survey year, all caves had been contaminated by *P. destructans* for at least two years. Although this could be due to the decrease in the capture of highly susceptible species (Table 1) caused by WNS related declines within each cave [72], climatic variation between years could also impact disease spread. Similar trends have been documented after the arrival of *Batrachochytrium dendrobatidis*, the pathogenic fungal agent of chytridiomycosis in frogs. Pathogen loads in naïve frog populations increased dramatically in the first year, causing a rapid rise in infection intensity and prevalence in densely populated habitats [73]. As the pathogen load on infected frogs increased, many populations suffered from high rates of mortality. However, the survival of infected individuals led to pathogen endemism and population persistence on the frogs. A similar

dynamic may be occurring in hibernacula contaminated by *P. destructans* [74, 75]. Individuals with high pathogen loads perish in the first year, perhaps allowing for individuals with minor *P. destructans* infections to return to the hibernacula the following winter. Further, some scientists hypothesize that increased incidences of chytridiomycosis are linked to increases in global temperatures creating optimal sites for the pathogen [76, 77]. Similar responses may occur with WNS if regions of the Southeast experience more extreme winters, creating more favorable conditions for *P. destructans* growth or limited opportunities for bats to replenish fat stores. Repeated low-level exposure to *P. destructans* or endemism of the pathogen, mild winters, and episodic feeding may allow for persistence of bat populations hibernating in the Southeast.

Recent evidence from the Northeast suggests some populations of bats have the ability to persist and reproduce despite continued exposure to WNS [42, 75, 78, 79]. Comparing *P. destructans* loads on the same species sampled while torpid in the Northeastern and active in the Southeastern U.S., we find the average fungal loads over the season were consistently lower on active bats (Table 2). *Eptesicus fuscus* and *M. lucifugus* captured in Tennessee had lower loads than those sampled in northeastern hibernacula, whereas fungal loads on *M. septentrionalis* and *M. sodalis* were similar towards the beginning and end of hibernation. *Perimyotis subflavus* in the Northeast, however, had higher loads than individuals sampled in the Southeast. Meaningful direct comparisons are lacking due to insufficient numbers sampled in the Northeast. Ultimately, we are seeing that Pd + bats captured in Tennessee have similar loads to torpid individuals sampled in more northern areas of their range, indicating that activity and survival in the Southeast may be more closely linked with short, mild winters and moderate prey levels during winter.

Conclusions

The depopulation of naïve bat hosts by WNS will likely lead to chronic population depression [2] due to the long-term persistence of *P. destructans* within cave environments. Whereas, mortality in the Northeast can reach 90% within two years of WNS confirmation [12, 13], population declines likely attributed to WNS in the Southeast occur four to five years after confirmation and tend to be less severe in some species [26, 27, 40, 80]. Our findings support the hypothesis that emergence from caves during winter may influence the variation seen in pathogen load and infection intensity among species. By understanding the species-specific dynamics of *P. destructans* within active winter populations, management strategies, such as regional area closures and bio-control treatments can be implemented more effectively. In the Southeast, mitigation measures, such as cave area closures used to minimize external cave disturbances (e.g. Great Smoky Mountains National Park) or bio-control agents (e.g. *Rhodococcus rhodochrous* [81] and chitosan), may work best when targeting hibernacula with small-bodied bats such as *M. lucifugus*, *M. septentrionalis*, and *P. subflavus*; species that are being hit the hardest by WNS in the Southeast [80, 61]. Finally, our study suggests that populations of some bats are persisting regardless of repeated exposure to *P. destructans*. Although the region is currently experiencing WNS-related mortality within highly affected species, mild temperatures and the persistent availability of prey during winter may allow the Southeast to serve as a refuge for surviving bat populations.

Additional file

Additional file 1: Figure S1. Peak load of *P. destructans* on bats captured leaving hibernacula. Maximum monthly load of *P. destructans* for seven species captured at five cave sites in Tennessee during the 2012–2013 and 2013–2014 hibernation period (October–April). Circles indicate months where Pd positive individuals were captured. Species acronym codes: EPFU – *Eptesicus fuscus*, MYGR – *Myotis grisescens*, MYLE – *Myotis leibii*, MYLU – *Myotis lucifugus*, MYSE – *Myotis septentrionalis*, MYSO – *Myotis sodalis*, PESU – *Perimyotis subflavus*. (PNG 264 kb)

Abbreviations

C^t: Cycle threshold; DNA: Deoxyribonucleic acid; IGS: Intergenic spacer; ng: Nanograms; PCR: Polymerase chain reaction; Pd-: *Pseudogymnoascus destructans* negative; Pd +: *Pseudogymnoascus destructans* positive; rRNA: Ribosomal ribonucleic acid; U.S.: United States; UV: Ultraviolet light; WDI: Wing damage index; WNS: White-nose syndrome

Acknowledgements

We would like to thank Anna Chow, Max Cox, Neil Giffen, Reilly Jackson, Devin Jones, Kitty McCracken, Mariah Patton, and Ana Reboredo-Segovia for help in the field; permitting and field help from Great Smoky Mountains National Park, Tennessee Wildlife Resources Agency, Tennessee chapter of The Nature Conservancy, Tennessee Department of Environmental Conservation, and the Tennessee regional office of the US Fish and Wildlife Service. We would also like to thank Colin Sobek for assisting with lab work at Northern Arizona University, and the White-nose syndrome and North American Society for Bat Research (NASBR) communities for countless discussions, advice, and continued support throughout this study. We would like to thank Melquisedec Gamba-Rios for graphic support and assistance in R, as well as keeping RFB sane while finishing her degree. Finally, we would like to thank all reviewers for their comments on earlier versions of the manuscript.

Funding

The research presented in this manuscript was funded by the White-nose syndrome research grant through Basically Bats Wildlife Conservation, Inc., University of Tennessee Institute of Agriculture Center for Wildlife Health, University of Tennessee Department of Ecology and Evolutionary Biology, and the US Geological Survey.

Authors' contributions

RFB designed the project, collected all field samples, analyzed all data, collaborated on obtaining research funding, and wrote the manuscript. EVW assisted with collection of field samples and collaborated on obtaining research funding. JTF and KLP performed the genetic analyses. GFM assisted with the project design, collaborated on obtaining research funding, and was a major contributor in writing the manuscript. All authors read and approved the final manuscript.

Competing interests

The authors declare that they have no competing interests.

Author details

[1]Department of Ecology and Evolutionary Biology, University of Tennessee, Knoxville, TN, USA. [2]Department of Ecosystem Science and Management, Pennsylvania State University, University Park, Pennsylvania, USA. [3]Department of Forestry, Wildlife, and Fisheries, University of Tennessee, Knoxville, TN, USA. [4]Center for Microbial Genetics and Genomics, Northern Arizona University, Flagstaff, AZ, USA. [5]Department of Molecular, Cellular, and Biomedical Sciences, University of New Hampshire, Durham, NH, USA.

References

1. Daszak P, Berger L, Cunningham AA, Hyatt AD, Green DE, Speare R. Emerging infectious diseases and amphibian population declines. Emerg Infect Dis. 1999;5:735–48. doi:10.3201/eid0506.990601.
2. Daszak P, Cunningham AA, Hyatt AD. Emerging infectious diseases of wildlife: threats to biodiversity and human health. Science. 2000;287:443–9. doi:10.1126/science.287.5452.443.
3. Rachowicz LJ, Hero J-M, Alford RA, Taylor JW, Morgan JAT, Vredenburg VT, et al. The novel and endemic pathogen hypotheses: competing explanations for the origin of emerging infectious diseases of wildlife. Conserv Biol. 2005;19:1441–8. doi:10.1111/j.1523-1739.2005.00255.x.
4. Mendelson JR, Lips KR, Gagliardo RW, Rabb GB, Collins JP, Diffendorfer JE, et al. Confronting amphibian declines and extinctions. Science. 2006;313:48.
5. Fisher MC, Henk DA, Briggs CJ, Brownstein JS, Madoff LC, McCraw SL, et al. Emerging fungal threats to animal, plant and ecosystem health. Nature. 2012;484:186–94. doi:10.1038/nature10947.
6. Blehert DS, Hicks AC, Behr MJ, Meteyer CU, Berlowski-Zier BM, Buckles EL, et al. Bat White-nose syndrome: an emerging fugal pathogen? Science. 2009;323:227.
7. Blehert DS, Lorch JMJM, Ballmann AE, Cryan PM, Meteyer CU. Bat White-nose syndrome in North America. Fungal diseases: an emerging threat to human, animal and plant health. Washington, D.C.: The National Academies Press; 2011. p. 167–76.
8. Zukal J, Bandouchova H, Brichta J, Cmokova A, Jaron KS, Pikula J, et al. White-nose syndrome without borders: *Pseudogymnoascus destructans* infection tolerated in Europe and Palearctic Asia but not in North America. Sci Rep. 2016;6:19829. doi:10.1038/srep19829.
9. Hoyt JR, Sun K, Parise KL, Lu G, Langwig KE, Jiang T, et al. Widespread bat White-nose syndrome fungus, Northeastern China. Emerg Infect Dis. 2016; 22:140–2.
10. Leopardi S, Blake D, Puechmaille SJ. White-nose syndrome fungus introduced from Europe to North America. Curr Biol. 2015;25:R217–9. doi:10.1016/j.cub.2015.01.047.
11. U.S. Fish and Wildlife Service. White-nose syndrome: the devastating disease of hibernating bats in North America. Hadley: Fish and Wildlife Service; 2014. p. 2.
12. Turner GG, Reeder DM, Coleman JTH. A five-year assessment of mortality and geographic spread of White-nose syndrome in North American bats and a look to the future. Bat Res News. 2011;52:13–27.
13. Frick WF, Pollock JF, Hicks AC, Langwig KE, Reynolds DS, Turner GG, et al. An emerging disease causes regional population collapse of a common North American bat species. Science. 2010;329:679–82. doi:10.1126/science.1188594.

14. Langwig KE, Frick WF, Bried JT, Hicks AC, Kunz TH, Marm KA. Sociality, density-dependence and microclimates determine the persistence of populations suffering from a novel fungal disease, White-nose syndrome. Ecol Lett. 2012;15:1050–7. doi:10.1111/j.1461-0248.2012.01829.x.

15. Maine JJ, Boyles JG. Bats initiate vital agroecological interactions in corn. Proc Natl Acad Sci. 2015:1–6. doi:10.1073/pnas.1505413112.

16. Kunz TH, Braun de Torrez E, Bauer D, Lobova T, Fleming TH. Ecosystem services provided by bats. Ann N Y Acad Sci. 2011;1223:1–38. doi:10.1111/j.1749-6632.2011.06004.x.

17. Meteyer CU, Buckles EL, Blehert DS, Hicks AC, Green DE, Shearn-Bochsler V, et al. Histopathologic criteria to confirm White-nose syndrome in bats. J Vet Diagnostic Investig. 2009;21:411–4. doi:10.1177/104063870902100401.

18. Warnecke L, Turner JM, Bollinger TK, Misra V, Cryan PM, Blehert DS, et al. Pathophysiology of White-nose syndrome in bats: a mechanistic model linking wing damage to mortality. Biol Lett. 2013;9:20130177.

19. Willis CKR, Menzies AK, Boyles JG, Wojciechowski MS. Evaporative water loss is a plausible explanation for mortality of bats from White-nose syndrome. Integr Comp Biol. 2011;51:364–73. doi:10.1093/icb/icr076.

20. Cryan PM, Meteyer CU, Boyles JG, Blehert DS. Wing pathology of White-nose syndrome in bats suggests life-threatening disruption of physiology. BMC Biol. 2010;8:135. doi:10.1186/1741-7007-8-135.

21. Verant ML, Meteyer CU, Speakman JR, Cryan PM, Lorch JM, Blehert DS. White-nose syndrome initiates a cascade of physiologic disturbances in the hibernating bat host. BMC Physiol. 2014;14:10. doi:10.1186/s12899-014-0010-4.

22. Warnecke L, Turner JM, Bollinger TK, Lorch JM, Misra V, Cryan PM, et al. Inoculation of bats with European *Geomyces destructans* supports the novel pathogen hypothesis for the origin of White-nose syndrome. Proc Natl Acad Sci. 2012;109:6999–7003. doi:10.1073/pnas.1200374109.

23. Field KA, Johnson JS, Lilley TM, Reeder SM, Rogers J, Behr MJ, et al. The White-nose syndrome transcriptome: activation of anti-fungal host responses in wing tissue of hibernating little brown Myotis. PLoS Pathog. 2015;11:e1005168. doi:10.1371/journal.ppat.1005168.

24. Bouma HR, Carey HV, Kroese FGM. Hibernation: the immune system at rest? J Leukoc Biol. 2010;88:619–24. doi:10.1189/jlb.0310174.

25. Lilley TM, Prokkola JM, Johnson JS, Rogers EJ, Gronsky S, Kurta A, et al. Immune responses in hibernating little brown myotis (*Myotis lucifugus*) with White-nose syndrome. Proc R Soc B. 2017;284:20162252.

26. Foley J, Clifford D, Castle K, Cryan PM, Ostfeld RS. Investigating and managing the rapid emergence of White-nose syndrome, a novel, fatal, infectious disease of hibernating bats. Conserv Biol. 2011;25:223–31. doi:10.1111/j.1523-1739.2010.01638.x.

27. Hayman DTS, Pulliam JRC, Marshall JC, Cryan PM, Webb CT. Environment, host, and fungal traits predict continental-scale White-nose syndrome in bats. Sci Adv. 2016;2:1–13. doi:10.1126/sciadv.1500831.

28. Frank CL, Michalski A, Mcdonough AA, Rahimian M. The resistance of a North American bat species (*Eptesicus fuscus*) to White-nose syndrome (WNS). PLoS One. 2014;9:e113958. doi:10.1371/journal.pone.0113958.

29. Geiser F. Metabolic rate and body temperature reduction during hibernation and daily torpor. Annu Rev Physiol. 2004;66:239–74. doi:10.1146/annurev.physiol.66.032102.115105.

30. Wojciechowski MS, Jefimow M, Tegowska E. Environmental conditions, rather than season, determine torpor use and temperature selection in large mouse-eared bats (*Myotis myotis*). Comp Biochem Physiol Part A Mol Integr Physiol. 2007;147:828–40. doi:10.1016/j.cbpa.2006.06.039.

31. Boyles JG, Boyles E, Dunlap RK, Johnson SA, Brack V. Long-term microclimate measurements add further evidence there is no "optimal" temperature for bat hibernation. Mamm Biol. 2017; doi:10.1016/j.mambio.2017.03.003.

32. Webb PI, Speakman JR, Racey PA. How hot is a hibernaculum? A review of the temperatures at which bats hibernate. Can J Zool. 1996;74:761–5. doi:10.1139/z96-087.

33. Bandouchova H, Bartonicka T, Berkova H, Brichta J, Cerny J, Kovacova V, et al. *Pseudogymnoascus destructans*: evidence of virulent skin invasion for bats under natural conditions, Europe. Transbound Emerg Dis. 2006;2014:1–5. doi:10.1111/tbed.12282.

34. Zukal J, Bandouchova H, Bartonička T, Berkova H, Brack V, Brichta J, et al. White-nose syndrome fungus: a generalist pathogen of hibernating bats. PLoS One. 2014;9:e97224.

35. Martínková N, Bačkor P, Bartonička T, Blažková P, Cervený J, Falteisek L, et al. Increasing incidence of *Geomyces destructans* fungus in bats from the Czech Republic and Slovakia. PLoS One. 2010;5:e13853. doi:10.1371/journal.pone.0013853.

36. Langwig KE, Frick WF, Reynolds R, Parise KL, Drees KP, Hoyt JR, et al. Host and pathogen ecology drive the seasonal dynamics of a fungal disease, White-nose syndrome. Proc R Soc London B. 2015;282:20142335.

37. Bernard RF, McCracken GF. Winter behavior of bats and the progression of White-nose syndrome in the southeastern United States. Ecol Evol. 2017:1–10. doi:10.1002/ece3.2772.

38. Weather Underground. 2016. https://www.wunderground.com. Accessed 6 July 2016.

39. Washington Dept. of Fish and Wildlife, U.S. Fish and Wildlife Service, U.S. Geological Survey. Bat with White-nose syndrome confirmed in Washington State; 2016. p. 2.

40. Turmelle AS, Allen LC, Jackson FR, Kunz TH, Rupprecht CE, McCracken GF. Ecology of rabies virus exposure in colonies of Brazilian free-tailed bats (*Tadarida brasiliensis*) at natural and man-made roosts in Texas. Vector-Borne Zoonotic Dis. 2010;10:165–75. doi:10.1089/vbz.2008.0163.

41. Dimitrov DT, Hallam TG, Rupprecht CE, McCracken GF. Adaptive modeling of viral diseases in bats with a focus on rabies. J Theor Biol. 2008;255:69–80. doi:10.1016/j.jtbi.2008.08.007.

42. Dobony CA, Hicks AC, Langwig KE, von Linden RI, Okoniewski JC, Rainbolt RE. Little brown Myotis persist despite exposure to White-nose syndrome. J Fish Wildl Manag. 2011;2:190–5. doi:10.3996/022011-JFWM-014.

43. Flock B. Tennessee bat population monitoring and white nose syndrome surveillance. 2013. Retrieved from: http://www.tnbwg.org/Files/13-22%202013%20Bat%20Population%20Monitoring%20and%20White%20Nose%20Syndrome%20Surveillance.pdf

44. Holliday C. White-nose syndrome disease surveillance and bat population monitoring report. 2012. Retrieved from: http://www.tnbwg.org/2012%20White%20Nose%20Syndrome%20Report.pdf.

45. Samoray S. White-nose syndrome monitoring and bat population survey of hibernacula in Tennessee. 2011. Retrieved from: http://www.tnbwg.org/2011%20Tennessee%20Hibernaculum%20Survey%20Report_FINAL.pdf

46. Carr JA, Bernard RF, Stiver WH. Unusual bat behavior during winter in Great Smoky Mountains National Park. Southeast Nat. 2014;13:N18–21.

47. Zahn A, Rupp D. Ectoparasite load in European vespertilionid bats. J Zool. 2004;262:383–91. doi:10.1017/S0952836903004722.

48. Reichard JD, Kunz TH. White-nose syndrome inflicts lasting injuries to the wings of little brown myotis (*Myotis lucifugus*). Acta Chiropterologica. 2009;11:457–64. doi:10.3161/150811009X485684.

49. Turner GG, Meteyer CU, Barton H, Gumbs JF, Reeder DM, Overton B, et al. Nonlethal screening of bat-wing skin with the use of ultraviolet fluorescence to detect lesions indicative of White-nose syndrome. J Wildl Dis. 2014;50:566–73. doi:10.7589/2014-03-058.

50. McGuire LP, Turner JM, Warnecke L, McGregor G, Bollinger TK, Misra V, et al. White-nose syndrome disease severity and a comparison of diagnostic methods. EcoHealth. 2016;13:60–71. doi:10.1007/s10393-016-1107-y.

51. USGS National Wildlife Health Center. Bat White-nose Syndrome (WNS)/Pd surveillance submission guidelines. Madison: USGS National Wildlife Health Center; 2013. p. 36.

52. Shuey MM, Drees KP, Lindner DL, Keim P, Foster JT. Highly sensitive quantitative PCR for the detection and differentiation of *Pseudogymnoascus destructans* and other Pseudogymnoascus species. Appl Environ Microbiol. 2014;80:1726–31. doi:10.1128/AEM.02897-13.

53. Muller LK, Lorch JM, Lindner DL, O'Connor M, Gargas A, Blehert DS. Bat White-nose syndrome: a real-time TaqMan polymerase chain reaction test targeting the intergenic spacer region of *Geomyces destructans*. Mycologia. 2013;105:253–9. doi:10.3852/12-242.

54. Shelley V, Kaiser S, Shelley E, Williams T, Kramer M, Haman K, et al. Evaluation of strategies for the decontamination of equipment for *Geomyces destructans*, the causative agent of the White-nose syndrome (WNS). J Cave Karst Stud. 2013;75:1–10. doi:10.4311/2011LSC0249.

55. Sikes RS, ACUC. 2016 Guidelines of the American Society of Mammalogists for the use of wild mammals in research and education. J Mammal. 2016;97:663–88. doi:10.1093/jmammal/gyw078.

56. Bates D, Maechler M, Bolker B, Walker S. lme4: Linear mixed-effects models using "Eigen" and S4. In: R package version 1.1–13. 2017.

57. R Core Team. 2017. R: a language and environment for statistical computing. Vienna: R Foundation for Statistical Computing. https://www.r-project.org/.

58. Bernard RF, Foster JTJT, Willcox EV, Parise KLKL, Mccracken GF. Molecular detections of the causative agent of White-nose syndrome on Rafinesque's big-eared bats (*Corynorhinus rafinesquii*) and two species of migratory bats in the Southeastern USA. J Wildl Dis. 2015;51:519–22. doi:10.7589/2014-08-202.

59. Reeder DM, Frank CL, Turner GG, Meteyer CU, Kurta A, Britzke ER, et al. Frequent arousal from hibernation linked to severity of infection and mortality in bats with White-nose syndrome. PLoS One. 2012;7:e38920. doi: 10.1371/journal.pone.0038920.

60. Jonasson KA, Willis CKR. Changes in body condition of hibernating bats support the thrifty female hypothesis and predict consequences for populations with White-nose syndrome. PLoS One. 2011;6:e21061. doi:10. 1371/journal.pone.0021061.

61. Reichard JD. Wing-damage index used for characterizing wing condition of bats affected by White-nose Syndrome. Retrieved from https://www.fws. gov/northeast/PDF/Reichard_Scarring%20index%20bat%20wings.pdf.

62. Begon M, Bennett M, Bowers RG, French NP, Hazel SM, Turner J. A clarification of transmission terms in host-microparasite models: numbers, densities and areas. Epidemiol Infect. 2002;129:147–53. doi:10.1017/ S0950268802007148.

63. Pikula J, Bandouchova H, Novotny L, Meteyer CU, Zukal J, Irwin NR, et al. Histopathology confirms White-nose syndrome in bats in Europe. J Wildl Dis. 2012;48:207–11.

64. U.S. Fish and Wildlife Service. Gray bat recovery plan. St. Louis: Fish and Wildlife Serivce; 1982. p. 143.

65. U.S. Fish and Wildlife Service. Gray Bat (*Myotis grisescens*) 5-year review: summary and evaluation. Columbia: Fish and Wildlife Service; 2009. p. 34.

66. Verant ML, Boyles JG, Waldrep W, Wibbelt G, Blehert DS. Temperature-dependent growth of Geomyces destructans, the fungus that causes bat White-nose syndrome. PLoS One. 2012;7:e46280. doi:10.1371/journal.pone.0046280.

67. Stevenson DE, Tuttle MD. Survivorship in the endangered gray bat (*Myotis grisescens*). J Mammal. 1981;62:244–57.

68. Tuttle MD. Population ecology of the gray bat (*Myotis grisescens*): factors influencing growth and survival of newly volant young. Ecology. 1976;57:587–95.

69. Perkins CE. Microclimate and bat occupation trends at Gorman cave, Colorado Bend State Park, San Saba county, Texas durin 2010–2011. 2011. Report to Texas Parks and Wildlife Department, Colorado Bend State Park, Texas, USA. 2011. p. 28.

70. Fujita MS, Kunz TH. *Pipistrellus subflavus*. Mammalian Species. 1984. pp. 1–6. doi:10.2307/3504021

71. Briggler JT, Prather JW. Seasonal use and selection of caves by the eastern pipistrelle bat (*Pipistrellus subflavus*). Am Midl Nat. 2003;149:406–12. doi: 0003/0003-0031(2003)149[0406:SUASOC]2.0.CO;2.

72. Campbell J. Tennessee winter bat population and White-nose syndrome monitoring report for 2014-2015 and 2015-2016. 2016. Retrieved from: http://www.tnbwg.org/2016%20Annual%20Monitoring%20Report.pdf

73. Briggs CJ, Knapp RA, Vredenburg VT. Enzootic and epizootic dynamics of the chytrid fungal pathogen of amphibians. Proc Natl Acad Sci. 2010;107: 9695–700. doi:10.1073/pnas.0912886107.

74. Frick WF, Cheng TL, Langwig KE, Hoyt JR, Janicki AF, Parise KL, et al. Pathogen dynamics during invasion and establishment of White-nose syndrome explain mechanisms of host persistence. Ecology. 2017;98:624–31. doi:10.1002/ecy.1706.

75. Langwig KE, Hoyt JR, Parise KL, Frick WF, Foster JT, Kilpatrick AM, et al. Resistance in persisting bat populations after White-nose syndrome invasion. Philos Trans R Soc B. 2017;372:20160044.

76. Pounds JA, Bustamante MR, Coloma LA, Consuegra JA, Fogden MPL, Foster PN, et al. Widespread amphibian extinctions from epidemic disease driven by global warming. Nature. 2006;439:161–7. doi:10.1038/nature04246.

77. Bosch J, Carrascal LM, Duran L, Walker S, Fisher MC. Climate change and outbreaks of amphibian chytridiomycosis in a montane area of Central Spain; is there a link? Proc R Soc B. 2007;274:253–60. doi:10.1098/rspb.2006.3713.

78. Reichard JD, Fuller NW, Bennett AB, Darling SR, Moore MS, Langwig KE, et al. Interannual survival of *Myotis lucifugus* (Chiroptera: Vespertilionidae) near the epicenter of White-nose syndrome. Northeast Nat. 2014;21:N56–9.

79. Lilley TM, Johnson JS, Ruokolainen L, Rogers EJ, Wilson CA, Schell SM, et al. White-nose syndrome survivors do not exhibit frequent arousals associated with *Pseudogymnoascus destructans* infection. Front Zool. 2016;13:12. doi:10. 1186/s12983-016-0143-3.

80. Lorch JM, Meteyer CU, Behr MJ, Boyles JG, Cryan PM, Hicks AC, et al. Experimental infection of bats with *Geomyces destructans* causes White-nose syndrome. Nature. 2011;480:376–8. doi:10.1038/nature10590.

81. Cornelison CT, Keel MK, Gabriel KT, Barlament CK, Tucker TA, Pierce GE, et al. A preliminary report on the contact-independent antagonism of *Pseudogymnoascus destructans* by *Rhodococcus rhodochrous* strain DAP96253. BMC Microbiol. 2014;14:246.

82. Langwig KE, Frick WF, Reynolds R, Parise KL, Drees KP, Hoyt JR, et al. Data from: host and pathogen ecology drive the seasonal dynamics of a fungal disease, White-nose syndrome. Dryad Digit Repos. 2014;282:20142335.

Effects of agrochemicals on disease severity of *Acanthostomum burminis* infections (Digenea: Trematoda) in the Asian common toad, *Duttaphrynus melanostictus*

Uthpala A. Jayawardena[1,2], Jason R. Rohr[3], Priyanie H. Amerasinghe[4], Ayanthi N. Navaratne[5] and Rupika S. Rajakaruna[2*]

Abstract

Background: Agrochemicals are widely used in many parts of the world posing direct and indirect threats to organisms. Xenobiotic-related disease susceptibility is a common phenomenon and a proposed cause of amphibian declines and malformations. For example, parasitic infections combined with pesticides generally pose greater risk to both tadpoles and adult frogs than either factor alone. Here, we report on experimental effects of lone and combined exposures to cercariae of the digenetic trematode *Acanthostomum burminis* and ecologically relevant concentrations of (0.5 ppm) four pesticides (insecticides: chlorpyrifos, dimethoate; herbicides: glyphosate, propanil) on the tadpoles and metamorphs of the Asian common toad, *Duttaphrynus melanostictus*.

Results: All 48 cercaraie successfully penetrated each host suggesting that the pesticides had no short-term detrimental effect on cercarial penetration abilities. When the two treatments were provided separately, both cercariae and pesticides significantly decreased the survival of tadpoles and metamorphs and induced developmental malformations, such as scoliosis, kyphosis, and skin ulcers. Exposure to cercariae and the two insecticides additively reduced host survival. In contrast, mortality associated with the combination of cercariae and herbicides was less than additive. The effect of cercariae on malformation incidence depended on the pesticide treatment; dimethoate, glyphosate, and propanil reduced the number of cercarial-induced malformations relative to both the control and chlorpyrifos treatments.

Conclusions: These results show that ecologically relevant concentrations of the tested agrochemicals had minimal effects on trematode infections, in contrast to others studies which showed that these same treatments increased the adverse effects of these infections on tadpoles and metamorphs of the Asian common toad. These findings reinforce the importance of elucidating the complex interactions among xenobiotics and pathogens on sentinel organisms that may be indicators of risk to other biota.

Keywords: Amphibians, Trematodes, Glyphosate, Chlorpyrifos, Dimethoate, Propanil, Malformation

* Correspondence: rupikar@pdn.ac.lk
[2]Department of Zoology, University of Peradeniya, Peradeniya, Sri Lanka
Full list of author information is available at the end of the article

Background

Amphibian populations in many parts of the world are experiencing declines and malformations owing to multiple causes, such as xenobiotics, diseases, radiation, habitat destruction, and climate change. [1, 2]. Among these causes, considerable attention has been paid to the effects of chemical contaminants on disease risk [3–7]. Amphibians prefer to live in littoral zones of wetland or aquatic ecosystems where there is a high potential exposure to agrochemicals [8]. Pesticides can travel over large expanses of about 1000 km [9, 10] and therefore can affect the aquatic life cycle stages of the amphibians.

Many studies conducted on effects of xenobiotic on amphibians have focused on direct mortality and developmental defects that might contribute to population declines [11–15]. For instance, the direct mortality of late stage larvae of green frogs (*Rana clamitans*) and spring peepers (*Pseudacris crucifer*)was studied by exposing them to 3 ppb of the insecticide carbaryl [16]. Relyea et al. [12] reported that exposure to 380 ppb of the herbicide glyphosate (Roundup) caused 40% reduction of survival of American toad (*Bufo americanus*), leopard frog (*Rana pipiens*), and gray tree frog (*Hyla versicolor*)-tadpoles. Other than effects on survival, growth reductions due to pesticide exposure can potentially reduce population growth rates of amphibians [12]. Furthermore, pesticides may delay or accelerate amphibian metamorphosis [17–19]; delays could cause mass mortality events if the water body dries up before metamorphosis and accelerated metamorphosis can compromise the immune capacity of metamorphs [20]. In addition to this indirect effect on immunity, pesticides can also be directly immunotoxic increasing susceptibility to infectious diseases [21].

Infectious diseases are particularly important because they are well-documented, widespread causative agents of amphibian population declines [22–24]. Among the amphibian infectious diseases, those caused by trematode infections have received much interest [23, 25, 26]. Deformed amphibians and associated mass mortality events became a major environmental issue during the late 1990's [4] and later on, trematode infections were identified as the major cause of many of these deformities [27–30]. By deforming their hosts, the trematodes are believed to enhance the chances that the intermediate host is depredated by a vertebrate definitive host, thus facilitating their life cycle completion [the handicapped frog hypothesis; 4, 31].

Agrochemicals consistently seem to affect interactions between amphibian hosts and trematode parasites [4, 32]. For example, *Echinostoma trivolvis* infection of cricket frogs has increased in areas with detectable levels of herbicides in Midwestern United States [32]. Similarly, *E. trivolvis* infection in *Rana clamitans* has increased in areas

closer to nutrient and where other chemical inputs were high [26]. To corroborate these findings, Rohr and colleagues [33] demonstrated that the trematode infections were higher in amphibians exposed to atrazine, glyphosate, carbaryl, and malathion. Furthermore, elevated levels of nitrogen and phosphorous associated with fertilizer use increased amphibian trematode infections [33–35].

In this study, we examined the effects of *Acanthostomum burminis* infections in the tadpoles and metamorphs of the Asian common toad, *Duttaphrynus melanostictus* in the presence of four pesticides: two herbicides (glyphosate and propanil), and two insecticides (chlorpyrifos and dimethoate). Individual effects of these pesticides on *A. burminis* infections in the same developmental stages of the hourglass tree frog, *Polypedates cruciger*, and *D. melanostictus* were previously reported [35–39]. In these species, *A. burminis* induced mainly axial and some limb malformations, increased mortality and time to metamorphosis, and decreased size at metamorphosis [35, 36, 39], whereas the four pesticides increased malformations, mortality, and time to metamorphosis [37, 38]. Many laboratory studies suggest that in the presence of pesticides, trematode-induced effects are enhanced [40–43]. Recently, exposure to the combination of cercariae of *A. burminis* and pesticides revealed that the two factors pose greater risks to frogs than either factor alone [44]. Even though, cercariae are often sensitive to xenobiotics [45–47], *A. burminis* cercariae in both the control and pesticide treatments penetrated the tadpoles showing no signs of toxicity before the infection [44]. Whereas previous work on *D. melanostictus* explored the effects of pesticides and *A.burminis* in isolation only, here we build upon work that suggests that pesticide exposure can enhance trematode infection by crossing the presence and absence of pesticides with the presence and absence of *A.burminis* to test whether pesticides increase or decrease risk from this infection in *D. melanostictus*. Consequently, this work will help move the field towards a more general conclusion regarding the risk that the combined effect of the pesticides and cercariae pose to amphibians.

Methods
Study animals

The Asian common toad, *Duttaphrynus melanostictus*is at least concerned species, distributed all over Sri Lanka, especially in human-altered habitats. The adults lay egg strands in slow-flowing streams or in water pools. Four newly spawned egg clutches of *D. melanostictus* were collected from ponds in the Peradeniya University Park (7°15′15″N 80°35′48″E / 7.25417°N 80.59667°E) and were transferred to the research laboratory in the Department of Zoology, University of Peradeniya, Sri Lanka. The egg strands were placed in a glass aquarium filled with dechlorinated tap water. Tadpoles were fed ground fish flakes twice a day (~10% body mass). The debris and faeces that collected at the bottom of the aquaria were siphoned out and water level was replenished daily. Water temperature was maintained around 27° -30 °C and pH was maintained around 6.5–7.0.

Adults of *Acanthostomum burminis* reproduce sexually in the common freshwater snake [39] and release eggs with the faeces of these hosts. A free-living larval stage, miracidiae comes out when the eggs encounter water and look for the first intermediate host, a snail. Once in the snail host, they reproduce asexually and a second free-living larval stage, cercaria, is released. Cercariae search for their second intermediate host, which is an amphibian. The cercariae encyst subcutaneously as metacercariae in amphibians. When an infected amphibian is consumed by a water snake, the life cycle is completed.

Pleurolophocercous cercariae of *A. burminis* released from the freshwater snail species *Thiara scabra* (Family: Thiaridae) were used for the trematode exposures in this experiment. *Thiara scabra* is a common freshwater snail, found associated with muddy/sandy bottom closer to riverine vegetation [48]. *Thiara scabra* were collected from the university stream and were placed in plastic vials containing 10–15 mL of dechlorinated tap water, under sunlight to induce cercarial shedding. The snails that were shedding cercariae were kept individually in separate vials to obtain a continuous supply of cercariae for the exposures. One infected snail was used for all the tadpole exposures per clutch. Thus, four source snails were used to expose the tadpoles from the four clutches of toads. This is advantageous because the blocking factor removes variation from the error term that is due to both the source of the tadpoles (clutch) and the source of the cercariae (snail), increasing statistical power to detect an effect of treatments.

Test chemicals

The tadpoles and cercariae were exposed to commercial formulations of four widely used agrochemicals; two organophosphorous insecticides (chlorpyrifos and dimethoate) and two herbicides (glyphosate and propanil). The concentration of the active ingredient (a.i.) tested and any known surfactants in the commercial formulation were given in Table 1. The test concentrations (0.5 ppm) for each pesticide were selected based on available literature

[49, 50] and information from Pesticide Registrar Office in Peradeniya on field concentrations of these chemicals.

Exposure of tadpoles to ceracraie and pesticides

Each tadpole (5 days post-hatch, Gosner stages 25–26 [51]) was placed in a separate specimen cup containing 15–20 mL of test solution (dechlorinated tap water/ 0.5 ppm- chlorpyrifos/ dimethoate/ glyphosate/ propanil). Tadpoles assigned to receive trematodes (Table 2) were exposed to12 cercariae per day for four consecutive days. Cercarial penetration was observed under a dissecting microscope and the containers were examined every half hour to ensure that no free swimming cercariae remained. A total of 800 tadpoles were tested requiring 20randomly selected tadpoles from each clutch for each treatment (20 tadpoles per clutch × 5 pesticide treatments × 2 trematode treatments ×4 clutches = 800 tadpoles). After exposure to the cercariae, 20 tadpoles of each treatment regime were assigned to one of 10glass aquaria (15 × 15 × 25 cm) containing 2 L of one of the five test solutions (dechlorinated tap water or 0.5 ppm of chlorpyrifos, dimethoate, glyphosate, or propanil). The tadpoles were raised in the same test medium until metamorphosis. The test solution was renewed once a week and temperature and pH were maintained between26 and 30° C and 6.5 and 7.0, respectively under a natural photoperiod of approximately 12:12 h.

Data collection and analyses

Tadpole mortality, forelimb emergence (stage 42, [52]), and metamorphosis were assessed daily. The dead tadpoles were removed and preserved in 70% alcohol. Snout vent length (SVL) to nearest 0.01 cm and body mass to nearest 0.001 g were recorded at metamorphosis. Malformations were reported at 10 and 30 days post hatching for larvae and at metamorphosis. Malformations were identified and categorized according to Meteyer [52], and severely malformed metamorphs were euthanized with MS-222 and preserved. All procedures described herein were approved by the Animal Ethical Review Committee (AERC/06/12) at the Postgraduate Institute of Science, University of Peradeniya.

Data were analyzed using Statistica (version 6) software (Statsoft, Tulsa, OK). We used binomial regression to test whether temporal block and the main and interactive effects of pesticide and cercarial treatments affected the

Table 1 Active ingredient, surfactant and commercial name of the pesticides used in the study

Active ingredient and strength	Surfactant/solvent	Trade name
Chlorpyrifos[O,O-DiethylO-(3,5,6-trichloro-2-pyridyl) phosphorothioate] 400 g/L	Xylene	Lorsban 40 EC®orPattas®
Dimethoate (O,O-Dimethyl phosporodithioate) 400 g/L	Water	Dimethoate 40EC®
Glyphosate [N-(Phosphonomethyl) glycine] 360 g/L	POEA	Round Up® or Glyphosate®
Propanil [N-(3,4-dichlorophenyl) propanamide] 360 g/L	Cyclohexanone & petroleum solvents	3, 4-DPA®

Table 2 Experimental design used to test the individual and combined exposure of *Acanthostomum*cercariae and pesticides on *D. melanostictus*tadpoles

Parasite	Exposure medium				
	Dechlorinated tap water (Control)	Chlorpyrifos	Dimethoate	Glyphosate	Propanil
No cercariae	–	–	–	–	–
Cercariae (12 × 4 = 48)	+	+	+	+	+

Note: Concentration of 0.5 ppm was used for all four chemicals. 20 tadpoles were tested in each treatment

proportion of frogs that survived in each tank. The binomial error distribution was further used to assess how the treatments affected malformation frequency of 10 days post-hatching. Because all 20 tadpoles were reared in a single tank in each temporal block, we used the mean of each tank as the replicate. Thus, each treatment had four replicates for these analyses. If any main effect or interaction were significant, a Fisher's LSD Posthoc test was conducted to evaluate which treatments differed from one another. Because all 20 tadpoles were reared in a single tank in each temporal block, we used the mean of each tank as the replicate and thus each treatment had four replicates total for these analyses. If any main effect or interaction were significant, a Fisher's least significant difference (LSD) Posthoc test was conducted to evaluate which treatments differed from one another. If temporal block was not significant, it was dropped from the statistical model.

Results

All the cercariae penetrated each tadpole because no cercariae were found in the exposure vials at the end of exposure. Thus, the number of infections per tadpole was the same throughout the pesticide treatments.

Survival of the tadpoles

Both the effect of pesticide treatment (Main effect: $\chi^2_4 = 54.53$, $p = 0.0001$) and cercariae (Main effect: $\chi^2_1 = 46.31$, $p = 0.0001$; Fig 1a) increased tadpole mortality (Fig. 1a). In the absence of cercariae, all four pesticides significantly increased mortality (Fig. 1a). Tadpoles exposed to chlorpyrifos, dimethoate, glyphosate, and propanil had 6.50, 7.75, 7.75, and 4.25 times the mortality as those exposed to the pesticide control (Fig. 1a). In the absence of pesticides, ceracriae exposed tadpoles had 6.75 times mortality as those not exposed to cercariae (Fig. 1a). An interaction between pesticide and cercarial treatments on the probability of death ($\chi^2_1 = 11.02$, $p = 0.026$, Fig. 1a) was also reported. This interaction was caused mostly by the combination of herbicides and cercariae having a less than additive effect on mortality (Fig. 1a).

Malformations of tadpoles and metamorphs

Significantly more malformations with than without pesticides (Main effect: $\chi^2_4 = 40.11$, $p = 0.0001$) and with

than without cercariae (Main effect: $\chi^2_1 = 138.33$, $p = 0.0001$; Fig. 1b) were recorded in ten days post hatch tadpoles. No malformations reported in the absence of both cercariae and pesticides (Fig. 1b). Without cercariae, chlorpyrifos, glyphosate, dimethoate, and propanil induced malformations in 24, 20, 16, and 16% of the tadpoles, respectively (Fig. 1b). In the absence of pesticides, cercariae induced malformations in 53% of the tadpoles (Fig. 1b). Notably, the pesticide treatment had a significant effect on the cercarial effect on malformation incidence (Pesticide x cercariae: $\chi^2_4 = 28.10$, $p < 0.001$). Dimethoate, glyphosate, and propanil reduced the number of cercarial-induced malformations relative to the control and chlorpyrifos treatments (Fig. 1b). Scoliosis (vertebral column curvature, laterally deviated spine), kyphosis (hunched back, abnormal convexed spine), and edema were observed as malformations.

Size at metamorphosis

Despite affecting toad survival and malformations, there was no any effects of pesticide or cercarial treatments on body mass (Pesticide: $F_{4,30} = 0.87$, $p = 0.494$, Cercariae: $F_{1,30} = 0.48$, $p = 0.492$, Interaction: $F_{4,30} = 0.24$, $p = 0.914$; Fig. 2a)or SVL at metamorphosis(Pesticide: $F_{4,30} = 0.16$, $p = 0.956$, Cercariae: $F_{1,30} = 0.42$, $p = 0.520$, Interaction: $F_{4,30} = 1.32$, $p = 0.284$; Fig. 2b) .

Developmental rate

Pesticide and cercariae exposure caused significant elevations in the TE_{50},-days until 50% of the metamorphs had forelimb emergence- and days to metamorphosis (Pesticides- Main effect: $F_{4,30} = 3.97$, $p = 0.011$; $F_{4,30} = 4.28$, $p = 0.007$, respectively; and cercariae- main effect: $F_{1,30} = 12.48$, $p = 0.001$; $F_{1,30} = 30.10$, $p < 0.001$, respectively, Fig. 3a,b). Without cercarial exposure, tadpoles exposed to chlorpyrifos, glyphosate, dimethoate, and propanil took 3.7, 5.1, 1.3, and 0.8 more days to metamorphose, respectively, compared to the tadpoles of the control group (Fig. 3b). Without pesticide exposure, those exposed to cercariae took 10.5 more days to complete the metamorphosis compared to those not exposed to cercariae (Fig. 3b). Moreover, cercarial effect on days until 50% of the frogs had forelimb emergence and days to metamorphosis were depended on the pesticide treatment (Pesticide x cercariae: $F_{4,30} = 2.95$,

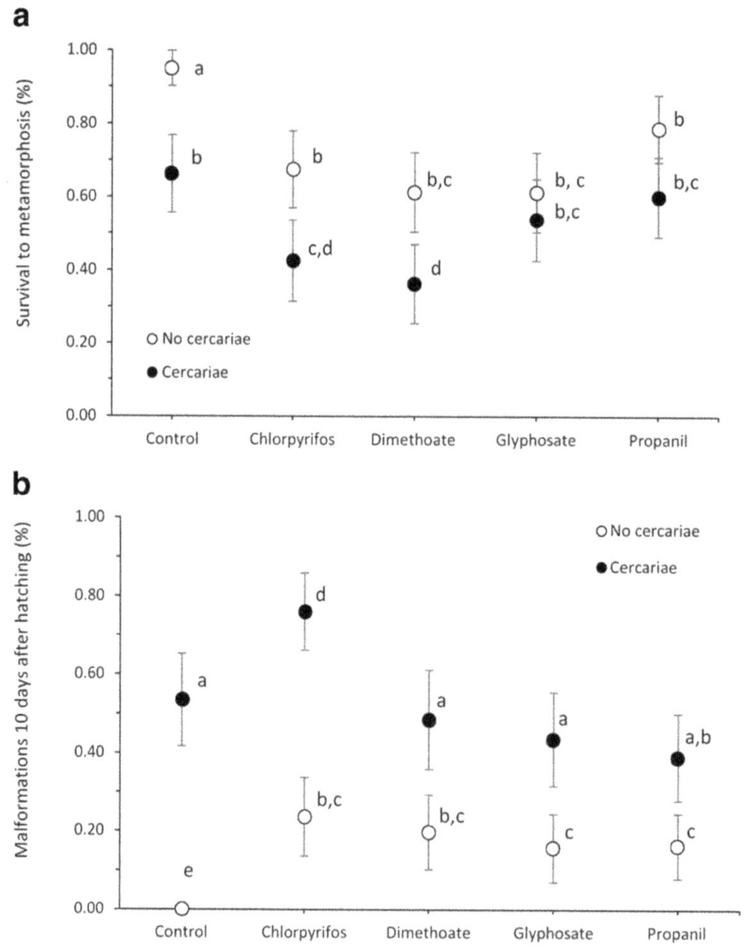

Fig. 1 Mean proportion (± 95% confidence interval, *n* = 4 tanks) of *D. melanostictus* tadpoles that survived until metamorphosis (**a**) and that had malformations approximately 10 days post-hatching (**b**) after the exposure to five treatments (control water, chlorpyrifos, dimethoate, glyphosate and propanil) along with the presence or absence of exposure to cercariae of the trematode *Acanthostomum burminis*. Treatments that do not share letters were deemed significantly different from one another based on none overlapping confidence intervals

$p = 0.036$; $F_{4,30} = 4.09$, $p = 0.009$, respectively). This was because cercariae increased and decreased development time in chlorpyrifos and dimethoate media, respectively, compared to the pesticide control (Fig. 3a,b).

Discussion

Exposure to cercariae of *A. burminis* alone and the four pesticides alone significantly increased mortality and malformations in the Asian common toad compared to the water control. However, individual chemicals interacted with the parasites in different ways. Exposure to the cercariae in the presence of the two insecticides (chlorpyrifos and dimethoate) additively enhanced the effects on mortality induced by either treatment alone. However, exposure to the cercariae in the presence of the herbicides resulted in an antagonistic interaction where survival in the combined treatment was less than additive. Moreover, the effect of cercariae on malformation incidence depended on the pesticide treatment.

Dimethoate, glyphosate, and propanil reduced the number of cercarial-induced malformations relative to the control and chlorpyrifos treatments.

In contrast to the current study, in a previous study on common hourglass tree frog tadpoles, the combined exposure of pesticides and cercariae resulted in a marked reduction in survival and significantly elevated levels of malformations compared to the lone exposures [44]. Differences in the traits of the Asian common toad and hourglass tree frog might explain these differences. The Asian common toad has thick, dry skin and the adults are nocturnal, terrestrial habitat generalists found frequently in human-altered agricultural and urban areas. The hourglass tree frog has thin skin and is an arboreal species found mostly in agricultural land, home gardens, houses, and other buildings. The differences in their skin are even visible at the tadpole stage, as tadpoles of the toad have thick dark skin and those of the frog have thin light skin.

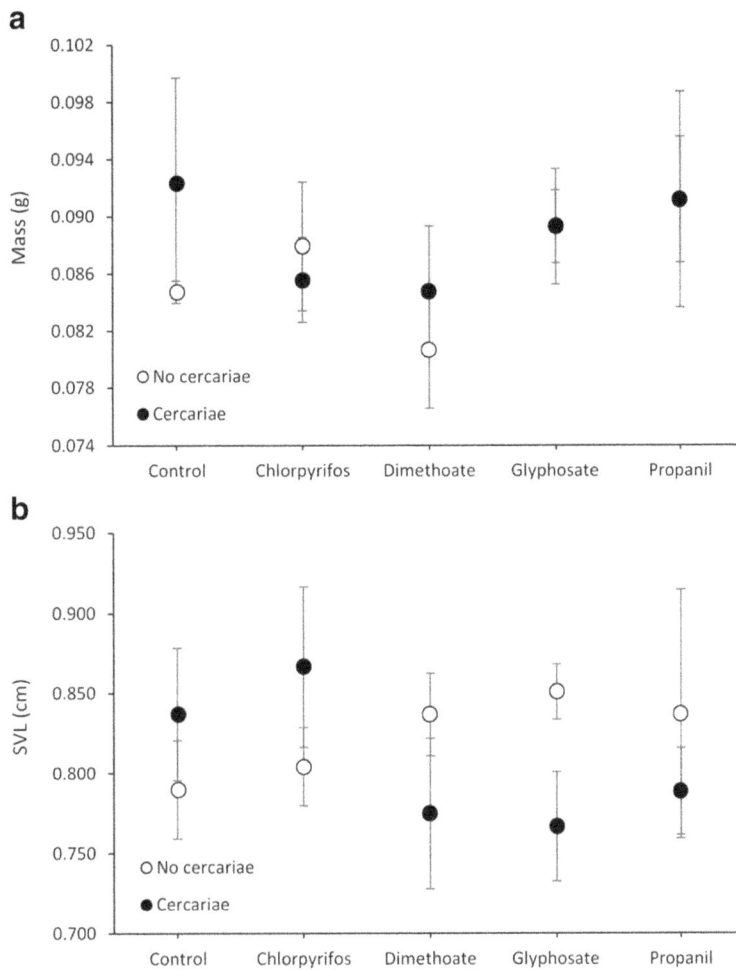

Fig. 2 Mean (±SE, n = 4 tanks) mass (**a**) and snout-vent length (SVL) (**b**) of *D. melanostictus* toads at metamorphosis when exposed to five treatments (control water, chlorpyrifos, dimethoate, glyphosate and propanil) along with the presence or absence of exposure to cercariae of the trematode *Acanthostomum burminis*. There were no significant main effects of pesticides or cercariae on mass or SVL, nor was there a significant interaction between these predictors

Pesticide and cercarial treatments affected developmental traits of toads. There was no difference in the size of the tadpoles exposed to either cercariae or pesticides or both compared to the size of tadpoles in the water control. However, significant lengthening of the developmental period (i.e. days until 50% of the frogs had forelimb emergence and days to metamorphosis) was observed for tadpoles exposed to pesticides, cercariae, or both compared to the water control. Moreover, the effect of cercariae on the growth period depended on the pesticide treatment. Relative to the control, cercariae increased the developmental period in the presence of chlorpyrifos and decreased development in the presence dimethoate. In contrast to our results, Jayawardena et al. [44] discovered that the same combined cercarial and pesticide treatments as we used here significantly lengthened the growth period and

reduced growth rates in the common hourglass tree frog relative to the two treatments alone.

The enhanced effect of pesticides on trematode disease severity might be due to impairment of the amphibian immune system. Immunosuppressive effects of pesticides have been reported in various studies [2, 39, 53–57]. Exact mechanisms of these immunosuppressive effects are unknown. However, Edge et al. [8] suggested that pesticides, particularly glyphosate, may affect skin peptides that can provide an important barrier to infections. Similarly, Gibble and Baer [58] reported that the in-vitro activity of antimicrobial peptides was reduced by agricultural runoff containing the herbicide atrazine [58]. In several other instances, pesticide exposure was associated with decreased melanomacrophage activity in the liver [59], reduced spleen cellularity [60], decreased lymphocyte proliferation [57], decreased white

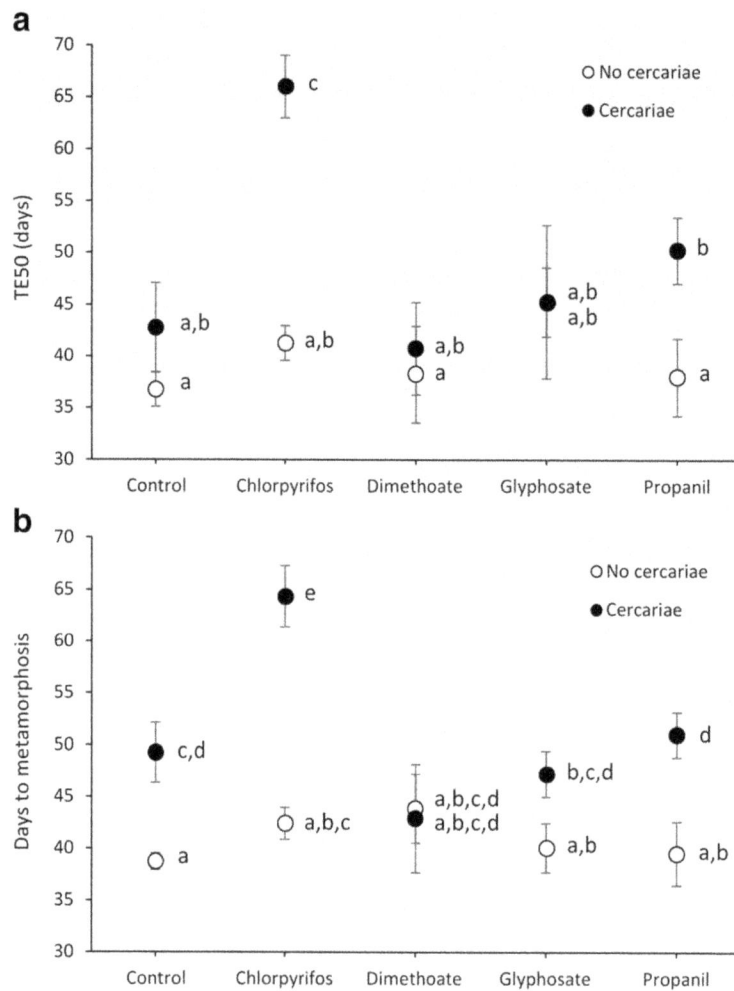

Fig. 3 Mean experimental day when 50% of *D. melanostictus* toads had forelimb emergence (TE50) (**a**) and mean days to metamorphosis (**b**) when exposed to five treatments (control water, chlorpyrifos, dimethoate, glyphosate and propanil) crossed with the presence or absence of exposure to cercariae of the trematode *Acanthostomum burminis*. Treatments that do not share letters were deemed significantly different from one another based on a Fisher's LSD post hoc multiple comparison test

blood cell counts [61], and elevated parasite prevalence [40, 41, 55, 59]. On the other hand, pesticide induced altered activity patterns may indirectly increase susceptibility to parasite infection [62] as the active tadpoles can avoid free-swimming larval stages such as cercariae by showing unusual swimming behavior [63, 64]. However, in the present study, tadpole ability to behaviorally avoid cercariae was controlled by exposing the tadpoles to parasites in a small volume of water where all the parasites successfully penetrated the tadpole, forcing them to rely primarily on physiological defenses, such as immune responses [65].

None of the pesticides tested in the present study had any detrimental effect on trematode survival. Similarly, ecologically relevant concentrations of atrazine, glyphosate, carbaryl, and malathion showed no apparent effect on embryo and miracidium (free-living stage) survival of

Echinostoma trivolvis, a common trematode of amphibians [65]. In addition, renicolid cercariae had an improved survival under increasing concentrations of glyphosate, with cercariae living about 50% longer in 3.6 mg a.i. L^{-1} of glyphosate than in control conditions [65]. In addition, several studies [67–70] have investigated the pollutants effect on either the cercarial surplus of the snails or their successive survival. For instance, Kelly et al. [59] recently showed that the New Zealand snail *Potamopyrgus antipodarum* released approximately three times more *Telogaster opisthorchis* cercariae per day when exposed to glyphosate than when kept in glyphosate-free water. In many cases, exposure to pollutants, such as metals, pesticides, and herbicides, reduces replication of trematodes within snails [66, 67] or their rate of emergence from snails [45, 68]. However, some studies report reduced virulence of trematode infections in the

Effects of agrochemicals on disease severity of Acanthostomum burminis infections...

113

presence of chemicals. For instance, Koprivnikar et al. [69] showed that trematode cercariae exposed to atrazine has less success in infesting the tadpoles than those in the control groups.

As described by Rohr et al. [4], the majority of *Acanthostomum* cercariae crawl towards the cloacal vent and form cysts in the crease between the body and tail, where limb buds are located. However, *A. burminis* does not appear to be as virulent as the more well-known *Ribeiroia ondatrae* that also causes amphibian limb deformities. Unlike *Ribeiroia*, *Acanthostomum* cysts are not visible as swollen lumps, perhaps because of their smaller size [70]. Apparently, *Ribeiroia* cysts average 300–350 μm in length (excysted metacercariae, 500–650 μm and adult 4160–5250 μm in length; [70]), whereas *Acanthostomum* cercariae average 216 μm in length. Size of cercariae has been suggested to affect the virulence of trematode infections [4], with larger metacercariae presumably causing more tissue damage, eliciting greater immune responses, and consuming more host resources.

In the field, combinations of pesticides and trematodes may have adverse or beneficial effects on amphibian populations. Pesticides may enhance snail population densities or immunosuppress hosts, thereby promoting deadly amphibian infections [40, 59, 71]. Mesocosm studies conducted by Rohr and colleagues [33] revealed that atrazine increases algal and snail biomass and increases trematode loads in immunosuppressed *Rana pipiens* tadpoles. In the present study, exposures to cercariae in the presence of the two insecticides further reduced host survival relative to the cercariae or insecticides alone. In contrast, herbicides had less than additive effects on mortality associated with cercarial exposures.

In many instances, pesticide concentrations in water bodies are too low to cause direct amphibian mortality. However, their interactions with other biotic and abiotic factors can induce substantial amphibian mortality, as shown in the current study. Hence, the effects of multiple stressors must be more thoroughly considered in ecological risk assessments of wildlife [72].

Conclusion

Although previous studies have shown that pesticides increase the adverse effects of cercariae infection on frogs, the results of the present study revealed that ecologically relevant concentrations of the four pesticides: chlorpyrifos, dimethoate, glyphosate and propanil caused only a slight effect on *A. burminis* infection in *D. melanostictus*. However, the importance of evaluating complex interactions of pollutants and infections on sensitive biota in our ecosystem is highlighted here.

Abbreviations
GLM: General linear model; SVL: Snout vent length; TE50: Time to metamorphosis

Acknowledgements
Authors thank V. Imbuldeniya and Y.G. Ariyaratne of the Department of Zoology, University of Peradeniya for technical assistance.

Funding
Financial support was provided by the National Science Foundation of Sri Lanka (NSF/2005/EB/02 to R.S.R.)

Authors' contributions
UA carried out the study under the guidance of RS, AN, and PH. JRR guided in analyses and interpreting the results. UA drafted the manuscript and RS, JRR, AN and PH reviewed it before the initial submission. All authors read and approved the final manuscript.

Competing interests
The authors declare that there is no Competing interest.

Author details
[1]Postgraduate Institute of Science, University of Peradeniyai, Peradeniya, Sri Lanka. [2]Department of Zoology, University of Peradeniya, Peradeniya, Sri Lanka. [3]Department of Integrative Biology, University of South Florida, Tampa, FL, USA. [4]International Water Management Institute, C/o ICRISAT, Patancheru – 502, Hyderabad, Andhra Pradesh 324, India. [5]Department of Chemistry, University of Peradeniya, Peradeniya, Sri Lanka.

References
1. Houlahan JE, Findlay CS, Schmidt BR, Meyer AH, Kuzmin SL. Quantitative evidence for global amphibian population declines. Nature. 2000;404:752–5.
2. Hayes TB, Falso P, Gallipeau S, Stice M. The cause of global amphibian declines: a developmental endocrinologist's perspective. J Exp Biol. 2010; 213:921–33.
3. Davidson C. Declining downwind: amphibian population declines in California and historical pesticide use. Ecol Appl. 2004;14:1892–902.
4. Rohr JR, Raffel TR, Sessions SK. Digenetic trematodes and their relationship to amphibian declines and deformities. In: Heatwole H, Wilkinson JW, editors. Amphibian biology, Amphibian decline: diseases, parasites, maladies and pollution, vol. Vol 8. Chipping Norton: Surrey Beatty & Sons; 2009. p. 3067–88.
5. Schotthoefer AM, Rohr JR, Cole RA, Koehler AV, Johnson CM, Johnson LB, Beasley VR. Effects of wetland vs. landscape variables on parasite communities of *Rana pipiens*: links to anthropogenic factors. Ecol Appl. 2011;21(4):1257–71.
6. McMahon TA, Brannelly LA, Chatfield MW, Johnson PT, Joseph MB, McKenzie VJ, et al. Chytrid fungus *Batrachochytrium dendrobatidis* has non-amphibian hosts and releases chemicals that cause pathology in the absence of infection. Proc Natl Acad Sci U S A. 2013;110(1):210–5.
7. Rohr JR, Raffel TR, Halstead NT, McMahon TA, Johnson SA, Boughton RK, Martin LB. Early-life exposure to a herbicide has enduring effects on pathogen-induced mortality. Proc Roy Soc Lond BBio. 2013;280(1772):20131502.
8. Edge CB, Gahl MK, Pauli BD, Thompson DG, Houlahan JE. Exposure of juvenile green frogs (*Lithobates clamitans*) in littoral enclosures to a glyphosate-based herbicide. Ecotox Environ Safe. 2011;74(5):1363–9.

9.   Fenelon J, Moore R. Transport of agrichemicals to ground and surface waters in a small central Indiana watershed. JEnvironQual. 1998;27:884–94.

10.   Vogel JR, Majewski MS, Capel PD. Pesticides in rain in four agricultural watersheds in the United States. J Environ Qual. 2008;37:1101–15.

11.   Davidson C, Mahony N, Struger J, Ng P, Pettit K. Spatial tests of the pesticide drift, habitat destruction, UV-B, and climate change hypothesis for California amphibian declines. Conserv Biol. 2002;16:1588–601.

12.   Relyea RA. The lethal impact of roundup on aquatic and terrestrial amphibians. Ecol Appl. 2005;15:1118–24.

13.   Relyea RA. A cocktail of contaminants: how mixtures of pesticides at low concentrations affect aquatic communities. Oecologia. 2009;159:363–76.

14.   Rohr JR, Crumrine PW. Effects of an herbicide and an insecticide on pond community structure and processes. Ecol Appl. 2005;15(4):1135–47.

15.   Rohr JR, Sager T, Sesterhenn TM, Palmer BD. Exposure, post exposure, and density-mediated effects of atrazine on amphibians: breaking down net effects into their parts. Environ Health Perspect. 2006;114:46–50.

16.   Storrs SI, Kiesecker JM. Survivorship patterns of larval amphibians exposed to low concentrations of atrazine. Environ Health Perspect. 2004:1054–7.

17.   Howe CM, Berrill M, Pauli BD, Helbing CC, Werry K, Veldhoen N. Toxicity of glyphosate-based pesticides to four north American frog species. Environ Toxicol Chem. 2004;23:1928–38.

18.   Sparling DW, Fellers GM. Toxicity of two insecticides to California, USA, anurans and its relevance to declining amphibian populations. Environ Toxicol Chem. 2009;28:1696–703.

19.   Rohr JR, Elskus A, Shepherd B, Crowley P, McCarthy T, Niedzwiecki J, Sager T, Sih A, Palmer B. Multiple stressors and salamanders: effects of an herbicide, food limitation, and hydroperiod. Ecol Appl. 2004;14:1028–40.

20.   Gervasi SS, Foufopoulos J. Costs of plasticity: responses to desiccation decrease post-metamorphic immune function in a pond-breeding amphibian. Funct Ecol. 2008;22(1):100–8.

21.   Carey C, Cohen N, Rollins-Smith L. Amphibian declines: an immunological perspective. Develop Comp Immunol. 1999;23:459–72.

22.   Berger L, Speare R, Daszak P, Green D, Cunningham A. Chytridiomycosis causes amphibian mortality associated with population declines in the rain forests of Australia and central America. Proc Natl Acad Sci U S A. 1998;95:9031–6.

23.   Daszak P, Berger L, Cunningham AA, Hyatt AD, Green DE, Speare R. Emerging infectious diseases and amphibian population declines. Emerg Infect Dis. 1999;5:735–48.

24.   Stuart S, Chanson J, Cox N, Young B, Rodrigues A, Fischman D, Waller R. Status and trends of amphibian declines and extinctions worldwide. Science. 2004;306:1783–6.

25.   Johnson PT, Lunde KB, Zelmer DA, Werner JK. Limb deformities as an emerging parasitic disease in amphibians: evidence from museum specimens and resurvey data. Conserv Biol. 2003;17(6):1724–37.

26.   Skelly DK, Bolden SR, Holland MP, Freidenburg LK, Friedenfelds NA, Malcolm TR. Urbanization and disease in amphibians. Disease Ecology: Community Structure and Pathogen Dynamics. 2006:153–67.

27.   Fried B, Pane PL, Reddy A. Experimental infection of Rana pipiens tadpoles with Echinostoma trivolviscercariae. Parasitol Res. 1997;83(7):666–9.

28.   Johnson PTJ, Sutherland DR. Amphibian deformities and Ribeiroia infection: an emerging helminthiasis. Trends Parasitol. 2003;19(8):332–5.

29.   Sessions SK, Franssen RA, Horner VL. Morphological clues from multilegged frogs: are retinoids to blame? Science. 1999;284(5415):800–2.

30.   Johnson PT, Lunde KB, Ritchie EG, Launer AE. The effect of trematode infection on amphibian limb development and survivorship. Science. 1999;284(5415):802–4.

31.   Sessions SK. What is causing deformed amphibians, Amphibian conservation. Washington: Smithsonian Press; 2003. p. 168–86.

32.   Beasley VR, Faeh SA, Wikoff B, Staehle C, Eisold J, Nichols D, Brown LE. Risk factors and declines in northern cricket frogs (Acriscrepitans). In: Lannoo MJ, editor. Amphibian declines: the status of United States species. Berkeley: University of California Press; 2004. p. 75–86.

33.   Rohr JR, Schotthoefer AM, Raffel TR, Carrick HJ, Halstead N, Hoverman JT, Johnson CM, Johnson LB, Lieske C, Piwoni MD. Agrochemicals increase trematode infections in a declining amphibian species. Nature. 2008a;455:1235–9.

34.   Johnson PT, Townsend AR, Cleveland CC, Glibert PM, Howarth RW, McKenzie VJ, et al. Linking environmental nutrient enrichment and disease emergence in humans and wildlife. Ecol Appl. 2010;20(1):16–29.

35.   Rajakaruna RS, Piyatissa PMJR, Jayawardena UA, Navaratne AN, Amerasinghe PH. Trematode infection induced malformations in the common hourglass treefrogs. JZool. 2008;275:89–95.

36.   Jayawardena UA, Rajakaruna RS, Navaratne AN, Amerasinghe PH. Toxicity of pesticides exposure on common hourglass tree frog, Polypedates Cruciger. Int JAgri Biol. 2010a;12:641–8.

37.   Jayawardena UA, Rajakaruna RS, Navaratne A, Amerasinghe PH. Trematode induced malformations in amphibians: effect of infection at pre limb bud stage tadpoles of Polypedates cruciger. J NatSci Found. 2010b;38:241–8.

38.   Jayawardena UA, Navaratne AN, Amerasinghe PH, Rajakaruna RS. Acute and chronic toxicity of four commonly used agricultural pesticides on the common toad, Bufo melanostictus. J Nat Sci Found. 2011;39(3):267–76.

39.   Jayawardena UA, Navaratne AN, Amerasinghe PH, Rajakaruna RS. Malformations and mortality in the Asian common toad induced by exposure to pleurolophocercous cercariae (Trematoda: Cryptogonimidae). Parasitol Int. 2013;62:246–52.

40.   Kiesecker JM. Synergism between trematode infection and pesticide exposure: a link to amphibian limb deformities in nature? Proc Natl Acad Sci. 2002;99(15):9900–4.

41.   Budischak SA, Belden LK, Hopkins WA. Effects of malathion on embryonic development and latent susceptibility to trematode parasites in ranid tadpoles. Environ Toxicol Chem. 2008;27:2496–500.

42.   Budischak SA, Belden LK, Hopkins WA. Relative toxicity of malathion to trematode-infected and noninfected Rana palustris tadpoles. Arch Environ Contam Toxicol. 2009;56(1):123–8.

43.   Rohr JR, Raffel TR, Sessions SK, Hudson PJ. Understanding the net effects of pesticides on amphibian trematode infections. Ecol Appl. 2008b;18(7):1743–53.

44.   Jayawardena UA, Rohr J, Nawaratne AN, Amerasinghe PH, Rajakaruna RS. Combined effect of pesticides and trematode infections on amphibian survival, growth and malformations in hourglass tree frog Polypedates cruciger. Eco Health. 2016;13:111–22.

45.   Morley NJ, Irwin SWB, Lewis JW. Pollution toxicity to the transmission of larval digeneans through their molluscan hosts. Parasitology. 2013;126:S5–S26.

46.   Pietrock M, Marcogliese DJ. Free-living Endo helminth stages: at the mercy of environmental conditions. Trend Parasitol. 2003;19:293–9.

47.   Blanar CA, Munkittrick KR, Houlahan J, Mac Latchy DL, Marcogliese DJ. Pollution and parasitism in aquatic animals: a meta-analysis of effect size. Aqua Toxicol. 2009;93(1):18–28.

48.   Jayawardena UA, Rajakaruna RS, Amerasinghe PH. Cercariae of trematodes in freshwater snails in three climatic zones in Sri Lanka. Ceylon Journal of Science (Biological Sciences). 2011;39(2):95–108.

49.   Aponso GLM, Magamage C, Ekanayake WM, Manuweera GK. Analysis of water for pesticides in two major agricultural areas of the dry zone. Annals of the Sri Lanka Department of Agriculture. 2003;5:7–22.

50.   Wijesinghe MR. Ecotoxicology: why is it a discipline of growing importance? Sri Lanka: Proc Inst Biol; 2012.

51.   Gosner KL. A Simplified table for staging anuran embryos and larvae with notes on identification. Herpetologica. 1960;16(3):183–90.

52.   Meteyer CU. Field guide to malformations of frogs and toads: with radiographic interpretations (no. 2000–0005). US Fish and Wildlife Service. 2000. pp. 1-20.

53.   Taylor SK, Williams ES, Mills KW. Effects of malathion on disease susceptibility in Woodhouse's toads. JWild Dis. 1999;35:536–41.

54.   Christin MS, Gendron AD, Brousseau P, Menard L, Marcogliese DJ, Cyr D, Ruby S, Fournier M. Effects of agricultural pesticides on the immune system of Rana pipiens and on its resistance to parasitic infection. Environ Toxicol Chem. 2003;22:1127–33.

55.   Gilbertson MK, Haffner GD, Drouillard KG, Albert A, Dixon B. Immunosuppression in the northern leopard frog (Rana pipiens) induced by pesticide exposure. Environ Toxicol Chem. 2003;22:101–10.

56.   Lewis J, Hoole D, Chappell LH. Parasitism and environmental pollution: parasites and hosts as indicators of water quality. Parasitology. 2003;126:S1–3.

57.   Christin MS, Menard L, Gendron AD, Ruby S, Cyr D, Marcogliese DJ, Rollins-Smith L, Fournier M. Effects of agricultural pesticides on the immune system of Xenopus laevis and Rana pipiens. Aqua Toxicol. 2004;67:33–43.

58.   Gibble RE, Baer KN. Effects of atrazine, agricultural runoff, and selected effluents on antimicrobial activity of skin peptides in Xenopus laevis. Ecotox Environ Safe. 2011;74(4):593–9.

Effects of agrochemicals on disease severity of Acanthostomum burminis infections...

115

59. Kelly DW, Poulin R, Tompkins DM. Townsend CR. Synergistic effects of glyphosate formulation and parasite infection on fish malformations and survival. J Appl Ecol. 2010;47:498–504.

60. Forson D, Storfer A. Effects of atrazine and iridovirus infection on survival and life-history traits of the long-toed salamander (*Ambystomama crodatylum*). Environ Toxicol Chem. 2006;25:168–73.

61. Bridges CM, Semlitsch RD. Variation in pesticide tolerance of tadpoles among and within species of Ranidae and patterns of amphibian decline. Conserv Biol. 2000;14:1490–9.

62. Thiemann GW, Wassersug RJ. Patterns and consequences of behavioral responses to predators and parasites in Rana tadpoles. Biol J Linn Soc. 2000;71:513–28.

63. Rohr JR, Civitello DJ, Crumrine PW, Halstead NT, Miller AD, Schotthoefer AM, Stenoien C, Johnson LB, Beasley VR. Predator diversity, intraguild predation, and indirect effects drive parasite transmission. Proc Natl Acad Sci U S A. 2015;112(10):3008–13.

64. Rohr JR, Swan A, Raffel TR, Hudson PJ. Parasites, info-disruption, and the ecology of fear. Oecologia. 2009;159:447–54.

65. Raffel TR, Sheingold JL, Rohr JR. Lack of pesticide toxicity to *Echinostoma trivolvis* eggs and miracidia. J Parasitol. 2009;95(6):1548–51.

66. Yescott RE, Hansen EL. Effect of manganese on *Biomphalaria glabrata* infected with *Schistosoma mansoni*. J Invert Path. 1976;28:315–20.

67. Stopper GF, Hecker L, Franssen RA, Sessions SK. How trematodes cause limb deformities in amphibians. J Exper Zool. 2002;294(3):252–63.

68. Hira P, Webbe G. The effect of sublethal concentrations of the molluscicide triphenyl lead acetate on *Biomphalaria glabrata* (say) and on the development of *Schistosoma mansoni* in the snail. J Helminthol. 1972;46:11–26.

69. Koprivnikar J, Forbes MR, Baker RL. Effects of atrazine on cercarial longevity, activity, and infectivity. J Parasitol. 2006;92:306–11.

70. Müller R, Baker JR. Medical parasitology: Lippincott Williams & Wilkins; 1990.

71. Johnson PTJ, Chase JM, Dosch KL, Hartson RB, Gross JA, Larson DJ, Sutherland DR, Carpenter SR. Aquatic eutrophication promotes pathogenic infection in amphibians. Proc Natl Acad Sci U S A. 2007;104:15781–6.

72. Rohr JR, Salice CJ, Nisbet RM. The pros and cons of ecological risk assessment based on data from different levels of biological organization. Crit Rev Toxicol. 2016;46:756–84.

High-resolution monitoring from birth to sexual maturity of a male reef manta ray, *Mobula alfredi*, held in captivity for 7 years: changes in external morphology, behavior, and steroid hormones levels

Ryo Nozu[1,2]* , Kiyomi Murakumo[2], Rui Matsumoto[1,2], Yosuke Matsumoto[2], Nagisa Yano[2], Masaru Nakamura[1], Makio Yanagisawa[1,2], Keiichi Ueda[1,2] and Keiichi Sato[1,2]

Abstract

Background: In this study, we report the first case of a male reef manta ray, *Mobula alfredi*, becoming sexually mature in captivity and present its reproductive characteristics.

Methods: We investigated changes in external morphology, behavior, and levels of steroid hormones in a male *M. alfredi* during its sexual maturation process.

Results: At 2 years and 6 months of age, the male exceeded 300 cm in disc width. Then, at around 3 years of age, the male started to chase a female in the tank and exhibited androgen levels similar to that of another matured male, indicating that the study specimen had begun sexual maturation endocrinologically. Its first copulation event was observed at 5 years and 4 months and appeared behaviorally similar to field observations. Seven months after copulation, we performed a biopsy to collect its semen, including any motile sperm.

Conclusion: Taken together, these results indicate that the captive male *M. alfredi* individual shows signs of sexual maturation (size of disc width, testosterone levels, reproductive behavior) already with the age of 2.5 to 3 years. As a first copulation was observed with 5 years 4 months of age and the presence of sperm were confirmed at the age of 5 years 11 months of age, the studied animal reached its full sexual maturity at the latest at the age of around 5 years.

Keywords: Reef manta ray, *Mobula Alfredi*, Male, Sexual maturation, In-captivity, Mating behaviors, Sex steroid hormones

Background

Manta rays comprise two species that occur in tropical, sub-tropical, and temperate waters around the world. As fully-grown adults, they are among the largest ray species. Due to their tendency to aggregate predictably in relatively shallow waters around islands and along coastlines, they have become an attraction for recreational

* Correspondence: r-nozu@okichura.jp
[1]Okinawa Churashima Research Center, Okinawa Churashima Foundation, 888 Ishikawa, Motobu, Okinawa 905-0206, Japan
[2]Okinawa Churaumi Aquarium, 424 Ishikawa, Motobu, Okinawa 905-0206, Japan

divers in many countries. Manta ray-based ecotourism potential can be important for coastal communities, particularly in developing countries, where it can generate great economic benefits [1, 2].

Until recently, the genus *Manta* was considered mono-specific, but was redescribed [3] in 2009 as comprising the reef manta ray, *Manta alfredi* (Krefft 1868) and the giant manta ray *M. birostris* (Walbaum 1792). In a recent study, the mitochondrial genome analysis showed that both *Manta* species are nested within the other *Mobula* species and sister to *M. mobular* [4]; the authors noted that *Manta* is an invalid generic name.

Therefore, in the present paper, the genus *"Mobula"* was adopted. Both species were listed as globally vulnerable on the IUCN Red List [5, 6].

Previous reports have provided some ecological information about manta ray species, such as distribution, movement patterns, courtship, mating behavior, gestation period, and reproductive periodicity [7–9]. However, there is little information on their reproductive biology. Observations of captive specimens in aquariums enable us to obtain accurate information of the same individual over time. At the Okinawa Churaumi Aquarium (OCA), *M. alfredi* individuals thrive since 1988; much information on the biology of this species has been generated in this aquarium. For example, the OCA recorded the first captive reproduction of a *M. alfredi* [10, 11]. On June 8, 2006, a male mated with a female and on June 16, 2007, the female gave birth to a female pup (193 cm disc width and 68.5 kg body weight). On June 17, 2008, the same pair gave birth to a male pup that was used in the present study. In addition, the copulating behavior was similar to that reported for Ogasawara wild mantas [12] and the duration of clasper insertion into the cloaca was approximately 11 to 18 s. Based on the observation, the authors concluded that the duration from copulation to parturition was approximately 1 year, which were consistent with field observations [7]. On the other hand, it is thought that *M. alfredi* usually takes a resting interval of at least 1 year between pregnancies [7]. The captive female in the OCA, however, mated with a male in the tank immediately after giving birth; the female manta ray became pregnant again and gave birth the next year [10]. This suggests that breaks in pregnancy in the wild are not due to the physical conditions, however, only environmental cues may be essential. Additionally, aspects of the embryonic respiratory system of *M. alfredi* were revealed by an ultrasonographic experiment on the pregnant individual [13]. This study provides the first direct evidence of the respiratory behavior of a *M. alfredi* embryo by demonstrating that the embryo acquires oxygen from the uterine fluid using gill ventilation. Thus, observations on *M. alfredi* in captivity can not only strengthen previous studies but also provide new biological information.

To achieve efficient captive breeding of *M. alfredi* in other facilities, it is essential to accumulate more information on its reproductive ecology and physiology. However, only few studies on these aspects exist because of the difficulty in closely following *M. alfredi* over its complete life-span and the high effort involved. Here, we confirm that a male born in the OCA reached sexual maturity in captivity. Moreover, we provide behavioral and physiological information on its captive reproduction obtained from detailed and continuous time series observations.

Methods

An observed individual and rearing conditions

A male (identification number: No10–2) was born in the "Kuroshio" tank (a 10 m deep, rectangular aquarium (35 × 27 m)) of the OCA on June 17, 2008, and was observed continuously (Fig. 1a). After birth, the neonate was immediately moved to the sea pen (diameter: 30 m, depth: 12–15 m) and kept there for approximately 10 months to naturalize feeding and allow the development of swimming ability without any influence from other individuals. The pup was then moved to the main tank again. Our study period ended at the end of November 2015 as the animal has been moved and been kept solo since then. The circulation rate of water in the tank was 16 turnovers per day (recycled seawater 12 turnovers and fresh seawater 4 turnovers) and water temperature was not under thermal control. The daily feed dosage was 0.5–0.8% of its body weight. The body mass was calculated using the allometry equation (Body weight (kg) = 0.000013563* disc width (cm) $^{2.889}$ (R^2 = 0.973)) that was derived from actual measurement values of 18 *M. alfredi* individuals (Matsumoto et al., personal communication). Feeding was performed twice a day, and the feed included *Euphausia superba*, *E. pacifica*, *Sergestes lucens*, and *Engraulis japonicus*.

Estimation of disc width

Disc width (DW) of specimen No10–2 was measured twice: at his birth (182 cm) and at 10 months of age (261 cm). Thereafter, its outer interorbital distance (IOD) was measured in June and/or December of each year. Estimated DW was calculated from IOD using the growth equation (LogDW (cm) = 0.84841*LogIOD (cm) + 0.88452, R^2 = 0.94) that derived from actual measurement values of 18 *M. alfredi* individuals (Matsumoto et al., personal communication).

Behavior observation and test for stiffness of claspers

During the day, aquarium staff visually monitored the individual but not continuously. When the individual exhibited specific behaviors, we recorded the behaviors through photographs and/or videos from inside and/or outside the tank. To confirm the claspers status, we hand-tested them for stiffness at the birth (June 2008), 5 year and 4 months (Oct 2013) and 5 years 11 months of age (May 2014) while No10–2 swam.

Semen collection

To confirm whether or not the individual produced mature sperm, we obtained semen from No10–2 via a biopsy while swam (Additional file 1: Movie S1). At 5 years and 11 months of age (on May 30, 2014), a plastic tube (All Silicone Foley Balloon Catheters, Product No. 001027 0120, Create Medic Co., Ltd., Kanagawa)

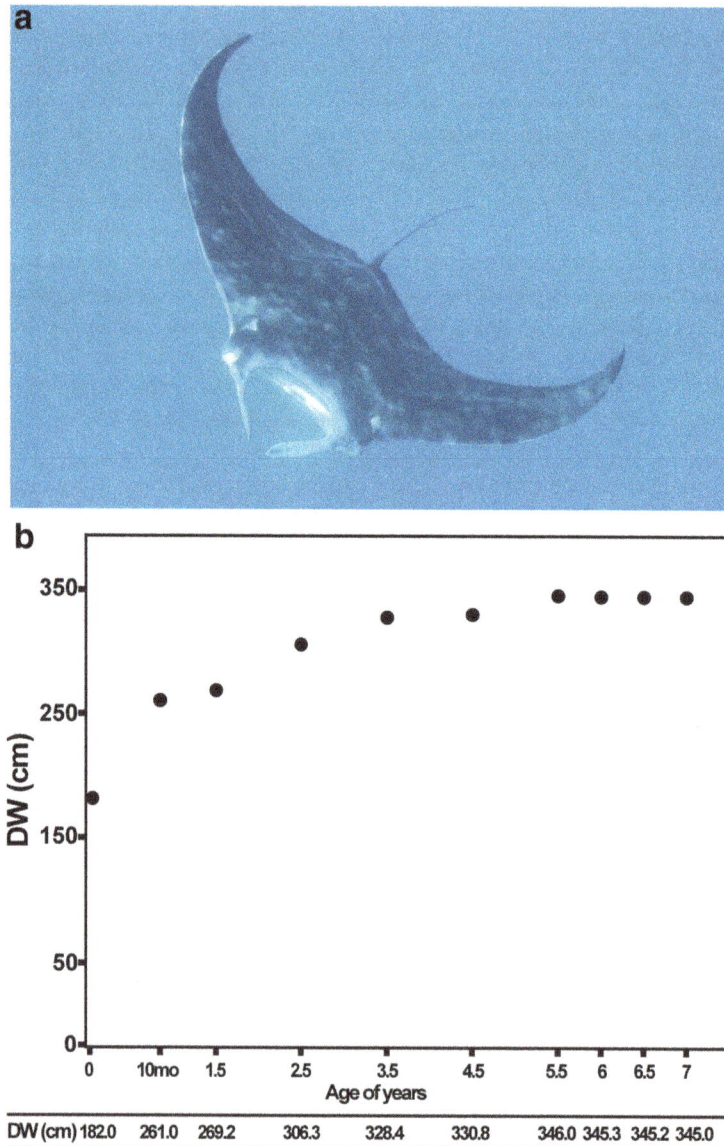

Fig. 1 Photograph of No10–2 soon after his birth and a chart of his growth of disc width (DW). **a** No10–2 swimming in the ocean sea pen. **b** Actual DW measurements were taken at birth and at 10 months of age. After that, estimated DW values were calculated by interorbital distance. mo; months

connected with syringes (5 ml for ballooning, 50 ml for collecting semen, Terumo Corporation, Tokyo) was slipped into his clasper and semen was obtained. The semen was observed under light microscopy and video-recorded; it was also observed under a light microscope after Giemsa staining. We estimated that the specimen did not suffer severe stress due to the biopsy because feed consumption was unaffected after the semen collection.

Blood collection and measurement of plasma steroid hormones levels
No10–2 had received husbandry training since it was approximately 1 year old, which enabled us to collect

blood samples regularly after 3 years and 3 months of age, from its left or right pectoral fin vein or artery, using a syringe (Terumo Corporation, Tokyo) with a 20 gauge needle (Terumo Corporation, Tokyo) while No10–2 swam. The time of blood collection was fixed in the morning. However, during certain months, we could not collect blood samples. The collected blood was placed into a heparinized tube and plasma was obtained by centrifugation (1400×g for 20 min. at room temperature). Obtained plasma was stored at –30 °C until analysis. To compare mature and immature male characteristics, the preserved plasma obtained from other males (i.e., No6 mature, 318-cm DW; No26 immature, 215-cm DW) was also analyzed. Both male individuals were caught from the wild. No6 was regarded as sexually mature because previous observations showed that No6 mated with a female in the OCA [10]. In addition, No26 was regarded as immature based on the previous report that DW at sexual maturity in wild was 270–300 cm [8].

Extraction of plasma steroid was performed according to a previous procedure [14]. Briefly, plasma steroids were extracted three times using 2.5 mL diethyl ether. The extracts were evaporated and the residue was reconstituted with 2 times its volume of assay buffer (0.05 M borate buffer, pH 7.8, containing 0.5% bovine serum albumin). Estradiol 17 beta (E2), testosterone (T), and dihydrotestosterone (DHT) were determined by ELISA, following the methods of Asahina et al. (1995) [15]. Additionally, progesterone (P4) was determined using the Progesterone ELISA kit (Item No. 582601, Cayman Chemical Company, MI) according to the manufacturer's instructions. We made it a top priority to measure testosterone levels based on the volume of preserved plasma sample. Samples and standards were applied in duplicate to each plate. Based on the information from the supplier (Cosmo Bio Co., Ltd., Tokyo), the major cross-reaction of the E2 antibody for estrone was 0.8%, and 0.5% for estriol. The cross-reaction of the T antibody for DHT and 11-ketotestosterone (11KT) was 7.0% and 0%, respectively; the cross-reaction of DHT antibody for T and 11KT was 48% and 0%, respectively; the cross reaction of the P4 antibody for E2 and T was 7.2% and <0.05%, respectively.

Results

Increase in disc width and elongation of claspers

DW at birth of the individual was 182 cm. Growth data of DW are shown in Fig. 1b: DW increased from the birth to 5 years and 6 months of age, after which it remained at ca. 345 cm until the end of the study. At birth, its claspers were shorter than its pelvic fins (Fig. 2a). The organ elongated during development. At an examination at 5 years and 4 months of age, its length exceeded the pelvic fins already (Fig. 2b).

However, we could not confirm when exactly the clasper started to increase in size and exceeded the pelvic fins during our observation period. Furthermore its stiffness at this examination was similar to that of the mature individual No6.

Appearance of reproductive behavior

At 3 years of age (in June 2011; * in Fig. 5), we observed that No10–2 slowly chased females; this behavior is known as a "mating train" (Additional file 2: Movie S2). This behavior was observed several times over several days. At 5 years and 4 months of age (on October 6, 2013; ** in Fig. 5), No10–2 copulated with a female for the first time (Fig. 3; Additional file 3: Movie S3). The male chased the female actively (Fig. 3a), which attempted to position immediately over the dorsal surface of the female and cause the female to rear up, after then bit the tip of the female left pectoral fin, and then inserted its clasper into the female cloaca (Fig. 3b). During copulation, both sank to the bottom of the tank; once they reached the bottom, they swam away separately. The duration from biting the female left pectoral fin to the separation was 40–50 s. After the copulation, vigorous chasing of No10–2 continued for several days and then quieted. Two months after copulation, we confirmed that the female was not pregnant, using ultrasonographic imaging. At 5 years and 11 months of age (May 2014; *** in Fig. 5), the male chased the female actively for several days, but no mating was observed. When the individual reached 6 years and 11 months (May 2015; *** in Fig. 5), we observed active chasing without copulation for several days. And then, we again observed active chasing at 7 years and 2 months of age (August 2015) for a couple of weeks, followed by copulation at 7 years and 3 months of age (on September 2, 2015; ** in Fig. 5). Accordingly, we confirmed that the female was not pregnant.

Production of motile spermatozoa

At 5 years and 11 months of age, mobility of spermatozoa in the semen collected from No10–2 was confirmed (Fig. 4a, b; Additional file 4: Movie S4). The sperm had a

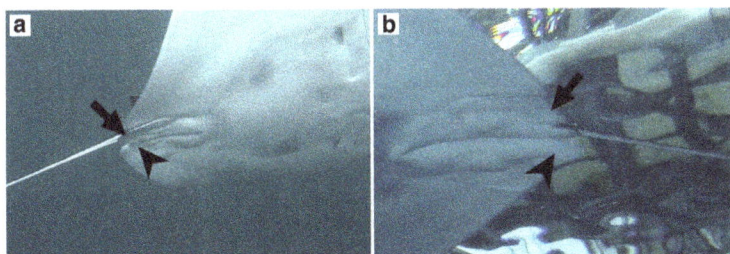

Fig. 2 Morphological changes in claspers. **a** Claspers at 4 months old (in October 2008). **b** Claspers exceeded the distal end of the pelvic fins at 5 years and 4 months of age (in October 2013). Arrowheads indicate claspers; arrows indicate pelvic fin

Fig. 3 Snapshots of mating behaviors in No10–2. a Active chasing behavior observed in October 2013. b Copulation observed in October 2013

typical chondrichthian spiral head structure and rotated along the long axis (Fig. 4c).

Changes in the levels of steroid hormones

Changes in the concentrations of steroid hormones of No10–2 are shown in Fig. 5 and Additional file 5: Table S1. Androgens levels were stable from 3 years of age.

Irregular peaks of testosterone levels were detected from age 6 onwards. Variation in dihydrotestosterone showed a pattern similar to that of testosterone. Estrogen showed basal level and did not range dramatically throughout the observation period. Progesterone level increased at 6 years of age, from February to June, then decreased. At 3 years of age, androgen levels of the

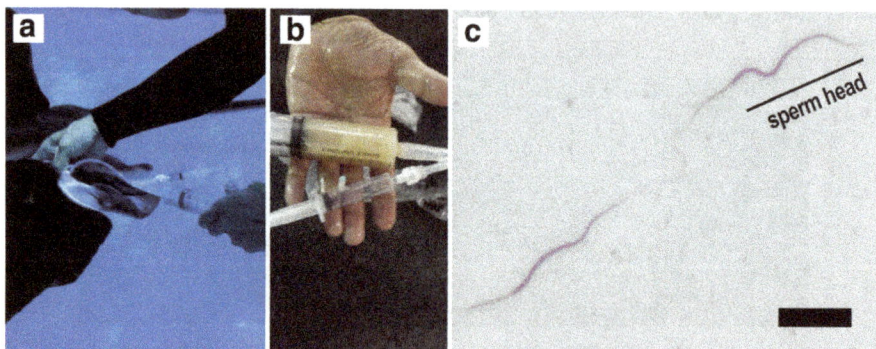

Fig. 4 Semen from No10–2. a Plastic tube inserted into the clasper. b Collected semen. c Giemsa-stained sperm. Scale bars = 20 µm

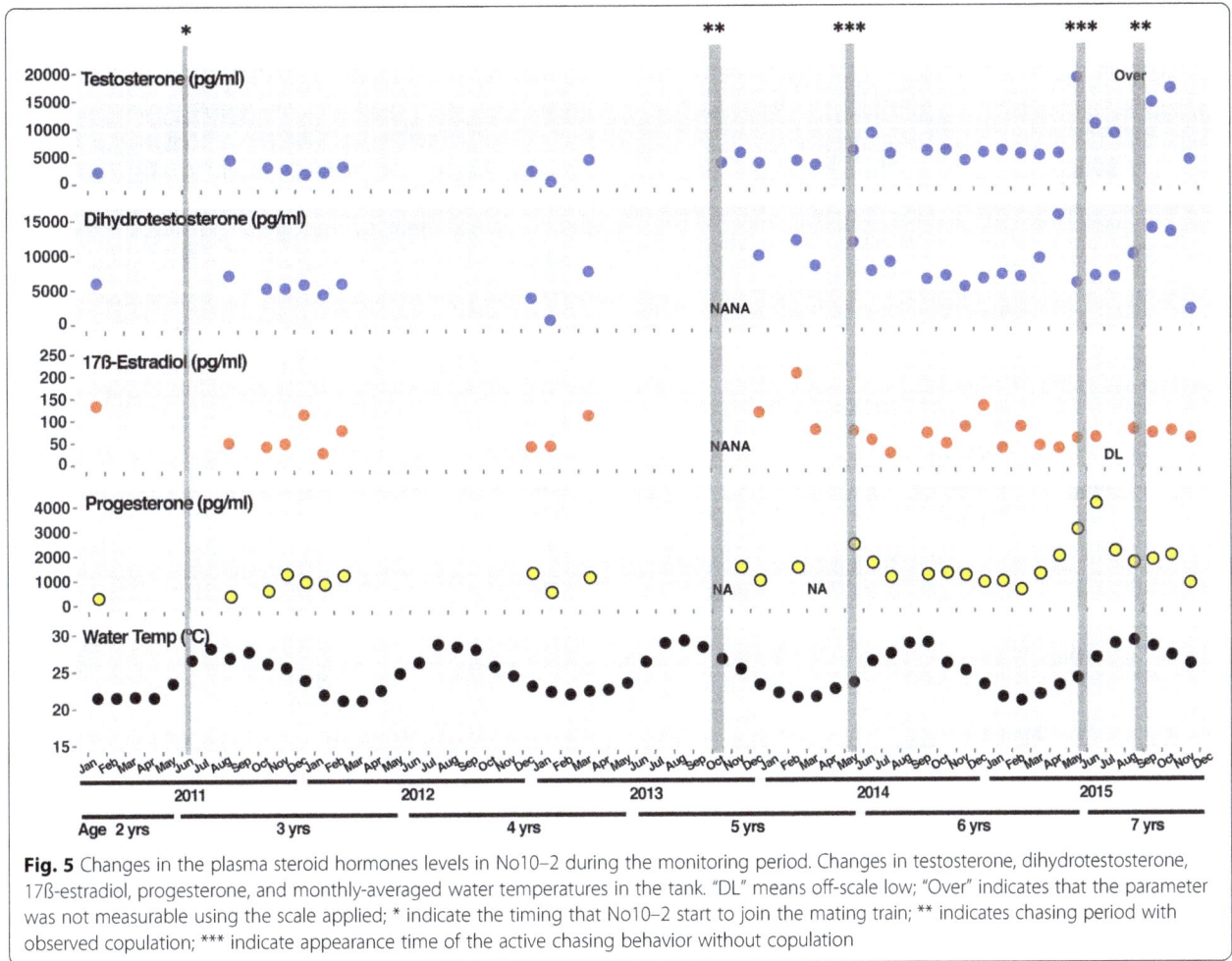

Fig. 5 Changes in the plasma steroid hormones levels in No10–2 during the monitoring period. Changes in testosterone, dihydrotestosterone, 17ß-estradiol, progesterone, and monthly-averaged water temperatures in the tank. "DL" means off-scale low; "Over" indicates that the parameter was not measurable using the scale applied; * indicate the timing that No10–2 start to join the mating train; ** indicates chasing period with observed copulation; *** indicate appearance time of the active chasing behavior without copulation

studied specimen was within the range of androgen levels of the matured male (No6; Fig. 6, Additional file 6: Table S2), which in turn was approximately 25–135 times higher than that of the immature male (No26; Fig. 6, Additional file 7: Table S3). Immediately prior to the chasing behaviors seen in May and August 2015, increases in testosterone levels were observed, especially for T levels in August that exceeded the upper threshold of the standard curve (Fig. 5).

Discussion

It has been previously reported that the age and DW at maturity in males was 3–6 years and 270–300 cm, respectively [8, 16]. Our observations of reproductive age in captivity are consistent with these field observations [8]. On the other hand, regarding DW, when No10–2 was only 2 years and 6 months old — that is, 6 months younger the minimum age for maturity than in a wild male — No10–2 already had exceeded the DW size at maturity (300 cm) of wild males. However, although the testosterone level is already higher than in an immature animal, the animal did not show chasing or

mating behavior at this time. These results suggest that age might be a more important factor than attaining sexual maturity size in *M. alfredi*. In contrast, it has been reported that the timing of pubertal development in male bonnethead sharks, *Sphyrna tiburo*, may be more associated with its size than its age [17]. On the basis of these disparities, it is likely that the limiting factor of sexual maturation in elasmobranchs depends on the particular species and/or its environment. Moreover, the results of the present study implied that that growth rates and possibly size at maturity might be quite plastic in *M. alfredi*, and very dependent on the individual's experience.

Clasper length and calcification are often-used criteria of male sexual maturity in elasmobranchs, in that a rapid increase in clasper length marks the onset of sexual maturity [18]. Under these criteria, the length of the claspers of No10–2 had already exceeded that of the pelvic fins, and its stiffness at 5 years and 4 months of age was similar to that of matured No6, this suggested that No10–2 had completed sexual maturation by this age. However, we could not confirm when exactly the clasper started to increase in size and exceeded the

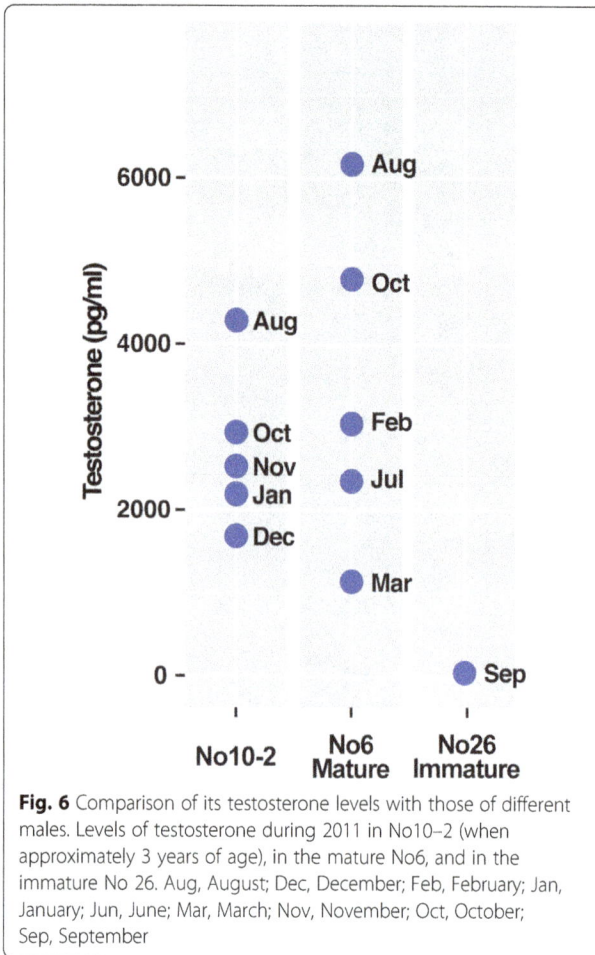

Fig. 6 Comparison of its testosterone levels with those of different males. Levels of testosterone during 2011 in No10–2 (when approximately 3 years of age), in the mature No6, and in the immature No 26. Aug, August; Dec, December; Feb, February; Jan, January; Jun, June; Mar, March; Nov, November; Oct, October; Sep, September

pelvic fins during our observation period. Therefore, in order to determine whether the timing of clasper elongation is consistent with other characteristics of sexual maturation (especially mating behavior and/or steroid hormone levels), further detailed observations are needed if a new male is born.

In this study, we observed that a *M. alfredi* born in captivity could exhibit mating behavior in the tank. Mating behavior of the giant manta ray, *M. birostris*, was observed in detail at Ogasawara Islands, Japan [12]; in this study, the serial mating behavior was classified into five steps: chasing, biting, copulating, post-copulating, and separating. In the case of *M. alfredi*, mating behavior observed in Mozambique also followed these five steps [7]. The present observations revealed that the captive male manta ray is able to engage in serial mating behavior typical of that seen in the field.

The first copulation of No10–2 was observed at 5 years and 4 months of age (in October 2013), a second one at 7 years and 3 months of age (in September 2015). Because this behavior was observed during visual monitoring in the day, we were unable to eliminate the

possibility that any other occasion may have been overlooked. However, in here, we discussed based on the copulations we observed actually. Most of the copulations in *M. alfredi* at the OCA were observed from May to July, indicating that their breading season in captivity lasted for several months [10]. However, copulation in 2015 occurred in September, which differed from that in the past. In Hawaii and the Maldives, *M. alfredi* exhibits two mating seasons in a given year [7, 19]. For Hawaii, it was suggested that females that are unsuccessful in getting pregnant during the first mating season may mate again during the second season [19]. In the present study, copulation occurred in September, which was an unusual time at the OCA, presumably because of the failure to copulate successfully during the ordinary breeding season (May to July).

The results of the present study showed the changes in steroid hormones over time in a male *M. alfredi* born in captivity. In particular, it is reported that the increase in androgens could be associated with puberty in male elasmobranch. In male bonnethead sharks, *S. tiburo*, androgen increased during puberty [17]. Additionally, in a male whale shark kept for over 20 years in the OCA, its testosterone levels definitely increased during the dramatic extension of his claspers (Matsumoto et al., personal communication), indicating that testosterone levels increased during sexual maturation in male whale sharks as well. On the other hand, in the present study, the androgen level of No10–2 did not exhibit such an upward trend during monitoring. Furthermore, testosterone levels in another mature male (No6) were the same range as those of No10–2 at 3 years of age (in 2011), whereas testosterone levels in the immature male (No26) were extremely lower than those in No10–2. On the basis of these facts, it is highly likely that No10–2 had already started the sexual maturation process endocrinologically before 3 years of age.

Conclusion

Although born in captivity, a male *M. alfredi* reached sexual maturity at the age of 5 years. This fact raises the possibility of complete breeding of *M. alfredi* in captivity. Importantly, the present study is the first to provide endocrinological data on a male *M. alfredi* in captivity. Since there have been limited published reports about captive breeding or copulation of *M. alfredi*, it is important that the present information obtained at the OCA be shared to achieve breeding of *M. alfredi* at other institutions. We hope to contribute to a better understanding of the reproductive biology of *M. alfredi* through long-term observations in captivity.

Additional files

Additional file 1: Movie S1. Semen collection. (MOV 9686 kb)

Additional file 2: Movie S2. Mating train. The top position is a female, the second is No10–2, and the third is another matured male. (MOV 15007 kb)

Additional file 3: Movie S3. First-time copulation of No10–2 recorded on October 6, 2013. (MOV 34609 kb)

Additional file 4: Movie S4. Motile sperm included in the semen. (MOV 760 kb)

Additional file 5: Table S1. Plasma steroid hormones levels and monthly average water temparature during the observation period in No10-2. (XLSX 13 kb)

Additional file 6: Table S2. Plasma testosterone levels in No6. (XLSX 9 kb)

Additional file 7: Table S3. Plasma testosterone level in No26. (XLSX 8 kb)

Abbreviations
DHT: Dihydrotestosterone; DW: Disc width; E2: Estradiol 17 beta; IOD: Interorbital distance; P4: Progesterone; T: Testosterone; The OCA: Okinawa Churaumi Aquarium

Acknowledgements
We are grateful to all of the Okinawa Churaumi Aquarium staff for their kind assistance.

Fundings
This work was supported in part by JSPS KAKENHI Grant Number 16 K21717 (RN) and 25,292,128 (MN).

Authors' contributions
RN, MN and KS conceived of the study, and participated in its design, and wrote the manuscript. RN performed steroid hormones assay. KM, RM, NY, YM, MY and KU collected blood samples, carried out the morphometry and observed behaviors. KM and RM carried out the semen collection and observation. All authors read and approved the final manuscript.

Competing interests
The authors declare that they have no competing interests.

References
1. Anderson RC, Adam MS, Kitchen-Wheeler A-M, Stevens G. Extent and economic value of manta ray watching in Maldives. Tourism Mar Environ. 2011;7(1):15–27.
2. O'Malley MP, Lee-Brooks K, Medd HB. The global economic impact of manta ray watching tourism. PLoS One. 2013;8(5):e65051.
3. Marshall AD, Compagno LJV, Bennett MB. Redescription of the genus Manta with resurrection of Manta alfredi (Krefft, 1868) (Chondrichthyes; Myliobatoidei; Mobulidae). Zootaxa. 2301;2009:1–28.
4. White WT, Corrigan S, Yang L, Henderson AC, Bazinet AL, Swofford DL, Naylor GJ. Phylogeny of the manta and devilrays (Chondrichthyes: mobulidae), with an updated taxonomic arrangement for the family. Zool J Linnean Soc. 2017; doi:https://doi.org/10.1093/zoolinnean/zlx018:.
5. Marshall A, Bennett MB, Kodja G, Hinojosa-Alvarez S, Galvan-Magana F, Harding M, Stevens G, Kashiwagi T. Manta birostris. In:The IUCN Red List of Threatened Species 2011. 2011. doi:https://doi.org/10.2305/IUCN.UK.2011-2.RLTS.T198921A9108067.en. Accessed Jan 2016.
6. Marshall A, Kashiwagi T, Bennett MB, Deakos M, Stevens G, McGregor F, Clark T, Ishihara H, Sato K. Manta alfredi. In:The IUCN Red List of Threatened Species 2011. 2011. doi:https://doi.org/10.2305/IUCN.UK.2011-2.RLTS.T195459A8969079.en. Accessed Jan 2016.
7. Marshall AD, Bennett MB. Reproductive ecology of the reef manta ray Manta alfredi in southern Mozambique. J Fish Biol. 2010;77(1):169–90.
8. Couturier LI, Marshall AD, Jaine FR, Kashiwagi T, Pierce SJ, Townsend KA, Weeks SJ, Bennett MB, Richardson AJ. Biology, ecology and conservation of the Mobulidae. J Fish Biol. 2012;80(5):1075–119.
9. Couturier LIE, Jaine FRA, Townsend KA, Weeks SJ, Richardson AJ, Bennett MB. Distribution, site affinity and regional movements of the manta ray, Manta alfredi (Krefft, 1868), along the east coast of Australia. Mar Freshw Res. 2011;62(6):628.
10. Matsumoto Y, Uchida S. Reproductive behaviour of manta rays (Manta birostris) in captivity. In: Proceedings of the 7th International Aquarium Congress: 2008; Shanghai, China; 2008. p. 123–37.
11. Uchida S, Toda M, Matsumoto Y: Captive records of manta rays in Okinawa Churaumi Aquarium. In: Joint Meeting of Ichthyologists and Herpetologists: 23-28 July 2008; Montreal, Can Underwrit.
12. Yano K, Sato F, Takahashi T. Observations of mating behavior of the manta ray, Manta birostris, at the Ogasawara Islands, Japan. Ichthyol Res. 1999;46(3):289–96.
13. Tomita T, Toda M, Ueda K, Uchida S, Nakaya K. Live-bearing manta ray: how the embryo acquires oxygen without placenta and umbilical cord. Biol Lett. 2012;8(5):721–4.
14. Nozu R, Nakamura M. Cortisol administration induces sex change from ovary to testis in the protogynous wrasse, Halichoeres trimaculatus. Sex Dev. 2015;9(2):118–24.
15. Asahina K, Kambegawa A, Higashi T. Development of a microtiter plate enzyme-linked immunosorbent assay for 17-alpha, 20-beta-21-trihydroxy-4-pregnen-3-one, a teleost gonadal steroid. Fisheries Sci. 1995;61(3):491–4.
16. Deakos MH. Paired-laser photogrammetry as a simple and accurate system for measuring the body size of free-ranging manta rays Manta alfredi. Aquat Biol. 2010;10(1):1–10.
17. Gelsleichter J, Rasmussen LEL, Manire CA, Tyminski J, Chang B, Lombardi-Carlson L. Serum steroid concentrations and development of reproductive organs during puberty in male bonnethead sharks, Sphyrna tiburo. Fish Physiol Biochem. 2002;26(4):389–401.
18. Clark E, von Schmidt K. Sharks of the central gulf coast of Florida. Bull Mar Sci. 1965;15(1):13–81.
19. Deakos MH. The reproductive ecology of resident manta rays (Manta alfredi) off Maui, Hawaii, with an emphasis on body size. Environ Biol Fish. 2012;94(2):443–56.

13

Indigenous house mice dominate small mammal communities in northern Afghan military bases

Christoph Gertler[1,11], Mathias Schlegel[1,12], Miriam Linnenbrink[2], Rainer Hutterer[3], Patricia König[4], Bernhard Ehlers[5], Kerstin Fischer[1], René Ryll[1], Jens Lewitzki[6], Sabine Sauer[7], Kathrin Baumann[1], Angele Breithaupt[8], Michael Faulde[9], Jens P. Teifke[8], Diethard Tautz[2] and Rainer G. Ulrich[1,10*]

Abstract

Background: Small mammals are important reservoirs for pathogens in military conflicts and peacekeeping operations all over the world. This study investigates the rodent communities in three military bases in Northern Afghanistan. Small mammals were collected in this conflict zone as part of Army pest control measures from 2009 to 2012 and identified phenotypically as well as by molecular biological methods.

Results: The analysis of the collected small mammals showed that their communities are heavily dominated by the house mouse *Mus musculus* and to a lesser extent *Cricetulus migratorius* and *Meriones libycus*. The origin of *M. musculus* specimens was analyzed by DNA sequencing of the mitochondrial cytochrome *b* gene and D-loop sequences. All animals tested belonged to the *Mus musculus musculus* subspecies indigenous to Afghanistan. The results were supported by detection of two nucleotide exchanges in the DNA polymerase gene of *Mus musculus* Rhadinovirus 1 (MmusRHV1), a herpesvirus, which is specific for all gene sequences from Afghan house mice, but absent in the MmusRHV1 sequences of German and British house mice. Studies of astrovirus RNA polymerase gene sequences did not yield sufficient resolution power for a similarly conclusive result.

Conclusions: House mouse populations in military camps in Northern Afghanistan are indigenous and have not been imported from Europe. Nucleotide sequence polymorphisms in MmusRHV1 DNA polymerase gene might be used as an additional phylogeographic marker for house mice.

Keywords: Military bases, Rodent, Shrew, *Mus musculus*, Phylogeography, Viruses, Public health, Afghanistan, ISAF, One health, Invasive species

Background

Modern warfare in the early 20st century is characterized by peacekeeping operations as well as counter-insurgency warfare [1]. Despite the advancements of logistics, medical support and technology, hygienic conditions in military bases in the Central Asian theatre or in Forward Operating Bases (FOBs) used for combat operations, still may support the spread of infectious agents [2–5]. Especially arthropod- or rodent-borne pathogens still pose a significant threat. Many of the current UN peacekeeping missions are located in areas that pose an infection risk with tropical viral, bacterial or protozoan pathogens [2–8]. The largest military operation of the last decade involving international personnel was the International Security Assistance Force (ISAF) in Afghanistan which started in December 2001. At the beginning of this study, it involved 56,420 military personnel from 50 countries in more than 700 military bases all over the country [1, 8, 9]. One of the largest bases in Northern Afghanistan was Camp Marmal near Mazar-e-Sharif, Balkh Province, Northern Afghanistan. The base did cater for up to 5500 military and non-military personnel from 16 countries.

* Correspondence: rainer.ulrich@fli.de
[1]Friedrich-Loeffler-Institut, Federal Research Institute for Animal Health, Institute of Novel and Emerging Infectious Diseases, Greifswald-Insel Riems, Greifswald, Germany
[10]German Center for Infection Research (DZIF), Partner Site Hamburg-Luebeck-Borstel-Insel Riems, Greifswald-Insel Riems, Greifswald, Germany
Full list of author information is available at the end of the article

Camp Marmal was set up as a Forward Support Base (FSB) and was extended to serve as a transportation hub for up to 112,000 personnel and 28,000 tons of supplies per year at peak times from 2005 to 2009. It was connected to the surrounding provinces via five stationary Provincial Reconstruction Teams (PRTs) and additional FOBs, built up temporarily if operationally needed [1]. Due to the large quantities of personnel and goods transported through Camp Marmal as well as the garrison size of this installation, it was an excellent habitat for rodents that may carry arthropod vectors and /or zoonotic pathogens.

Northern Afghanistan is an endemic region for a large variety of human pathogens that may be carried by rodents or arthropod vectors, e.g. *Salmonella typhi*, *S. typhimurium*, *Rickettsia* spp., *Leptospira* spp., *Coxiella burnetii*, *Leishmania* spp., *Giardia* spp., *Plasmodium* spp. as well as Crimean-Congo hemorrhagic fever virus and West Nile virus [2]. A major outbreak of a zoonotic disease occurred in the Camp Marmal area between 2004 and 2006. More than 4200 cases of Zoonotic Cutaneous Leishmaniasis (ZCL) were recorded within the local Afghan population and occurred mostly from August to November of every year within this period [3]. ISAF personnel were also affected with more than 200 ZCL cases diagnosed between 2004 and 2005 [4]. To prevent further infections of ISAF as well as employed civil personnel with *Leishmania* spp., a parasite with a rodent reservoir host, both major reconstruction works and pest control operations have been conducted throughout the further operation time of Camp Marmal [5] by a dedicated pest control unit. As Afghanistan is home to a wide range of rodents which may harbor pathogens [10, 11], a seasonal monitoring for rodents and mosquitos was implemented to verify the effectiveness of used pest control measures and to reassess the risk of infectious diseases present inside military camps of German responsibility. In addition to Camp Marmal, pest control measurements were set in place at the PRTs Kunduz and Fayzabad as well as inside smaller FOBs if feasible concerning the security situation.

Transmission of small mammal-borne pathogens within military bases such as Camp Marmal could take place through ubiquitous rodents found in almost any human settlement, such as *Rattus* spp. or *Mus* spp.. Most prominent within these genera is the house mouse *Mus musculus*, a highly mobile, omnivorous and oligotrophic commensal with birth rates of 5–10 litters and with an average of six to eight young per litter [12]. Because of these high birth rates, house mouse populations are characterized by a strong resilience towards predation and pest control measures. These animals are considered an invasive species especially on island locations and caused devastations

to the biodiversity of New Zealand as well as several island ecosystems of the South Atlantic and Indian Oceans during European exploration in the 19th century [13, 14]. This may be due to quick genotypic and phenotypic adaptations (e.g. adaptation to local pathogens or the local climate and food spectrum) in relatively short periods of time after invading an island [15].

Personnel and goods were initially transported into Camp Marmal from Germany and other international locations primarily via air transport from Termez (Uzbekistan) or Kabul (Afghanistan) as well as by train especially during the initial work-up phase until 2009. Following the construction of a runway and throughout the collection time of this study, most goods were airlifted from Germany (Cologne Airport, Leipzig Airport) directly to Camp Marmal [1]. Although pest control measures were in place in Germany before loading, delays during consignment in Kabul or Termez and "stand stills" during train rides could not be avoided. Therefore, it cannot be excluded that commensalic rodents such as *M. musculus* were transferred into Camp Marmal within the actual cargo and subsequently carry pathogens into the country.

This study aims at analyzing the small mammal communities available during pest control in Camp Marmal and the PRT Kunduz and PRT Fayzabad bases within the Northern Afghan theatre of military operations. The three bases were chosen as collection sites as they were visited by the pest control unit of the German Army. Collections were conducted as regular as possible as permitted by the military situation and necessity for pest control. The small mammal communities were analyzed by both phenotype and several molecular biological methods such as PCR-based analysis of the mitochondrial cytochrome *b* gene and D-loop region. Finally, we compared DNA polymerase (*DPOL*) gene sequences of a herpesvirus, *Mus musculus* rhadinovirus 1 (MmusRHV1, a double stranded (ds) DNA virus) as well as RNA-dependent RNA polymerase (*RdRp*) gene sequences of murine astroviruses (single stranded (ss) RNA viruses) in *M. musculus* specimens collected at Camp Marmal in Afghanistan and several locations in Europe. In the absence of serious zoonotic viruses in the Afghan specimens, these viruses were selected to investigate their suitability as additional phylogeographic markers, as they were detected in a significant number of Afghan house mouse specimens, have a high host-specificity and low or lacking zoonotic potential.

Methods
Small mammal trapping and transfer procedures
Small mammal collections were conducted by a dedicated Army pest control team (Sanitäts-Hygiene-Trupp)

if feasible concerning the security situation, as described before [16]. Briefly, a total of 751 small mammals were trapped using mouse and rat snap traps or collected as rodenticide-poisoned carcasses with a daily control of all sampling sites from January 2009 to October 2012 after the initial work-up of the three German Military Camps based on the implemented pest control management. Trapping sites were disseminated inside the camps with special focus on sites with reported or suggested occurrence of rodents as well as vulnerable places like kitchens or food storage facilities. The size and location of the trap used was based on apparent environmental factors, the local habitat and the knowledge of small mammal behavior as decided by veterinary personnel. For the collection site Hazrat-e Sultan, a statistically insignificant number of five animals were collected. For ten animals, no collection site could be determined during dissections. Hence, these 15 animals were omitted in the in-depth analysis of this study. All collected small mammals were registered by the veterinary medical staff of Bundeswehr, stored in aluminum containers at −20 °C prior to transportation and dissection at the Friedrich-Loeffler-Institut, Greifswald-Insel Riems, Germany. Dissections were conducted according to standardized protocols in a biosafety level 3 containment laboratory. Weight, length (body and tail) and sex of each specimen and number of embryos of pregnant females were determined for future analysis on their association with pathogen prevalence. The phenotypical identification of small mammal species was accompanied by photo documentation of each animal. Tissues were removed in specific order as described before [16], placed in 1.5 ml reaction tubes and stored at −20 °C until further experimentation. Control animals were selected to reflect the diversity of European house mice with specific regard to the presence of both *Mus musculus domesticus* in the Western part of Germany and *Mus musculus musculus* in the Eastern part of Germany. As airlifts to Afghanistan were conducted from both areas (Cologne airport and Leipzig airport), animals from Western Germany and North Western Czech Republic (representing the Central European *M. m. musculus* population) were selected. The trapping of control animals for astrovirus and herpesvirus investigations at five sites in Germany and one site in the United Kingdom have been reported previously [13, 17, 18]. As no morphological vouchers could be preserved in this study, some preliminary species identifications were based on the photographic documentation and comparisons with museum specimens.

Nucleic acid extraction, PCR amplification of cytochrome *b* gene and DNA sequence analysis

Molecular species identification of small mammals was conducted with a modified protocol based on a method described before [19]. Briefly, frozen small mammal tail or ear pinna tissue was cut into slices of 0.5 mm by 0.5 mm with a sterile scalpel, transferred into a 1.5 ml reaction tube and immersed in 300 µl of tissue lysis buffer (50 mM KCl, 10 mM Tris-HCl pH 9.0, 0.45% (v/v) Nonidet P40, 0.45% (v/v) Tween 20 (Sigma-Aldrich, Munich, Germany)). Three microliters of a 10 mg ml^{-1} proteinase K (Sigma-Aldrich) solution were added and the reaction mixture was incubated over night at 56 °C with mixing intervals of 10 s and 1000 rpm and pauses of 3 min in a thermomixer (Eppendorf, Hamburg, Germany). Residual tissue was pelleted by centrifugation at 16,000×g for 1 min in a tabletop centrifuge (Eppendorf) and lysates were transferred to another reaction tube. Remaining proteinase K was inactivated by incubation at 95 °C for 10 min. An additional tenfold dilution of each lysate was prepared in sterile water. Each 22.5 µl of PCR master mix contained 17.65 µl of sterile water, 2.5 µl of 10× PCR buffer, 250 µM of each dNTP, 1.5 mM of $MgCl_2$ (all Qiagen, Hilden, Germany), 200 nM of each primer cytochrome b-F (5`-TCA TCM TGA TGA AAY TTY GG-3`) and cytochrome b-R (5`-ACT GGY TGD CCB CCR ATT CA-3`) (MWG-Eurofins, Munich, Germany), and 0.1 µl of Platinum Taq Polymerase (Life Technologies, Darmstadt, Germany). For each template, 2.5 µl of undiluted extracts were applied in separate reactions. PCR reaction programs contained an initial denaturation step of 94 °C for 3 min, followed by 40 cycles of 94 °C for 1 min, 53 °C for 30 s and 72 °C for 1 min. A final elongation step was conducted at 72 °C for 10 min. All PCR products were visualized on 1.5% agarose gels after electrophoresis at 100 V for 30 min and ethidium bromide staining on an INTAS Classic gel documentation system (INTAS, Göttingen, Germany). For all reactions that failed to amplify, tenfold diluted DNA extracts were used for an additional PCR amplification trial as described above. Due to poor cytochrome *b* sequence quality in some *Meriones* specimens, the DNA extraction was repeated from kidney tissue using EURx GeneMATRIX Tissue DNA Purification Kit (Roboklon, Potsdam, Germany) and followed the protocol of the manufacturer. The PCR protocol presented above was modified by replacing buffer and polymerase by 5 µl of Colorless GoTaq Flexi buffer (Promega, Mannheim, Germany) and 1 U of GoTaq Polymerase (Promega) while reducing the amount of sterile water to 15.15 µl.

PCR products were purified with the Nucleospin kit (Macherey-Nagel, Düren, Germany). Sequencing reactions were performed using the ABI Big Dye v3.1 kit (Applied Biosystems, Foster City, CA, USA) according to the manufacturer's guidelines. Products of the sequencing reactions were purified with a NucleoSeq kit (Macherey-Nagel) and sequencing was performed on an AP 310 Genetic Analyzer (Applied Biosystems) in both forward and reverse direction. Resulting sequences were assembled and edited with BioEdit (version 7.2.5) using the Contig Alignment Program subroutine [20] and tested for chimeric sequences with the Pintail 1.1 program [21]. Molecular species determination was conducted by comparison of sequences obtained in this study to published sequences using the Megablast algorithm of the BLAST program [22]. Phylogenetic analysis of all small mammal specimens was conducted by sequence alignment in BioEdit (version 7.2.5) and ClustalW [20]. Phylogenetic tree construction was conducted via CIPRES portal, using the FastTreeMP on XSEDE (2.1.9) routine including 1000 bootstraps; the Jukes-Cantor substitution model with Gamma distribution was used [23].

Sequence determination and analysis of the mitochondrial D-loop

A 647 base pair (bp) fragment of the mitochondrial D-loop was either sequenced as described before [24] or in several cases with an alternative primer pair (forward 5′-CTG AAT CCT AGT AGC CAA CC-3′ and reverse 5′-AGT GTT TTT GGG GTT TGG C-3′) using the same conditions. DNA sequencing reactions were conducted with the ABI Big Dye v3.1 Kit (Applied Biosystems) according to the manufacturer's guidelines and analyzed with an ABI 3730 automated sequencer. Raw sequence data were edited using the software CodonCodeAligner 4.04 (CodonCode Corporation, Centerville, MA, USA) and aligned using the ClustalW algorithm implemented in the program MEGA 6.06 [25]. Sequences of *M. m. domesticus* (Norway, Denmark, Germany and France) [26, 27] and *M. m. musculus* (Kazakhstan and Czech Republic) [26] were included as references. Gaps and indels were excluded from the analysis. A Median Joining Network [28] was constructed using the program PopART (http://popart.otago.ac.nz).

Generic PCR for detection of herpesviruses

DNA was extracted from spleen tissue of 18 house mouse specimens each from an urban area in Cologne (Germany), from a city park in Saarbrücken (Germany) and a rural setting in Radolfzell (Germany) as well as from 36 specimens from Camp Marmal (Afghanistan). DNA samples were obtained with the EURx GeneMATRIX Tissue DNA Purification Kit (Roboklon) according to the protocol "A" described in the manufacturer's guidelines. DNA was visualized on 0.8% agarose gels to assess DNA quality and quantity. PCR amplification of the cytochrome *b* gene was conducted for each sample as described above to determine DNA amplificability. Amplification of the *DPOL* gene was conducted as described before [17]. Briefly, PCR was performed in nested format using a mixture of six degenerated/deoxyinosine substituted primers and delivering an expected

600 bp DNA fragment. The PCR reaction mix of 25 µl contained 5.25 µl of DNase/RNase free sterile water, 5 µl of Colorless GoTaq Flexi buffer (Promega), 2 µl of 25 mM $MgCl_2$, 5 µl of 5 M betaine (Sigma-Aldrich), 0.5 µl of a dNTP mix containing 10 µM of each dNTP (Promega), 1.25 µl of dimethyl sulfoxide (DMSO), 2 µl of a primer mix containing 100 nM of each primer DFA[n/i] f (GAY TTY GC(N/i) AGY YT(N/i) TAY CC, ILK[n/i] f (TCC TGG ACA AGC AGC AR(N/i) YSG C(N/i)M T(N/i) AA as well as KG1[n/i] r (GTC TTG CTC ACC AG(N/i) TC(N/i) AC(N/i) CCY TT, 1 U of GoTaq Polymerase (Promega) and 2 µl template DNA. PCR conditions involved an initial denaturation step of 94 °C for 10 min followed by 40 cycles of 95 °C for 20 s, 46 °C for 30 s and 72 °C for 30 s. A final extension step of 72 °C was conducted for 7 min.

Two microliters of each reaction were applied in a nested PCR reaction using the PCR conditions described above but applying the primers TGV [n/i] f (TGT AAC TCG GTG TAY GG(N/i) TTY AC(N/i) GG(N/i) GT and IYG [n/i] r (CAC AGA GTC CGT RTC (N/i)CC RTA (D/i)AT), a reduced extension time of 20 s and total amount of 30 cycles. Resulting PCR products were subjected to gel electrophoresis in a 3% agarose gel in 1× TAE buffer at 100 V for 45 min. Bands were excised with a scalpel blade and purified using the QiaEx II gel extraction kit (Qiagen) according to the manufacturer's guidelines. Sequencing reactions using the Big Dye v3.1 kit (Applied Biosystems) and DNA sequence analysis were performed as described above. Resulting nucleotide and in silico translated amino acid sequences were amended by sequences from further sites in Germany and the United Kingdom from a previous study as described therein [17] and a reference sequence (accession number AY854167). Alignments of viral nucleotide and amino acid sequences were conducted with BioEdit (version 7.2.5) and ClustalW [20].

Generic RT-PCR for detection of astroviruses

RNA was extracted from the liver tissue of a total of 120 house mice that originated from Camp Marmal (36 specimens), Saarbrücken, Cologne and Radolfzell (12 specimens each), Gelsenkirchen (urban), Lünen (rural), Stuttgart (animal husbandry) (8 specimens each) as well as from a mouse breeding facility (24 specimens) using the Genematrix DNA/RNA/Protein Universal Purification kit (Roboklon). RNA quality and quantity was assessed as described for DNA extracts above. Amplification of the *RdRp* gene was conducted with a modified hemi-nested RT-PCR protocol [29]. Briefly, a master mix of 25 µl each containing 12.5 µl of Superscript® III One-step PCR Buffer, 400 mM $MgSO_4$, 10 µM each of the forward primers AS1-f (5′-GARTTYGATTGGRCKCGK TAYGA-3′) and AS2-f (5′-GARTTYGATTGGRCKAG

GTAYGA-3') and 20 μM of the reverse primer AS3-r (5'-GGYTTKACCCACATNCCRAA-3') as well as 1 μl of the Superscript® III One-step PCR reagent and 2.5 μl of the RNA extracts were applied in each reaction using the PCR program described before [29].

A second PCR was conducted using a master mix of 25 μl each containing 2.5 μl of Peqlab Pwo Complete buffer (VWR International, Erlangen, Germany), 50 μM of each dNTP, 10 μM each of the forward primers AS4-f (5'-CGKTAYGATGGKACKATHCC-3) and A5-f (5'-AGGTAYGATGGKACKATHCC-3'), 20 μM of primer AS3-r and 1 U of Peqlab Pwo polymerase (VWR International) and 2.5 μl of the initial PCR mix using the PCR program described before [29]. Resulting PCR products were visualized on a 3% agarose gel. All samples delivering the expected 422 bp PCR products were subjected to an additional hemi-nested PCR using 50 μl reactions containing identical final concentrations of all chemicals and 5 μl of RNA extracts to generate additional amounts of PCR product. The complete reactions were subjected to gel electrophoresis on a 3% agarose gel. All bands of approximately 422 bp were extracted from the gel and purified using the QiaEx II kit (Qiagen) according the manufacturer's guidelines. Sequencing of the resulting bands was conducted with the Big Dye v3.1 kit (Applied Biosystems) and DNA sequence analysis was performed as described above. For the sequencing reactions, final concentrations of 1 μM each for a mix of the primers AS4-f and A5-f as well as 2 μM of primer AS3-r were applied. Alignment of viral sequences was conducted as described for herpesviruses. Phylogenetic analysis was performed via CIPRES portal, using the FastTreeMP on XSEDE (2.1.9) routine including 1000 bootstraps with the Jukes-Cantor substitution model and ClustalW, including a discrete Gamma distribution [23].

Results

Small mammal trapping

A total of 751 small mammals were trapped in Northern Afghanistan from January 2009 to October 2012. The vast majority of small mammals in this study originated from Camp Marmal (648) while lower numbers were trapped in PRT Fayzabad (27) and PRT Kunduz (61). A total of around 220 mammals were trapped each in 2009 and 2011, while around 100 animals were trapped in each 2010 and 2012. In total, small mammal communities in all three camps consisted of no more than four members of the order Rodentia and two representatives of the order Soricomorpha (Fig. 1). The house mouse (*Mus musculus*) comprised more than 80% of trapped animals overall, ranging from 67 to 98% in all three locations. A total of 629 house mouse specimens were collected which comprised 377 males and 248 females. The

sex of the four remaining specimens could not be determined. Thirty-three females (13.3%) were pregnant at the time of trapping and carried a total of 205 embryos (an average of 6.2 embryos per female, ranging from 1 to 10 embryos per female). Among the remaining rodents trapped, the gray dwarf hamster (*Cricetulus migratorius*) was most prominent with an average percentage of 12.8% of trapped animals which comprised 47 females and 48 males. Six females (12.8%) were pregnant and carrying an average of 6.5 (ranging from 5 to 9) embryos. The abundance of *C. migratorius* specimens ranged from 0 to 33.3% of trapped animals in all three sites. Remaining trapped animals could be identified as members of the rodent genera *Meriones* (2.1%), *Rattus* (0.1%), and the shrew genera *Suncus* (0.1%) and *Crocidura* (1.1%). Despite the different numbers of trapped animals in all three sites and a general trend towards a *Mus musculus* (*C. migratorius*) dominated small mammal community in Northern Afghan military bases, there were subtle differences between individual bases. In contrast to Camp Marmal, small mammal communities at PRT Fayzabad contained a higher percentage of *C. migratorius* specimens but no other small mammals. In contrast to this, the mammal community at PRT Kunduz was entirely dominated by *M. musculus* but contained a single trapping of *Crocidura* cf. *suaveolens*. Despite a variation in trapping success (Additional file 1: Figure S1) percentages of genera collected at individual annual quarters represent the average community composition. In contrast to all other small mammal species in this study both *M. musculus* and *C. migratorius* specimens could be collected throughout the whole period from 2009 to 2012.

Identification of small mammals by DNA sequencing of the mitochondrial cytochrome *b* gene

Specimens initially determined as *M. musculus* showed the highest average identity of the partial cytochrome *b* gene of 99.1% (99% to 100%) to reference *M. musculus musculus* sequences from Central Asia (KF697060; AB649551) and a lower average identity of 97.7% (97% to 98%) to various sequences of *M. musculus domesticus* (Additional file 2: Table S1). Animals identified as gray dwarf hamsters showed an average identity of 96.8% (91% to 98%) to a reference *C. migratorius* sequence (AY288508) originating from Northeast Pakistan (Additional file 2: Table S2). All 16 jirds subjected to DNA sequencing displayed an average identity of the mitochondrial cytochrome *b* gene of 96.9% (93% to 99%) to that of the Libyan Jird (*Meriones libycus*; AJ851266; see Additional file 2: Table S2). This high sequence similarity was also reflected in a corresponding phylogenetic tree (data not shown). The single *Rattus* specimen collected in this study showed 93% sequence identity to *R. tanezumi* (JQ793907; Additional file 2: Table S2).

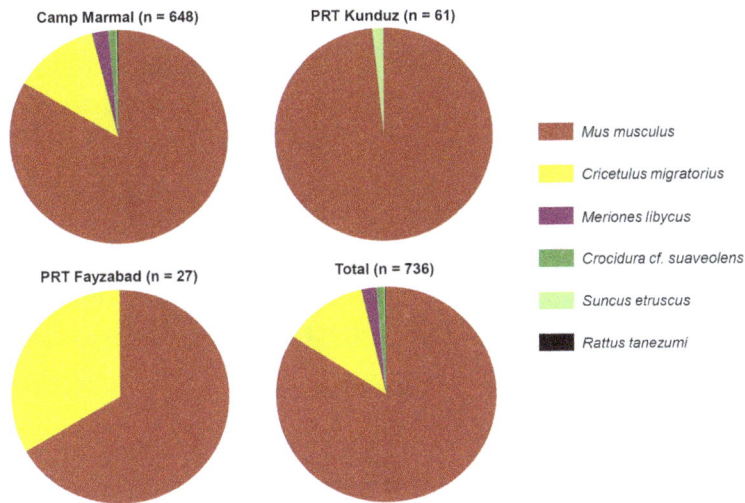

Fig. 1 Percentages of individual species within small mammal communities in Northern Afghan military bases Camp Marmal (Mazar-e-Sharif) as well as PRTs (Provincial Reconstruction Team bases) Kunduz and Fayzabad. Total numbers for each site were *n* = 648 for Camp Marmal, *n* = 27 for PRT Fayzabad and *n* = 61 for PRT Kunduz. Five additional animals were collected at Hazrat-e-Sultan; for 10 animals no trapping site information was available

Among *Crocidura* specimens, members of *Crocidura* cf. *suaveolens* were detected (DQ630062; Additional file 2: Table S3). The *Suncus* species showed high identity of the cytochrome *b* gene to *Suncus etruscus* (FJ716836; Additional file 2: Table S3).

Analysis and comparison of the mitochondrial D-loop from *Mus musculus* specimens

The subspecies status of house mice was determined by sequencing of the mitochondrial D-loop region of animals trapped in this study and reference sequences of the Western house mouse (*M. m. domesticus*) and Eastern house mouse (*M. m. musculus*) which differ in both genotype and phenotype. A phylogeographic analysis including 214 mitochondrial D-loop sequences revealed mice trapped in this study belong to the *M. m. musculus* clade, and therefore cluster with samples from Kazakhstan and the Czech Republic which serve as a reference for mice from neighboring Germany (e.g. the Leipzig area; Fig. 2). The "*musculus*"-haplogroup is clearly distinct from *M. m. domesticus* (from various Western European countries, Africa and South West Asia). Furthermore, Fig. 2 gives evidence for separation of Afghan house mice D-loop sequences from sequences of house mice from the Czech Republic.

Comparison of DNA sequences of *Mus musculus* rhadinovirus 1 detected in *M. musculus* of different geographic regions

Using generic *DPOL* gene PCR for herpesvirus detection in house mouse samples from Camp Marmal and three locations in Germany, a total of 14 out of 114 reactions produced products of approximately 250 bp length each

(Fig. 3). Five out of 60 samples from Camp Marmal, six out of 18 samples from Cologne, two out of 18 samples from Saarbrücken and one out of 18 samples from Radolfzell delivered positive results. All 14 viral sequences showed 97% similarity to a single MmusRHV1 reference sequence described before (AY854167) [17]. A comparison of the novel sequences from Camp Marmal/ Afghanistan (MeS), Cologne (COL), Saarbrücken (SBR) and Radolfzell (RAZ), and those for the Leverkusen (LEK), Roklum (ROK) and Liverpool (LIV) sites generated in an earlier study [17] indicated that all sequences from any given site were identical to each other, with exception of one sequence from the Leverkusen site (number LEK_1; Fig. 3). Viral sequences for each trapping location showed a unique *DPOL* DNA sequence with the exception of identical sequences from the sites Cologne (COL_1 – COL_6)/Radolfzell (RAZ_1) and Liverpool (LIV_1 – LIV_4)/Leverkusen (LEK_2 – LEK_9). The other European MmusRHV1 DNA sequences showed a difference in a single nucleotide (Roklum, ROK, position 148; Saarbrücken, SBR, position 158). Similarly, sequences obtained from Camp Marmal/ Afghanistan differed from all European sequences in two positions (10 and 157). Partial amino acid sequences of the deduced DPOL protein show that most of the nucleotide differences are silent and do not translate into amino acid substitutions (Additional file 3: Figure S2). DPOL amino acid sequences from MmusRHV1 detected in house mice from Camp Marmal and all other sites were almost identical, however, all sequences from Saarbrücken showed a substitution of isoleucine for valine in position 53 and all sequences from Roklum including the reference sequence displayed a substitution

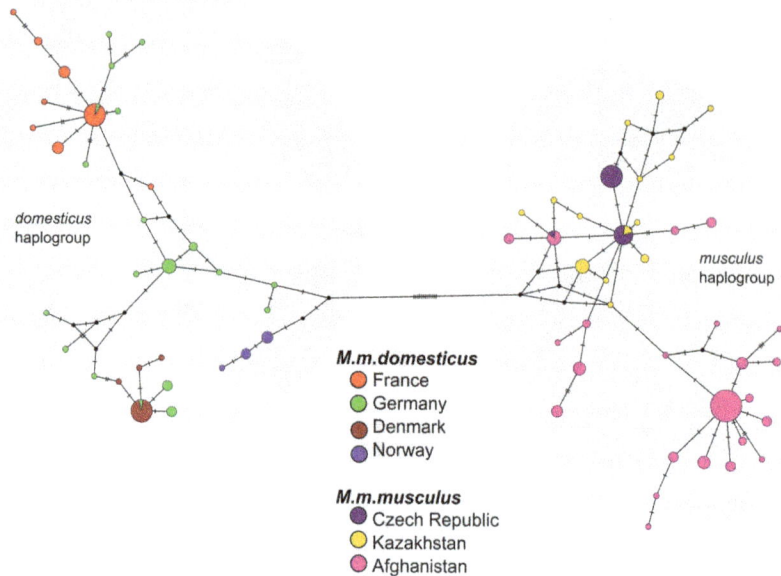

Fig. 2 Median Joining Network of 214 mitochondrial D-loop sequences of house mice. The relationship between individuals of seven populations, representing the two most prominent subspecies of house mice *M. m. musculus* and *M. m. domesticus* are shown. Mice collected in Afghanistan could be clearly attributed to *M. m. musculus*. The number of mutation steps is given as hatch marks

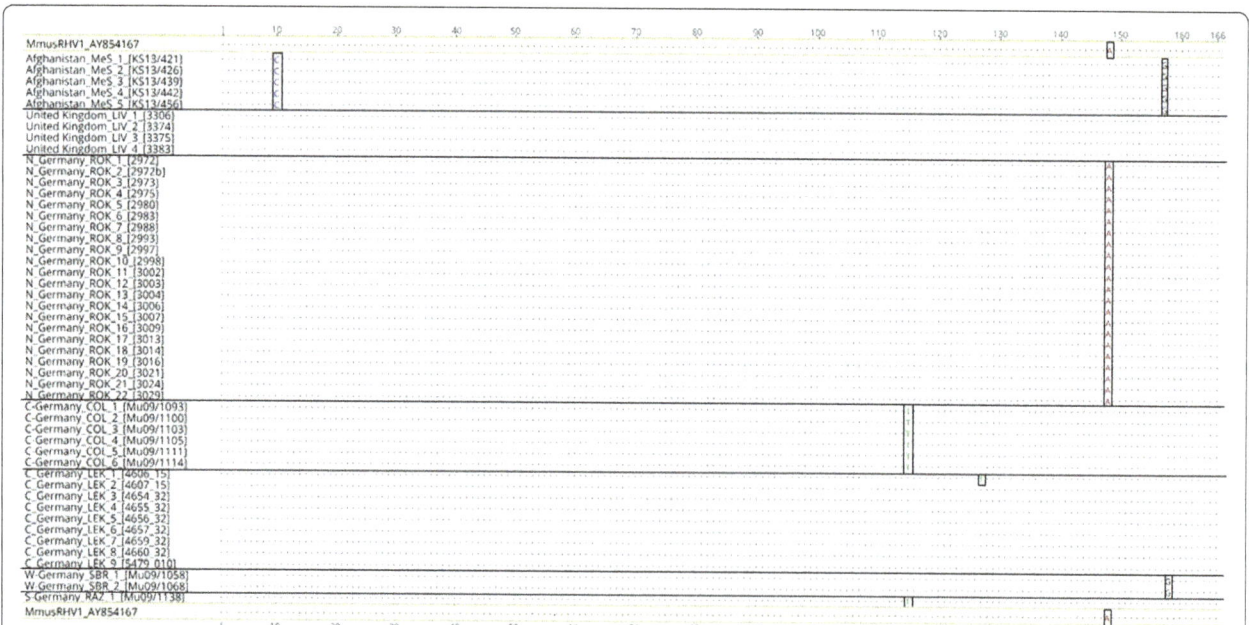

Fig. 3 Comparison of partial DNA sequences of the *DPOL* gene of MmusRHV1 from *Mus musculus* specimens collected in Afghanistan, Germany and the United Kingdom to a reference sequence from Germany. All sequences from Afghanistan originated from Camp Marmal (Mazar-e-Sharif (MeS)), whereas all sequences from the United Kingdom originated from Liverpool (LIV). Sequences from Germany resulted from specimens trapped in Northern (N) Germany (Roklum (ROK)), Central (C) Germany (Cologne (COL) and Leverkusen (LEK)), Southwest (W) Germany (Saarbrücken (SBR)) as well as South (S) Germany (Radolfzell (RAZ)). Nucleotide exchanges in the *DPOL* sequence of MmusRHV1 strains are marked. Included sequences were generated in this study (Afghanistan: 5, Cologne: 6, Saarbrücken: 2, Radolfzell: 1; *n* = 14) or have been determined in a reference study (Leverkusen: 9, Roklum: 22, Liverpool: 4; *n* = 35; [17])

of glutamic to aspartic acid in position 49 compared to all other sequences.

Detection of astrovirus RNA sequences in *M. musculus* from Afghanistan and Germany

The generic astrovirus RT-PCR showed for 13 out of 120 samples products of about 400 bp. Seven out of 36 samples from Camp Marmal, four out of eight samples from Gelsenkirchen and Lünen, one out of eight samples from Stuttgart and one out of 12 samples from Radolfzell delivered positive results. No positive samples were obtained from the mouse breeding facility. Two distinct types of astroviruses were observed: a sample from Radolfzell as well as one out of seven samples from Camp Marmal displayed 84% identity to rat astroviruses (e.g. rat astrovirus RS297, accession number HM450385), whereas all other astrovirus sequences detected in this study showed an average of 92% (range: 89% to 93%) identity to house mouse astroviruses (e.g. astrovirus TF20LM, accession number JQ408742). A phylogenetic analysis of the 13 novel sequences and 33 astrovirus reference sequences highlights this apparent difference (Fig. 4).

Discussion

Composition of the small mammal community of FSB camp Marmal and PRTs Fayzabad and Kunduz

Military bases of the 21st century offer a multitude of ecological niches for commensal rodents within artificial human environments. This is especially true for temporary or makeshift bases set up in countries with infrastructure deteriorated by war such as Afghanistan, which offer a unique combination of high accessibility and large amounts of resources for commensalic small mammals capable of exploiting them. As a result, rodent communities determined in this study were strongly dominated by *M. musculus*. The overall small mammal diversity in all trapping sites was very low with a total of six small mammal species, four of which belonged to the order Rodentia. The highest diversity of small mammals was observed for the Camp Marmal site. In contrast, only two species each were encountered at the PRT Kunduz and the PRT Fayzabad site, albeit the total amount of animals trapped at both PRTs was significantly lower for PRT Kunduz (n = 61) and PRT Fayzabad (n = 27) than for Camp Marmal (n = 648). Among the small mammal fraction were sporadic trappings of insectivores, such as *Crocidura* cf. *suaveolens* and *Suncus etruscus* which could have been attracted by invertebrates feeding on such bait. Only a single specimen of the genus *Rattus* was collected during the study time of almost three years. This specimen had a high identity of the cytochrome *b* gene to *R. tanezumi*, a species commonly encountered in Central Asia. Gray dwarf hamsters (*Cricetulus migratorius*) and jirds (*Meriones libycus*) proved to be the second and

third most abundant mammals trapped. Gray dwarf hamsters accounted for 12.8% of trappings. This omnivorous (granivorous/insectivorous) rodent is widespread through Central Asia and can inhabit a large variety of grassland habitat, especially at high altitudes. In contrast to *Mus musculus* this species prevails in very dry habitats and was reported to have a significantly lower litter size and birth rate [30]. However, the results of our study show that both the possible litter sizes and the percentage of pregnant females of both species were similar. Similar to *C. migratorius*, *Meriones libycus* trapped in this study are relatively widespread burrowing rodents. These animals are strongly adapted to arid environments and can entirely cover their water supply through their food sources [31].

The small mammal species trapped in this study however only represents a snapshot of the true small mammal diversity of Afghanistan. While the predominant species detected in this study, *Mus musculus*, *Meriones libycus* and *C. migratorius* have been observed in previous studies from Afghanistan [10, 11] and a study from Northeastern Iran [32], the total small mammal diversity of this region also comprises members of the genera *Calomyscus*, *Alticola*, *Blanfordimys*, *Ellobius*, *Microtus*, *Allactaga*, *Jaculus*, *Dryomys*, *Salpingotus*, *Gerbillus*, *Apodemus*, *Nesokia*, *Rattus*, *Tatera* and *Sorex*.

The limited small mammal diversity detected at the three sampling sites Mazar-e-Sharif, Fayzabad and Kunduz within the dataset however is not unusual for urban settings. Previous studies from Vancouver, Canada [33], and Buenos Aires, Argentina [34], have shown that urban small mammal communities are dominated by *Rattus norvegicus*, *Rattus rattus* or *Mus musculus*. A pivotal factor determining which of the three species dominates an urban environment is the structure and microenvironment provided by the specific location. These species excel in microenvironments that contain man-made structures as well as rural habitats, e.g. shanty towns or parklands, but struggle to compete with other rodents in less urbanized regions. The three trapping sites investigated in this study were semi-temporary urban environments with an unusually high transition rate of human personnel and goods. Due to the military use of the trapping sites, density of human personnel per square meter was very high and accommodation was available in the shape of container architecture. For logistical and strategic reasons, all three trapping sites were located adjacent to local air strips or at the outskirts of the actual cities of Mazar-e-Sharif, Fayzabad and Kunduz and were embedded in a predominantly rural surrounding. This setting therefore may select small mammal communities for commensals such as *M. musculus* due to both easy accesses in and out of the base but also offers easy access to food resources. In

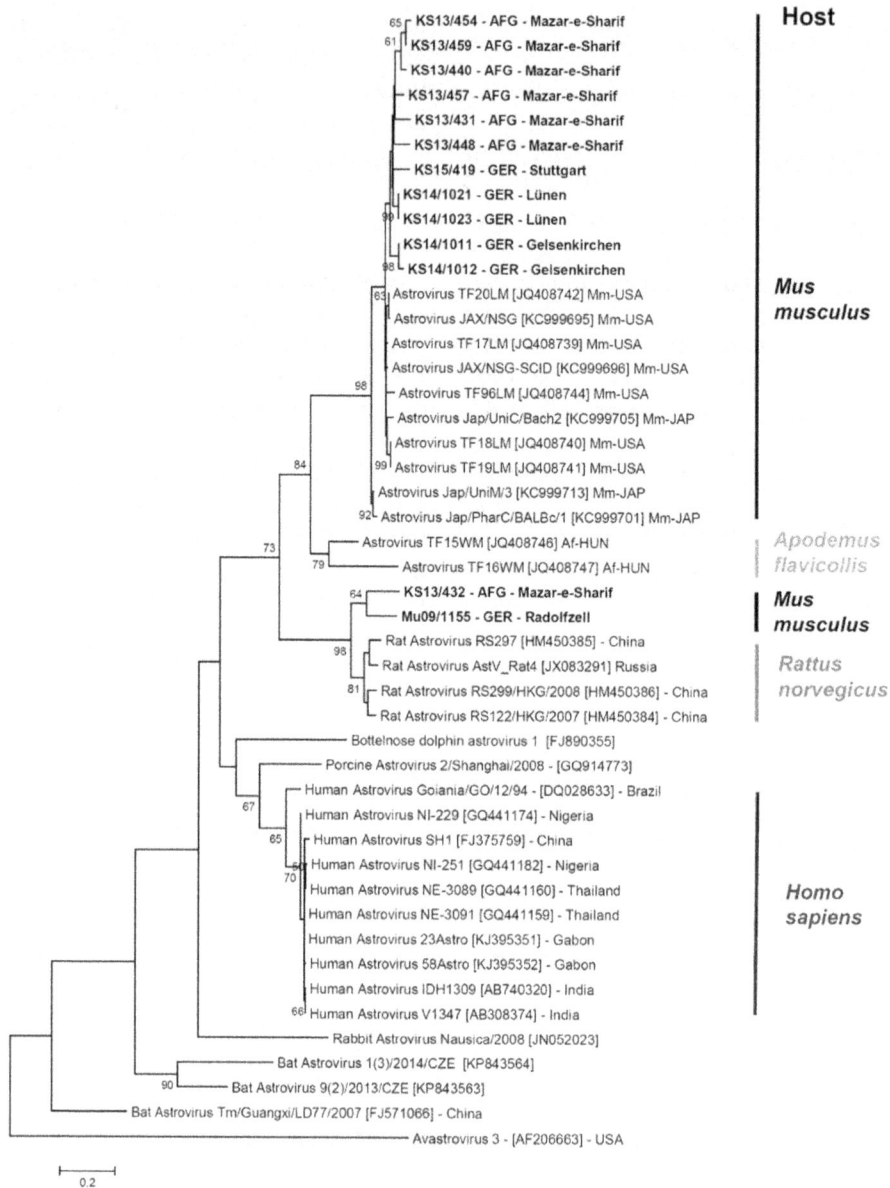

Fig. 4 Molecular phylogenetic analysis of astrovirus sequences from house mice collected in Afghanistan and Germany as well as reference sequences from the United States and Japan by Maximum Likelihood method. The evolutionary history was inferred by using the Maximum Likelihood method based on the Jukes-Cantor Model. Bootstrapping with 1000 replications was performed and support values below 50% were omitted in the figure for clarity. A discrete Gamma distribution was used to model evolutionary rate differences among sites (5 categories (+G). The analysis involved 46 nucleotide sequences. All positions containing gaps and missing data were eliminated. There were a total of 231 positions in the final dataset. Evolutionary analyses were conducted via CIPRES portal using the FastTreeMP on XSEDE (2.1.9) routine [23]

contrast to the patchy rodent population composition and density in an urban environment observed in Vancouver [33], the communities in Camp Marmal, PRT Kunduz and PRT Fayzabad showed a relatively similar composition concerning the dominance of house mice and occurrence of gray dwarf hamsters. This could be due to the modular and standardized design of makeshift military bases developed for missions such as ISAF. Similar to rodent communities observed in Buenos Aires [34], Camp Marmal and the PRTs Kunduz and Fayzabad also contained hemerophobic rodent species, e.g. *C. migratorius* and *Meriones libycus*. A previous study conducted in habitats similar to the Afghan trapping sites in Uzbekistan showed presence of both species and indicated preferences for similar microenvironments and food sources [30]. While *C. migratorius* is a granivorous and insectivorous rodent that prefers rocky or gravely surfaces, *Meriones libycus* was primarily trapped on sandy ground and prefers green vegetation and seeds [30, 31].

The results of this study greatly vary from an earlier report on rodent sightings in Camp Marmal in the period from 2004 to 2006. Within this time, a major ZCL outbreak occurred in this area that involved more than 4200 cases within a period of 36 months and was attributed to a high abundance of great gerbils (*Rhombomys opimus*) and transmission by *Phlebotomus papatasi* sandflies [3]. Interestingly, no specimens of *R. opimus* could be detected among the rodents collected in the present study. The presence of *R. opimus* in the vicinity of humans in Northern Afghanistan is associated with infrastructure degradation due to the Afghan Civil War and agricultural changes e.g. the construction of irrigation canals for intensified crop in the 1980's. In case of Camp Marmal, the massive increase of the local *R. opimus* population was attributed to banks of loose soil erected during construction works in the vicinity of the base. By 2006, the soil banks were removed, graveled or replaced by a concrete wall surrounding parts of the base [5]. These modifications in combination with pest control measures, the reconstruction of infrastructure as well as the introduction of food sources more suitable to commensalic rodents (e.g. *Mus musculus*) therefore may have changed the habitat to the disadvantage of *R. opimus* while simultaneously preventing from *Leishmania* spp. infection by interrupting the local main transmission mode [5].

Origins of house mice (*M. musculus*) trapped in military bases of northern Afghanistan

An essential question resulting from our data is the origin of the small mammals trapped in this study. As Camp Marmal received large quantities of goods via air transport from Leipzig, Germany, to Kabul and successive land transport during the trapping period [1] most supplies for the base hence originated from Germany or nations sharing the base (e.g. Denmark and Norway). It is not unlikely that within such a large logistic operation, an ubiquitous and invasive species such as *M. musculus* could have been accidentally introduced into the military base through supply crates or containers from Europe. Similarly, indigenous populations of *M. musculus* could have been recruited into the Northern Afghan bases by providing local animals with an environment and food sources suitable for them.

Even though the phylogeography of the house mouse is not fully resolved yet, it is clear that the house mouse is split up into at least three distinct subspecies, *M. musculus domesticus*, *M. musculus musculus* and *M. musculus castaneus* [35] which are already well known. Furthermore, the clade *M. musculus bactrianus* has been recently confirmed [36, 37]. Among these, two predominant subspecies, *M. musculus domesticus* (the Western house

mouse) and *M. musculus musculus* (the Eastern house mouse) are spread globally due to human colonization. While *M. m. domesticus* is predominant in Western Europe and Northern Africa, Australia and the Americas, *M. m. musculus* can be found in the continental climate of Eastern Europe and Asia [38–41]. A hybridization zone of both subspecies has been observed in Eastern Germany and Austria extending through the Adriatic coast [38–41]. However, hybrids of Eastern and Western house mice in the European hybridization zone are considered significantly less competitive than members of each subspecies through a series of genetic mechanisms [42–45].

In this study, we investigated the origin of house mice trapped at three sites in Afghanistan with several techniques: First, amplification and sequencing of the cytochrome *b* gene enabled a quick detection of subspecies of *M. musculus*. All analyzed specimens of house mice trapped at the three Afghan sites showed a high similarity of the cytochrome *b* gene to *M. m. musculus* but a lower similarity to *M. m. domesticus*. These findings were supported by the analysis of the mitochondrial D-loop region. All 85 sequences of animals trapped in this study used for the analysis of the mitochondrial D-loop region did unequivocally cluster with sequences obtained from *M. m. musculus* specimens originating from Central Asia in an earlier study [41], which also confirms the finding of Prager et al. (1998) [46].

Further, this analysis also shows that transports from airports close to the European *M. m. musculus* populations (i.e. the airports in Eastern Germany, Denmark and Norway) were less likely the source of the mice sampled in Afghanistan (Fig. 2). In case of Eastern German Airports, D-loop sequences of *M. m. musculus* from the Czech Republic were used as reference due to their geographic proximity. The results presented above are supported furthermore by a recent study that compared wild populations of house mice from eight Eurasian locations (including specimens from Camp Marmal) by whole genome sequencing [47]. These results clearly show substantial differences between Eastern European and Central Asian specimens of *M. musculus musculus* on the level of the whole genome. Furthermore, that study included female and male individuals and there is no evidence of male import from Western Europe to Afghanistan [47].

A limitation of both methods that are based on mitochondrial DNA that is solely inherited from the maternal lineages is their incapability of detecting hybrid offspring of male *M. m. musculus* specimens that could have been introduced from Eastern German airports into Afghanistan. In contrast to conventionally used methods to overcome this limitation (e.g. microsatellite analysis), this study investigated viral DNA and RNA sequences of the polymerase genes of MmusRHV1 and murine astroviruses as alternative markers to test their suitability for this very purpose.

MmusRHV1 has so far exclusively been detected in *Mus musculus* and the *DPOL* gene is highly conserved in contrast to other genes within the viral genome (e.g. the gene coding for glycoprotein B) [17]. The results of this study confirm the conserved nature of the *DPOL* gene of MmusRHV1, but show distinct local patterns of nucleotide variations. Out of the sequence dataset obtained from Afghanistan and six different locations in Europe, unique nucleotide exchanges were detected at two sites in Germany (Saarbrücken and Roklum), whereas identical sequences were found at locations Cologne/Radolfzell and Liverpool/Leverkusen. Interestingly, different sequences for the *DPOL* gene were observed in specimens from Cologne and Leverkusen, two neighboring cities on opposing banks of the river Rhine. Afghan MmusRHV1 sequences displayed two unique and silent nucleotide exchanges that did not occur in European samples. Nucleotide substitutions in the DNA sequence of MmusRHV1 proved to affect the amino acid sequence of the actual DPOL protein in only two positions within the whole dataset analyzed in this study (positions 49 and 53). In the first case aspartic acid was substituted for glutamic acid, whereas in the second case isoleucine was substituted for valine and thus involved amino acids with similar chemical properties and functional groups. This combination of conserved DNA sequence but location-specific nucleotide variations makes the partial *DPOL* gene of MmusRHV1 a promising candidate for a novel phylogeographic marker for the analysis of geographic origins of *Mus musculus* specimens. Yet, a limiting factor of this method could be the percentage of infected animals. In this study, an average of 12.3% (14/114) of Afghan house mice was infected with MmusRHV1 which was slightly lower than in a previous study [17]. As it cannot *a priori* be assumed that a high enough percentage of a house mouse population at each trapping site is infected with MmusRHV1, the current procedure requires trapping and processing of large numbers of house mice to provide a statistically valid data set for each trapping site. Further problems could be the relatively low sensitivity of the nested pan-PCR used in this study or the fact that herpesviruses are more complicated to be detected during latency.

To back up the results obtained for MmusRHV1, a dsDNA virus, another virus that shows high prevalence in house mouse populations was investigated in this study. For that purpose, astroviruses were selected as an example of a ssRNA virus frequently detected in rodents and more specifically in house mice [48–50]. As they are RNA viruses, murine astroviruses are in general prone to a high sequence variability driven by the lacking proof-reading function of the RdRp as well as specific features of the astrovirus genome. In addition, these viruses are also capable of genomic recombination which has already been described for human astroviruses [50, 51]. In the case of this study, two different genotypes of astroviruses were detected each in house mice from Afghanistan and Germany. While most of the novel astrovirus sequences have shown high similarities to other murine astrovirus sequences, a single sequence each from Afghanistan and Germany did group with rat astroviruses. Within the murine astroviruses, no apparent grouping into location-specific groups could be detected. The apparent high sequence variability of the region analyzed in this study in combination with a relatively small database of murine astrovirus genomes available at this moment of time therefore currently make astroviruses a less suitable candidate for virus-based phylogeographical analysis than gammaherpesviruses, i.e. MmusRHV1.

Finally, the combined results of three of the four methods tested indicate that the establishment of an invasive house mouse population originating from Northern, Central or Western Europe after a hypothetical transportation is unlikely. These findings are supported by a previous study conducted on the Kerguelen Archipelago and other Southern Ocean Islands [14]. D-loop analysis indicated that the house mouse population that conducts the primary invasion of the islands cannot be superseded by a secondary invasion of the same species. The mechanism for this resilience against either takeover or interbreeding of primary colonizers by secondary colonizers is presumed to be due to quick adaptations to biotic and abiotic factors of the new habitat. Primary driving forces for these adaptations may be climate, food sources and microhabitat structure [14, 15, 27]. In case of the Kerguelen Archipelago, Western house mice from Europe performed the primary invasion and quickly adapted to a colder climate, a diet mainly based on insects as well as a less structured habitat due to the high number of offspring and low generation time. Therefore, secondary colonizers invading the islands faced an already adapted population of competitors. In case of the Afghan military camps, house mice must be adapted to a significantly harsher and drier continental climate at elevated altitude and to a desert habitat with significantly less cover and vegetation in comparison to habitats in Central Europe.

Outlook - *Mus musculus* as dominant commensal of ISAF garrisons in northern Afghanistan

The results of previous research [33, 34] as well as this study imply that rodent communities dominated by house mice may form spontaneously in sites like Camp Marmal due to the habitat structure and food sources introduced by the construction of the military bases. While these results urgently require further investigation in military bases of similar design, they may indicate an effect of this specific rodent species on life in Northern Afghanistan military bases. From 2004 to 2006, high numbers of *Rhombomys opimus* were detected in the vicinity of Camp Marmal due to earth works, yet for the period from 2009 to 2012 after having implemented continuous *R. opimus* control including environmental changes, none of the trapped mammals belonged to this species. While another possible reservoir host for *Leishmania* spp., *Meriones libycus* [52], was trapped within the Camp Marmal site, it only represented 2.1% of total trapped animals. This may indicate that *R. opimus*-dominated communities inside Camp Marmal site from 2004 to 2006 were superseded by house mouse-dominated communities by 2010. In contrast to this species, house mice are known to be carriers of other pathogens such as lymphocytic choriomeningitis virus, *Leptospira* spp., *Rickettsia* spp. and *Salmonella* spp. [53]. The presence of a house mouse population in the military camps in this study therefore may have provided a surrounding with significantly lower number of reservoir hosts of zoonotic pathogens such as *Leishmania* spp. for both soldiers and civilian personnel. Hence, house mice in Camp Marmal and the PRTs Kunduz and Fayzabad have successfully colonized a habitat previously occupied by reservoir hosts of such pathogens. However, further studies could provide an insight if this is a common process occurring throughout Afghanistan (or in fact other military bases in development on other continents) and more importantly, if this process is reversed after the end of the military operations, e.g. by investigating current rodent communities in now abandoned Northern Afghan military bases and determining the prevalence of zoonotic pathogens within them.

Conclusions

The results of this study highlight that changes in the architecture and design of the makeshift military bases investigated affected the resident rodent communities as the population of *R. opimus* encountered before 2009 disappeared from the bases and a rodent population dominated by Eastern house mice could be found after this point in time. Albeit a relatively low probability of undetected hybridization of introduced and local house mice, both maternal mitochondrial DNA markers and viral marker genes of a DNA virus infecting these Afghan rodents indicate that an introduction of *M. musculus musculus* from Europe seems unlikely.

Additional files

Additional file 1: Percentages of individual species within small mammal communities in Northern Afghan military bases as well as Provincial Reconstruction Team bases with regard to the time of trapping. (DOCX 130 kb)

Additional file 2: Results of DNA sequencing of the cytochrome *b* gene of animals phenotypically identified as house mice (*Mus musculus*), gray dwarf hamsters (*Cricetulus migratorius*), jirds (*Meriones* spp.), rat (*Rattus* spp.) and as members of the shrew genera *Suncus* and *Crocidura*. (DOC 206 kb)

Additional file 3: Comparison of partial amino acid sequences of the deduced DPOL protein of MmusRHV1 from *Mus musculus* specimens trapped in Afghanistan, Germany and the United Kingdom to a reference sequence from Germany. (DOCX 472 kb)

Abbreviations

bp: base pair; COL: Cologne; DMSO: Dimethyl sulfoxide; DPOL: DNA polymerase; ds: double stranded; FOB: Forward Operating Base; FSB: Forward Support Base; ISAF: International Security Assistence Force; LEK: Leverkusen; LIV: Liverpool; MeS: Mazar-e-Sharif, Camp Marmal/Afghanistan; MmusRHV1: *Mus musculus* Rhadinovirus 1; PRT: Provincial Reconstruction Team; RAZ: Radolfzell; RdRp: RNA-dependent RNA polymerase; ROK: Roklum; SBR: Saarbrücken; ss: single stranded; ZCL: Zoonotic Cutaneous Leismaniasis

Acknowledgements

The authors thank the ISAF staff of Camp Marmal, PRT Kunduz and PRT Fayzabad and in particular all veterinarian teams and numerous preventive medicine technicians. We acknowledge LtCol Alfred Binder for his contributions to veterinary science which form the bases of our study, and Dr. Ulrich Schotte for critical reading the manuscript and helpful comments. The authors thank Anke Mandelkow, Marie Luisa Schmidt, Maria Justiniano Suarez and Dörte Kaufmann for their excellent technical assistance, and Ulrike M. Rosenfeld, Sabrina Schmidt, Hanan Sheikh Ali, Elisa Heuser, Julia Schneider, Franziska Thomas, Maysaa Dafalla and Stefan Fischer for their help during dissections and laboratory work, Stephan Drewes for assistance in dissections and data analysis as well as Carina Spahr and Richard Kruczewski for kindly providing house mouse specimens.

Funding

These investigations were funded by contract-research-projects for the Bundeswehr Medical Service M/SABX/005 and E/U2 AD/CF512/DF557.

Authors' contributions

Conceived and designed the study: CG MS JPT DT SS RGU. Performed research: CG MS ML JL KB AB RR. Analyzed data: CG MS RH ML. Contributed new methods or models: KF BE PK. Wrote the paper: CG RH MF SS BE ML RR KF DT RGU. All authors read and approved the final manuscript.

Competing interests

The authors declare that they have no competing interests.

Author details

[1]Friedrich-Loeffler-Institut, Federal Research Institute for Animal Health, Institute of Novel and Emerging Infectious Diseases, Greifswald-Insel Riems, Greifswald, Germany. [2]Max Planck Institute of Evolutionary Biology, Plön, Germany. [3]Stiftung Zoologisches Forschungsmuseum Alexander Koenig, ZFMK, Bonn, Germany. [4]Friedrich-Loeffler-Institut, Federal Research Institute for Animal Health, Institute of Diagnostic Virology, Greifswald-Insel Riems, Greifswald, Germany. [5]Robert Koch-Institut, Division 12 "Measles, Mumps, Rubella, and Viruses Affecting Immunocompromised Patients", Berlin, Germany. [6]Landratsamt Weilheim-Schongau Veterinäramt, Weilheim i. OB, Germany. [7]Bundeswehr Medical Academy, Military Medical Research and Development, Division E, Munich, Germany. [8]Friedrich-Loeffler-Institut, Federal Research Institute for Animal Health, Department of Experimental Animal Facilities and Biorisk Management, Greifswald-Insel Riems, Greifswald, Germany. [9]Zentrales Institut des Sanitätsdienstes der Bundeswehr Koblenz, Abteilung I Medizin, Koblenz, Germany. [10]German Center for Infection Research (DZIF), Partner Site Hamburg-Luebeck-Borstel-Insel Riems, Greifswald-Insel Riems, Greifswald, Germany. [11]Present Address: RWTH Aachen, Institute for Biotechnology, Aachen, Germany. [12]Present Address: Seramun Diagnostica GmbH, Heidesee, Germany.

References

1. Chiari B. From Venus to Mars? Provincial reconstruction teams and the European military experience in Afghanistan, 2001–2014. Freiburg i. Breisgau: Rombach Verlag AG; 2014.

2. Wallace MR, Hale BR, Utz GC, Olson PE, Earhart KC, et al. Endemic infectious diseases of Afghanistan. Clin Infect Dis. 2002;34:171–207.

3. Faulde M, Schrader J, Heyl G, Amirih M, Hoerauf A. Zoonotic cutaneous leishmaniasis outbreak in Mazar-e Sharif, northern Afghanistan: an epidemiological evaluation. Internat. J Med Microbiol. 2008;298:543–50.

4. Faulde M, Heyl G, Amirih ML. Zoonotic cutaneous leishmaniasis, Afghanistan. Emerg Inf Dis. 2006;12:1623–4.

5. Faulde M, Schrader J, Heyl G, Hoerauf A. High efficacy of integrated preventive measures against zoonotic cutaneous leishmaniasis in northern Afghanistan, as revealed by quantified infection rates. Acta Trop. 2009;110:28–34.

6. Lee HW, Baek LJ, Johnson KM. Isolation of the etiologic agent of Korean hemorrhagic fever. J Infect Dis. 1978;137:298–308.

7. Bugert JJ, Welzel TM, Zeier M, Darai G. Hantavirus infection–haemorrhagic fever in the Balkans–potential nephrological hazards in the Kosovo war. Nephrol Dial Transplant. 1999;14:1843–4.

8. Press and Media Section, North Atlantic Treaty Organization (NATO). Afghanistan report 2009. NATO, public diplomacy division, press and media section, media operation center (MOC)2009. Available: www.nato.int.

9. Press and Media Section, North Atlantic Treaty Organization (NATO). International Security Assistance Force (ISAF). Key facts and figures. NATO, Public Diplomacy Division, Press and Media Section, Media Operation Center (MOC). 2009. Available: www.nato.int.

10. Hassinger JD. Street expedition to Afghanistan. A survey of the mammals of Afghanistan, resulting from the 1965 street expedition (excluding bats) [by] Jerry D. Hassinger. [Chicago]: Field Museum of Nat Hist; 1973.

11. Mohammadian H. An introduction to mammals of Afghanistan. Teheran, Iran: Shabpareh Publishing Institute; 2011.

12. Tattershall FN, Smith RH, Nowell F. Experimental colonization of contrasting habitats by house mice. Z Säugetierkunde. 1997;62:350–8.

13. Pelz H-J, Rost S, Müller E, Esther A, Ulrich RG, et al. Distribution and frequency of VKORC1 sequence variants conferring resistance to anticoagulants in Mus musculus. Pest Manag Sci. 2012;68:254–9.

14. Hardouin EA, Chapuis J-L, Stevens MI, van Vuuren J, Quillfeldt P, et al. House mouse colonization patterns on the sub-Antarctic Kerguelen archipelago suggest singular primary invasions and resilience against re-invasion. BMC Evol Biol. 2010;10:325.

15. Babiker H, Tautz D. Molecular and phenotypic distinction of the very recently evolved insular subspecies Mus musculus helgolandicus Zimmermann, 1953. BMC Evol Biol. 2015;15:160. doi:10.1186/s12862-015-0439-5.

16. Schlegel M, Baumann K, Breithaupt A, Binder A, Schotte U, et al. What about the role of rodents as vectors for zoonotic pathogens in mission areas of the Bundeswehr? Wehrmed Monatsschrift 2012;56:203–207. [in German].

17. Ehlers B, Kuchler J, Yasmum N, Dural G, Voigt S, et al. Identification of novel rodent herpesviruses, including the first gammaherpesvirus of Mus musculus. J Virol. 2007;81:8091–100.

18. Hasenkamp N, Solomon T, Tautz D. Selective sweeps versus introgression - population genetic dynamics of the murine leukemia virus receptor Xpr1 in wild populations of the house mouse (Mus musculus). BMC Evol Biol. 2015; 15:248. doi:10.1186/s12862-015-0528-5.

19. Schlegel M, Ali HS, Stieger N, Groschup MH, Wolf R, et al. Molecular identification of small mammal species using novel cytochrome b gene-derived degenerated primers. Biochem Genet. 2012;50:440–7.

20. Hall T. BioEdit: a user-friendly biological sequence alignment editor and analysis program for windows 95/98/NT. Nucleic Acid Symp. 1999;41:95–8.

21. Ashelford KE, Chuzhanova NA, Fry JC, Jones AJ, Weightman AJ. At least 1 in 20 16S rRNA sequence records currently held in public repositories is estimated to contain substantial anomalies. Appl Environ Microbiol. 2005;71:7724–36.

22. Altschul SF, Gish W, Miller W, Myers EW, Lipman DJ. Basic local alignment search tool. J Mol Biol. 1990;215:403–10.

23. Miller MA, Pfeiffer W, Schwartz T. Creating the CIPRES Science Gateway for inference of large phylogenetic trees. In: Proceedings of the Gateway Computing Environments Workshop (GCE), 14 Nov. 2010, New Orleans, LA, pp 1–8.

24. Prager EM, Sage RD, Gyllensten U, Thomas WK, Hübner R, et al. Mitochondrial DNA sequence diversity and the colonization of Scandinavia by house mice from East Holstein. Biol J Linnean Soc. 1993;50:85–122.

25. Tamura K, Stecher G, Peterson D, Filipski A, Kumar S. MEGA6:Molecular Evolutionary Genetics Analysis version 6.0. Mol Biol Evol. 2013;30:2725–9.

26. Ihle S, Ravaoarimanana I, Thomas M, Tautz D. An analysis of signatures of selective sweeps in natural populations of the house mouse. Mol Biol Evol. 2006;23:790–7.

27. Bonhomme F, Orth A, Cucchi T, Rajabi-Maham H, Catalan J, Boursot P, et al. Genetic differentiation of the house mouse around the Mediterranean basin: matrilineal footprints of early and late colonization. Proc R Soc B. 2011;278:1034–43.

28. Bandelt H, Forster P, Röhl A. Median-joining networks for inferring intraspecific phylogenies. Mol Biol Evol. 1999;16:37–48.

29. Chu DKW, Poon LLM, Guan Y, Peiris JSM. Novel astroviruses in insectivorous bats. J Virol. 2008;82:9107–14.

30. Shenbrot GI. Spatial structure and niche patterns of a rodent community in the south Bukhara desert (middle Asia). Ecography. 1992;15:347–57.

31. Alagaili AN, Mohammed OB, Bennett NC, Oosthuizen MK. A tale of two jirds: the locomotory activity patterns of king jird (Meriones rex) and Lybian jird (Meriones libycus) from Saudi Arabia. J Arid Environm. 2013;88:102–12.

32. Darvish J, Siahsarvie R, Mirshamsi O, Kayvanfar N, Hashemi N, et al. Diversity of the rodents of northeastern Iran. Iran J Animal Biosystem. 2006;2:57–76.

33. Himsworth CG, Jardine CM, Parsons KL, Feng AYT, Patrick DM. The characteristics of wild rat (Rattus spp.) populations from an inner-city neighborhood with a focus on factors critical to the understanding of rat-associated zoonoses. PLoS One. 2014;9:e91654.

34. Cavia R, Cueto GR, Suárez OV. Changes in rodent communities according to the landscape structure in an urban ecosystem. Landscape Urban Plan. 2009;90:11–9.

35. Boursot P, Auffray JC, Britton-Davidian J, Bonhomme F. The evolution of house mice. Annu Rev Ecol Syst. 1993;24:119–52.

36. Hamid HS, Darvish J, Rastegar-Pouyani E, Mahmoudi A. Subspecies differentiation of the house mouse Mus musculus Linnaeus, 1758 in the center and east of the Iranian plateau and Afghanistan. Mammalia. 2017;81:147–68.

37. Adhikari P, Han S-H, Kim Y-K, Kim T-W, Thapa TB, Subedi N, Adhikari P, Oh H-S. First molecular evidence of Mus musculus bactrianus in Nepal inferred from the mitochondrial DNA cytochrome B gene sequences. Mitochondrial DNA Part A > DNA Mapping, Sequencing, and Analysis. 2017; doi:10.1080/24701394.2017.1320994

38. Guénet J-L, Bonhomme F. Wild mice: an ever-increasing contribution to a popular mammalian model. TIG. 2003;19:24–31.

39. Hardouin EA, Orth A, Teschke M, Darvish J, Tautz D, Bonhomme F. Eurasian house mouse (Mus musculus L.) differentiation at microsatellite loci identifies the Iranian plateau as a phylogeographic hotspot. BMC Evol Biol. 2015;15:26. doi:10.1186/s12862-015-0306.

40. Mikula O, Auffray J-C, Macholan M. Asymmetric size and shape variation in the central European transect across the house mouse hybrid zone: asymmetric variation in mouse hybrids. Biological J Linnean Soc. 2010;101:13–27.

41. Darvish J, Orth A, Bonhomme F. Genetic transition in the house mouse, Mus musculus of eastern Iranian plateau. Folia Zool. 2006;55:349–57.

42. Baird SJE, Macholán M. What can the Mus musculus musculus/M. m. domesticus hybrid zone tell us about speciation? In: Evolution of the house mouse. Cambridge, UK: Cambridge University Press; 2012. p. 334–61.

43. Teeter KC, Payseur BA, Harris LW, et al. Genome-wide patterns of gene flow across a mouse hybrid zone. Genome Res. 2008;18:67–76.

44. Turner LM, Schwahn DJ, Harr B. Reduced male fertility is common but highly variable in form and severity in a natural house mouse hybrid zone. Evolution. 2012;66:443–58.

45. Suzuki TA, Nachman MW. Speciation and reduced hybrid female fertility in house mice. Evolution. 2015;69:2468–81.

46. Prager EM, Orrego C, Sage RD. Genetic variation and phylogeography of central Asian and other house mice, including a major new mitochondrial lineage in Yemen. Genetics. 1998;150(2):835–61.

47. Harr B, Karakoc E, Neme R, Teschke M, Pfeifle C, Pezer Ž, Babiker H, Linnenbrink M, Montero I, Scavetta R, Reza Abai M, Puente Molins M, Schlegel M, Ulrich RG, Altmüller J, Franitza M, Büntge A, Künzel S, Tautz D. Genomic resources for wild populations of the house mouse, Mus musculus and its close relative Mus spretus. Sci Data. 2016;3:160075.

48. Kjeldsberg E, Hem A. Detection of astroviruses in gut contents of nude and normal mice. Arch Viro. 1985;84:135–40.

49. Chu DK, Chin AW, Smith GJ, Chan KH, Guan Y, Peiris JS, Poon LL. Detection of novel astroviruses in urban brown rats and previously known astroviruses in humans. J Gen Virol. 2010;91:2457–62. doi:10.1099/vir.0.022764-0.

50. Ng TF, Kondov NO, Hayashimoto N, Uchida R, Cha Y, Beyer AI, Wong W, Pesavento PA, Suemizu H, Muench MO, Delwart E. Identification of an astrovirus commonly infecting laboratory mice in the US and Japan. PLoS One. 2010;8:E66937.

51. De Benedictis P, Schultz-Cherry S, Burnham A, Cattoli G. Astrovirus infections in humans and animals – molecular biology, genetic diversity, and interspecies transmissions. Inf Gen Evol. 2011;11:1529–44.

52. Parvizi P, Moradi G, Akbari G, Farahmand M, Ready PD, et al. PCR detection and sequencing of parasite ITS-rDNA gene from reservoir host of zoonotic cutaneous leishmaniasis in central Iran. Parasitol Res. 2008;103:1273–8.

53. Meerburg BG, Singleton GR, Kijlstra A. Rodent-borne diseases and their risks for public health. Crit Rev Microbiol. 2009;35:221–70.

Demography of a small, isolated tiger (*Panthera tigris tigris*) population in a semi-arid region of western India

Ayan Sadhu[1], Peter Prem Chakravarthi Jayam[1,4], Qamar Qureshi[2], Raghuvir Singh Shekhawat[3], Sudarshan Sharma[3] and Yadvendradev Vikramsinh Jhala[1*]

Abstract

Background: Tiger populations have declined globally due to poaching, prey depletion, and habitat loss. The westernmost tiger population of Ranthambhore in India is typified by bottlenecks, small size, and isolation; problems that plague many large carnivore populations worldwide. Such populations are likely to have depressed demographic parameters and are vulnerable to extinction due to demographic and environmental stochasticity. We used a combination of techniques that included radio telemetry, camera traps, direct observations, and photo documentation to obtain 3492 observations on 97 individually known tigers in Ranthambhore between 2006 and 2014 to estimate demographic parameters. We estimated tiger density from systematic camera trap sampling using spatially explicit capture-recapture (SECR) framework and subsequently compared model inferred density with near actual density.

Results: SECR tiger density was same as actual density and recovered from 4.6 (SE 1.19) to 7.5 (SE 1.25) tigers/100km^2 over the years. Male: female ratio was 0.76 (SE 0.07), and cub: adult tigress ratio at 0.48 (SE 0.12). Average litter size was estimated at 2.24 (SE 0.14). Male recruitment from cub to sub-adult stage (77.8%, SE 2.2) was higher than that of females (62.5%, SE 2.4). But male recruitment rate as breeding adults from the sub-adult stage (72.6%, SE 2.0) was lower than females (86.7%, SE 1.3). Annual survival rates, estimated by known-fate models, of cubs (85.4%, $CI_{95\%}$ 80.3–90.5%) were lower than that of juvenile (97.0%, $CI_{95\%}$ 95.4–98.7%) and sub-adult (96.4%, $CI_{95\%}$ 94.0–98.9%) tigers. Adult male (84.8%, $CI_{95\%}$ 80.6–89.2%) and female (88.7%, $CI_{95\%}$ 85.3–92.2%) annual survival rates were similar. Human-caused mortality was 47% in cubs and 38% in adults. Mean dispersal age was 33.9 months (SE 0.8), males dispersed further (61 Km, SE 2) than females (12 Km, SE 1.3). Higher age of first reproduction (54.5 months, SE 3.7) with longer inter-birth intervals (29.6 months, SE 3.15) was likely to be an effect of high tiger density.

Conclusion: Demographic parameters of Ranthambhore tigers were similar to other tiger populations. With no signs of inbreeding depression there seems to be no eminent need for genetic rescue. The best long-term conservation strategy would be to establish and manage a metapopulation in the Ranthambhore landscape.

Keywords: Camera traps, Dispersal, Inter-birth interval, Known fate, Litter size, Mortality, Radio-telemetry, Ranthambhore, Spatially explicit capture-recapture, Survival

* Correspondence: jhalay@wii.gov.in
[1]Department of Animal Ecology and Conservation Biology, Wildlife Institute of India, Chandrabani, Dehra Dun, Uttarakhand 248001, India
Full list of author information is available at the end of the article

Background

At the onset of the nineteenth century, India was home to nearly 40,000 tigers (*Panthera tigris tigris*, Linnaeus) [1], while currently there are around 2200 left [2]. The decline in tigers was primarily due to hunting, prey depletion, followed by habitat loss [3]. A timely and proactive conservation measure, in the form of Project Tiger initiated in 1973 by the Indian Government [4], initially halted the rapid decline caused by trophy hunting. But an increased demand for tiger body parts in China and Southeast Asia in the past 25 years has severely impacted wild tiger populations. Demand driven poaching resulted in the local extinction of tigers in Sariska and Panna Tiger Reserves in India [5, 6]. Most tiger populations currently are small, isolated, and highly structured [3, 7, 8]. Such populations are vulnerable to extinction events caused by environmental and demographic stochasticity [9, 10]. Ranthambhore was a famous hunting reserve for the *Maharajas* of Jaipur, and numerous *shikar* (hunting) camps were organized in pre and post-independence era [11]. Subsequent to India's independence, intensity of hunting increased since tiger *shikar* was considered a social status symbol. This unregulated hunting caused a severe decline in Ranthambhore tiger population, and before the onset of Project Tiger (1973), there were around 14 tigers left in Ranthambhore [12]. After an initial recovery in 1980's, rampant poaching in 1992 and 2005, caused Ranthambhore tiger population to decline below 15 individuals from about 40 [13, 14]. Local extinctions in the last five decades suggest that tigers of the semi-arid region of western India are most vulnerable [15]. The tiger population of Ranthambhore is the only population that survives in western India. It typifies the problems many large carnivore populations face globally i.e. small founder population and lack of connectivity with other source populations. Small isolated populations like Ranthambhore are susceptible to loss of genetic variability caused by genetic drift and inbreeding depression [16, 17]. Such populations often lose their ability to adept in response to environmental changes and some manifest deleterious effects in the form of morphological abnormalities and depressed population vigour [18, 19]. Hence, understanding demographic parameters of a potentially genetically compromised population to determine the need for genetic rescue is important for developing appropriate conservation strategies [20]. Quantifying demographic parameters needs long-term data over multiple generations and for long-lived carnivores such datasets are rare [21]. Till date most population studies conducted on tigers aim at estimating abundance [22–29], while studies on demographic parameters (like survival rates, litter size, sexual maturity, reproductive success) have been sparse (but see [30–32] for *P. t. tigris* and, [33, 34] for *P. t. altaica*).

Herein, we report demographic parameters of free ranging tigers from Ranthambhore Tiger Reserve from a nine-year study where 97 individually known tigers were monitored, and annual density estimated by spatially explicit capture-recapture using camera traps. By 2012 due to intensive monitoring, we had photo-captured almost all tigers of Ranthambhore and had developed a catalogue for individually identifying them. We use this information to compare snapshot density estimated by model-based inference with near reality. We also compare demographic parameters of Ranthambhore tigers with those of other tiger populations and conclude that though Ranthambhore tigers have undergone population bottlenecks, with limited gene flow and small population size, their demographic parameters do not seem to be compromised.

Methods
Study area

The study was conducted from 2006 to 2014 in Ranthambhore Tiger Reserve (hereafter RTR, latitudes $25^0 41'$ N to $26^0 22'$ N and longitudes $76^0 16'$ E to $77^0 14'$ E) which is situated at the junction of two ancient mountain ranges, the Aravalli and the Vindhya. RTR is part of the western Indian landscape that has Sariska Tiger Reserve in the north, Kuno Wildlife Sanctuary and Madhav National Park in the east, Ramgarh Visdhari Wildlife Sanctuary and Mukundara Hills Tiger Reserve in the south-western part (Fig. 1). The core area of RTR was composed of Ranthambhore National Park (392 km^2), Sawai Mansingh Sanctuary (290 km^2) while Kailadevi Wildlife Sanctuary (630 km^2) was designated as the buffer zone of RTR. Within this western Indian landscape, tigers were only present in Ranthambhore NP during the commencement of this study and subsequently colonised Sawai Mansingh Sanctuary in 2008–09. These together comprise the only source population of tigers in the landscape. During this study, tigers from this population were reintroduced in Sariska [35] and six tigers dispersed into northern as well as south-eastern and eastern part of the landscape (Fig. 1).

The sub-tropical dry climate of RTR experiences three distinct seasons: mostly dry winters (October–February, minimum average temperature 5 °C, relative humidity ~10%), hot summers (March–June, mean maximum temperature 45 °C, relative humidity 10–15%), and humid monsoons (July–September, average rainfall 700 mm, relative humidity >60%). RTR primarily comprises of steep hills, gentle slopes, plateaus, and narrow valleys dotted with shallow man-made perennial lakes. The area is representative of dry deciduous *Anogeissus pendula* forests in association with *Acacia, Butea, Capparis, Zizyphus* and *Prosopis* species (5B/C_2 - Northern Tropical Dry

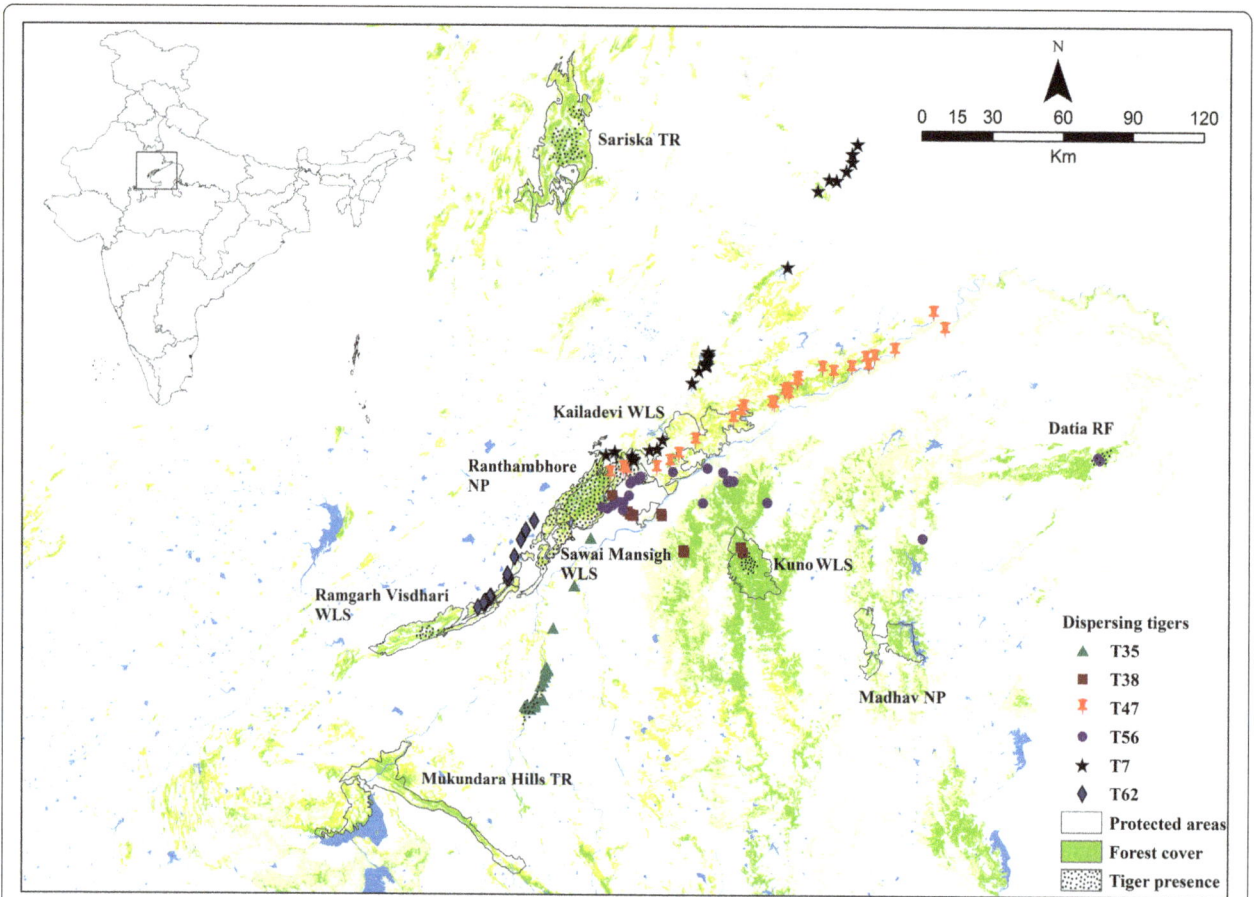

Fig. 1 Landscape of Ranthambhore showing the tiger reserve and other tiger occupied forest patches in western India. Locations of six individual tigers that dispersed out of the tiger reserve are shown. The map inset shows the location of Ranthambhore Tiger Reserve within IndiaTR - Tiger Reserve; NP - National Park; WLS- Wildlife Sanctuary; RF – Reserve Forest; Ranthambhore NP, Sawai Mansingh WLS and Kailadevi WLS together constitute the Ranthambhore Tiger Reserve

Deciduous forests. 6B/DS1 - *Zizyphus* scrub, DS1 - Dry deciduous scrub and 5/DS4 - Dry Grasslands, of Champion & Seth [36] classification). A diverse assemblage of carnivore species (17 species from 7 different families) were recorded during the course of the study, which include tiger, leopard (*Panthera pardus*, Linnaeus), sloth bear (*Melursus ursinus*, Shaw), striped hyena (*Hyaena hyaena*, Linnaeus), caracal (*Caracal caracal*, Schreber), fishing cat (*Prionailurus viverrinus*, Bennett), jungle cat (*Felis chaus*, Schreber), desert cat (*Felis silvistris*, Schreber), rusty-spotted cat (*Prionailurus rubuginosa*, I. Geoffroy Saint-Hilaire), golden jackal (*Canis aureus*, Linnaeus), Indian fox (*Vulpes bengalensis*, Shaw), honey badger (*Mellivora capensis*, Schreber), common palm civet (*Paradoxurus hermaphorditus*, Pallas), small Indian civet (*Viverricula indica*, É. Geoffroy Saint-Hilaire), Indian gray mongoose (*Herpestes edwardsii*, É. Geoffroy Saint-Hilaire), small Indian mongoose (*Herpestes auropunctatus*, Hodgson), and ruddy mongoose (*Herpestes smithi*, Gray). Tiger prey species present in the study area were chital (*Axis axis*, Erxleben), sambar (*Rusa unicolor*,

Kerr), nilgai (*Boselaphus tragocamelus*, Pallas), chinkara (*Gazella bennetti*, Skyes), wild pig (*Sus scrofa*, Linnaeus), common langur (*Semnopithecus entellus*, Dufresne) and rhesus macaque (*Macaca mulata*, Zimmermann).

Field methods: Monitoring of tigers

We monitored 97 individual tigers during the study (2006 to 2014) through camera traps, radio-telemetry and routine patrolling (for direct sightings and photo-documentation) by researchers and forest staff. We developed criteria for classifying tigers into age groups by observing known-age individuals and use teeth eruption, ware, and body characteristics (see Additional file 1 for age estimation of tigers) similar to that of lions [37, 38]. We classified tigers into six age classes, namely, cubs (< 12 months), juveniles (12-24 months), sub-adults (2–3 years), young adults (4–5 years), prime adults (6–10 years) and old adults (> 10 years). Of the 97 tigers, 74 were known since cub stage, and their age was known to the exact month. Sex of the individuals was ascertained by the time cubs were 6 months old.

Camera trapping

Camera traps were used a) in a systematic grid-based design ($4km^2$ in 2006, 2009, 2012 and 2013, and $2km^2$ in 2014) for a short duration to estimate tiger abundance, and b) to target specific areas so as to determine the presence and use of the area by particular individual tigers throughout the study period.

a) The entire tiger occupied part of RTR was sampled in a systematic manner during the study by placing a pair of camera traps at each selected location within a grid. After conducting a reconnaissance survey, camera traps were placed on dirt roads, animal trails, fire lines and dry river beds at locations that maximized the chances of photo-capturing tigers. A pair of camera traps (TrailMaster® Lenexa KS USA, Cuddeback™ Green Bay USA, MOULTRIE® Alabama USA, or Stealth Cam® LLC Grand Prairie USA), facing each other, were placed at each location to get both flank photos of tigers at the same time. Each Camera was programmed with unique trap ID, time and date stamp on each photograph. Location of each camera trap was recorded by handheld GPS unit (Garmin 72™ and Garmin Etrex® 10, Kansas, USA) and plotted on a digitized map of RTR in GIS domain to ensure no sampling holes were present. Cameras were checked every 2–3 days to ensure proper functioning and recovery of data. Each photo captured tiger was identified to individuals by comparing their stripe patterns and given a unique id (e.g. T1, T2, and so on). The entire study area was sampled simultaneously

within a period of 28 to 51 days so as to adhere to the assumption of population closure [39]. The camera trapped area sampled each year as estimated by joining the outer most camera traps ranged between 139 to $492 km^2$ (Table 1).

b) Besides systematic camera trapping conducted once in a year, cameras were also used in a need based manner to record photographs of specific tigers. Areas of tigers that were not seen for over 60 days were specifically targeted. These special efforts were carried out by placing multiple camera traps and conducting an intensive ground search in most probable locations of that tiger. The effort continued till the fate of that tiger was ascertained by locating it or confirming its death or dispersal. Since the landscape outside of the tiger reserve was human-dominated, the presence of dispersing tigers was quickly detected by reports of livestock kills and sighting of tigers or their signs by villagers. The research team along with forest department staff subsequently tried to locate each such tiger and ascertain its identity through sighting, camera traps or hand held photography.

Radio telemetry

Eight tigers (three adult males, one adult female, three sub-adult males and one sub-adult female) were radio-collared between April 2007 to May 2009. Tigers were anesthetized with ketamine hydrochloride in combination with medetomidine injected intramuscularly using a gas-powered projectile dart delivery system [40]. Tigers were collared with a Very High-Frequency transmitter (Telonics, Arizona, USA) and in most cases with a Global Positioning System with ground download facility

Table 1 Sampling details and parameters estimates of tiger density from camera trap based spatial capture-recapture analysis in Ranthambhore Tiger Reserve

Year	No. of camera locations	Camera trap Polygon (km²)	Trap Nights	M_t $+_1^a$	N^b(SE)	P^e	Known tiger population > 1 Year ♂	♀	Cub	D (SE)^d /100km² (CI$_{95\%}$)	Known density^c /100km²	g0 (SE)	σ (SE) Km
2006	40	139	48	16	16 (0.73)	1.0	5^f	15^f	15^f	4.62 (1.19) (2.81–7.59)	–	♀: 0.06 (0.009)	2023.0 (131.6)
												♂: 0.03 (0.007)	4162.6 (430.4)
2009	48	162	28	25	26 (1.52)	0.96	19^f	17^f	2^f	8.75 (1.79) (5.88–13.02)	–	♀: 0.05 (0.009)	1547.0 (110.2)
												♂: 0.05 (0.009)	1564.4 (104.3)
2012	60	223	45	22	30 (3.9)	0.73	12	16	6	5.68 (1.22) (3.74–8.64)	5.60	♀: 0.08 (0.008)	1529.0 (81.7)
												♂: 0.08 (0.007)	2136.9 (97.2)
2013	76	464	51	37	40 (2.62)	0.92	18	20	14	7.56 (1.25) (5.47–10.44)	7.60	♀: 0.05 (0.005)	1480.7 (85.5)
												♂: 0.05 (0.005)	1948.7 (83.3)
2014	182	492	48	39	39 (1.38)	1.0	21	21	12	7.22 (1.16) (5.27–9.88)	7.63	♀: 0.05 (0.006)	1496.5 (73.3)
												♂: 0.04 (0.003)	2161.7 (78.4)

^{a}M_{t+1}: Unique adult tigers photo-captured in camera traps
^{b}N: Population estimates derived from spatially explicit capture recapture technique (regional population size using 'region.N');
^{c}Known density: Known tiger population/ tiger occupied area ♂: Male; ♀: Female
^{d}D: Density estimates (model averaged); g0: detection probability at home range center; σ: movement parameter
^{e}P: Proportion of the population detected by camera trap survey
^{f}Since all tigers were not photo-captured, here we have reported the minimum number of known individuals

(HABIT, British Columbia, Canada). Soon after the radio-collaring operation, Atipamezole was administered for reversing the effect of the tranquilizer [40]. Animals were left after natural reflexes and behaviour returned and subsequently monitored through telemetry. Radio-collared tigers were tracked and regularly monitored (>three times a week) from a vehicle or on foot throughout the functional period of those collars. Tigers were tracked with the help of a hand-held directional 3- element Yagi antenna and Telonics and HABIT receivers. After the battery life of the radio-collars, the surviving collared tigers were monitored through camera traps and visual sightings.

Routine patrolling

Due to intensive camera trapping over the years and photo documentation by researchers and park officers, we were reasonably certain that almost all tigers of RTR were photo-captured by 2012 and were individually known. Our claim of knowing almost all RTR tigers was substantiated by the fact that in subsequent years no unknown adult tiger was recorded either by camera traps or by any other means. All additions to RTR population from 2012 onward till date were from known cubs. A photo album was developed and shared with the staff and officers of RTR in 2012 with additions and deletions done every 6 months. Since adult tiger numbers ranged between 20 to 40 individuals, it was possible to identify each tiger on most occasions it was sighted and often photographed with digital cameras by forest department staff and researchers. When in doubt these photographs were compared with the photo-album or on the computer to ascertain the tiger's identity. A daily record on all tiger observations was maintained in a register which was subsequently transferred to a database.

Analytical methods: Estimating demographic parameters of tigers

Tiger abundance

We used likelihood based spatially explicit capture-recapture (SECR, [41, 42] in package 'secr' on R platform [43, 44] to estimate tiger density from the systematically sampled camera trap data over the years (2006, 2009, 2012, 2013, and 2014; Table 1). SECR consists of two sub-models. The distribution sub-model depicts the spatial distribution of detectors and animal captures in the landscape. The detection sub-model (g(x)) declines with increasing distance between the animal's activity centre, and this spatial scale of detection is parameterized by sigma (σ). A spatial capture history matrix, a trap layout matrix, and a habitat mask that excluded non-habitat areas from the SECR model space were prepared and used in secr. Home-range size (as indexed by σ) is

often correlated with density [45]. Since the tiger population of RTR increased during our study period, we, parameterized σ and g0 (capture probability at the activity centre) separately for each year. Male and female tigers were likely to differ in their ranging patterns. Hence we used gender as a covariate to model heterogeneity in movement parameter (σ). Half-normal detection function was used to model σ. We used AIC_c (Akaike Information Criterion corrected for sample size) [46] to compare models with the null model (where g0 and σ were constant) and amongst themselves. Models with less than five delta AICc values were considered probable and the parameter estimates were obtained by AICc weighted model averages [47]. Cubs (<12 months) were excluded from density estimation since this cohort is underrepresented in camera traps and has relatively high mortality [22, 48]. Since by 2012 we were reasonably certain that almost all tigers of RTR were individually known to us, we use this information to compare estimates obtained by SECR with actual known density.

Sex ratio and female reproductive parameters

Since sex of all tigers was known, we have calculated sex ratio (adult males: Adult females) for each year (sampling without replacement) by counting the total number of males and females in the population, and an estimate of its versatility between years [49].

Before giving birth, females restricted themselves to a small area of their territory [30, 50] and were often detected through frequent photo-captures and higher intensity of use at a particular site. Births were confirmed from photo-captures or direct sighting of lactating females (Additional file 1), while most cubs were recorded (photo-captured or sighted) with their mother only after they were about 2 months old. Since litter size at birth was rarely known, our reported litter size could be an underestimate as mortality before 2 months' age was not known. The ratio of cub: adult tigress was computed. Age at first reproduction was determined by recording first birth of tigresses that were monitored since they were cubs. Inter-birth intervals were recorded from intensively monitored tigresses that littered more than once during the study period. We recorded intervals between two successive litters when all cubs of the previous litter died before reaching dispersal age (more than 2 years) and compare these with intervals between two successive litters when cubs of the previous litter survived beyond dispersal age.

Reproductive success and recruitment

We recorded reproductive success of tigresses as the number of cubs that survived to the age of independence (24 months). Individually identified tiger cubs that were monitored up to young adult stage allowed us to determine the age at which these tigers acquired territories either by displacing established tigers or in vacant habitats through dispersal. We calculated recruitment as the proportion of cubs that survived to the sub-adult stage ($\geq$ 24 month). We also computed recruitment of sub-adults to successful breeders as the proportion of sub-adults that subsequently established territories. Territoriality was inferred when a young-adult exclusively used an area that was earlier used by other adult tigers of the same gender or dispersed to a vacant habitat and lived there for several months.

Survivorship

We estimated stage specific (cub, juvenile, sub-adult, young adult, prime adult, and old adult) survival probabilities of tigers by using known-fate model [51] in program MARK (*ver.*7, [52]). Known-fate models use Kaplan-Meier estimator [53, 54] to estimate survival which requires that fate of the individual is known with certainty during the period a particular individual is monitored. All individuals do not enter the study simultaneously but are added in a staggered manner known as the staggered entry design [54]. Radio-telemetered individuals provide the best data for such analyses [51]. However, in our case, besides the eight collared tigers that were located several times a week, each known individual tiger was observed (camera trap photo-captured, directly observed or photographed from hand-held cameras) at least once every month, and its fate recorded (see Additional file 2). In rare cases when an individual was not observed during an interval the known-fate model allows its observation to be 'censored' from the analysis [51]. Data on survival/death of tigers was compiled on a monthly basis. We subsequently pooled this data for a three-month period to have a good number of observations on each tiger as well as to have an interval that was meaningful for survivorship analysis of tigers that are reasonably long-lived. Live-dead encounter history matrix for tigers was made by pooling encounter histories of 3 months into one interval where *10* represented survival of the individual throughout the interval, *11* represented mortality of the individual during the interval, and *00* represented censoring the individual when that individual was not observed during the three-month interval. Only 59 observations on seven tigers out of 3492 observations from 97 tigers were censored during the study (see Additional file 2), forming a very small proportion (1.6%) of observation where fate could not be ascertained for that interval. All of these seven tigers were observed in subsequent periods, but by censoring them for intervals they were not recorded, we add to the uncertainty in our estimates [51]. Seven tigers (two juveniles, one sub-adult male, one young adult

male, and three old adult females) out of 97 that we monitored went 'missing' during the study period. We considered two extreme scenarios for these individuals in our survival analysis: a) a highly likely conservative scenario, that these tigers were poached, and b) a more optimistic but less likely scenario that these tigers dispersed and we were unable to trace them. In case of scenario 'a', we modelled these tigers as dead, and in case of scenario 'b', we censored them from our analysis. The resultant estimates of survival would encompass the true estimate. For our analysis, the matrix was right censored for all the surviving tigers at the end of December 2014. Tigers that survived and were monitored from cub to adulthood were included in subsequent stages with the assumption that survival rates were independent for different stages. As part of tiger reintroduction program, seven tigers were translocated from RTR to Sariska Tiger Reserve during the study period [55]. These individuals were censored in our survival analysis at the time of their translocation since these tigers would be exposed to a different set of environmental factors in Sariska that determine their survival probability. We formulated and run different candidate models which were ecologically plausible and compare them using AICc values. Model averaged survival estimates were obtained for models with less than five delta AICc values [47].

Mortality events were confirmed when carcasses were recovered, and causes of mortality were recorded on the basis of post-mortem report compiled by experienced veterinary personnel and/or by questioning eyewitnesses (if any). Cub mortalities were confirmed when found dead. However, not all cub carcasses could be recovered. A cub was considered dead if it was not detected (not photo-captured or sighted) with its mother for more than a month [34]. There were no incidents where a cub that was not recorded for over a month (considered dead) was ever seen again. We considered seven tiger cubs (ranging from 2 to 10 months) that were provisioned by park managers after the death of their mother, as dead since they were unlikely to have survived in the wild on their own.

The small and isolated nature of RTR made it possible for us to follow and ascertain dispersal events of tigers. Tigers that left the National Park had to traverse human-dominated areas where their presence was detected readily by local communities who were always on a high vigil for large carnivores. The presence of tigers was detected from signs and examination of livestock kills (which were compensated by the Government). Our research team and park managers ascertained the identity of these tigers by targeted effort of camera trapping and visual sighting. Due to the high profile nature of RTR as well as the small tiger population that was vulnerable to poaching, the park management along with the research team made a concentrated effort to locate each tiger individually to ascertain its wellbeing. Mortality events were categorized into natural mortality (death of the mother, infanticide, old age, disease, intra-specific strife), human-caused mortality (poaching, poisoning, accidents due to human causes), and unknown.

Dispersal age and distances

We considered tigers to have reached dispersal age once they became ≥24 month old (considering the lowest age of dispersal). At this age most tigers no longer moved with their mothers and were capable of hunting on their own. Dispersals from the natal area were confirmed through telemetry, camera trap photographs, direct observations and hand-held photography of tigers. Dispersal distances (Euclidean distance) of these tigers were measured from the centre of their natal area to the most extreme location of that tiger.

Results

Tiger abundance

On the average an effort of 3715 (SE 1324) camera trap nights were invested each year (Table 1). In most years the model having g0 as constant (.) and movement parameter (σ) having sex-specific responses was selected as the best-fit model. In year 2009 the null model (g0(.), σ(.)) was selected as the best model (Additional file 3: Table S1). Annual model averaged density estimates varied from 4.6 to 8.7 tigers per 100 km^2 (Table 1). The movement parameter (σ) was consistently larger for males compard to females and was found to decline asymptotically with an increase in density for males (Fig. 2). The density estimate obtained from SECR did not differ from near actual density (Table 1).

Sex ratio and female reproductive parameters

Adult sex ratio (male: female) in the initial years (2006–07) was female biased (0.38, SE 0.04) and became marginally female biased (0.91, SE 0.04, 2008–14) in subsequent years. Overall male: female ratio was 0.76 (SE 0.07) during the study period. The total number of breeding females in RTR ranged between 12 to 15 (see Additional file 4: Figure S1). We recorded litter size from 33 litters of 18 females, and the mean litter size was 2.24 (SE 0.14; range = 1–4 cubs). Most of the litters were of two (50%) or three (31%) cubs (Fig. 3). Overall cub: adult female ratio ($n = 9$ years) was 0.48 (0.12 SE). We could ascertain the age at first reproduction for 11 tigresses that were monitored since they were cubs as 54.5 months (3.7 SE; range 33–68 months). Eight out of these 11 tigresses produced their first litter after 4 years of age, while only two tigresses gave birth before 3 years (see Additional file 5: Figure S2). The average interval between two successive litters (inter-birth interval) was 29.6 months (SE 3.15; range = 7–51 months, $n = 14$ intervals from 8 tigresses). Intervals were shorter when all cubs of previous litter died before reaching independence (24 months, 15.0, SE 4.04 months, $n = 3$) than when cubs of previous litters survived till independence (33.64, SE 2.83 months, $n = 11$, see Additional file 6: Figure S3).

Reproductive success and recruitment

More than 50% of the intensively monitored breeding females ($n = 18$) successfully raised all their litters to independence, while around 10% of the females failed to raise any of their cubs during the study period (Fig. 4). Data on 51 individuals indicated that male recruitment rate from cub to the sub-adult stage was higher (77.8%, SE 2.2) than females (62.5%, SE 2.4). But male recruitment rate as breeding adults in the population from the

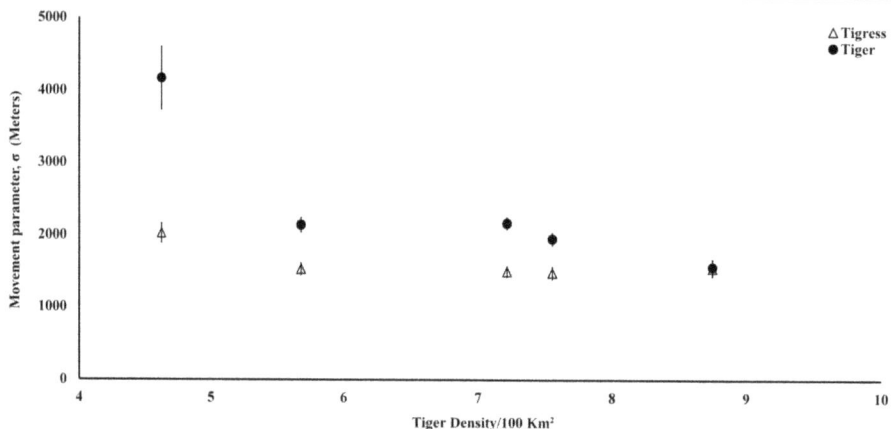

Fig. 2 The relationship between the movement parameter σ estimated using likelihood based spatially explicit capture recapture models and tiger density. Movement parameter σ for tigers declined asymptotically with increasing density while σ for tigresses remained relatively constant with increasing density

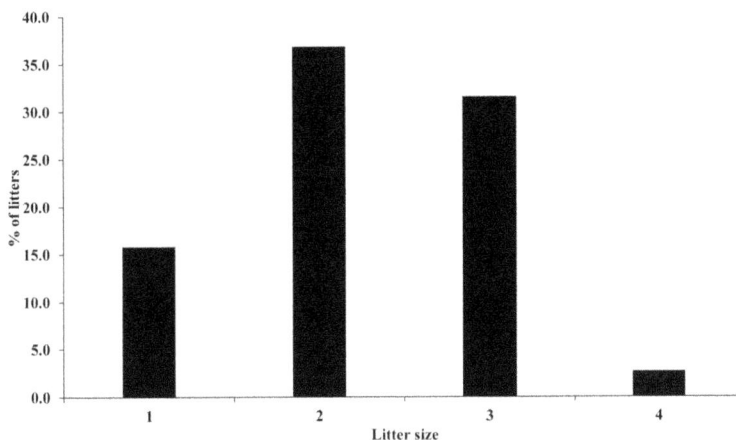

Fig. 3 Percent frequency of different sized litters ($n = 33$) of tigresses observed in Ranthambhore Tiger Reserve

sub-adult stage was lower (72.6%, SE 2.0) than females (86.7%, SE 1.3, Table 2).

Survivorship

The average annual survival rate of cubs (85.4%) were comparatively lower than that of other age classes (Table 3). Annual survival rate increased after cub stage, remained constant till prime adult stage after which it declined (Table 3). Survival rates for males and females were similar for younger stages while females had marginally higher survival in older stages (Table 3). The survival estimates with our conservative approach of considering seven missing tigers as dead did not statistically differ from survival estimates when these tigers were censored from the analysis (Additional file 7: Table S2).

We recorded 25 mortality events (17cubs and juveniles, and 8 adults) during our study. Amongst all cub and juvenile mortality, 41% were natural (infanticide and death of mother due to natural causes), 47% were human caused (mother poached, accidents due to human

causes and poisoning) and 12% of these could not be ascertained (Fig. 5). Amongst all adult mortality, 50% were natural (old age, disease, and intra-specific strife), and 38% were human-caused (poisoning and poaching, Fig. 5), the cause of 12% adult mortality could not be determined.

Dispersal age and distances

We recorded dispersal of 29 tigers, of these six were long distance dispersal out of RTR (Fig. 1). Mean dispersal age of tigers in RTR was 33.9 months (SE 0.8, range = 24–42 months). Mean male dispersal distances (60.6, SE 2.1 Km) were larger than that of females (12.1, SE 1.3 Km, Fig. 6). Most of the females established their territory near their natal area while the majority of the males dispersed further from their natal areas. Five out of 16 males and one out of 13 females dispersed outside RTR and settled in forest patches within the larger landscape. These long distance dispersal movements ranged from 56 to 220 km Euclidean distance from their natal areas.

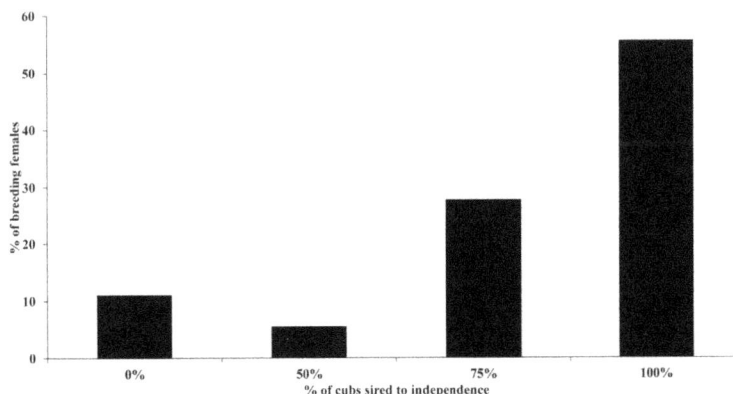

Fig. 4 Reproductive success of breeding females in Ranthambhore Tiger Reserve measured as a percentage of cubs that survived to recruitment age (24 months) ($n = 74$ cubs from 33 litters of 18 females)

Table 2 Recruitment of tigers with known fate (from ~2 months of age to age of independence >2 years and as territorial adults >3 years) in Ranthambhore Tiger Reserve

Classes	No. of cubs	Recruitment to adult stage (%)	Recruitment as territorial adults (%)
Male	27	21 (77.8%, $CI_{95\%}$: 73.5–82.0%)	16 (76.2%, $CI_{95\%}$: 72.4–80.0%)
Female	24	15 (62.5%, $CI_{95\%}$: 57.8–67.1%)	13 (86.7%, $CI_{95\%}$: 84.1–89.2%)
All	51	36 (70.6%, $CI_{95\%}$: 64.2–76.9%)	25 (69.4%, $CI_{95\%}$: 64.0–74.8%)

Discussion

Tigers are conservation dependent species and require substantial investments in terms of management and protection for their long-term persistence [56]. Information on survival, recruitment, litter size, reproductive parameters, dispersal, sex ratio and density that we provide in this paper are the basis of estimating population viability and planning management interventions.

We mostly use standard methodology, but adept some and develop a few approaches required for procuring data from endangered carnivores. The subsequent analyses of these data do not violate any underlying assumptions of the analytical procedures. We believe that we were justified in using Known-Fate model for estimating survival due to the high frequency (every month) of observations of 97 individually known tigers for determining their fate (see Additional file 2). As explained in our

Table 3 Survival rates of tigers (n = 97) in Ranthambhore between 2006 to 2014

Age class	Gender	Sample size	Average annual survival rate[a] ($CI_{95\%}$)
Cubs (< 12 months)	Male	39	85.35 (80.3–90.4) %
	Female	35	85.40 (80.3–90.5) %
Juveniles (1–2 years)	Male	33	97.05 (95.4–98.7) %
	Female	26	97.06 (95.4–98.7) %
Sub adults (2–3 years)	Male	28	96.46 (94.0–98.9) %
	Female	19	96.49 (94.1–98.9) %
Young adults (3–5 years)	Male	20	93.87 (88.0–99.8) %
	Female	18	94.26 (89.0–99.6) %
Prime adults (5–10 years)	Male	15	82.53 (74.6–90.4) %
	Female	20	86.43 (80.7–92.1) %
Old adults (> 10 years)	Male	3	82.78 (76.9–88.7) %
	Female	12	84.52 (79.1–90.0) %
Adults (> 3 years)	Male	38	84.88 (80.6–89.2) %
	Female	50	88.74 (85.3–92.2) %
All adults (> 3 years)	Male and Female	88	86.99 (84.3–89.7) %

[a]Conservative estimates, where we have considered seven 'missing' tigers as dead; a more likely scenario

methods, we 'censored' surviving individuals when the study was completed and for the few occasions in-between when we could not determine the fate of individuals [51]. Open population capture-mark-recapture models cannot distinguish between emigration and mortality, and are confounded by the nuisance parameter of imperfect detection [51, 57]. Therefore, known-fate models though being data intensive, provide more precise and more informative parameter estimates.

During the initial phase of the study (i.e. 2006) tiger density in RTR was low (Table 1) as the population was recovering from a recent decline caused by poaching. In subsequent years, density increased with good protection and fluctuated between 5.6 to 8.7 tigers/ 100km^2 with a mean density of 7.5 (SE 2.7) tigers/ 100km^2 (Table 1). This fluctuation in tiger density was likely due to synchrony in breeding by several females and recruitment of a large cohort of sub-adults that became available for camera trap sampling at an approximate interval of 2 years. As tiger density increases within a limited area, we would expect home-range to either decrease and/or show an increase in overlap. The movement parameter σ is an index of home-range radius over short time duration [42]. Efford et al. [45] show an asymptotic decline in σ with increase in tiger density from data across India. Herein, we demonstrate a similar relationship between σ and tiger density within a single population which we believe is ecologically more meaningful (Fig. 2). The asymptotic nature of σ for males and a relatively constant σ for females suggests that male tigers' home range decline to some extent with increase in tiger density, while home range size in females, which is based primarily on food availability [31], does not change with density. This could be interpreted to suggest that either tigresses do not invest energy in acquiring a home range larger than required to rear cubs or that Ranthambhore National Park was already near carrying capacity density with little scope of reduction in home-range sizes for tigresses. The increase in tiger numbers in RTR was accompanied with an increase in tiger occupancy while the number of breeding females remained relatively constant (12 to 15) during our study period (see Additional file 4: Figure S1). Camera trap data suggests that the entire core area of RTR (Ranthambhore National Park and Sawai Mansingh Sanctuary) was occupied by tigers by 2014. We also witnessed intense competition between mothers and daughters for breeding territories. This was another indication that tiger density was at or near carrying capacity within the core of RTR. For high density populations near carrying capacity regulatory mechanisms may include delayed age of first reproduction, increased inter-birth intervals, smaller litter sizes and depressed survival [58]. Delayed female age of first reproduction and longer intervals between two

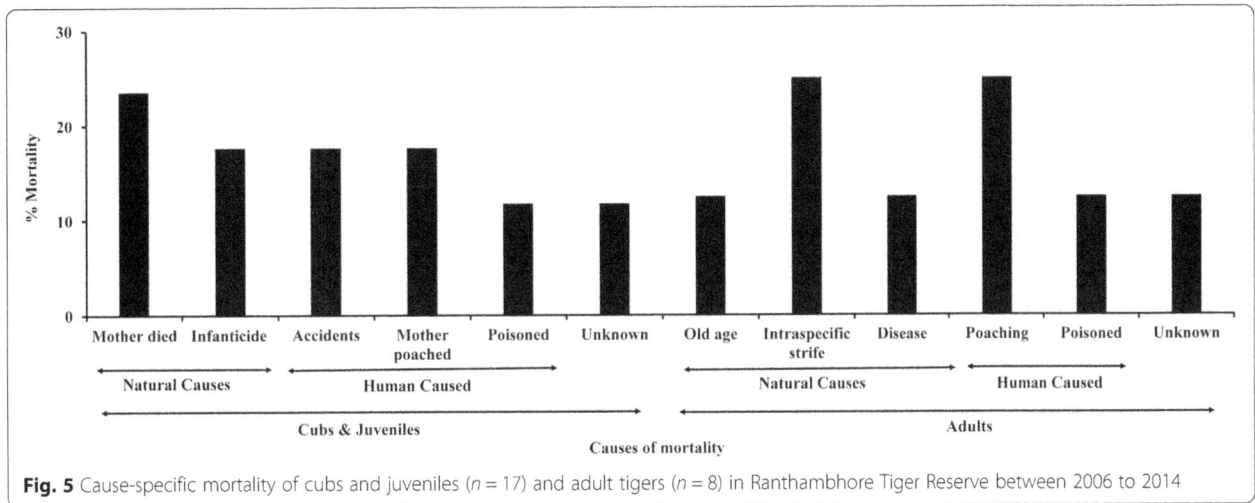

Fig. 5 Cause-specific mortality of cubs and juveniles (n = 17) and adult tigers (n = 8) in Ranthambhore Tiger Reserve between 2006 to 2014

successive litters observed in RTR tigers were likely an effect of high tiger density at or nearing carrying capacity inside the core of RTR. Parameters that could potentially be depressed by inbreeding depression like litter size, cub survival, and disease caused mortality were either similar or better in comparison to other tiger populations (Table 4).

In highly inbred populations morphological abnormalities are often observed and resistance to disease is often compromised [18, 59]. During our study we did not encounter cubs with abnormalities or skeletal defects and mortality of tigers attributed to disease was only one. The morphometric measurement of RTR tigers captured for radio-collaring were among those of the largest recorded for tigers in India (YV Jhala, unpublished data). These observations along with comparable demographic parameters to other tiger populations, suggest that there were no deleterious effects that had as yet manifested

among RTR tigers due to population bottlenecks, small size and isolation [18]. This seems likely, since, till recent times the semiarid zone tiger population of central India was large and well connected [15] and the population bottlenecks that RTR tigers passed through did not remain very small for long periods of time (less than one generation time) [60]. However, we lack data on early infant mortality and foetal loss during pregnancy, which are some of the parameters that would be influenced by inbreeding depression. Reddy et al. [61] conclude that RTR tigers had reasonable genetic diversity comparable to other central Indian tiger populations. Studies that link genetic variability with population size, connectivity, and ultimately with demography are required for conservation management of large carnivores.

Females that had lost cubs before their recruitment age gave birth within a smaller time interval as also reported in tigers [62] and in Asiatic lions [38]. Due to a high level of protection, poaching was rare inside RTR during the study period. Therefore, long and stable tenures of territorial males were recorded (mean 5.6 years, range 4 to 11 years, n = 13). Only three cases of infanticide were observed during this study; this contrasts with studies on tigers in Chitwan National Park in Nepal [30] and Asiatic lions in Gir [38], where infanticide was a major cause of cub mortality. Mean dispersal age of RTR tigers (~33 months) was higher than reported for Chitwan (~23 months) [30], and Amur tigers (~19 months, Kerley et al. 2003). Young tigers were not compelled to leave their natal territories due to new male takeovers [30]. This stability in territorial male tenures allowed young tigers, especially males, to continue to live longer within their natal area enhancing their survival and achieving rapid growth to breeding body size. On four occasions we observed sub-adult male tigers occupy small ranges in peripheral areas of resident male territories. Once such tigers become sufficiently

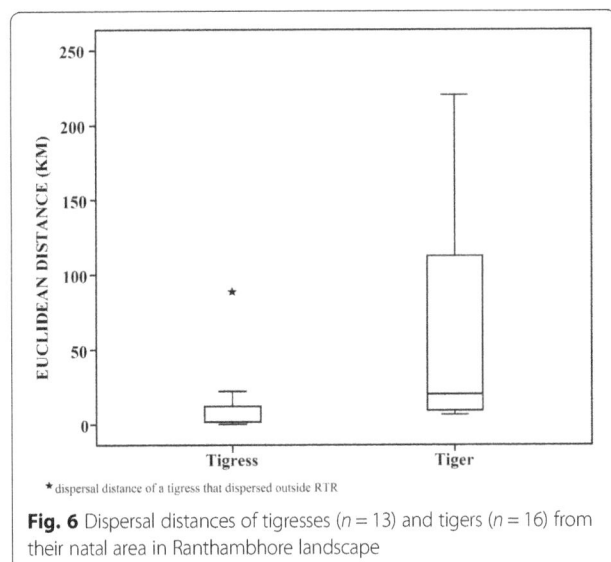

Fig. 6 Dispersal distances of tigresses (n = 13) and tigers (n = 16) from their natal area in Ranthambhore landscape

Table 4 Comparison of demographic parameters of tigers from Ranthambhore Tiger Reserve with demographic parameters from published studies

Sources	Area (sub-species)	Litter size	Female age at 1st reproduction (years)	Inter-birth intervals (months)	Cub survival probability	Adult survival	
						Male	Female
Sankhala 1978 [67]	Zoo, India (*Panthera tigris tigris*)	2.9 (*n* = 49)	3–6 years	24–36	NA	NA	
Smith & McDougal 1991 [30]	Wild, Nepal (*P. T. tigris*)	2.98 (n = 49)	3.4 years (n = 7)	21.6 (n = 7)	0.65	NA	
Chundawat et al. 2002 [68]	Wild, India (*P. t. tigris*)	2.3 (*n* = 12)	NA	21.6 (n = 14)	NA	NA	
Kerley et al. 2003 [33], Goodrich et al. 2008 [34]	Wild, Russia (*P. T. altaica*)	2.4 (± 0.6) (n = 16)	4 (± 0.4) years (*n* = 4)	21.4 (± 4.4) (*n* = 7)	0.53–0.59	0.63 (±0.2)	0.81 (±0.1)
Singh et al. 2013 [32]	Wild, India (*P. t. tigris*)	2.9 (± 0.2) (n = 18)	NA	25.2 (± 1.8) (n = 9)	NA	NA	
Singh et al. 2013 [32]	Wild, India (*P. t. tigris*)	2.3 (±0.1) 2(*n* = 22)	NA	33.4 (± 3.7) (n = 7)	NA	NA	
Present study	Wild, India (*P. t. tigris*)	2.24 (±0.14) (n = 33)	4.54 (± 0.3) years (n = 11)	29.6 (± 3.1) (*n* = 14)	0.85 (±0.02)	0.84 (±0.02)	0.88 (±0.01)

NA not available

large so as to challenge resident males, they expand their range into that of the residential males' territory. By following this strategy relatively smaller and inexperienced sub-adult males avoid lethal encounters with larger and experienced males during their initial dispersal stage [31].

Ten out of 13 sub-adult females established territories near their natal areas, two females established their territory inside RTR but ~20 km away from their natal area, and one female dispersed outside the reserve (~90 km). We observed that females either established their territory beside their mothers (*n* = 3) or occupied a part of their mother's territory and gradually pushed their mother off (*n* = 5). One female that initially occupied her mother's territory by displacing her mother produced her first litter in this territory, but subsequently shifted her territory with her 12-month-old cubs (of her second litter) ~10 km away from her natal territory. This shift coincided with the takeover of her natal territory by a new male tiger. Her natal territory was subsequently occupied by her daughter from her first litter.

The dispersing sex in tigers is known to be males [31]. Amongst RTR tigers, males did disperse larger distances than females, but opportunities to disperse were restricted in this landscape. Large dispersal distances travelled by tigers (once out of RTR), and older age of dispersal reflects the difficulty a tiger faces in locating appropriate vacant habitat to settle. The small reserve size combined with very little and disjunct tiger habitat available outside the reserve system within a hostile human-dominated habitat matrix restricted dispersal. Males that managed to disperse and locate habitat patches tried to settle there, but due to lack of female tigers in these patches, were unable to breed. If areas of Kailadevi Wildlife Sanctuary (part of RTR buffer), Kuno Wildlife

Sanctuary, Ramgarh Visdhari Wildlife Sanctuary and Mukundara Hills Tiger Reserve (Fig. 1) are made free of human settlements through incentivised voluntary relocation scheme (WPA 1972; 2006 amendment), these areas could harbour breeding population of tigers. Simultaneously dispersal corridors between these tiger habitats and with core of RTR need to be secured and restored to promote a metapopulation in this landscape. Initially, once these areas have been appropriately restored and have sufficient prey base, this tiger dispersal could be aided by translocating tigers. Establishing and managing the tiger population in the larger Ranthambhore landscape as a metapopulation [63] would be desirable for long-term conservation.

Park authorities often intervene by treating injured tigers in-situ or supplementing food resources for orphan/sick cubs. These activities, although well intentioned, hinder the natural process of selection and the social dynamics of the species [38, 64]. Such interventions also reduce the genetic fitness of the population over time by ensuring the survival of unfit individuals especially if sick animals are treated, saved and allowed to breed [18]. Interfering with the natural process may be necessary for highly endangered populations where every living individual counts. However, RTR is now a high tiger density area [65], and therefore, management interference of health care should be extremely selective, if any.

Conclusion

Our study did not find any evidence of detrimental effects resulting from a small population that could potentially be inbred and there seems to be no current need for genetic rescue [20] of RTR tigers. Currently RTR has about 15 breeding units and adult female survival of about 88%, these are bare minimal requirements to

Demography of a small, isolated tiger (Panthera tigris tigris) population in a semi-arid region of western...

149

ensure long-term persistence [66]. Managing the Ranthambhore landscape by restoring habitat patches and connectivity with RTR so as to promote a metapopulation of tigers would enhance the potential of long-term persistence of this last remaining semi-arid tiger population in western India.

Additional files

Additional file 1: Field Guide for Aging Tigers. (PDF 8068 kb)

Additional file 2: Individual tiger monitoring data used for "Known Fate" survival analysis. (PDF 98 kb)

Additional file 3: Table S1. Sampling details and parameters estimates of annual tiger density from camera trap based spatial capture-recapture analysis in Ranthambhore Tiger Reserve. (PDF 84 kb)

Additional file 4: Figure S1. Number of breeding tigresses in each year observed in Ranthambhore Tiger Reserve during the study period. (TIFF 258 kb)

Additional file 5: Figure S2. Age at first reproduction of tigresses (n = 11) observed in Ranthambhore Tiger Reserve. (TIFF 515 kb)

Additional file 6: Figure S3. Inter-birth intervals of tigresses in Ranthambhore Tiger Reserve when all cubs of previous litter died before reaching independence (n = 3 litters) and when cubs of previous litters survived beyond 24 months (n = 11). (TIFF 344 kb)

Additional file 7: Table S2. Survival rates of tigers (n = 97) estimated in an optimistic (where we have censored seven 'missing' tigers) as well as a conservative (a more likely scenario where we have considered seven 'missing' tigers as dead) approach in Ranthambhore between 2006 to 2014. (PDF 8 kb)

Abbreviations

AIC$_c$: Akaike Information Criterion, corrected for sample size; CI$_{95\%}$: 95% Confidence Interval; g0: Capture probability at activity centre; n: Sample size; NP: National Park; RTR: Ranthambhore Tiger Reserve; SE: Standard Error; SECR: Spatially Explicit Capture Recapture; TR: Tiger Reserve; σ: Sigma, movement parameter

Acknowledgements

Rajesh Gopal is specially thanked for encouragement and facilitation. Director, Dean and Research Coordinator of the Wildlife Institute of India are acknowledged for logistic support. We thank G. V. Reddy, Rajesh Gupta, and Y. K. Sahu for support. We thank all frontline staff of Ranthambhore Tiger Reserve and our field assistants Ram Prasad, Mujahid, Kannaiya, and Javed for their sincere effort. S. Dutta, U. Kumar, and K. Banerjee are acknowledged for assistance with data analysis. G. S. Bhardwaj and D. S. Shaktawat are acknowledged for providing photographs.

Funding

This work was funded by the National Tiger Conservation Authority for the research project "Monitoring Source Population of Tiger in Ranthambhore Tiger Reserve".

Authors' contributions

YVJ and QQ conceived the study, and raised funds, YVJ & QQ supervised the study, AS and PPC, YVJ conducted the field work, YVJ did the radio-collaring, QQ, RSS and SS coordinated the field work, assisted in radio-collaring, RSS and SS provided logistic support, assisted in monitoring tigers and locating dispersing tigers, AS and YVJ analysed the data and wrote the manuscript. All authors have read, provided inputs, and given their consent to publish the final manuscript.

Competing interests

The authors declare that they have no competing interests.

Author details

[1]Department of Animal Ecology and Conservation Biology, Wildlife Institute of India, Chandrabani, Dehra Dun, Uttarakhand 248001, India. [2]Department of Population Management, Capture and Rehabilitation, Wildlife Institute of India, Chandrabani, Dehra Dun, Uttarakhand 248001, India. [3]Rajasthan Forest Department, Jaipur, Rajasthan 302004, India. [4]Tamilnadu Forest Department, Udumalpet, Tamil Nadu 642126, India.

References

1. Gee EP. The wildlife of India. London: UK. Collins; 1964.
2. Jhala YV, Qureshi Q, Gopal R. The status of tigers in India 2014. New Delh and Dehradun: National Tiger Conservation Authority and The Wildlife Institute of India; 2015.
3. Dinerstein E, Loucks C, Wikramanayake E, Ginsberg J, Sanderson E, Seidensticker J, Forrest J, Bryja G, Heydlauff A, Klenzendorf S, Leimgruber P. The fate of wild tigers. Bioscience. 2007;57:508–14.
4. Panwar HS. Project Tiger: the reserves, the tigers and their future. In: Tilson RL, Seal US, editors. Tigers of the world: the biology, biopolitics, management and conservation of an endangered species. Park Ridge: Noyes Publications; 1987. p. 110–7.
5. Check E. The tiger's retreat. Nature. 2006;441:927–30.
6. Gopal R, Qureshi Q, Bhardwaj M, Singh RKJ, Jhala YV. Evaluating the status of the endangered tiger Panthera Tigris and its prey in Panna Tiger Reserve, Madhya Pradesh, India. Fauna and Flora International, Oryx. 2010;44:383–9.
7. Ranganathan J, Chan KMA, Karanth KU, Smith JLD. Where can tiger persist in the future? A landscape-scale, density-based population model for the Indian subcontinent. Biol Conserv. 2008;141:67–77.
8. Yumnam B, Jhala YV, Qureshi Q, Maldonado JE, Gopal R, Saini S, Srinivas Y, Fleischer RC. Prioritizing Tiger conservation through landscape genetics and habitat linkages. PLoS One. 2014;9(11):e111207.
9. Caughley GC. Directions in conservation biology. J Anim Ecol. 1994;63:215–44.
10. Purvis A, Gittleman JL, Cowlishaw G, Mace GM. Predicting extinction risk in declining species. Proc Biol Sci. 2000;267:1947–52.
11. Singh K. Shikar Camps. In: Rangarajan M, editor. The Oxford anthology of Indian wildlife volume I: hunting and shooting. New Delhi: Oxford University Press; 1999. p. 35–50.
12. IBWL (Indian Board for Wild Life). Project Tiger, a planning proposal for the preservation of the Tiger (Panthera tigris tigris Linn.) in India. Dehradun: F.R.I. Press; 1972.
13. Jackson P. Fifty years in the tiger world: an introduction. In: Tilson RL, Seal US, editors. Tigers of the world: the biology, politics, management and conservation of an endangered species. Park Ridge: Noyes Publications; 2010. p. 1–15.
14. Sharma S, Wright B. Monitoring tigers in Ranthambhore using digital pugmark technique: Technical Report. New Delhi: Wildlife Protection Society of India; 2005.
15. Chundawat RS, Sharma K, Gogate N, Malik PK, Vanak AT. Size matters: scale mismatch between space use patterns of tigers and protected area size in a tropical dry Forest. Biol Conserv. 2016;197:146–53.

16. Allendorf FW, Leary RF. Heterozygosity and fitness in natural populations of animals. In: Conservation biology: the science of scarcity and diversity. M. E. Smith, editor, Sinauer, Sunderland, MA; 1986. p. 57–76.

17. Frankham R, Briscoe DA, Ballou JD. Introduction to conservation genetics. Cambridge: Cambridge University Press; 2002.

18. Keller LK, Waller DM. Inbreeding effects in wild populations. Trends Ecol Evol. 2002;17:230–41.

19. O'brien SJ. Tears of the cheetah: the genetic secrets of our animal ancestors. St Martin's Press New York: Thomas Dunne Books; 2003.

20. Pimm SL, Dollar L, Bass OL. The genetic rescue of the Florida Panther. Anim Conserv. 2006;9:115–22.

21. Balme GA, Batchelor A, Britz NDEW, Seymour G, Grover M, Hes L, Macdonald DW, Hunter LTB. Reproductive success of female leopards Panthera pardus: the importance of top-down processes. Mammal Rev. 2012;43:221–37.

22. Karanth KU. Estimating tiger (Panthera tigris) populations from camera-trap data using capture-recapture models. Biol Conserv. 1995;71(3):333–8.

23. Karanth KU, Nichols JD. Estimation of tiger densities in India using photographic captures and recaptures. Ecology. 1998;79:2852–62.

24. Karanth KU, Nichols JD, Kumar NS, Link WA, Hines JE. Tigers and their prey: predicting carnivore densities from prey abundance. PNAS. 2004;101:4854–8.

25. Kawanishi K, Sunquist ME. Conservation status of tigers in a primary rainforest of peninsular Malaysia. Biol Conserv. 2004;120:329–44.

26. Barlow ACD, Ahmed MIU, Rahman MM, Howlader A, Smith AC, Smith JLD. Linking monitoring and intervention for improved management of tigers in the Sundarbans of Bangladesh. Biol Conserv. 2008;141:2031–40.

27. Jhala YV, Gopal R, Qureshi Q. Status of tigers, co-predators, and prey in India. New Delhi and Dehradun: National Tiger Conservation Authority, Govt. of India and Wildlife Institute of India; 2008.

28. Jhala YV, Qureshi Q, Gopal R. Can the abundance of tigers be assessed from their signs? J Appl Ecol. 2011;48:14–24.

29. Harihar A, Prasad DL, Ri C, Pandav B, Goyal SP. Losing ground: tigers Panthera Tigris in the north-western Shivalik landscape of India. Oryx. 2009; 43:35–43.

30. Smith JLD, McDougal C. The contribution of variance in lifetime reproduction to effective population size in tigers. Conserv Biol. 1991;5:484–90.

31. Smith JLD. The role of dispersal in structuring the Chitwan tiger population. Behaviour. 1993;124:165–95.

32. Singh R, Mazumdar A, Sankar K, Qureshi Q, Goyal SP, Nigam P. Interbirth interval and litter size of free-ranging Bengal tiger (Panthera tigris tigris) in dry tropical deciduous forests of India. Eur J Wildlife Res. 2013;59:629–36.

33. Kerley LL, Goodrich JM, Miquelle DG, Smirnov EN, Quigley H, Hornocker MG. Reproductive parameters of wild female Amur (Siberian) tigers (Panthera tigris altaica). J Mammal. 2003;84:288–98.

34. Goodrich JM, Kerley LL, Smirnov EN, Miquelle DG, McDonald L, Quigley HB, Hornocker MG, McDonald T. Survival rates and causes of mortality of Amur tigers on and near the Sikhote-Alin biosphere Zapovednik. J Zool. 2008;276:323–9.

35. Sankar K, Goyal SP, Qureshi Q. Assessment of status of tiger (Panthera tigris) in Sariska Tiger Reserve, Rajasthan. Dehra Dun: A Report submitted to the Project Tiger, Ministry of Environment & Forests, Govt. of India, New Delhi Wildlife Institute of India; 2005. p. 1–26.

36. Champion HG, Seth SK. A revised survey of the Forest types of India. Dehli: Manager of Publications; 1968.

37. Schaller GB. The Serengeti lion. Chicago: University of Chicago Press; 1972.

38. Banerjee K, Jhala YV. Demographic parameters of endangered Asiatic lions (Panthera leo persica) in Gir forests. India Journal of Mammalogy. 2012;93: 1420–30.

39. Chao A, Huggins RM. Classical closed-population capture-recapture models. In: Amstrup SC, TL MD, Manly BF, editors. Handbook of capture-recapture analysis. Princeton: Princeton University Press; 2010. p. 22–35.

40. Kreeger TJ. Handbook of wildlife chemical immobilization. 1st ed. Laramie: International Wildlife Veterinary Services Inc; 1996. p. 175.

41. Efford MG. Estimation of population density by spatially explicit capture-recapture analysis of data from area searches. Ecology. 2011;92:2202–7.

42. Borchers DL, Efford MG. Spatially explicit maximum likelihood methods for capture-recapture studies. Biometrics. 2008;64:377–85.

43. Efford MG. Secr: spatially explicit capture-recapture models. R package version 2.9.3. 2015. http://CRAN.R-project.org/package=secr.

44. R Core Team. R: a language and environment for statistical computing. Vienna: R foundation for Statistical Computing. http://www.R-project.org/. Accessed 15 Mar 2017.

45. Efford MG, Dawson DK, Jhala YV, Qureshi Q. Density-dependent home-range size revealed by spatially explicit capture–recapture. Ecography. 2015; 39:676–88.

46. Akaike H. A new look at the statistical model identification. IEEE Trans Autom Control. 1974;19:716–23.

47. Burnham KP, Anderson DR. Model selection and multimodel inference- a practical information-theoretic approach. 2nd ed. New York: Springer; 2002.

48. Sharma RK, Jhala Y, Qureshi Q, Vattakaven J, Gopal R, Nayak K. Evaluating capture–recapture population and density estimation of tigers in a population with known parameters. Anim Conserv. 2010;13(1):94–103.

49. Skalski JR, Ryding KE, Millspaugh JJ. Wildlife demography: analysis of sex, age, and count data. Burlington: Elsevier Academic Press; 2005.

50. Sunquist ME. The social organization of tigers (Panthera tigris) in Royal Chitawan National Park, Nepal. Washington: Smithsonian Institution Press; 1981.

51. Williams BK, Nichols JD, Conroy MJ. Analysis and management of animal populations. Oxford: Academic Press; 2002.

52. Cooch W, White GC. Program MARK: a gentle introduction. 17th ed; 2017. p. 16-1–16-25. http://www.phidot.org/software/mark/docs/book/.

53. Kaplan EL, Meier P. Non-parametric estimation from incomplete observations. J Am Stat Assoc. 1958;53:457–81.

54. Pollock KH, Winterstein SR, Bunck CM, Curtis PD. Survival analysis in telemetry studies: the staggered entry design. J Wildlife Manage. 1989;53:7–15.

55. Sankar K, Nigam P, Malik PK, Qureshi Q, Bhattarcharjee S. Monitoring of reintroduced tigers (Panthera tigris tigris) in Sariska Tiger Reserve, Rajasthan. Technical report –1. Dehradun: Wildlife Institute of India; 2013.

56. Sanderson EW, Forrest J, Loucks C, Ginsberg J, Dinerstein E, Seidensticker J, Leimgruber P, Songer M, Heydlauff A, O'Brien T, Bryja G. Setting priorities for tiger conservation: 2005–2015. In: Tilson RL, Seal US, editors. Tigers of the world: the science, politics, and conservation of Panthera tigris. Park Ridge, New Jersey: Noyes Publications; 2010. p. 143–61.

57. Nichols JD. Modern open-population capture-recapture models. In: Amstrup SC, TL MD, Manly BF, editors. Handbook of capture-recapture analysis. Princeton: Princeton University Press; 2010. p. 88–123.

58. Derocher AE, Stirling I. The population dynamics of polar bears in western Hudson Bay. In: McCullough D, Barrett R, editors. Wildlife 2001: populations. England: H. Elsevier Science Publishers Ltd; 1992. p. 1150–9.

59. Räikkönen J, Vucetich JA, Peterson RO, Nelson MP. Congenital bone deformities and the inbred wolves (Canis Lupus) of isle Royale. Biol Conserv. 2009;142:1025–31.

60. Reddy GV, Tyagi RK, Bhatnagar D, Soni RG, Daima ML, Sen A. Management plan of Ranthambhore Tiger Reserve (2002–2003 to 2011–2012). Rajasthan, India: Forest Department; 2002. p. 363–6.

61. Reddy PA, Gour DS, Bhavanishankar M, Jaggi K, Hussain SM, Harika K, Shivaji S. Genetic evidence of Tiger population structure and migration within an isolated and fragmented landscape in Northwest India. PLoS One. 2012;7:e 29827.

62. Karanth KU. Tiger ecology and conservation in the Indian subcontinent. J Bombay Nat Hist Soc. 2003;100:169–89.

63. Hanski I. Metapopulation dynamics. Nature. 1998;396:41–9.

64. Packer C, Brink H, Kissui BM, Maliti H, Kushnir H, Caro T. Effects of trophy hunting on lion and leopard populations in Tanzania. Conserv Biol. 2011; 25(1):142–53.

65. Sadhu A, Gupta D, Latafat K, George S, Jhala YV, Qureshi Q. Ranthambhore Tiger Reserve. In: Jhala YV, Qureshi Q, Gopal R, editors. The status of tigers in India 2014. New Delhi and Dehradun: National Tiger Conservation Authority and The Wildlife Institute of India; 2015. p. 167–9.

66. Chapron G, Miquelle DG, Lambert A, Goodrich JM, Legendre S, Clobert J. The impact on tigers of poaching versus prey depletion. J Appl Ecol. 2008; 45:1667–74.

67. Sankhala K. Tiger! The story of the Indian tiger. London: Collins; 1978.

68. Chundawat RS, Gogate N, Malik PK. Understanding tiger ecology in the tropical dry deciduous forests of Panna Tiger Reserve. Final report. Dehradun: Wildlife Institute of India; 2002.

Species-level divergences in multiple functional traits between the two endemic subspecies of Blue Chaffinches *Fringilla teydea* in Canary Islands

Jan T. Lifjeld[1]*(iD), Jarl Andreas Anmarkrud[1], Pascual Calabuig[2], Joseph E. J. Cooper[3], Lars Erik Johannessen[1], Arild Johnsen[1], Anna M. Kearns[1,4], Robert F. Lachlan[3], Terje Laskemoen[1], Gunnhild Marthinsen[1], Even Stensrud[1] and Eduardo Garcia-del-Rey[5]

Abstract

Background: One of the biggest challenges in avian taxonomy is the delimitation of allopatric species because their reproductive incompatibility cannot be directly studied in the wild. Instead, reproductive incompatibility has to be inferred from multiple, divergent character sets that indicate a low likelihood of allopatric populations amalgamating upon secondary contact. A set of quantitative criteria for species delimitation has been developed for avian taxonomy.

Results: Here, we report a broad multi-trait comparison of the two insular subspecies of the Blue Chaffinch *Fringilla teydea*, endemic to the pine forests of Tenerife (ssp. *teydea*) and Gran Canaria (ssp. *polatzeki*) in the Canary Islands. We found that the two taxa were reciprocally monophyletic in their whole mitogenomes and two Z chromosome introns. The genetic distance in mitogenomes indicates around 1 Mya of allopatric evolution. There were diagnostic differences in body morphometrics, song and plumage reflectance spectra, whose combined divergence score (=11) exceeds the threshold level (=7) set for species delimitation by Tobias et al. (Ibis 152:724–746, 2010). Moreover, we found a marked divergence in sperm lengths with little range overlap. Relatively long sperm with low intra- and intermale CV compared to other passerines suggest a mating system with high levels of sperm competition (extrapair paternity) in these taxa.

Conclusion: The large and diagnostic divergences in multiple functional traits qualify for species rank, i.e., Tenerife Blue Chaffinch (*Fringilla teydea*) and Gran Canaria Blue Chaffinch (*Fringilla polatzeki*). We encourage a wider use of sperm traits in avian taxonomy because sperm divergences might signal reproductive incompatibility at the postcopulatory prezygotic stage, especially in species with sperm competition.

Keywords: Integrative taxonomy, Mitogenomes, Reproductive barriers, Plumage colour, Song, Speciation, Sperm size

* Correspondence: j.t.lifjeld@nhm.uio.no
[1]Natural History Museum, University of Oslo, PO Box 1172 Blindern, 0318 Oslo, Norway
Full list of author information is available at the end of the article

Background

When two populations of the same species become spatially separated, they will start to diverge as a product of the combined processes of mutation, selection and drift, and eventually become separate species [1–3]. Such allopatric evolution is thought to be the main pathway to the formation of new species [2, 4–7]. However, since evolutionary divergence is a gradual process, it is operationally difficult to define the actual threshold at which two allopatric populations can be classified as separate species. Moreover, there is no consensus about the criteria for delimiting species or the definition of species [8–10]. The controversies over species concepts basically come down to the role of evolutionary history versus population processes, i.e., the criteria of monophyly (phylogenetic species concepts, e.g., [11]) and reproductive incompatibility (the biological species concept [12, 13]).

In avian taxonomy there is a general consensus that species must have diagnostic characteristics and an independent evolutionary history [11, 14]. Those two features are tightly linked, as diagnostic, heritable characters take time to evolve. However, there is disagreement over the emphasis of reproductive incompatibility. Whereas proponents of the phylogenetic species concept argue that it is irrelevant whether two independently evolving species will merge or remain distinct upon secondary contact [11, 15] advocates of the biological species concept argue that reproductive incompatibility is the key criterion for species delimitation [14]; allopatric or parapatric populations that show diagnostic differences but have not yet attained reproductive incompatibility are better classified as subspecies within a polytypic species. On the other hand, Gill [16] recently argued for a reversal of current taxonomic practice in which "splitting" rather than "lumping" taxa should be the null hypothesis, and the burden of proof placed on "lumping" rather than on "splitting" taxa at the species level.

A recent review on avian species-level taxonomy since 1950 has shown a steady increase in the number of bird species, due to a trend towards splitting polytypic species [17]. This trend is caused by an eclectic taxonomic practice, in which multiple criteria for species delimitation have been applied [17, 18]. The trend can also be seen as a necessary correction to a taxonomic bias during the first half of the 20th century, when the number of recognized species was reduced by more than 50 % due to "default lumping" of taxa into polytypic species without any critical assessment of their diagnosability, monophyly or reproductive incompatibility [17, 19]. Hence, it follows that many subspecies still retained in current classifications are ghosts of the past and should undergo taxonomic revision with a pluralistic set of species criteria [16, 18].

While diagnosability and monophyly are relatively easy to assess by phenotypic and genotypic traits, the assessment of reproductive incompatibility seems very challenging for allopatric taxa [15], and must therefore be inferred from divergences in characters that are supposed to be functionally important in reproduction. The criteria for deciding when two allopatric or parapatric taxa have attained enough reproductive incompatibility to justify species rank have been outlined by Helbig et al. [14]. In principle, reproductive incompatibility is seen as a hypothesis derived from comparative evidence, where the magnitude of multiple character divergences matches that of related species that coexist in sympatry, and therefore are undoubtedly reproductively incompatible. Tobias et al. [20] proposed a quantitative scoring system for species assignment based on a set of phenotypic and behavioural traits derived from a data set of well-recognised sympatric or parapatric species pairs. These traits likely play a role in habitat and niche adaptations, or in sexual attraction among mates, and can thus be regarded as premating isolation mechanisms.

Here we argue that reproductive incompatibility can also be inferred from traits that have a functional role after copulation, viz. sperm traits. Spermatozoa are extremely diverse in size and form across the animal kingdom [21] and are often divergent among closely related species [22]. Sperm morphology traits have therefore proved useful for taxonomy and phylogenetic inference in many animal groups, including birds [23]. While there are generally marked structural differences in avian spermatozoa among avian orders and families [24], recent comparative studies on passerine birds have revealed considerable variation in sperm length among related species [25, 26]. There is also evidence of sperm length variation among geographically structured populations or subspecies [27–30], which suggests that sperm length can be a useful taxonomic marker also for incipient species.

In this paper, we report a comparative study of multiple character divergences between the two island populations of the Blue Chaffinch *Fringilla teydea*, which are currently ranked as subspecies [31–33]. The two taxa, endemic to the central islands of Tenerife (ssp. *teydea*) and Gran Canaria (ssp. *polatzeki*) in the Canarian archipelago, are phenotypically distinct in male plumage [34], as well as in morphometrics and vocalizations [35, 36]. A recent study confirmed the statistical significance of these measures and suggested lifting the taxa to species rank [37]. There is also some evidence that the taxa are divergent in mtDNA [38, 39] and nuclear microsatellites [40]. A timely question is therefore whether the two taxa are fully diagnosable in multiple genotypic and phenotypic traits to an extent that would merit species rank. Here we present quantitative evidence in a broad array of traits, including mitogenomes, nuclear introns, biometrics, male plumage colour, song and sperm morphology. We assess the magnitude of these divergences in relation to divergences between undisputed sister species, following the current guidelines for species delimitation in avian taxonomy [14, 20].

Methods
Study species
The Blue Chaffinch *Fringilla teydea* is a medium-sized to large finch (~30 g), larger than the two other congeneric species, the Common Chaffinch *F. coelebs* and the Brambling *F. montifringilla*. The distribution of the Blue Chaffinch is restricted to the high-elevation (1200–2300 m) forests of Canary Pine *Pinus canariensis* on Tenerife and Gran Canaria [35]. Pine seeds constitute the staple food throughout the year [41, 42], but the diet also includes a significant proportion of arthropods [34, 35]. Like most finches, *F. teydea* is sexually dimorphic in size and plumage, with males being larger and more colourful (leaden-blue) than females (olive-brown). The two island populations are taxonomically recognized as subspecies and show morphological differences predominantly in adult males [34]: 1) *polatzeki* is slightly duller, more ashy-olive grey than *teydea*, 2) the black band on the lower forehead is considerably more pronounced in *polatzeki*, 3) the tips of median and greater coverts are light bluish-grey in *teydea* and distinctly broader and contrasting white in *polatzeki*, and 4) *teydea* is generally larger (bill, wing and body size) than *polatzeki*. Figure 1 depicts the adult male of the two taxa.

The male territorial song is about a 2 s long strophe in both taxa, which consists of a first phrase of a falling series of soft, disyllabic notes in *polatzeki* and a more same-pitch series of harder, monosyllabic notes in *teydea*. The second phrase is a prolonged syllable in both taxa, but it is markedly softer or subdued in *polatzeki*, and more like a crescendo in *teydea*. These differences are notable in the sonograms in Fig. 1.

According to the IUCN Red List, the total population size of *F. teydea* is 1800–4500 individuals, with the majority on Tenerife (*teydea*) and only less than 250 birds (*polatzeki*) left in the wild on Gran Canaria [43]. A more recent survey estimated a population size of about 16 000 individuals on Tenerife [44]. The *polatzeki* population has declined severely during the last century due to habitat loss from logging and fragmentation of the pine forest [34]. In 2007, a wildfire further destroyed much of the core habitat in the Inagua area, and the population size dropped to only 122 individuals the following year [45]. Although the species seems to cope well with wildfires of mild and moderate severity [46], access to high-quality pine forest habitat in combination with stochastic population fluctuations seems to be a critical factor to the survival of the small Gran Canaria population.

Data and sample collection
Data for analysis originate from museum specimens (plumage coloration, American Museum of Natural History, New York), from measurements and samples (sperm and blood) of birds caught in the wild (Tenerife) or in captivity (Gran Canaria; the wildlife recuperation center in Tafira), and from song recordings of wild birds on both islands. The Gran Canaria captive breeders were either wild-caught or the first generation of wild-caught birds.

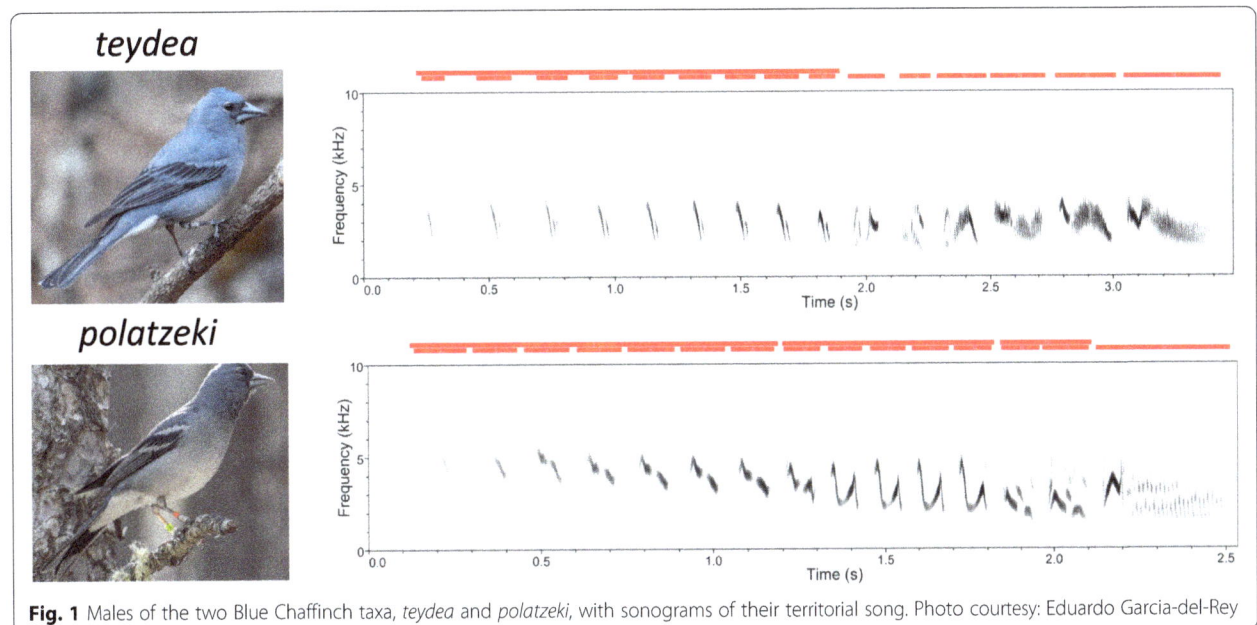

Fig. 1 Males of the two Blue Chaffinch taxa, *teydea* and *polatzeki*, with sonograms of their territorial song. Photo courtesy: Eduardo Garcia-del-Rey

One person (EGDR) measured their wing length (straightened and flattened chord, nearest 1 mm), tail length (from base between the central pair of rectrices to the tip, nearest 1 mm), tarsus length (between extreme bending points, nearest 0.1 mm), bill length (from skull, nearest 0.1 mm), bill depth (at distal edge of nostrils, nearest 0.1 mm) and body mass (0.1 g). Wing length was measured with a stopped ruler, tail length with an unstopped ruler, tarsus and bill with a digital calliper, and body mass with a Pesola 50 g balance. Since first-year birds have significantly shorter wings and tail than older birds [47], we excluded first-year birds for the analyses of these characters. About 10–30 µL blood was collected in a capillary tube after brachial venepuncture and stored in absolute ethanol for subsequent DNA analyses in the lab. Ejaculate samples (1–3 µL) were collected in a capillary tube after cloacal massage, diluted in 20–30 µL phosphate-buffered saline and fixed in 300 µL 5 % formaldehyde [48] for subsequent measurements of sperm morphometrics in the microscopy lab.

Songs were recored from nine individual male *teydea* on three locations on Tenerife (Vilaflor, La Guancha and Las Lagunetas) and eight individual *polatzeki* on Gran Canaria (Llanos de la Pez) during May 2015. All individuals were in adult plumage. All recordings were made by EGDR using a Fostex recorder with a parabolic Telinga microphone.

Genetic analyses

Genomic DNA was extracted using a commercial spin column kit (E.Z.N.A. DNA Kit; Omega Bio-Tek) or a GeneMole® automated nucleic acid extraction instrument (Mole Genetics), following the manufacturers' protocols.

Mitochondrial DNA was amplified from high molecular weight DNA extracts using two primer pairs: MtCorvus531F (GGATTAGATACCCCACTATGC) and mt Corvus9431R (GTCTACRAAGTGTCAGTATCA), and mtCorvus8031F (CCTGAWCCTGACCATGAACCTA) and mtCorvus926R (GAGGGTGACGGGCGGTATGTA) designed for study on Ravens *Corvus corax* (JAA, AJ and AMK, unpublished data). These two primer pairs yielded amplicons of ~8900 bp (Amplicon 1) and ~9700 bp (Amplicon 2). Annealing sites and overlapping regions are illustrated in Fig. 2. The following PCR conditions were utilized for amplification: 1X reaction buffer, 200 µM of each dNTP, 0.5 µM of each primer, ~20 ng template DNA, 0.02 U/µl Q5 High-Fidelity DNA polymerase (New England Biolabs) and dH$_2$O to a final volume of 25 µl. The following thermal profiles were employed: Amplicon 1 – Initial denaturation 98 °C in 30 s, 35 cycles with denaturation 98 °C for 10 s, annealing 59 °C for 20 s and elongation 72 °C for 7.5 min, and a final elongation step for 2 min. Amplicon 2 - Initial

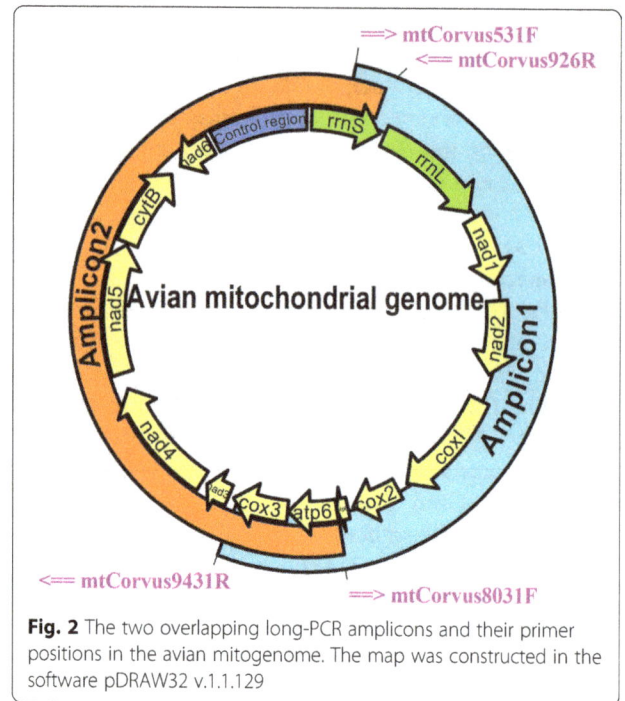

Fig. 2 The two overlapping long-PCR amplicons and their primer positions in the avian mitogenome. The map was constructed in the software pDRAW32 v.1.1.129

denaturation 98 °C in 30 s, 5 cycles with denaturation 98 °C for 10 s following a touch-down profile starting at 72 °C with 1 °C/cycle reduction, 30 cycles with denaturation 98 °C for 10 s, annealing 67 °C for 20 s and elongation 72 °C for 7.5 min, and a final elongation step for 2 min.

The complete PCR reactions were transferred to a 0.8 % agarose gel and ran at 90 V. When completely separated, the respective amplicons were cut from gel and purified using the GenJet Gel Extraction Kit (Thermo-Fischer Scientific). Concentrations of the purified amplicons were measured on a Qbit instrument (ThermoFischer Scientific) and equimolar amounts of each amplicon were pooled. Twenty ng of pooled amplicons where sheared using a Covaris M220 Focused-ultrasonicator (Covaris, Inc.), running the pre-programmed DNA shearing protocol for 800 bp twice. To generate barcoded libraries for sequencing, we employed the NEBNext library prep kit for Ion Torrent (New England Biolabs) on the sheared amplicons, using the IonXpress barcode adapter kit (ThermoFischer Scientific). Barcoded libraries were pooled and size selected (440–540 bp) using a Bluepippin instrument (Sage Science). Concentration of the final library was measured on a Fragment Analyzer (Advanced Analytical) using the DNF-474 High Sensitivity NGS Fragment Analysis kit. The sheared, size selected and barcoded amplicons were sequenced on an IonPGM instrument (ThermoFischer Scientific). The samples were sequenced on two different 314 chips.

Trimming and removal of low quality reads were performed on the Torrent Suite ™ software (ThermoFisher Scientific). A Common Chaffinch mitochondrial genome

(GenBank acc NC025599) was used as a reference in the Torrent Suite ™ software for coverage estimates, using the plugin coverageAnalysis (v4.4.2.2). Mitochondrial genomes were reconstructed using MITObim v1.8 [49]. The complete mitochondrial genome of Common Chaffinch was used as reference in the initial mapping. Mitochondrial genes were first automatically annotated using MITOs [50], and thereafter manually inspected. Distance estimates were calculated in MEGA6 [51] using the Maximum Composite Likelihood model for nucleotides [52] and the Poisson Correction model for amino acids [53].

For a more quantitative test of possible admixture of mitochondrial haplotypes between taxa, we sequenced the first part of the cytochrome oxidase subunit 1 (COI) gene (655 bp) for 175 individuals; 14 *polatzeki* and 161 *teydea* (see Additional file 1). We used the primer pair PasserF1 and PasserR1 [54].

Eight Z-chromosome introns (ALDOB-7, BRM-15, CHDZ-15, CHDZ-18, PTCH-6, VLDLR-7, VLDLR-8, VLDLR-12; [54]) were sequenced and screened for variation between the two taxa. Two loci (PTCH-6 and VLDLR-7) were found to have polymorphic sites and were sequenced for respectively 44 and 48 individuals [see Additional file 1]. The primer sequences for all introns are given by Borge et al. [55].

For the COI marker and the eight Z-chromosome introns, the PCR reaction volumes were 12.5 µL, containing 0.5 mM dNTPs, 0.3 U Platinum Taq DNA Polymerase (Invitrogen), 1x buffer solution (20 mM Tris-HCl, 50 mM KCl; Invitrogen), 2.5 mM MgCl$_2$, 0.1 µM forward and reverse primer and 2 µL DNA extract. The reactions were carried out under the following conditions: 2 min at 94 °C, 35 cycles of [30 s at 94 °C, 30 s at 50 °C (PTCH-6), 55 °C (COI) or 57 °C (VLDLR-7), and 45 s at 72 °C], and a final extension period of 10 min at 72 °C. Cycle-sequencing reactions were carried out using the BigDye Terminator v3.1 Cycle Sequencing Kit (Applied Biosystems), and the sequencing products were run on an ABI Prism 3130xl Genetic Analyzer (Applied Biosystems). The COI fragment was sequenced in both directions, whereas Z-introns were only sequenced with forward primers. Sequences were proofread in CodonCode Aligner v3.7.1 (CodonCode Corporation) and aligned in MEGA5.1 [56].

Divergences in mtDNA between the taxa were visualized in haplotype networks using the PopArt software [57]. All sequence data with their voucher information and Genbank accession numbers are given in Additional file 1.

Plumage colour measurements

We measured spectral reflectance of adult male plumage from five patches, i.e., back, crown, upper breast, rump and wing bar (median covert) on 15 specimens (10 *teydea*, 5 *polatzeki*) at the American Museum of Natural History (for voucher information see Additional file 2). Unfortunately, we were not able to measure the black band on the lower forehead because of poor feather structure in the study skins. Spectral reflectance was measured with an Ocean Optics USB2000 reflectance spectrophotometer, a PX-2 pulsed xenon light source (Ocean Optics, Dunedin, Florida) and a fiber-optic probe equipped with a 'probe pointer' to ensure measurements were taken at a constant distance and angle from the specimen. The following settings were used integration time = 20 msec, spectra average = 40 with a multiple strobe setting. All measurements were calibrated against a Spectralon white standard (Labsphere, North Sutton, New Hampshire) and a dark standard (no light). Five measurements were taken from the same spot on each plumage patch. The spectrometer was re-calibrated using the dark and white standards after every second plumage patch was measured (after ten measurements).

Raw reflectance spectra were imported in five nanometer (nm) bins between 300 and 700 nm using CLR: Colour Analysis Programs v1.05 [58], Brightness, chroma and hue colour variables [59] were calculated in CLRv1.05 using the formulae most appropriate for the slaty greyish-blue plumage of Blue Chaffinches [58]. These were B1 for brightness = R320–700), S5a for chroma = $S_5 = \sqrt{\left(B_r - B_g\right)^2 + \left(B_y - B_b\right)^2}$ and H4a for hue = arctan ([(By – Bb)/B1] / [(Br – Bg)/B1]). In both H4a and S5a, b (blue) = 400–475 nm, g (green) = 475–550 nm, y (yellow) = 550–625 nm, r (red) = 625–700 nm. We visually screened for outliers and mis-measurements by comparing the five reflectance curves taken for each individual at each plumage patch using Excel (2010 Microsoft Corporation). One mis-measurement was removed from the dataset (AMNH-788194, *F. t. teydea*, back measure #2). Mean spectral reflectance curves and colour variables were then calculated from the independent measures taken for each individual using JMP 11 (SAS Inc., Cary, NC).

Principle components analysis (PCA) was used to condense the reflectance spectra (81 measures between 300 and 700 nm) into two principle components (PCs). PC1 explained the majority of the variation in all five plumage patches (percent variation: 86–92 %; eigenvalue: 70–75). PC2 explained only a small fraction of the variation at the five plumage patches (percent variation: 6–8.5 %; eigenvalue: 4.8–6.8). We used two-tailed *t*-tests assuming equal variances to test for differences in brightness, chroma, hue and reflectance PCs of male plumage between *teydea* and *polatzeki*. All statistics were performed in JMP 11 (SAS Inc., Cary, NC).

Song analyses

Songs were analyzed using Luscinia (https://github.com/rflachlan/Luscinia/). First, the repertoire size of individuals (within the sample recorded) was determined by visual inspection of spectrograms. Song types were highly stereotypic within an individual's repertoire, leading us to have confidence in this method (note: the sample size of songs recorded per male was not sufficient to have confidence that all song types in each male's repertoire had been recorded). An exemplar was chosen for each song-type in each male's repertoire, and was measured using Luscinia. Spectrogram settings were as follows: frame length – 5 ms; time step 1 ms; maximum frequency – 10 kHz; dynamic range – variable between 40–50 dB depending on recording quality; dereverberation parameter – 100 %. A high-pass filter was applied before spectrograms were created with a threshold of 1.0 kHz. Measurement involved identifying trajectories of acoustic parameters for each element within the song, and classification of elements into repeated syllables and phrases. These methods have been described in previous studies, e.g., [60].

Next, songs were compared using Luscinia's implementation of the dynamic time warping algorithm (DTW). This algorithm searches for an optimal alignment of acoustic features between syllables, and then allows pairs of syllables to be compared by measuring Euclidean distances along this alignment. The DTW method takes into account all of the considerable and variable frequency modulation within syllables and in so doing allows a more holistic comparison of syllable structure than methods based on a small number of structural parameters (maximum frequency, syllable length, etc.). Of particular relevance for this study, Luscinia's DTW implementation has been successfully applied to a large dataset of songs recorded from the congeneric Common Chaffinch in which detailed descriptions of population divergence were found [60].

A key step in carrying out the DTW comparison is to normalize the various acoustic features relative to each other. In this analysis, the parameter weightings were as follows. Weighting of parameters: Time: 10; Fundamental Frequency: 2.755; Fundamental Frequency Change: 8.850; Vibrato Amplitude: 0.338; all others: 0. These values are the inverse of standard deviations across the Common Chaffinch database for the chosen parameters except for time, which is normalized in a different way (see [60] for further explanation). Compression factor was set to 0.2 (with a minimum element length of 10), Time SD was set to 1, a syllable repetition weighting of 0.2 was applied, and a maximum warp of 100 %. The weight by relative amplitude, log transform frequencies, interpolate in time warping, and dynamic warping options were selected. The comparison used the Stitch syllables method with 5 alignment points.

The DTW algorithm generated a dissimilarity matrix between syllables that was converted to dissimilarity matrices between songs (using the DTW method of integrating syllable dissimilarities) and between individuals (using the option of finding the best match between songs within individual repertoires). These dissimilarities formed the basis of further analysis.

First, we carried out a UPGMA clustering analysis of songs and individuals from the two Blue Chaffinch populations. We also clustered songs using the PAM k-medoid clustering algorithm and calculated the Global Silhouette Index for each k value as a way of searching for natural clusters in the data.

Next, we quantified divergence between populations. To do this we first calculated the spatial median of each population (based on an NMDS ordination of the data, using 20 dimensions). We then measured the acoustic distance between each song or individual data point to its own population's spatial median, as well as that of other populations. To quantify a measure of divergence, d_{AB}, between the two populations, A and B, we then calculated:

$$d_{AB} = \frac{\sum_{i=1}^{n_A} d_{i,SB} - d_{i,SA}}{n_A} + \frac{\sum_{i=1}^{n_B} d_{j,SA} - d_{i,SB}}{n_B}$$

SA and SB are the spatial medians of individuals or songs for populations A and B (with sample sizes n_A and n_B, respectively), and d_{iSA} and d_{iSB} represent the acoustic dissimilarity between a data point (individual or song) i and the spatial median for population A and B, respectively. The metric therefore quantifies the degree to which songs were closer to the spatial median of their own population rather than to the spatial median of the other population.

Local populations with no genetic differentiation may nevertheless be differentiated in song structure as a consequence of cultural divergence. Because cultural evolution is believed to occur at a much faster rate than genetic evolution, populations may become culturally differentiated in the face of high levels of gene flow between populations. It is important therefore to interpret levels of divergence in the context of how other genetically differentiated and undifferentiated populations have diverged. In this case, we used a large-scale analysis of Common Chaffinch, previously compared using the same DTW methods [60] and analyzed whether the differences seen between the two Blue Chaffinch populations matched those found between other Common Chaffinch populations.

Sperm morphometrics

A small aliquot of approximately 15 µl of the formaldehyde/sperm solution was applied onto a microscope slide and allowed to air-dry before inspection, digital imaging and measurement under light microscopy. We took digital images of spermatozoa at magnifications of 200× or 400×, using a Leica DFC420 camera mounted on a Leica DM6000 B digital light microscope. The morphometric measurements were conducted using Leica Application Suite (version 2.6.0 R1). We measured the length of head (i.e., acrosome and nucleus), the midpiece and the tail (i.e., the midpiece-free end of the flagellum) of ten intact spermatozoa per male, except for three males with very few measureable sperm (one *teydea* with 4 sperm, and two *polatzeki* with 2 and 1 sperm, respectively). Total sperm length was calculated as the sum of head, midpiece and tail length. We have previously shown that sperm length measurements have very low measurement error and high repeatability [61, 62]. In this study, sperm were measured by two different persons (TL and ES) who differed consistently in the way they measured sperm components, but were in close agreement over the measure of total sperm length (i.e., the sum of components). We therefore used their combined measurements for sperm total length, but only the measurements from one of them (TL) for the analyses of component lengths. Standardized values of intra- and intermale variation in sperm total length was expressed as the coefficient of variation ($CV = SD/mean \times 100$). The intermale variation in mean sperm length (CV_{bm}) is negatively correlated with the frequency of extrapair paternity across passerine birds [26] and is thus an indicator of the level of sperm competition. For this measure we applied a correction factor for variation in sample size (N), viz. $CV = SD/mean \times 100 \times (1+ 1/4\ N)$, as recommended by Sokal and Rohlf [63].

Estimation of phenotypic divergence

The quantitative estimation of phenotypic divergence was expressed by Cohen's *d* [64] as recommended by Tobias et al. [20]. We also adapted the taxonomic scoring system proposed by Tobias et al. [20] for the relevant subset of recommended characters. It includes two biometric variables (the largest positive and the largest negative divergence), two acoustic variables (strongest temporal and spectral character divergences), and the three strongest plumage characters. Effect sizes were transferred to a taxonomic score of 0–4; e.g., Cohen's *d* in the range of 0.2–2 gives a score of 1, *d* in the range of 2–5 gives a score of 2, *d* in the range 5–10 gives a score of 3, and *d* >10 equals a score of 4. Scores are then summed for all variables, and a total score of 7 or above qualifies for species status.

Results

Genetic divergences

The mitogenome sequencing in the two Ion PGM runs yielded 706,008 and 743,375 reads, respectively. In total,

37.3 % of reads filtered as polyclonals and 15.9 % as low quality reads. Sample-specific information regarding total number of reads, mapped reads, coverage and uniformity is provided in Table 1.

The mitogenomes contained 16784-16786 nucleotides. Gene annotation analyses revealed 13 protein coding genes, 2 rRNAs and 22 tRNAs. The gene arrangement followed the standard avian mitochondrial genome [65].

We found altogether 12 haplotypes among the 25 sequenced birds; nine among the 21 *teydea* and three among the 4 *polatzeki*. The haplotypes clustered in two distinct groups corresponding to the two taxa, as shown in a median-joining haplotype network (Fig. 3) with the mitogenome of a Common Chaffinch included as an outgroup. The mean substitution rate averaged 2.3 % for the entire mitogenome (Table 2). The table also shows the divergence for each gene region for nucleotides and amino acids, respectively.

To test for the possible introgression of mitochondrial haplotypes, we sequenced a larger number of presumably unrelated individuals of the two taxa for the standard DNA barcode region, i.e., the first part of the COI gene [66]. We found four haplotypes among 161 *teydea* and two haplotypes among 14 *polatzeki*. As with the full mitogenomes, the COI sequences clustered perfectly into two groups concordant with the two taxa (Fig. 4). Hence, there was no evidence of mitochondrial mismatch in this large sample. We are therefore confident that the two taxa are reciprocally monophyletic in their mtDNA.

We also found evidence of reciprocal monophyly in two Z-chromosome introns, PTCH-6 (559 bp) and VLDLR-7 (586 bp). There was one fixed point mutation in PTCH-6 between 24 *polatzeki* and 20 *teydea* sequences, and two fixed point mutations in VLDLR-7 between 26 *polatzeki* and 22 *teydea* sequences (Table 3). All three were G – A transitions. There were no other polymorphic sites in these two introns, or in any of the six other Z-linked introns sequenced from the two subspecies (see Additional file 1).

Biometrics

Since both subspecies are sexually size dimorphic in adults, we analysed the biometric measurements separately for each sex (Table 4). The nominate subspecies was significantly larger than *polatzeki* for all traits in both sexes, except for female tarsus length (Table 3). The difference was especially large for wing length and bill length, for which Cohen's *d* > 2 in both sexes (Table 4). A principal component analysis based on wing length, tail length and bill length clustered adult males in two distinct groups corresponding to the two subspecies (Fig. 5a, Table 5). For females, a similar tendency was found, though with some overlap (Fig. 5b, Table 4).

Table 1 Sequence read quality of the assembled mitogenomes of the *teydea* (*N* = 21) and *polatzeki* (*N* = 4) Blue Chaffinches

Taxon	NHMO Acc.no	Total reads	Mapped reads	Percent mapped	Mean coverage	Uniformity
teydea	27311	91589	82591	90 %	1346.0	83 %
teydea	27312	51724	38844	75 %	609.9	86 %
teydea	27314	14274	11660	82 %	185.3	63 %
teydea	36318	27983	25761	92 %	393.0	93 %
teydea	36336	4707	3530	75 %	53.68	89 %
teydea	36401	40339	38577	96 %	602.5	91 %
teydea	36403	26512	25730	97 %	393.6	92 %
teydea	36416	20220	19513	97 %	303.3	88 %
teydea	36439	16040	15522	97 %	243.8	90 %
teydea	36450	5257	4832	92 %	75.4	61 %
teydea	68704	29785	27615	93 %	251.4	93 %
teydea	68711	39708	36436	92 %	334.0	95 %
teydea	68716	33980	32278	95 %	302.1	96 %
teydea	68718	29386	26132	89 %	236.9	95 %
teydea	68720	30963	29409	95 %	265.1	95 %
teydea	68724	16511	15403	93 %	140.1	95 %
teydea	68726	19251	15951	83 %	144.4	90 %
teydea	68730	16774	15973	95 %	147.4	93 %
teydea	68737	6786	6460	95 %	59.63	95 %
teydea	68738	5002	4628	93 %	43.02	91 %
teydea	68739	5894	5637	96 %	52.28	94 %
polatzeki	30994	20982	19927	95 %	318.7	76 %
polatzeki	30997	20428	18435	90 %	287.0	92 %
polatzeki	30998	16961	13898	82 %	218.9	90 %
polatzeki	31001	11589	10507	91 %	167.9	89 %

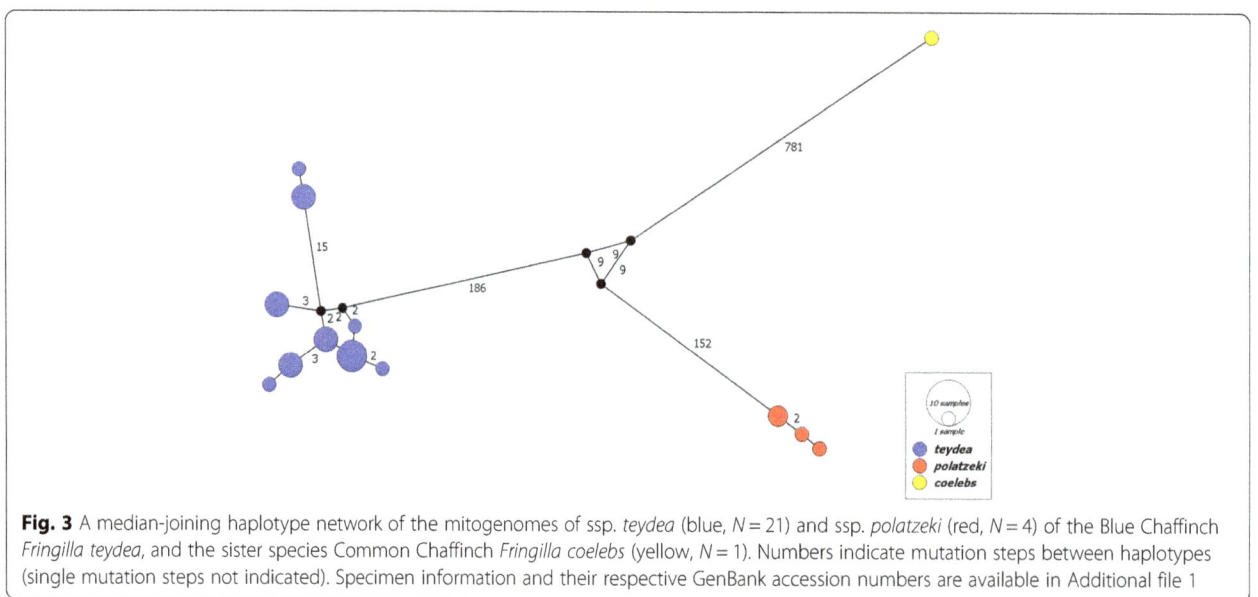

Fig. 3 A median-joining haplotype network of the mitogenomes of ssp. *teydea* (blue, *N* = 21) and ssp. *polatzeki* (red, *N* = 4) of the Blue Chaffinch *Fringilla teydea*, and the sister species Common Chaffinch *Fringilla coelebs* (yellow, *N* = 1). Numbers indicate mutation steps between haplotypes (single mutation steps not indicated). Specimen information and their respective GenBank accession numbers are available in Additional file 1

Table 2 Estimates of sequence divergence across the mitogenomes between *teydea* (N = 21) and *polatzeki* (N = 4) Blue Chaffinches

Genes	Nucleotides		Amino acids	
	Sequence	Divergence (%)	Sequence	Divergence (%)
12S	975	0.619 ± 0.254	-	-
16S	1598	0.695 ± 0.194	-	-
NAD1	973	2.530 ± 0.813	324	0.933 ± 0.475
NAD2	1038	2.725 ± 0.467	346	1.752 ± 0.711
COI	1548	1.968 ± 0.552	516	0.134 ± 0.097
COII	675	2.813 ± 0.814	225	0.893 ± 0.647
ATP8	168	3.776 ± 1.673	55	3.707 ± 2.547
ATP6	680	2.754 ± 0.609	226	0.443 ± 0.378
COIII	784	2.221 ± 0.829	261	1.046 ± 0.592
NAD3	342	3.194 ± 0.994	114	2.032 ± 1.226
NAD4l	290	2.125 ± 0.782	96	1.047 ± 1.006
NAD4	1369	1.955 ± 0.747	456	0.251 ± 0.024
NAD5	1816	3.155 ± 1.075	605	2.172 ± 0.656
Cyt *b*	1144	2.147 ± 0.602	380	1.058 ± 0.551
NAD6	518	3.427 ± 0.939	171	1.051 ± 0.737
CR	1311	2.944 ± 0.715	-	-
Whole mitogenome	16788	2.309 ± 0.716	-	-
Synonymous substitutions[a]			3523	8.072 ± 0.522
Nonsynonymous substitutions[a]			3523	0.950 ± 0.133

[a]NAD6 excluded

Divergences (mean ± SE) were calculated as the percentage of substitutions per site (nucleotide or amino acid) from averaging over all individual pair combinations between taxa. All ambiguous positions were excluded for each sequence pair. SE of the mean was calculated by a bootstrap procedure with 100 replicates

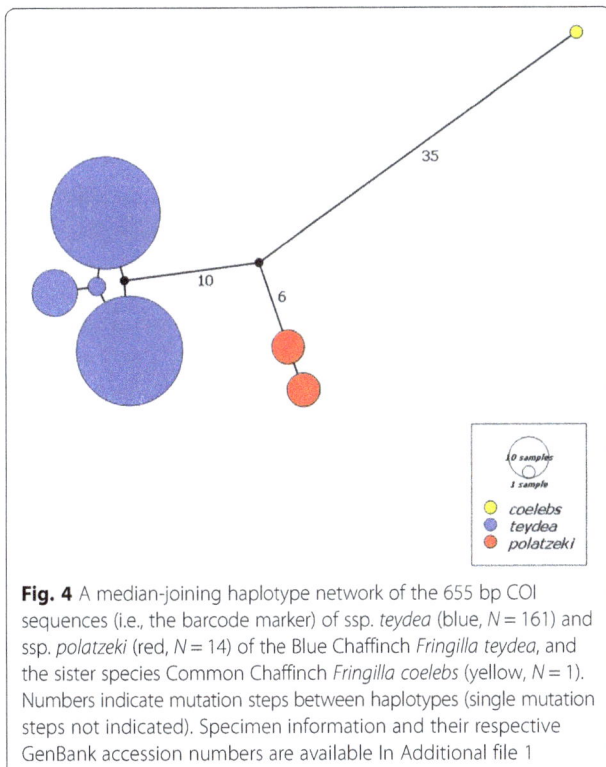

Fig. 4 A median-joining haplotype network of the 655 bp COI sequences (i.e., the barcode marker) of ssp. *teydea* (blue, N = 161) and ssp. *polatzeki* (red, N = 14) of the Blue Chaffinch *Fringilla teydea*, and the sister species Common Chaffinch *Fringilla coelebs* (yellow, N = 1). Numbers indicate mutation steps between haplotypes (single mutation steps not indicated). Specimen information and their respective GenBank accession numbers are available In Additional file 1

We can therefore conclude that among adult males, the two subspecies have diagnostic, non-overlapping body-size distributions.

Plumage

Pictures of study skins from AMNH are shown in Fig. 6, and average reflectance spectra for the five body parts are shown in Fig. 7. For three of the body parts (crown, back and rump), *teydea* males had significantly higher brightness and PC1 scores (which to a large extent reflects brightness) than *polatzeki* males (Table 6). In contrast, the wing bar of the median coverts was significantly brighter and showed a higher chroma in *polatzeki* males than in *teydea* males. The effect on our estimate of chroma was due to a steeper slope in the UV/blue parts of the spectrum for *polatzeki* males (Fig. 7). There was no difference in the reflectance of the upper breast. In summary,

Table 3 Fixed nucleotide substitutions between *teydea* and *polatzeki* Blue Chaffinches in two Z-chromosome introns

Taxon	VLDLR-7			PTCH-6	
	N	Pos. 148	Pos. 305	N	Pos. 464
teydea	22	G	G	20	A
polatzeki	26	A	A	24	G

Table 4 Sex-specific morphological divergences between the *teydea* and *polatzeki* Blue Chaffinches

Sex	Trait	*teydea* (Tenerife)	*polatzeki* (Gran Canaria)	Test of difference between means	Effect size Cohen's *d*
Males	Wing length (mm)	102.8 ± 2.1 (118)	96.8 ± 0.7 (16)	$t_{132} = 11.00$, $P < 0.001$	2.95
	Tail length (mm)	83.9 ± 1.9 (58)	78.2 ± 1.7 (15)	$t_{71} = 10.54$, $P < 0.001$	3.09
	Bill length (mm)	20.37 ± 1.09 (83)	16.76 ± 0.97 (22)	$t_{103} = 14.16$, $P < 0.001$	3.43
	Bill depth (mm)	10.48 ± 0.37 (83)	10.18 ± 0.28 (22)	$t_{103} = 3.54$, $P < 0.001$	0.86
	Tarsus length (mm)	26.31 ± 0.75 (83)	25.87 ± 0.30 (22)	$t_{103} = 2.69$, $P < 0.01$	0.65
	Body mass (g)	32.24 ± 1.70 (109)	28.33 ± 1.06 (22)	$t_{129} = 10.71$, $P < 0.001$	2.52
Females	Wing length (mm)	93.5 ± 2.0 (87)	89.2 ± 1.2 (16)	$t_{101} = 8.18$, $P < 0.001$	2.25
	Tail length (mm)	75.9 ± 2.0 (42)	72.8 ± 1.4 (15)	$t_{55} = 5.64$, $P < 0.001$	1.73
	Bill length (mm)	20.43 ± 1.30 (61)	16.88 ± 0.82 (16)	$t_{75} = 10.39$, $P < 0.001$	2.96
	Bill depth (mm)	10.30 ± 0.37 (61)	9.94 ± 0.27 (16)	$t_{75} = 3.63$, $P < 0.001$	1.03
	Tarsus length (mm)	25.82 ± 0.79 (61)	25.51 ± 0.40 (17)	$t_{76} = 1.65$, $P = 0.102$	0.46
	Body mass (g)	28.8 ± 1.5 (83)	27.1 ± 1.4 (17)	$t_{98} = 4.38$, $P < 0.001$	1.18

Mean values are given with their $\pm$ SD (*N*). Statistical tests are two-sample *t*-tests with unequal variances

teydea males display a brighter blue plumage in most body parts, whereas *polatzeki* display lighter wing bars.

Song

The male territorial song is about a 2 s long strophe in both taxa, which consists of a first section with one or two descending phrases of soft, disyllabic syllables in *polatzeki* and a more constant-pitch series of harder, monosyllabic syllables in *teydea*. The second section of the song consists of unrepeated syllables (or syllables only repeated once) with a buzzy, 'vibrato' characteristic. They are markedly softer or subdued in *polatzeki*, and more like a crescendo in *teydea*. These differences are notable in the sonograms in Fig. 1.

When we clustered songs (Fig. 8a) and individual repertoires (Fig. 8b) using the UPGMA algorithm, the two populations were clearly separated with no exceptions in either case. Similarly, *k*-medoid clusterings with songs and individuals classified both according to their population with 100 % accuracy with *k* = 2. Notably, both types of analysis represented unsupervised clustering analyses of the data (unlike, e.g., DFA), suggesting a clear divergence between the two populations. Furthermore, the Global Silhouette Index for songs (the only data set with sufficient sample size) showed a clear peak with *k* = 2 (Fig. 9), suggesting that a natural partition of the data set was into two clusters.

The divergence score between *polatzeki* and *teydea* was larger than that found between any pair of Common Chaffinch populations (Table 7). The Blue Chaffinch populations both had lower divergence scores with at least one Common Chaffinch population than they did with each other (Table 7).

These results demonstrate that Blue Chaffinch songs have diverged considerably between *polatzeki* and *teydea*. This can be illustrated in a Neighbor-Joining phylogram of all the populations included in the above analyses, based on inter-population differences in song structure (Fig. 10). This analysis successfully reconstructed much of the known topology of the relationships between these populations [60], suggesting a high phylogenetic signal for song structure. The only unusually placed population is Madeira (ssp. *maderensis*), which might have been predicted to be found in the Azorean – Canarian clade. In this phylogeny, the Blue Chaffinches connect near the root of the Atlantic Island and European Common Chaffinch populations. Moreover, and strikingly, the two Blue Chaffinch populations show substantial differentiation from one another.

Sperm

Spermatozoa were significantly longer in *polatzeki* than *teydea* males (Fig. 11, Table 8), i.e., opposite to the contrast seen in body size dimensions. The difference was mostly explained by the length of the midpiece, which made up about 88 % of the sperm total length in both taxa (Table 8). Sperm heads were also significantly longer in *polatzeki* than in *teydea*. The variation in sperm total length within males, as expressed by the CV_{wm} index, was exceptionally low in both taxa; i.e., close to 1 % (Table 8). Likewise, the variation among males in mean sperm total length (CV_{bm}) was low in both taxa, and variances did not differ significantly between them (Table 8). The low CV_{bm} values indicate a relatively high risk of sperm competition, and yield estimates of 39 % and 44 % extrapair young in broods of *teydea* and *polatzeki*, respectively, using the linear regression equation given in Lifjeld et al. [26] for passerine species.

Taxonomic scoring

We adopted the standardized taxonomic scoring system proposed by Tobias et al. [20] for species delimitation,

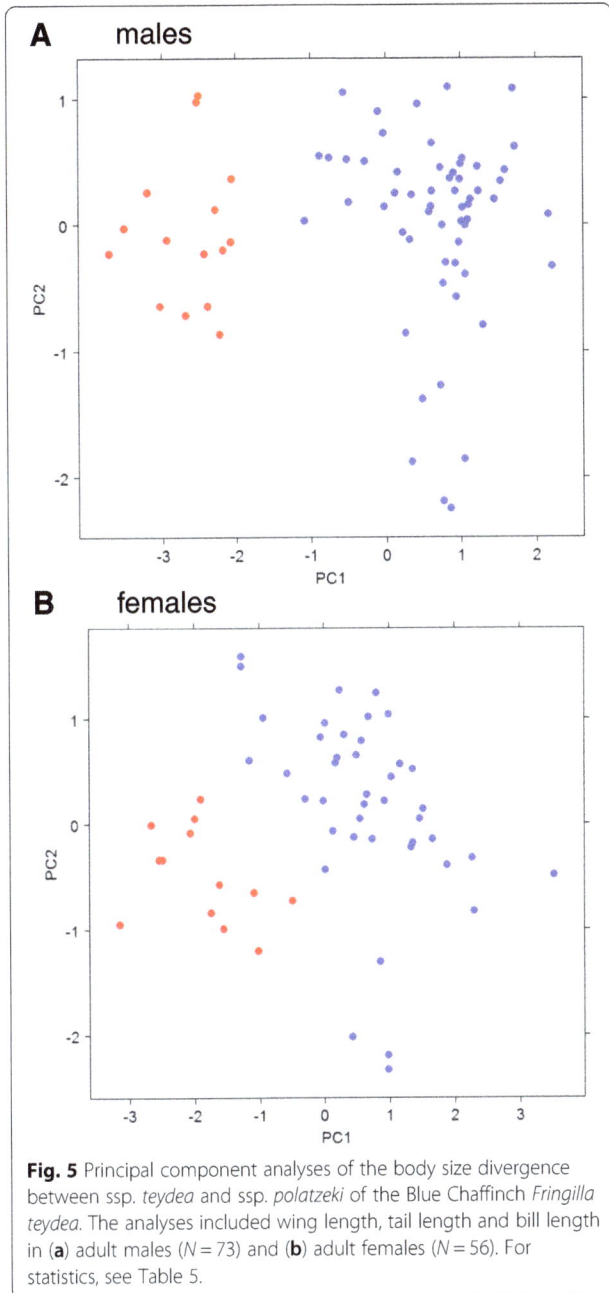

Fig. 5 Principal component analyses of the body size divergence between ssp. *teydea* and ssp. *polatzeki* of the Blue Chaffinch *Fringilla teydea*. The analyses included wing length, tail length and bill length in (**a**) adult males (N = 73) and (**b**) adult females (N = 56). For statistics, see Table 5.

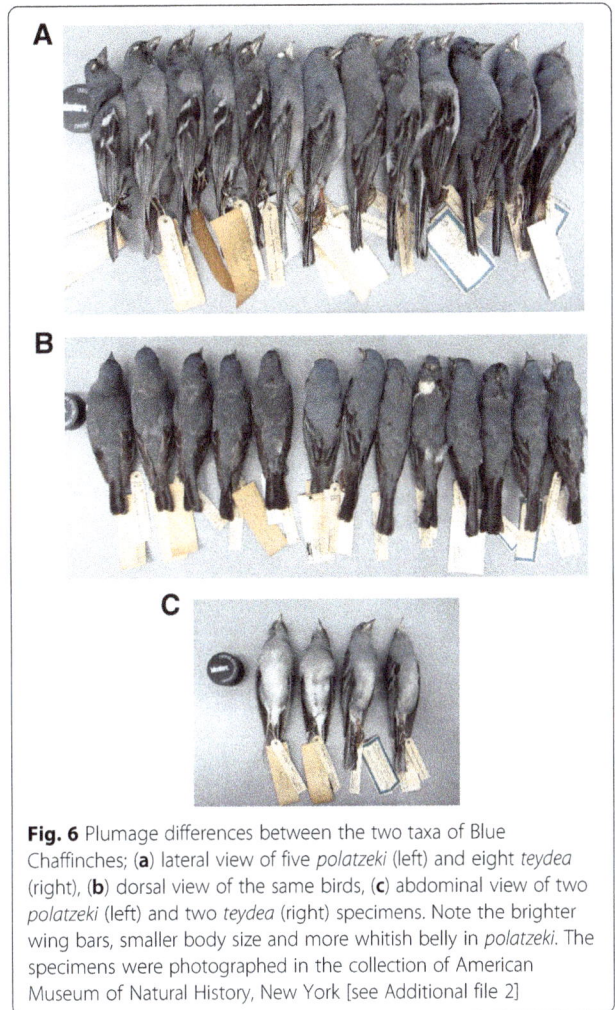

Fig. 6 Plumage differences between the two taxa of Blue Chaffinches; (**a**) lateral view of five *polatzeki* (left) and eight *teydea* (right), (**b**) dorsal view of the same birds, (**c**) abdominal view of two *polatzeki* (left) and two *teydea* (right) specimens. Note the brighter wing bars, smaller body size and more whitish belly in *polatzeki*. The specimens were photographed in the collection of American Museum of Natural History, New York [see Additional file 2]

and the scores are given in Table 9. For biometrics, the scoring system only allows for the largest increase and the strongest decrease among multiple variables, but since *teydea* is the larger in all characters, we could only select one. The largest effect size was estimated for bill length of males (Table 4), which scores as a medium divergence (score = 2). For scoring the acoustics, we used

Table 5 Factor loadings of three body size variables on the first two principal components in Principal Component Analyses of a) male and b) female *teydea* and *polatzeki* Blue Chaffinches, and their eigenvalues, percentage of variance explained, and *F*-statistics

Variable	Males		Females	
	PC1	PC2	PC1	PC2
Wing length	0.616	−0.189	0.636	−0.286
Tail length	0.582	−0,544	0.626	−0.351
Bill length	0.530	0.817	0.451	0.892
Eigenvalue	2.285	0.520	2.076	0.727
Variance explained	76.2 %	17.3 %	69.2 %	24.2 %
Test of difference between taxa	$F_{1,71} = 298.04, P < 0.001$	$F_{1,71} = 0.17, P = 0.68$	$F_{1,54} = 82.13, P < 0.001$	$F_{1,54} = 6.58, P = 0.013$

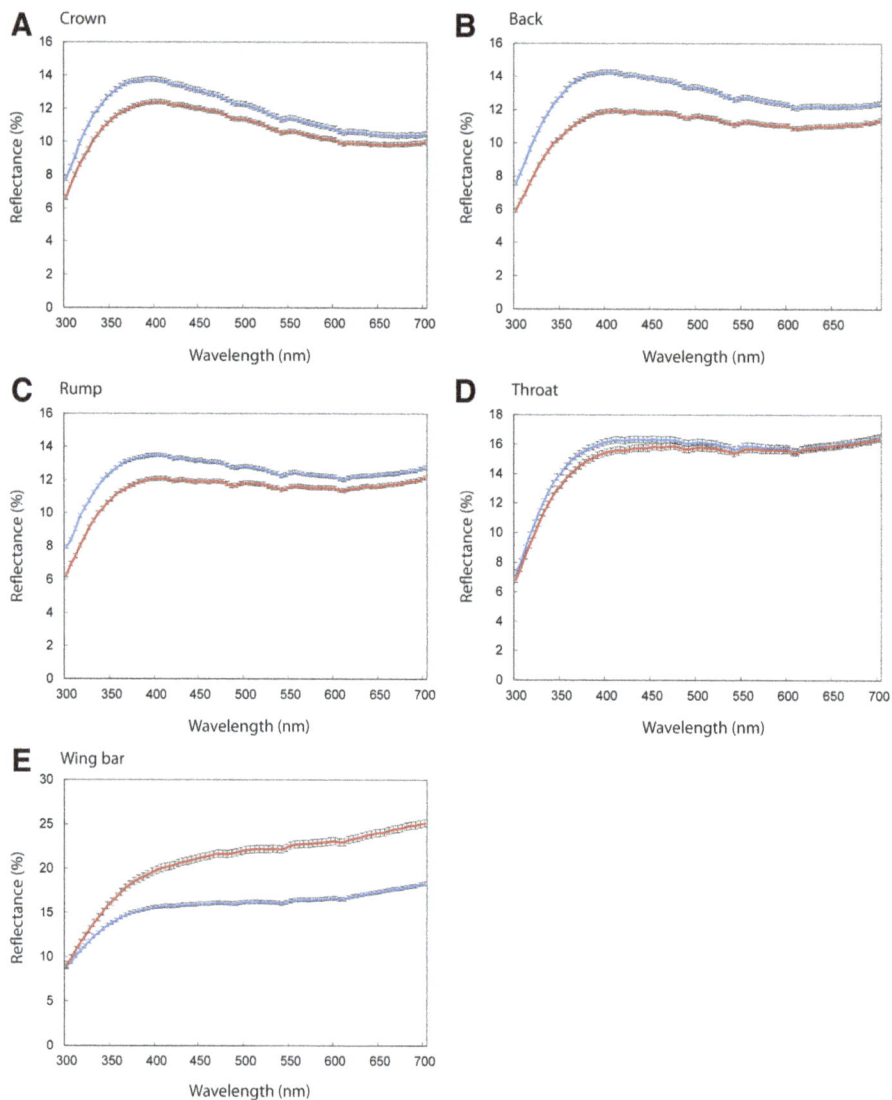

Fig. 7 Average reflectance (+ SE) from five plumage regions (**a**: crown, **b**: back, **c**: rump, **d**: upper breast, **e**: wing bar) of male *teydea* (blue lines, N = 10) and male *polatzeki* (red lines, N = 5) skins in the AMNH collection [Additional file 2]. Spectra were averaged for five scans of each plumage region

the quantitative measurements of temporal and spectral variables presented in Sangster et al. [37]; Table 4. The stable pitch in the first phrase of the *teydea* song versus the gradually decreasing pitch in *polatzeki* qualifies for a score of 3, whereas the increased amplitude of phrase 2 (crescendo) in *teydea* and reduced amplitude in *polatzeki*, gives a score of 2 (Table 9). We scored two diagnostic plumage traits, i.e., male body plumage colouration and male wing bars, as medium divergence. The male *teydea* has a brighter plumage, which gives a more bluish impression than the more greyish male *polatzeki*. The two wing bars formed by the tips of the median and greater coverts are white and sharply edged in the male *polatzeki*, resembling the conspicuous wing bars in the Common Chaffinch, whereas in the male *teydea* the wing bars are light bluish-grey with much less

contrast. The black band on the lower forehead is thicker and more distinct in male *polatzeki* than in male *teydea*, and was scored as minor divergence (score 1; Table 9). The total score obtained was 11, which is above the threshold of 7 for species assignment recommended by Tobias et al. [20].

Discussion
Modern taxonomy emphasizes an integrative approach in species delimitation, which means that multiple trait divergences should be assessed in a comparative framework [67–69]. This is the approach we have followed here. The Blue Chaffinches on Tenerife (*teydea*) and Gran Canaria (*polatzeki*) are clearly divergent in multiple traits. The distinct differences in male plumage and size formed the basis for their original description as two

Table 6 Plumage colour differences between *polatzeki* (N = 5) and *teydea* (N = 10) Blue Chaffinches

Plumage patch	Colour variable	*teydea*	*polatzeki*	Test of difference between means	Effect size (Cohens *d*)
Back	Brightness	2.424 ± 0.190	2.082 ± 0.176	$t_{13} = 3.36, P = 0.005$	1.98
	Chroma	0.093 ± 0.010	0.091 ± 0.002	$t_{13} = 0.45, P = 0.66$	0.26
	Hue	−1.045 ± 0.321	−1.103 ± 0.481	$t_{13} = 0.28, P = 0.78$	0.17
	PC1	1.992 ± 3.468	−3.985 ± 3.170	$t_{13} = 3.23, P = 0.007$	1.90
	PC2	−0.169 ± 1.426	0.338 ± 0.221	$t_{13} = -0.78, P = 0.45$	−0.46
Crown	Brightness	2.248 ± 0.159	2.056 ± 0.078	$t_{13} = 2.50, P = 0.026$	1.48
	Chroma	0.085 ± 0.015	0.076 ± 0.002	$t_{13} = 1.31, P = 0.21$	0.77
	Hue	−0.555 ± 0.250	−0.681 ± 0.191	$t_{13} = 0.98, P = 0.34$	0.58
	PC1	1.574 ± 4.249	−3.149 ± 2.088	$t_{13} = 2.32, P = 0.037$	1.36
	PC2	−0.279 ± 1.595	0.559 ± 0.853	$t_{13} = -1.09, P = 0.30$	−0.64
Rump	Brightness	2.358 ± 0.152	2.147 ± 0.112	$t_{13} = 2.72, P = 0.018$	1.61
	Chroma	0.095 ± 0.009	0.102 ± 0.008	$t_{13} = -1.46, P = 0.17$	−0.86
	Hue	−0.949 ± 0.691	−0.674 ± 1.154	$t_{13} = -0.58, P = 0.57$	−0.34
	PC1	1.691 ± 3.993	−3.382 ± 3.043	$t_{13} = 2.49, P = 0.027$	1.46
	PC2	0.420 ± 1.246	−0.839 ± 0.649	$t_{13} = 2.09, P = 0.056$	1.23
Upper breast	Brightness	2.896 ± 0.208	2.816 ± 0.201	$t_{13} = 0.71, P = 0.49$	0.42
	Chroma	0.143 ± 0.018	0.154 ± 0.011	$t_{13} = -1.22, P = 0.25$	−0.73
	Hue	−0.033 ± 1.223	1.136 ± 0.513	$t_{13} = -2.02, P = 0.064$	−1.19
	PC1	0.493 ± 4.429	−0.985 ± 4.149	$t_{13} = 0.62, P = 0.55$	0.37
	PC2	0.365 ± 1.308	−0.730 ± 1.130	$t_{13} = 1.59, P = 0.14$	0.94
Wing bar (median covert)	Brightness	2.962 ± 0.229	3.898 ± 0.334	$t_{13} = -6.43, P < 0.001$	3.78
	Chroma	0.193 ± 0.025	0.309 ± 0.035	$t_{13} = -7.52, P < 0.001$	4.38
	Hue	1.284 ± 0.069	1.268 ± 0.014	$t_{13} = 0.51, P = 0.62$	0.30
	PC1	−2.604 ± 1.995	5.209 ± 2.874	$t_{13} = -6.20, P < 0.001$	−3.65
	PC2	0.170 ± 1.229	−0.340 ± 1.065	$t_{13} = 0.79, P = 0.45$	0.46

separate taxa more than hundred years ago [70]; see also [34]. Here we have analysed these characters more quantitatively and show that males are diagnostically different in colour reflectance curves in multiple plumage characters and in body size dimensions. We have also shown that the two Blue Chaffinch taxa are reciprocally monophyletic in mitochondrial and nuclear DNA, and are divergent in sexual traits, like song and sperm, that are presumably important traits in mate choice and fertilization success.

In a recent paper, Sangster et al. [37] indicated a qualitatively similar divergence in male plumage characters and documented non-overlapping body size distributions in adult males, using a similar PCA approach as we have presented here. They also analysed song and calls, and found strong divergences between the two taxa. Moreover, they showed in a playback experiment that *polatzeki* males responded more aggressively to *polatzeki* songs than to *teydea* songs. Our two studies are therefore congruent in demonstrating multiple character divergences in the two taxa. When we employed the

quantitative scoring system for multiple phenotypic characters proposed by Tobias et al. [20], which specifically consider effect sizes in biometrics, plumage and vocalizations, we found that the divergence score (=11) for *teydea* and *polatzeki* exceeded the threshold (= 7) set for the species level.

The quantitative scoring system proposed by Tobias et al. [20] is designed to infer a meaningful assessment of reproductive isolation for allopatric taxa, which cannot be tested directly. This is a key criterion under the biological species concept. Species are seen as hypotheses in which divergences are tested against a reference material drawn from a number of sympatric or indisputably good sister species. The system is a helpful guide, but it also has some shortcomings. First of all, it does not include genetic divergences, which are essential for unravelling a species' evolutionary history and uniqueness. Second, the emphasis of biometry, plumage and vocalization may not always be sufficient or relevant to predict reproductive isolation, and their relative weights in the scoring system may seem somewhat subjective. In

A

| Tenerife:BCTeydea9 |
| Tenerife:BCTeydea8 |
| Tenerife:BCTeydea6 |
| Tenerife:BCTeydea1 |
| Tenerife:BCTeydea5 |
| Tenerife:BCTeydea4 |
| Tenerife:BCTeydea7 |
| Tenerife:BCTeydea7 |
| Tenerife:BCTeydea3 |
| Tenerife:BCTeydea2 |
| Gran Canaria:BCPol_8 |
| Gran Canaria:BCPol_7 |
| Gran Canaria:BCPol_6 |
| Gran Canaria:BCPol_8 |
| Gran Canaria:BCPol_8 |
| Gran Canaria:BCPol_5 |
| Gran Canaria:BCPol_8 |
| Gran Canaria:BCPol_8 |
| Gran Canaria:BCPol_8 |
| Gran Canaria:BCPol_4 |
| Gran Canaria:BCPol_3 |
| Gran Canaria:BCPol_2 |
| Gran Canaria:BCPol_1 |
| Gran Canaria:BCPol_7 |
| Gran Canaria:BCPol_7 |
| Gran Canaria:BCPol_6 |
| Gran Canaria:BCPol_5 |
| Gran Canaria:BCPol_6 |

B

| Tenerife:BCTeydea9 |
| Tenerife:BCTeydea8 |
| Tenerife:BCTeydea6 |
| Tenerife:BCTeydea1 |
| Tenerife:BCTeydea5 |
| Tenerife:BCTeydea4 |
| Tenerife:BCTeydea7 |
| Tenerife:BCTeydea3 |
| Tenerife:BCTeydea2 |
| Gran Canaria:BCPol_8 |
| Gran Canaria:BCPol_4 |
| Gran Canaria:BCPol_3 |
| Gran Canaria:BCPol_2 |
| Gran Canaria:BCPol_1 |
| Gran Canaria:BCPol_7 |
| Gran Canaria:BCPol_6 |
| Gran Canaria:BCPol_5 |

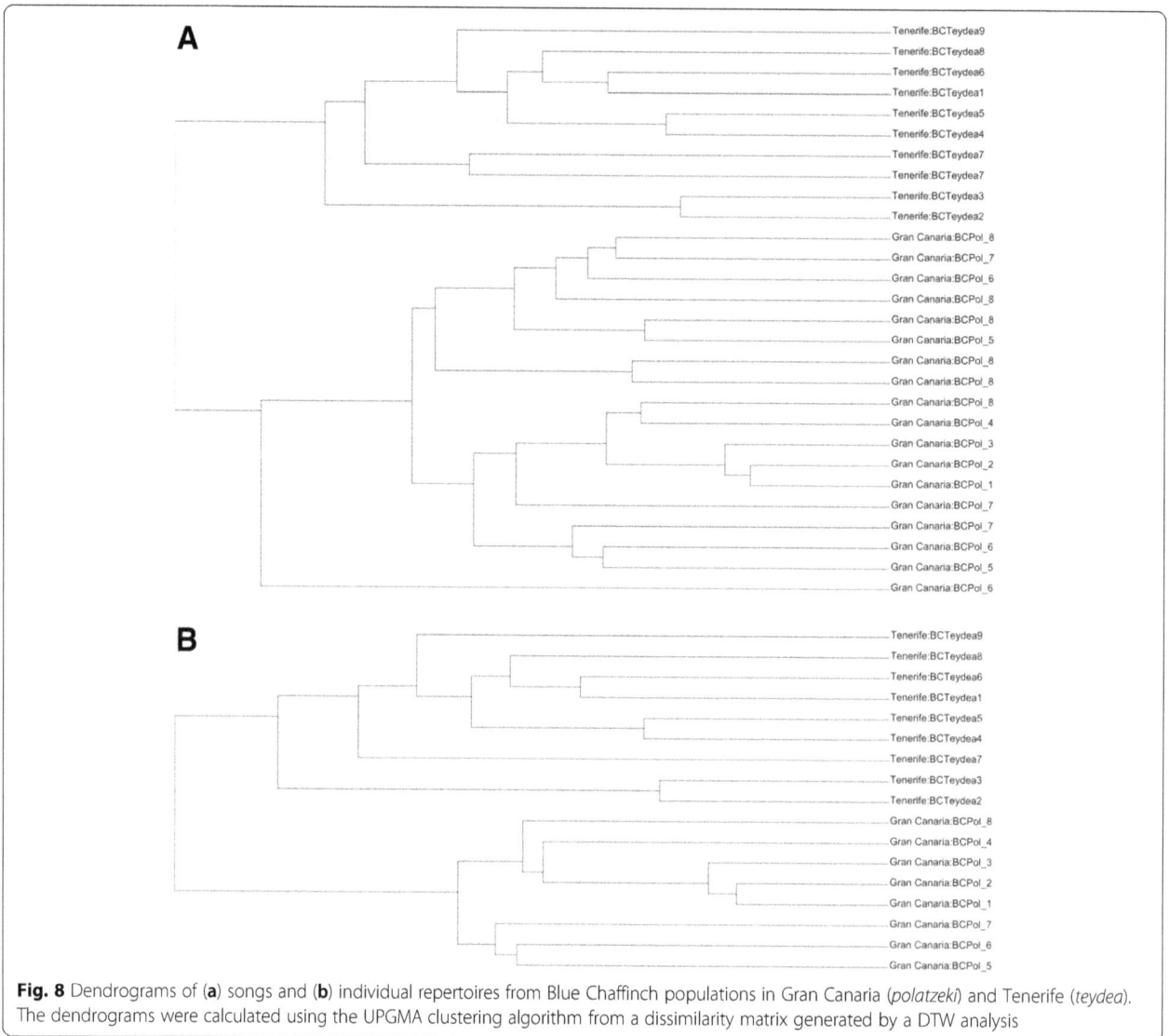

Fig. 8 Dendrograms of (**a**) songs and (**b**) individual repertoires from Blue Chaffinch populations in Gran Canaria (*polatzeki*) and Tenerife (*teydea*). The dendrograms were calculated using the UPGMA clustering algorithm from a dissimilarity matrix generated by a DTW analysis

the following we therefore discuss these aspects in more detail for the case of the Blue Chaffinch.

The multiple trait divergences observed between the two island populations of the Blue Chaffinch are the results of a long history of allopatric evolution. Their mitogenomes show an overall genetic distance of 2.3 % with some variation among the various genes (Table 2). Previously, Suarez et al. [39] reported a genetic distance of 2.3 % for the cytochrome *b* gene between *teydea* and *polatzeki*, which is similar to our value for the same gene (Table 2). In birds, mitochondrial genes seem to evolve in a clock-like manner, with around a 2 % sequence divergence per million years as estimated for the cyt *b* gene across multiple avian orders and with multiple calibration points over the past 12 million years [71]. Accordingly, the two populations may have evolved independently for as long as one million years. Our COI sequencing of a

fairly high number of individuals revealed no cases of introgression. Hence, their mitogenomes seem to be completely sorted and there is no evidence of recent or past gene flow between the two populations. It must be noted that mitochondrial DNA is inherited only through the maternal line, so the lack of mitochondrial gene flow suggest no female dispersal. However, we also found fixed mutations in nuclear DNA, i.e., two Z-chromosome introns (Table 2), which suggests that there are two distinct nuclear gene pools with no evidence of gene flow in either sex. Nuclear genes generally sort more slowly and have longer coalescence times [72]. Six other Z-chromosome introns showed no sequence divergence and should be considered incomplete lineage sorting. We have also shown in a previous study [40] that there are significant differences in allele size ranges for several microsatellite DNA markers.

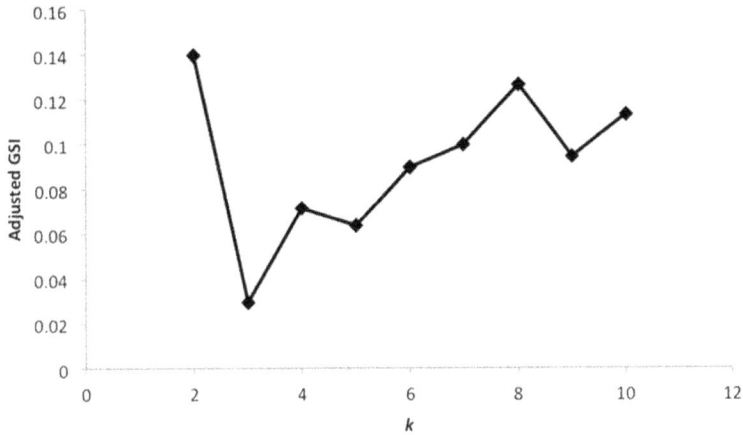

Fig. 9 Global Silhouette Index values for different clustering solutions produced by the *k*-medoids clustering algorithm applied to songs of *polatzeki* and *teydea*. The GSI has a value greater than 0 when data are clustered more than expected by chance. Higher values represent a greater clustering tendency. The GSI tends to produce higher values with smaller values of *k*, so we corrected the GSI by comparing its output with simulated data-sets. The peak with *k* = 2 corresponds to the division of songs between *polatzeki* and *teydea*

Even though species limits cannot be inferred directly from sequence divergence in mtDNA, it is interesting to note that there are many avian sister species with shorter genetic distances than what we see among the two subspecies of the Blue Chaffinch. A survey of the DNA barcode database BOLD [73] reveals that even within the same family (Fringillidae), there are sister species with a divergence less than 2 % in the genera *Carduelis*, *Spinus*, *Acanthis*, *Pyrrhula* and *Loxia* [74, 75]. Some of them, e.g., *Acanthis* and *Loxia*, are not even monophyletic, and their species taxonomy may be doubtful [76], but others are undisputedly good species.

Many subspecies on oceanic islands are distinct evolutionary units [77] and should undergo taxonomic revision. In the Canary Islands, the Blue Chaffinch is not unique in showing strong evolutionary divergence among islands. There is a lot of endemism in the fauna and flora of Canary Islands [78], and among birds there are several distinct island-specific taxa with subspecies status. In particular, several forest-dwelling passerines show distinct differentiation between the islands of Tenerife and Gran Canaria as for example the Afrocanarian Blue Tit *Cyanistes teneriffae* [79, 80], the Common Chaffinch [39] and the European Robin *Erithacus rubecula* [81]. Their genetic distances (cytochrome *b*) between Tenerife and Gran Canaria populations range from 1.0 % in Common Chaffinch [39] to 3.7 % in the European Robin [81]. These taxa should undergo further taxonomic assessment with respect to their species limits.

The habitat of the Blue Chaffinch is the forest of the Canary Pine, and pine seeds constitute a major food resource [41, 42]. The pine forest on Tenerife has not been so extensively logged as the one on Gran Canaria, and contains in general more older and larger trees, with larger cones and seeds than the reduced and mostly replanted forest in Gran Canaria [34]. The larger bill and

Table 7 Estimates of pairwise population divergence in song structure between populations of Common Chaffinches and Blue Chaffinches

Population A	Population B	Divergence
Iberia - Montseny	North Europe - Holland	0.0770
Iberia - Montseny	Britain - Hampshire	0.0642
Azores - Faial	Azores - São Miguel	0.0096
Iberia - Montseny	Azores - Faial	0.0960
Iberia - Montseny	Common Chaffinch - Gran Canaria	0.1126
Blue Chaffinch - Tenerife	Blue Chaffinch - Gran Canaria	0.1440
Blue Chaffinch - Tenerife	Azores - Faial	0.1250
Blue Chaffinch - Gran Canaria	Iberia - Montseny	0.0652

The divergence score represents the degree to which songs were closer to their own population's spatial median than to that of the other population. The table illustrates that the two Blue Chaffinch populations were more divergent than any other pair of populations

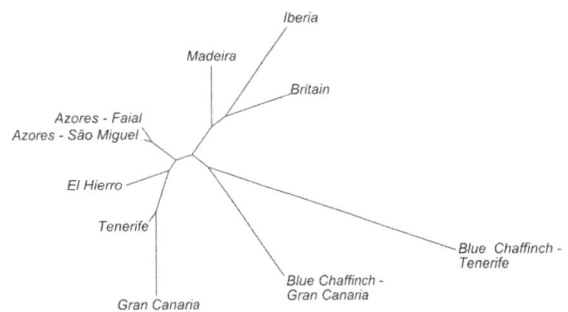

Fig. 10 Neighbor-Joining Phylogram showing evolutionary relationships between Common Chaffinch and Blue Chaffinch populations on the basis of song structure. For Common Chaffinch populations, only island/region is indicated

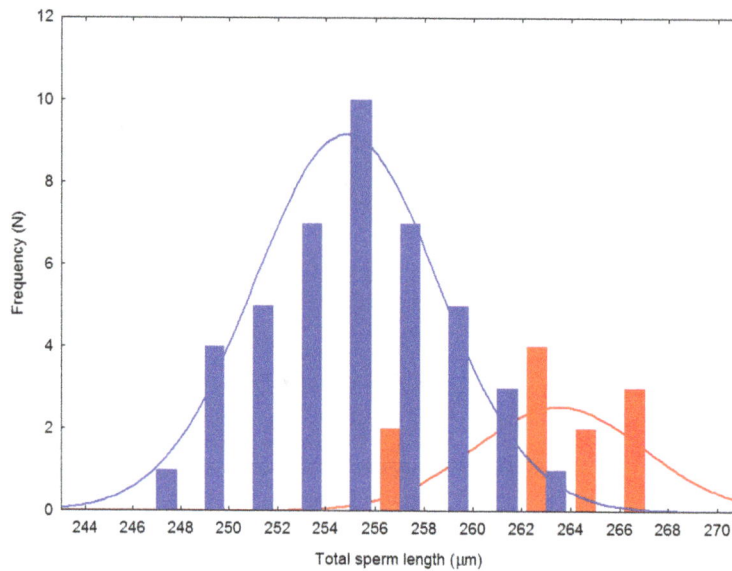

Fig. 11 Frequency distributions of total sperm length in the two Blue Chaffinch taxa *teydea* (blue) and *polatzeki*. (red). Fitted normal curves are indicated

body size in *teydea* in Tenerife than in *polatzeki* on Gran Canaria, therefore possibly reflect adaptations to different environments. It is an interesting parallel to the differentiation in bill and body size seen in Darwin finches on Galapagos, which are interpreted as adaptations to different seed sizes [82]. However, species limits in Darwin's finches is a highly contentious issue because bill and body size seem to evolve fast in these birds without any strong evidence of reproductive isolation [83]. It has therefore been argued that differentiation in these phenotypic characters reflect local adaptations within the same species rather than reproductive isolation between species. The same argument may apply to the Blue Chaffinch case. The differentiation in biometrics observed between *teydea* and *polatzeki* may be of rather recent origin due to human-induced habitat differences, and may not be a good indicator of reproductive isolation.

Traits that are more directly involved in reproduction, like sexual characters used in mate choice and mating competition, and gametic traits with a function in fertilization, may be more suitable as indicators of reproductive isolation. Elaborate plumage colour and feather ornaments, and display behaviours like song in passerine birds, are generally believed to evolve by sexual selection [84] and thereby function as premating isolation mechanisms [7].

Sperm traits, on the other hand, have a reproductive function after copulation and may represent a postmating or postcopulatory isolation mechanism. Our reasoning is that divergent sperm traits may be particularly important in passerine birds with sperm competition, because postcopulatory processes may be relatively more important for individual fitness in these systems. Across bird species, sperm length is positively correlated with the length of female sperm storage tubules [85, 86]. It has been demonstrated in insects that the form of the

Table 8 Sperm morphometrics of ssp. *teydea* and ssp. *polatzeki* of the Blue Chaffinch *Fringilla teydea*

Sperm trait	*teydea* (Tenerife)	*polatzeki* (Gran Canaria)	Test of difference between means	Test of difference between variances	Effect size Cohen's *d*
Head length (µm)	17.05 ± 0.70 (22)	18.01 ± 0.60 (11)	$t_{31} = -3.86$, $P < 0.001$	F-ratio = 1.34, $P = 0.65$	−1.47
Midpiece length (µm)	224.10 ± 5.78 (22)	232.92 ± 4.41 (11)	$t_{31} = -4.44$, $P < 0.001$	F-ratio = 1.71, $P = 0.38$	−1.69
Tail length (µm)	13.63 ± 5.21 (22)	12.49 ± 4.41 (11)	$t_{31} = 0.60$, $P = 0.55$	F-ratio = 1.16, $P = 0.84$	0.23
Total sperm length (µm)	254.75 ± 3.74 (43)	263.42 ± 3.46 (11)	$t_{52} = -6.96$, $P < 0.001$	F-ratio = 1.17, $P = 0.83$	−2.40
CV_{wm} total length (%)	1.24 ± 0.36 (43)	1.06 ± 0.26 (10)	$t_{51} = 1.44$, $P = 0.16$	F-ratio = 1.88, $P = 0.31$	0.60
CV_{bm} of mean total length (%)[a]	1.48	1.34			

[a]Adjusted for variation in sample size (N) by the formula $CV_{bm} = SD/mean \times 100 \times (1 + 1/4 N)$

Ten sperm cells were measured per male, except for one *teydea* and two *polatzeki* with very few sperm cells in the sampled ejaculate. Mean values are given with their ± SD (N)

Table 9 Quantitative scores for species delimitation of the two Blue Chaffinch taxa *teydea* (Tenerife) and *polatzeki* (Gran Canaria), following the scoring system of Tobias et al. [20]

Trait type	Most divergent character *teydea* vs. *polatzeki*	Effect size (Cohen's *d*)	Score
Morphology	Bill length (males)	3.43	2
Song	Frequency (MHz) change in 1st phrase	−9.08[a]	3
	Change in max amplitude from 1st to 2nd phrase	3.58[a]	2
Male colouration	Colouration wing bar (PC1)	−3.65	2
	Colouration back (PC1)	1.90	1
	Black band on forehead (faint vs. marked)	Minor	1
	Total score		11

[a]calculated from Sangster et al. ([37]; Table 3)

female reproductive tract drives the evolution of sperm morphology [87]. This is consistent with the idea that sperm length in a given species is adapted to its specific female environment. Nevertheless, there is also considerable variation in sperm length among males within the same population [60, 88], and this variation has a strong genetic basis [89]. Typically, species with large intraspecific variation in sperm length are characterized by little or no sperm competition, whereas species with a narrow sperm length variation have high levels of sperm competition [26, 28, 48, 90]. All of this suggests that sperm competition acts as an evolutionary force of stabilizing or purifying selection that reduces variation in sperm length among males in the population. That is, sperm competition should favour sperm lengths close to the population mean, and disfavour those that are particularly long or short. It further implies that two populations with divergent sperm length distributions may be reproductively incompatible under sperm competition. Hence, sperm traits might predict the presence of a postcopulatory prezygotic barrier [91, 92], or "competitive gametic isolation" (sensu Coyne & Orr [4]), and thus be taxonomically informative in delimitation of allopatric species that exhibit a mating system with sperm competition. It must be emphasized that the role of sperm competition as a selection force on sperm size variation in passerine birds is at present only a hypothesis, which needs empirical testing.

Conclusions

We conclude that the two Blue Chaffinch subspecies fulfil all the major criteria for species delimitation, and should therefore be assigned species status: the Tenerife Blue Chaffinch *Fringilla teydea* and the Gran Canaria Blue Chaffinch *Fringilla polatzeki*. This is in agreement with the recommendation by Sangster et al. [37]. The population on Gran Canaria is critically endangered with just a few hundred birds left in the wild [45]. Recognition of the population as a full species will presumably increase attention to its conservation as one of the most critically endangered birds in Europe [37]. However, we emphasize that conservation arguments should have no weight in the taxonomic assessments of species limits, and they have not affected our analyses and recommendation here either.

Additional files

Additional file 1: List of all blood and sperm samples with their collection number at Natural History Museum, Oslo, BOLD processID for COI sequences, and GenBank accession numbers for COI, whole mitogenomes and five nuclear intron sequences. Sequence data for three additional introns (<200 bp, i.e., too short for GenBank) are also included. (XLSX 28 kb)

Additional file 2: List of study skins used in the plumage colour spectrophotometry analyses and their voucher information in American Museum of Natural History, New York. (DOCX 15 kb)

Acknowledgements
We are grateful to Paul Sweet and Lydia Garetano for access to the skin collection in American Museum of Natural history, New York. Financial support was received from the Research Council of Norway (grant numbers 170853 and 196554 to JTL and 213592 to AJ).

Funding
The study was funded by three grants from the Research Council of Norway (grant numbers 170853 and 196554 to JTL and 213592 to AJ).

Authors' contributions
JTL and EGDR conceived and designed the study. JTL, PC, LEJ, AJ, TL, ES and EGDR conducted field work and collected data and samples for analysis. PC was responsible for keeping the captive population on Gran Canaria and supervised all handling and sampling of these birds. JAA conducted the mitogenome sequencing and GM conducted the other sequencing work, and both analysed the sequence data and drafted their respective parts of the Methods and Results sections. EGDR conducted the song recordings and JEJC and RFL analysed the song data and drafted the song-related parts of the manuscript. AMK conducted the spectrophotometric analyses of the study skins in American Museum of Natural History and AMK and AJ drafted the plumage-related parts of the manuscript. TL and ES prepared and measured the sperm, and EGDR conducted all the biometric measurements. AJ conducted the PCA analyses of the morphological data and JTL conducted the remaining statistical analyses. JTL drafted the entire manuscript. All co-authors read, commented and approved the final version.

Competing interests
The authors declare that they have no competing interests.

Author details
[1]Natural History Museum, University of Oslo, PO Box 1172 Blindern, 0318 Oslo, Norway. [2]Wildlife Recovery Center "Tafira", Vivero Forestal, Cabildo de Gran Canaria, 35017 Las Palmas de Gran Canaria, Canary Islands, Spain. [3]Department of Biological and Experimental Psychology, Queen Mary University of London, London E1 4NS, UK. [4]Department of Biological

Sciences, University of Maryland, Baltimore County, 1000 Hilltop Circle, Baltimore, MD 21250, USA. [5]Macaronesian Institute of Field Ornithology, C/ Enrique Wolfson 11-3, 38004 Santa Cruz de Tenerife, Canary Islands, Spain.

References

1. Lande R. Genetic variation and phenotypic evolution during allopatric speciation. Am Nat. 1980;116:463–79.
2. Mayr E. Animal Species and Evolution. Cambridge: Harvard University Press; 1963.
3. Templeton AR. The theory of speciation via the founder principle. Genetics. 1980;94:1011–38.
4. Coyne JA, Orr HA. Speciation. Sunderland: Sinauer; 2004.
5. Grant PR. Reconstructing the evolution of birds on islands: 100 years of research. Oikos. 2001;92:385–403.
6. Newton I. The Speciation and Biogeography of Birds. London: Academic; 2003.
7. Price T. Speciation in Birds. Greenwood Village: Roberts and Co; 2008.
8. De Queiroz K. Species concepts and species delimitation. Syst Biol. 2007;56:879–86.
9. Hey J. On the failure of modern species concepts. Trends Ecol Evol. 2006;21:447–50.
10. Wheeler Q, Maier R. Species concepts and phylogenetic theory: a debate. New York: Columbia University Press; 2000.
11. Cracraft J. Species concepts and speciation analysis. In: Johnston RF, editor. Curr Ornithol, vol. 1. New York: Plenum Press; 1983. p. 159–87.
12. Mayr E. Systematics and the Origin of Species from the Viewpoint of a Zoologist. New York: Columbia University Press; 1942.
13. Mayr E. The Growth of Biological Thought: Diversity, Evolution, and Inheritance. Cambridge: Harvard University Press; 1982.
14. Helbig AJ, Knox AG, Parkin DT, Sangster G, Collinson M. Guidelines for assigning species rank. Ibis. 2002;144:518–25.
15. Zink RM. Rigor and species concepts. Auk. 2006;123:887–91.
16. Gill FB. Species taxonomy of birds: Which null hypothesis? The Auk. 2014;131:150–61.
17. Sangster G. Increasing numbers of bird species result from taxonomic progress, not taxonomic inflation. Proc R Soc Lond B. 2009;276:3185–91.
18. Sangster G. The application of species criteria in avian taxonomy and its implications for the debate over species concepts. Biol Rev. 2014;89:199–214.
19. Haffer J. The history of species concepts and species limits in ornithology. Bull Br Ornithol Club Centenary Suppl. 1992;112A:107–58.
20. Tobias JA, Seddon N, Spottiswoode CN, Pilgrim JD, Fishpool LDC, Collar NJ. Quantitative criteria for species delimitation. Ibis. 2010;152:724–46.
21. Cohen J. Reproduction. London and Boston: Butterworths; 1977.
22. Pitnick S, Hosken DJ, Birkhead TR. Sperm morphological diversity. In: Birkhead TR, Hosken DJ, Pitnick S, editors. Sperm Biology: An Evolutionary Perspective. Oxford: Elsevier; 2009. p. 69–149.
23. Jamieson BGM, Ausió J, Justine J-L, editors. Advances in Spermatozoal Phylogeny and Taxonomy. Paris: Muséum national d'Histoire naturelle; 1995.
24. Jamieson BGM. Avian spermatozoa: structure and phylogeny. In: Jamieson BGM, editor. Reproductive Biology and Phylogeny of Birds Part A. Enfield: Science Publishers Inc; 2006. p. 249–511.
25. Immler S, Birkhead TR. Sperm competition and sperm midpiece size: no consistent pattern in passerine birds. Proc R Soc B. 2007;274:561–8.
26. Lifjeld JT, Laskemoen T, Kleven O, Albrecht T, Robertson RJ. Sperm length variation as a predictor of extrapair paternity in passerine birds. PLoS ONE. 2010;5:e13456.
27. Hogner S, Laskemoen T, Lifjeld JT, Pavel V, Chutný B, García J, Eybert M-C, Matsyna E, Johnsen A. Rapid sperm evolution in the bluethroat (Luscinia svecica) subspecies complex. Behav Ecol Sociobiol. 2013;67:1205–17.
28. Laskemoen T, Albrecht T, Bonisoli-Alquati A, Cepak J, Lope F, Hermosell I, Johannessen L, Kleven O, Marzal A, Mousseau T, et al. Variation in sperm morphometry and sperm competition among barn swallow (Hirundo rustica) populations. Behav Ecol Sociobiol. 2013;67:301–9.
29. Lüpold S, Linz G, Birkhead T. Sperm design and variation in the New World blackbirds (Icteridae). Behav Ecol Sociobiol. 2009;63:899–909.
30. Rowe M, Griffith S, Hofgaard A, Lifjeld J. Subspecific variation in sperm morphology and performance in the Long-tailed Finch (Poephila acuticauda). Avian Res. 2015;6:23.
31. Gill F, Donsker D, IOC World Bird List (v. 6.1). 2016. Available at http://www.worldbirdnames.org/.
32. BirdLife International. Species factsheet: Fringilla teydea. 2016. Available at http://www.birdlife.org.
33. Dickinson EC, Christidis L, editors. The Howard and Moore complete checklist of the birds of the world. Volume 2: Passerines. 4th ed. London: Aves Press; 2014.
34. Bannerman DA. Birds of the Atlantic Islands. A history of birds of the Canary Islands and of the Salvages. Edinburgh and London: Oliver & Boyd; 1963.
35. del Hoyo J, Elliott A, Christie DA, editors. Handbook of the Birds of the World - Vol. 15. Weavers to New World Warblers. Barcelona: Lynx Edicions; 2010.
36. Cramp S, Perrins CM, editors. The birds of the western Palearctic. Vol. 8. Oxford: Oxford University Press; 1994.
37. Sangster G, Rodríguez-Godoy F, Roselaar CS, Robb MS, Luksenburg JA. Integrative taxonomy reveals Europe's rarest songbird species, the Gran Canaria Blue Chaffinch Fringilla polatzeki. J Avian Biol. 2016;47:159–66.
38. Pestano J, Brown RP, Rodriguez F, Moreno A. Mitochondrial DNA control region diversity in the endangered blue chaffinch, Fringilla teydea. Mol Ecol. 2000;9:1421–5.
39. Suárez NM, Betancor E, Klassert TE, Almeida T, Hernández M, Pestano JJ. Phylogeography and genetic structure of the Canarian common chaffinch (Fringilla coelebs) inferred with mtDNA and microsatellite loci. Mol Phyl Evol. 2009;53:556–64.
40. Garcia-del-Rey E, Marthinsen G, Calabuig P, Estévez L, Johannessen L, Johnsen A, Laskemoen T, Lifjeld J. Reduced genetic diversity and sperm motility in the endangered Gran Canaria Blue Chaffinch Fringilla teydea polatzeki. J Ornithol. 2013;154:761–8.
41. Garcia-del-Rey E, Gil L, Nanos N, Lopez-de-Heredia U, Munoz PG, Fernandez-Palacios JM. Habitat characteristics and seed crops used by blue chaffinches Fringilla teydea in winter: implications for conservation management. Bird Study. 2009;56:168–76.
42. Garcia-del-Rey E, Nanos N, López-de-Heredia U, Muñoz P, Otto R, Fernández-Palacios J, Gil L. Spatiotemporal variation of a Pinus seed rain available for an endemic finch in an insular environment. Eur J Wildl Res. 2011;57:337–47.
43. The IUCN Red List of Threatened Species: Fringilla teydea. Volume 2015-4. 2016. Availabale at www.iucnredlist.org.
44. Garcia-del-Rey E. Birds of the Canary Islands. Spain: Sociedad Ornitologica Canaria; 2015.
45. Suarez NM, Betancor E, Fregel R, Rodríguez F, Pestano J. Genetic signature of a severe forest fire on the endangered Gran Canaria blue chaffinch (Fringilla teydea polatzeki). Conserv Genet. 2012;13:499–507.
46. Garcia-Del-Rey E, Otto R, Fernandez-Palacios JM, Munoz PG, Gil L. Effects of wildfire on endemic breeding birds in a Pinus canariensis forest of Tenerife, Canary Islands. Ecoscience. 2010;17:298–311.
47. Garcia-del-Rey E, Gosler AG. Biometrics, ageing, sexing and moult of the Blue Chaffinch Fringilla teydea teydea on Tenerife (Canary Islands). Ringing Migr. 2005;22:177–84.
48. Kleven O, Laskemoen T, Fossøy F, Robertson RJ, Lifjeld JT. Intraspecific variation in sperm length is negatively related to sperm competition in passerine birds. Evolution. 2008;62:494–9.
49. Hahn C, Bachmann L, Chevreux B. Reconstructing mitochondrial genomes directly from genomic next-generation sequencing reads—a baiting and iterative mapping approach. Nucleic Acids Res. 2013;41:e129.
50. Bernt M, Donath A, Jühling F, Externbrink F, Florentz C, Fritzsch G, Pütz J, Middendorf M, Stadler PF. MITOS: Improved de novo metazoan mitochondrial genome annotation. Mol Phyl Evol. 2013;69:313–9.
51. Tamura K, Stecher G, Peterson D, Filipski A, Kumar S. MEGA6: Molecular Evolutionary Genetics Analysis Version 6.0. Mol Biol Evol. 2013;30:2725–9.
52. Tamura K, Nei M, Kumar S. Prospects for inferring very large phylogenies by using the neighbor-joining method. Proc Natl Acad Sci U S A. 2004;101:11030–5.
53. Zuckerlandl E, Pauling L. Evolutionary divergence and convergence in proteins. In: Bryson V, Vogel HJ, editors. Evolving Genes and Proteins. New York: Academic; 1965. p. 97–166.
54. Lohman DJ, Prawiradilaga DM, Meier R. Improved COI barcoding primers for Southeast Asian perching birds (Aves: Passeriformes). Mol Ecol Resour. 2009;9:37–40.

55. Borge T, Webster MT, Andersson G, Saetre G-P. Contrasting patterns of polymorphism and divergence on the Z chromosome and autosomes in two *Ficedula* flycatcher species. Genetics. 2005;171:1861–73.

56. Tamura K, Peterson D, Peterson N, Stecher G, Nei M, Kumar S. MEGA5: Molecular evolutionary genetics analysis using maximum likelihood, evolutionary distance, and maximum parsimony methods. Mol Biol Evol. 2011;28:2731–9.

57. Leigh JW, Bryant D. popart: full-feature software for haplotype network construction. Methods in Ecology and Evolution. 2015;6:1110–6.

58. Montgomerie R. CLR: Colour Analysis Programs v1.05. CLR: Colour Analysis Programs v1.05. 2008. Available at http://post.queensu.ca/~mont/color/analyze.html.

59. Hill GE, McGraw KJ, editors. Bird Coloration. Mechanisms and measurements. Cambridge: Harvard University Press; 2006.

60. Lachlan Robert F, Verzijden Machteld N, Bernard Caroline S, Jonker P-P, Koese B, Jaarsma S, Spoor W, Slater Peter JB, ten Cate C. The progressive loss of syntactical structure in bird song along an island colonization chain. Curr Biol. 2013;23:1896–901.

61. Laskemoen T, Kleven O, Fossøy F, Lifjeld JT. Intraspecific variation in sperm length in two passerine species, the bluethroat *Luscinia svecica* and the willow warbler *Phylloscopus trochilus*. Ornis Fennica. 2007;84:131–9.

62. Laskemoen T, Kleven O, Fossøy F, Robertson RJ, Rudolfsen G, Lifjeld JT. Sperm quantity and quality effects on fertilization success in a highly promiscuous passerine, the tree swallow *Tachycineta bicolor*. Behav Ecol Sociobiol. 2010;64:1473–83.

63. Sokal RR, Rohlf FJ. Biometry. San Francisco: W. H. Freeman and co.; 1981.

64. Cohen J. Statistical Power Analysis for the Behavioral Sciences. 2nd ed. Hillsdale: Lawrence Erlbaum Associates; 1988.

65. Desjardins P, Morais R. Sequence and gene organization of the chicken mitochondrial genome: a novel gene order in higher vertebrates. J Mol Biol. 1990;212:599–634.

66. Hebert PDN, Stoeckle MY, Zemlak TS, Francis CM. Identification of birds through DNA barcodes. PLoS Biology. 2004;2:1657–63.

67. Yeates DK, Seago A, Nelson L, Cameron SL, Joseph LEO, Trueman JWH. Integrative taxonomy, or iterative taxonomy? Syst Entomol. 2011;36:209–17.

68. Dayrat B. Toward integrative taxonomy. Biol J Linn Soc. 2005;85:407–15.

69. Padial J, Miralles A, De la Riva I, Vences M. The integrative future of taxonomy. Front Zool. 2010;7:16.

70. Hartert E. Eine neue subspecies von *Fringilla teydea*. Ornithol Monatsber. 1905;13:164.

71. Weir JT, Schluter D. Calibrating the avian molecular clock. Mol Ecol. 2008;17:2321–8.

72. Zink R, Barrowclough G. Mitochondrial DNA under siege in avian phylogeography. Mol Ecol. 2008;17:2107–21.

73. Ratnasingham S, Hebert PDN. bold: The Barcode of Life Data System (http://www.barcodinglife.org). Mol Ecol Notes. 2007;7:355–64.

74. Johnsen A, Rindal E, Ericson PGP, Zuccon D, Kerr KCR, Stoeckle MY, Lifjeld JT. DNA barcoding of Scandinavian birds reveals divergent lineages in trans-Atlantic species. J Ornithol. 2010;151:565–78.

75. Kerr KCR, Lijtmaer DA, Barreira AS, Hebert PDN, Tubaro PL. Probing evolutionary patterns in Neotropical birds through DNA barcodes. PLoS ONE. 2009;4:e4379.

76. Lifjeld JT. When taxonomy meets genomics: lessons from a common songbird. Mol Ecol. 2015;24:2901–3.

77. Phillimore AB, Owens IPF. Are subspecies useful in evolutionary and conservation biology? Proc R Soc B. 2006;273:1049–53.

78. Juan C, Emerson BC, Oromi P, Hewitt GM. Colonization and diversification: towards a phylogeographic synthesis for the Canary Islands. Trends Ecol Evol. 2000;15:104–9.

79. Dietzen C, Garcia-del-Rey E, Castro GD, Wink M. Phylogeography of the Blue Tit (*Parus teneriffae*-group) on the Canary Islands based on mitochondrial DNA sequence data and morphometrics. J Ornithol. 2008;149:1–12.

80. Gohli J, Leder EH, Garcia-del-Rey E, Johannessen LE, Johnsen A, Laskemoen T, Popp M, Lifjeld JT. The evolutionary history of Afrocanarian blue tits inferred from genomewide SNPs. Mol Ecol. 2015;24:180–91.

81. Dietzen C, Witt HH, Wink M. The phylogeographic differentiation of the European robin *Erithacus rubecula* on the Canary Islands revealed by mitochondiral DNA sequence data and morphometrics: evidence for a new robin taxon on Gran Canaria? Avian Sci. 2003;3:115–31.

82. Grant PR, Grant BR. How and Why Species Multiply. The Radiation of Darwin's Finches. Princeton: Princeton University Press; 2008.

83. McKay BD, Zink RM. Sisyphean evolution in Darwin's finches. Biol Rev. 2015; 90:689–98.

84. Andersson M. Sexual selection. Princeton: Princeton University Press; 1994.

85. Briskie JV, Montgomerie R, Birkhead TR. The evolution of sperm size in birds. Evolution. 1997;51:937–45.

86. Kleven O, Fossøy F, Laskemoen T, Robertson RJ, Rudolfsen G, Lifjeld JT. Comparative evidence for the evolution of sperm swimming speed by sperm competition and female sperm storage duration in passerine birds. Evolution. 2009;63:2466–73.

87. Higginson DM, Miller KB, Segraves KA, Pitnick S. Female reproductive tract form drives the evolution of complex sperm morphology. Proc Natl Acad Sci USA. 2012;109:4538–43.

88. Cramer ERA, Laskemoen T, Kleven O, Lifjeld JT. Sperm length variation in House Wrens *Troglodytes aedon*. J Ornithol. 2013;154:129–38.

89. Simmons LW, Moore AJ. Evolutionary quantitative genetics of sperm. In: Birkhead TR, Hosken DJ, Pitnick S, editors. Sperm Biology: An Evolutionary Perspective. Oxford: Elsevier; 2009. p. 405–34.

90. Calhim S, Immler S, Birkhead TR. Postcopulatory sexual selection is associated with reduced variation in sperm morphology. PLoS ONE. 2007;2:e413.

91. Birkhead TR, Brillard J-P. Reproductive isolation in birds: postcopulatory prezygotic barriers. Trends Ecol Evol. 2007;22:266–72.

92. Cramer ERA, Stensrud E, Marthinsen G, Hogner S, Johannessen LE, Laskemoen T, Eybert M-C, Slagsvold T, Lifjeld JT, Johnsen A. Sperm performance in conspecific and heterospecific female fluid. Ecol Evol. 2016;6:1363–77.

16

Body condition scoring of Bornean banteng in logged forests

Naomi S. Prosser[1,2*], Penny C. Gardner[1,2], Jeremy A. Smith[2,3], Jocelyn Goon Ee Wern[1], Laurentius N. Ambu[4] and Benoit Goossens[1,2,4,5*]

Abstract

Background: The Bornean banteng (*Bos javanicus lowi*) is an endangered subspecies that often inhabits logged forest; however very little is known about the effects of logging on their ecology, despite the differing effects this has on other ungulate species. A body condition scoring system was created for the Bornean banteng using camera trap photographs from five forests in Sabah, Malaysia, with various past and present management combinations to establish if banteng nutrition suffered as a result of forest disturbance.

Results: One hundred and eleven individuals were photographed over 38,009 camera trap nights from April 2011 to June 2014 in five forests. Banteng within forests that had a recent history of reduced-impact logging had higher body condition scores than banteng within conventionally logged forest. Conversely, when past logging was conducted using a conventional technique and the period of forest regeneration was relatively long; the banteng had higher body condition scores.

Conclusion: The body condition scoring system is appropriate for monitoring the long-term nutrition of the Bornean banteng and for evaluating the extent of the impact caused by present-day reduced-impact logging methods. Reduced-impact logging techniques give rise to individuals with the higher body condition scores in the shorter term, which then decline over time. In contrast the trend is opposite for conventional logging, which demonstrates the complex effects of logging on banteng body condition scores. This is likely to be due to differences in regeneration between forests that have been previously logged using differing methods.

Keywords: Body condition scoring, Camera trap, Habitat degradation, Reduced-impact logging, Sabah, Tropical forest

Background

Body condition scores (BCS) measure the amount of soft tissue an animal has relative to its size and are useful as a general guide to the health and fitness of an animal [1, 2]. Usually BCS systems allocate a number that is associated with condition, with lower scores given to animals in poorer condition [3]. Many BCS systems for domestic mammals use palpitation to more accurately assess the condition [1, 2], however this is not practical in wild animal studies as they would first have to be captured, which is both costly and stressful for the animal. There are BCS techniques developed for visual assessment of wild animals in the field; for example on Indian elephants (*Elephas maximus indicus*) by Ramesh et al. [4], and from photographs of mule deer (*Odocoileus hemionus eremicus*) by Marshal et al. [5] and Sri Lankan elephant (*Elephas maximus maximus*) by Fernando et al. [6]. Visual BCS systems are useful for animals that cannot be handled [7]. Non-invasive scoring of mammal body condition using camera traps furthers the scope for BCS applications in wildlife management and for assessing the health of highly elusive species such as the Bornean banteng (*Bos javanicus lowi*) that are not directly observable in the wild. BCS systems are straightforward to conduct, although variability can arise between observer [7, 8]. Clear and systematic BCS systems are beneficial because they can be implemented by wildlife managers [9] and because they are reliable in tracking changes in soft tissue carried by an animal over time (Edmonson et al. 1989 cited in [3]).

* Correspondence: N.Prosser@warwick.ac.uk; goossensbr@cardiff.ac.uk
[1]Danau Girang Field Centre, c/o Sabah Wildlife Department, Wisma Muis, 88100 Kota Kinabalu, Sabah, Malaysia
[2]Organisms and Environment Division, School of Biosciences, Cardiff University, Sir Martin Evans Building, Museum Avenue, Cardiff CF10 3AX, UK
Full list of author information is available at the end of the article

BCS provide invaluable information on the health of the animal on an individual scale that can be related to the strength of their immune system (e.g. [10]), age at which they first breed (Carrion et al. 2007 cited in [8]), fertility [8] and mortality [11, 12]. Monitoring individual health allows the tracking of the health at the population scale [13] and BCS can be used to evaluate the factors limiting population growth [5]. If body condition is low, nutrition is likely to be the key limiting factor in population growth (Bowyer 2005 cited in [5]) but it may also indicate that the population has reached its carrying capacity [5, 8]. For these reasons it is possible that comparing results of BCS across different habitats may show optimal management techniques and be an indication of the longer term effects of different treatments.

Banteng arc classified as endangered by the IUCN Red List with the rate of their population decline being greater than 80 % in the last three generations in parts of their range [14]. The population size and structure in Sabah is unknown due to a lack of data [15]. The home range of banteng in Borneo is also unknown, however a bull has been observed travelling 23 km (P. Gardner unpublished observations cited in [15]) and herds of *Bos javanicus birmanicus* can occupy home ranges of up to 44.8 km^2 (Prayurasithi 1997 cited in [15]). The subpopulations studied here were not recaptured in any of the other forests studied. Banteng are crepuscular in their activities, spending more time on foraging and social activity at dawn and dusk, while the mid-part of the day is largely spent ruminating (P. Gardner unpublished observations cited in [15]). They feed on a wide variety of plant material, opting to graze in open areas and are more frequently found in open dipterocarp forests when available [15–17]. The Bornean subspecies is recorded as living in secondary forests and that logging, which opens up the forest floor, may benefit banteng due to the increased understory growth [14, 17, 18]. This suggestion is supported by Meijaard and Sheil [19] who observed that, with the exception of frugivores, ungulates are more successful in logged forests. Logging however removes timber and alters the habitat that may provide vital food sources for the banteng [18] and creates extensive disturbance [20]. Ancrenaz et al. [20] found it was possible to maintain populations of orangutans (*Pongo pygmaeus morio*) within commercial forests that adopt reduced-impact logging (RIL) techniques, whereas conventional logging was more damaging and resulted in localized extinctions. This pattern is also likely to be true for the Bornean banteng; Deramakot Forest Reserve in Sabah utilizes RIL techniques and observations suggest it may support a denser population of banteng [21]. It is possible that past and present logging techniques and management agendas will differ in their effect upon banteng populations, and that the most

destructive types of management should result in reduced body condition.

The body condition of banteng in Sabah should reflect the habitat suitability of the forests and the health of the banteng populations (Adamczewski 1993 cited in [8]). Logging alters the vegetation composition of a forest [18], therefore it is important to know how logging practices affect banteng body condition. This information will be especially important as there is currently very little knowledge of the impact of logging on banteng [18]. At present there is very little unlogged forest remaining in Sabah [22] suggesting that it is likely that banteng will be confined to commercial forests in the very near future. Comparisons of banteng BCS from forests that differ in management may indicate the suitability of the management techniques and the most effective approach to conserve the banteng in commercially managed forest.

We conducted the first identification of individual banteng in Sabah using unique natural markings. We then created the first BCS system using non-invasive camera trap images to identify the impact of the implementation of conventional logging, RIL and protection from logging, both in the past and present, upon the health of banteng in Sabah. It was expected that banteng living in forests with longer post-logging regeneration times and which had been subjected to RIL instead of conventional techniques would have higher BCS.

Methods
Study sites
Camera trap surveys were conducted in five forests in Sabah, Malaysian Borneo: Tabin Wildlife Reserve, Malua Forest Reserve, the buffer zone of Maliau Basin Conservation Area, Sipitang Forest Reserve and Sapulut Forest Reserve, located in east, east-central, south-central, south and west of Sabah, respectively (Fig. 1). Within these locations the habitat that is inhabited by banteng is predominately lowland and hill dipterocarp forest and freshwater swamp forest. The forests experience uniform temperatures and very little variation in rainfall across the year [23], however extensive climatic data is not available. These forests have undergone different past and present logging management methods (Table 1).

Data collection
Camera trap images of banteng were obtained from the five forest reserves at differing time periods and camera trap stations were distributed in a grid format and/or on an ad-hoc basis (Table 2). The differing sampling schemes were due to two different prior studies of banteng with different objectives: a PhD project by Gardner [24] and a state-wide survey of the remnant banteng populations. With both sampling schemes the cameras

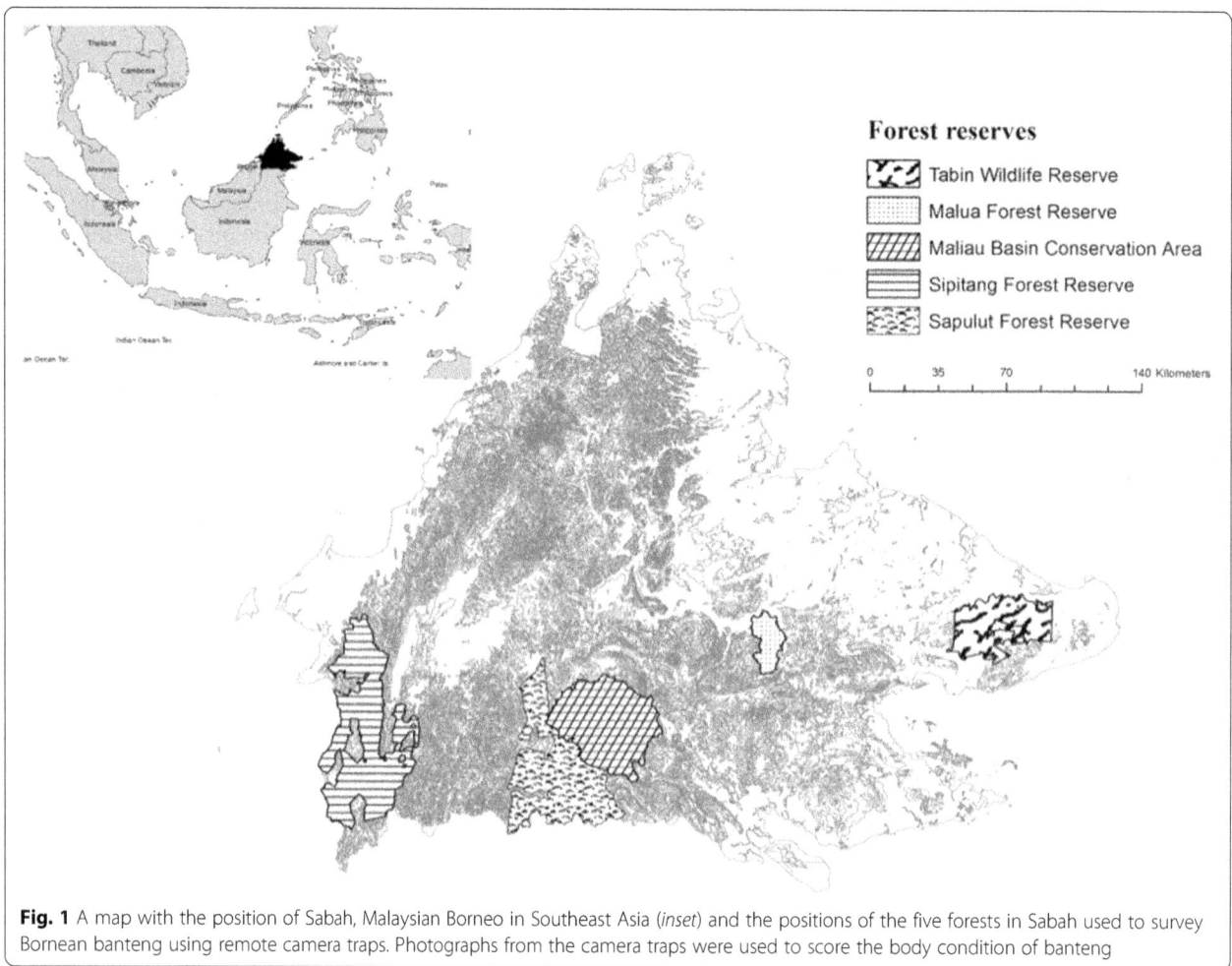

Fig. 1 A map with the position of Sabah, Malaysian Borneo in Southeast Asia (*inset*) and the positions of the five forests in Sabah used to survey Bornean banteng using remote camera traps. Photographs from the camera traps were used to score the body condition of banteng

were widely dispersed throughout the forest in question, making it reasonable to assume that the health of banteng individuals captured on camera are representative of the entire population. Cameras in a grid layout were positioned overlooking banteng or large mammal signs (or in the absence of these an animal trail) within a 50 m radius of a predetermined GPS position. Grid positions were determined primarily on access by vehicle, boat or on-foot, with a minimum distance of 500 m from the nearest unsealed road. Ad-hoc cameras were placed near or overlooking banteng signs (tracks or dung), direct sightings or in habitat suitable for banteng (*i.e.* internal openings or grassland). A camera trap station was comprised of two cameras facing each other, fixed to trees at approximately 10 m apart and 1 m high. Cameras were programmed to take three photographs at one-second intervals every time they were triggered. In low-light conditions images were taken in monochrome

Table 1 Logging categorisation scheme

Study Site	Past Logging Management	Present Logging Management	Year of Most Recent Logging
Tabin Wildlife Reserve	Conventional techniques [32]	Protected from logging	1989 [32]
Malua Forest Reserve	Conventional and RIL techniques [33]	Protected from logging for 50 years	2007 [33]
Buffer zone of Maliau Basin Conservation Area	RIL techniques [34]	Protected from logging during time of data collection, but logging has now recommenced.	1997 [34]
Sipitang Forest Reserve	Conventional techniques [35]	Areas logged with RIL techniques, managed as plantation or unlogged [35]	2010–2014 [35]
Sapulut Forest Reserve	Conventional techniques (Sabah Forestry Department staff, pers. obs.)	Logged using RIL techniques or managed as plantation (Sabah Forestry Department staff, pers. obs.)	2005–2014 (Sabah Forestry Department staff, pers. obs.)

The descriptions of the past and present logging managements of each of the five study forests in Sabah, Malaysia, showing the categories that were used in the data analysis

Table 2 Camera trap sampling scheme

Location	Study Period	No. of Stations	Camera make/model	Sampling scheme
Tabin Wildlife Reserve	April 2011–October 2012	130	Reconyx H500	2 grids: 4 km² (500 m between each station)
				2 grids: 6.25 km² (500 m between each station)
				8 ad-hoc stations placed in areas of banteng signs
Malua Forest Reserve	April 2011–June 2014	118	Reconyx H500 & PC800	3 grids: 6.25 km² (500 m between each station)
				10 ad-hoc stations placed in areas of banteng signs
Maliau Basin Conservation Area	June 2013–June 2014	21	Reconyx H500, PC800 & PC850	Cameras in areas of banteng signs
Sipitang Forest Reserve	September 2013–March 2014	30	Reconyx PC800	Cameras in areas of banteng signs
Sapulut Forest Reserve	November 2013–April 2014	30	Reconyx H500 & PC800	Cameras in areas of banteng signs

The camera trapping methods used to capture banteng in each forest reserve studied in Sabah, Malaysia

using an in-built infrared light with the exception of one camera in the buffer zone of Maliau Basin Conservation Area (Reconyx PC850), which operated with a white flash for colour pictures. Cameras were checked every 4 weeks for tampering and functionality, and to change the batteries and SD cards. Surveys were conducted for a minimum of 90 days in each forest. This study only utilized non-invasive remote camera trapping for data collection in order to minimize the disturbance caused to this shy and nervous large bovid. No experimental work and no direct handling of banteng individuals in the wild or in captivity was conducted. This study was executed in compliance with the Sabah Wildlife Department: no ethics statement was required and no ethics committee exists for wildlife studies within Sabah, Malaysia. Research permits were granted by the Sabah Biodiversity Council, reference numbers: JKM/MBS.1000-2/12(156) and JKM/MBS.1000-2/2 JLD.3 (18).

Body condition scoring

Individual profile cards of each banteng were created using camera trap images of the head, rear and both sides of the animal to create a reference allowing the banteng to be identified in the future if recapture occurred. Each individual was given a score for every day that it appeared in the camera traps using the five-point pictorial scoring system developed specifically for the Bornean banteng (Additional file 1). All the scores given to each banteng were included in the analysis and in cases where not all of the scored parts of the banteng were visible; the parts that were in sight were used to create the score. The five-score pictorial BCS system covers the full range of conditions observed in banteng captured on camera within the research period. Individuals were assigned a score of 1, 2, 3, 4 or 5, with high values corresponding to more soft tissue; a score of 1 was given to animals with prominent skeletal features such as the ribs, 2 when the hips protruded but the ribs

were not obvious, 3 when there was some rounding to the body and the most prominent skeletal features such as the hooks and pins were covered by some soft tissue, 4 when there was a relatively high degree of soft tissue covering the prominent skeletal features and the body was quite rounded, and 5 when all skeletal features on the body were covered in soft tissue giving a very rounded appearance. It is possible that body conditions both above and below this range exist in individuals not captured during this study and future users of this method could extend the scale to 0 or 6 if necessary, as with the elephant BCS system developed by Fernando et al. [6]. Predominantly only colour images were used in the profiling and for scoring the body condition. In exceptional circumstances, monochrome photographs were included when the light levels shifted and caused the camera to change from colour to monochrome photographs (or vice versa) while the herd was still present. These images retained sufficient light to permit individual recognition and scoring. Images where it was not possible to allocate an accurate score were not included in the dataset for analysis.

Logging management analysis

Past logging was defined as the logging that took place before the current logging management and present logging was defined as the logging being practiced at the time the images were captured. The logging history of each forest was classified into four categories that best describe the past and present logging methods: (1) RIL [25]; (2) industrial tree plantation: the area is clear felled and replanted; (3) conventional: traditional logging without RIL techniques; and (4) none: no logging. The time since logging in years was calculated from the year that that part of the forest was last harvested. When the exact date of logging cessation for the compartment was unknown, the last harvesting activity for the entire forest was used. When an individual was recaptured in different areas of the study forest that had had differing logging treatments on the same day, the photos were discounted. However every instance where an individual was only captured in an area of the same logging treatment in a day was included in the analysis.

R statistical software version 3.0.2 [26] and the package Ordinal [27] was used in the data analysis. Only the first BCS given to each individual was used in the analysis to remove bias for those individuals who were scored multiple times. The independent variables were Past Logging Type (categorical), Present Logging Type (categorical) and Years Since Logging (continuous) and these were included in a cumulative link model (CLM) to explain the dependent categorical variable, BCS. Density was not included in the analysis, as the nature of the camera trap deployment did not allow this to be estimated. An initial CLM of BCS explained by Past Logging, Present Logging, Years Since Logging and all possible two-way interactions was created and run with every possible link. Akaike information criterion (AIC) comparison was used to choose the best link. In order to remove insignificant terms and arrive at the most efficient model the initial CLM was reduced by AIC comparison by automatic stepwise refinement.

Results
Profiling
There was a differential surveying effort in each forest due to the nature of the primary objectives for the data collection and this is laid out in Table 3. Out of the total 681 daytime banteng captures there were 100 instances where the individual could not be reliably assigned a profile (Table 3). In total 111 individuals were identified and scored from the five forests (Additional file 2). Individuals were captured during the first 46 survey days. Two individuals were excluded from the analysis because no reliable scores could be allocated for any of the occasions they appeared on the camera. In Tabin Wildlife Reserve eight bulls, six cows and two juveniles were studied whereas 16 bulls, 17 cows and seven juveniles were studied for Malua Forest Reserve, seven bulls, 10 cows and

one juvenile for Maliau Basin Conservation Area, seven bulls, 12 cows and five juveniles for Sipitang Forest Reserve and six bulls, five cows and two juveniles for Sapulut Forest Reserve. The main natural features used to identify individuals were the shape of the horns, tears in the ears, the pattern at the edges of the stockings and scars. Occasionally other features that were used in identification were the length of tail, ear hairiness and pairing up of cows and calves (Gardner PC: Individual recognition and profiling of Bornean banteng, in preparation).

Forest management
The most efficient link in the initial model was a loglog link and after automatic stepwise deletion the final most efficient model was BCS explained by Years Since Logging in interaction with Past Logging. The interaction between years since logging and past logging type was significant ($p = 0.0192$), meaning that the relationship between the timing of logging and the BCS of the banteng differs depending on which logging method (RIL or conventional) was previously used in the area (Table 4; Fig. 2). BCS also differed significantly between past logging types ($p = 0.0238$) however time since logging did not significantly affect BCS ($p = 0.9907$) (Table 4). The BCS of banteng living in forest previously logged by RIL was significantly higher by 1.07 ± 0.47 than those living in forest that was previously logged using conventional methods. The BCS of banteng living in previously logged forests using the RIL method had declined over time, whereas banteng in the conventionally logged forests now have a higher BCS, after a longer regeneration time (Fig. 2). In the short term banteng have higher body condition in areas where RIL was used in the past, however after approximately 11 years of regeneration banteng BCS are higher in areas that have had conventional logging techniques in the past (Fig. 2).

Discussion
Profiling
Banteng can be reliably identified using differences in their coat colour, size and shape of horns, differences in the edges of the stockings, cuts and scars and any other

Table 3 Camera trap sampling results

	Camera Trap Nights	Daytime Captures	Unscored captures	Number of Individuals	Elevation of Captures (m) above sea level
Tabin Wildlife Reserve	14784	26	4	16	22–164
Malua Forest Reserve	12400	486	113	40	29–293
Maliau Basin Conservation Area	4971	77	18	18	267–398
Sipitang Forest Reserve	3714	56	11	24	781–1143
Sapulut Forest Reserve	2140	36	5	13	407–482
Total	38009	681	151	111	22–1143

The camera trapping survey and profiling results for the Bornean banteng in Sabah, Malaysia, showing how long cameras were left up, at what elevations, how many times individual banteng appeared on camera and how many captures could not be reliably given a score

Table 4 The results of the final CLM

Term	Coefficient	SE	Test Statistic	p-value
Past Logging * Years Since Logging	−0.09	0.040	−2.341	0.9907
Past Logging (RIL)	1.06	0.473	2.260	0.0238
Years Since Logging	−0.0001	0.017	−0.012	0.9907

The statistical results of the cumulative link model (CLM) of Banteng BCS ~ Past Logging*Years Since Logging for banteng in Sabah, Malaysia

permanent or semi-permanent distinguishing feature (Gardner PC: Individual recognition and profiling of Bornean banteng, in preparation). Profiling juveniles was more difficult, particularly the very young calves that had neither horns nor stockings and rapidly developed between photographic recaptures that were often infrequent. Furthermore, recognition was difficult when multiple calves in the same herd were born over the same time. It was necessary to be very flexible with the features used to recognize the individuals, as different lights and angles showed different features. When profiling it is important to keep in mind the changes that occur over time: new scars may be acquired, horns will curve more and the coat may change colour slightly. For forests where camera trapping was carried out over multiple years, instances of all stages of an individual's growth were included in their profile to allow their recognition at all ages. Although individuals would change in many ways; for example horn development and shape, number of scars and colour; features such as the pattern at the edge of their stockings would not change, allowing an individual to be successfully identified a number of years after it was last captured.

Body condition scoring

As body condition scoring has been successfully applied to elephants using only visual observations [4, 6, 9] it was expected that a scoring system for banteng using photos would be straightforward to implement. Camera

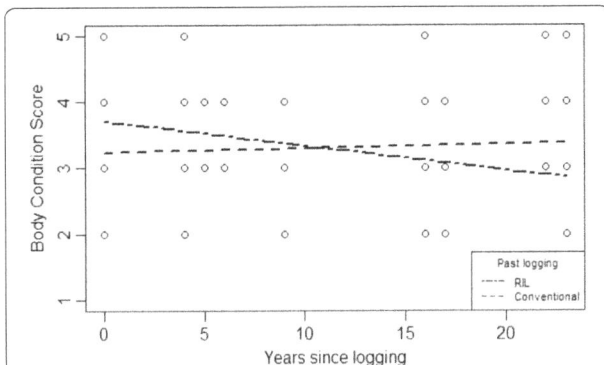

Fig. 2 The body condition scores (BCS) of banteng inhabiting forests in Sabah, Malaysia, with increasing time since the forest was last logged by either reduced-impact logging (RIL) or conventional techniques for an ordinal regression model with the raw data (circles) and predicted values (dashed lines). The body condition scores were obtained from camera trap images

trap captures are often limited in their photographic aspect and do not always permit the animal to be viewed from all angles to assess the condition, however this technique does allow a much closer observation of a wild animal than would normally be possible. Occasionally the camera functionality or weather conditions gave rise to poorer quality photos, which were more difficult to score. As stated by Wemmer et al. [9], the skeletal features are much more visible when an animal is stretched out mid-stride whereas they are much less visible when bunched up, each of which decreases or increases the perceived body condition respectively. This would be less of a problem with live observations. Fernando et al. [6] found that lighting and posture might cause a different score to be assigned to an animal, but the error is usually only small (equating to ± 0.5 BCS in this case). It was suggested that the use of a series of photographs would be beneficial to obtain more reliable scores of an animal [6]. As the camera traps used in this study provided a short series of photographs of each individual, we maximized the potential for capturing the features used for scoring. For these reasons, we believe our scores of each banteng are accurate given the harsh environmental conditions and the inability to conduct direct observations.

Forest logging effects

It was expected that the BCS would be higher in forests that have been marginally opened up by recent and small amounts of logging. This is because it would promote the growth of pioneer grasses, which are suitable forage for banteng, but that extensive logging would remove some of the important feeding habitats and therefore the banteng would be in poorer condition there. We found evidence in this study that supports this theory.

The results of this study indicated that forests that were logged using RIL in the recent past contained banteng with higher BCS than forests that were logged using only conventional logging methods. RIL techniques minimize the impact on the forest by removing timber using helicopter and/or long-distance cranes to avoid non-target stem destruction, soil erosion and compaction [21], however all logging activity still causes much disturbance and habitat degradation [28].

Most of Sabah's forests have been logged previously [29], therefore it is necessary to conserve banteng in logged areas. All logging in Sabah must now follow RIL techniques if the forest is not going to be converted into plantation [25], which will be less detrimental to banteng habitat than old-style conventional logging. The higher BCS in forests that have been logged recently using RIL may be due to the short-term benefits of increased understory growth of pioneer species in the first few years after logging [29]. This is due to the increased light intensity on this layer [30]. Pioneer species such as grasses were found in greater abundance when logging had occurred within the last ten years (S. Ridge, pers. obs.) making banteng forage more abundant. The nature of RIL techniques is less damaging than conventional techniques [25] because they allow the vegetation to recover more quickly [31]. Past logging was found to significantly influence banteng BCS but our results show that current disturbance or present logging does not manifest in different BCS. It is therefore possible that a negative lag effect exists and that banteng BCS may decline in the future. For this reason, it is important to minimize the impact of logging. Long-term monitoring of banteng BCS in an area undergoing changes in management will be required in order to assess the full consequences. The effects of past conventional logging techniques are evident today as low BCS, with the more recent the conventional logging the lower the BCS (Fig. 2). It was not possible to document banteng BCS prior to logging but we believe the lower scores to be due to conventional logging techniques decreasing the rate of vegetation regeneration [31]. We observed a gradual increase in BCS with time in forests that were regenerating from conventional logging; however we observed the opposite trend for RIL whereby BCS decreased at a steep gradient over time following RIL activity.

How recently a forest is logged and the type of logging undertaken significantly affect BCS and these factors are indicative of forest regeneration; forests logged using RIL will vary in the extent and speed of regeneration compared with conventional logging, which may take significantly longer. It is likely that the effects of logging on banteng are actually very complex with many more facets that were not covered in this study. A note of caution however regarding the interpretation of the significance of time since logging; although there was overlap in the time since logging between the three most recently logged forests (Malua Forest Reserve, Sipitang Forest Reserve and Sapulut Forest Reserve), the time of the two most distantly logged forests (Tabin Wildlife Reserve and the buffer zone of Maliau Basin Conservation Area) do not overlap with any other forest, therefore the increase in BCS with recent logging could actually be due to other factors not measured in this study within these latter two forests.

Conclusions

The BCS system is appropriate for monitoring the long-term nutrition of the Bornean banteng and for evaluating

the extent of the impact caused by present-day RIL methods. RIL techniques give rise to individuals with higher BCS in the shorter term, which then decline over time. In contrast the trend is opposite for conventional logging, demonstrating the complex effects of logging on banteng BCS. This is likely to be due to differences in regeneration between forests that have been previously logged using differing methods. This study highlights the lack of information there is into the effects of logging on the Bornean banteng and provides new methods for allowing research into this elusive mammal. It is imperative that this knowledge gap is closed and corrective measures are implemented to facilitate the persistence of this endangered species.

Abbreviations
AIC, Akaike Information Criterion; BCS, Body Condition Score; CLM, Cumulative Link Model; RIL, Reduced-Impact Logging

Acknowledgements
We would like to thank the local counterpart, the Sabah Wildlife Department for supporting this research, and the Sabah Biodiversity Centre for giving permission to carry out research in Sabah. Also, the Sabah Forestry Department (Datuk Dr. S. Mannan) and Sabah Foundation for permission to conduct our work in the different protected areas. The Leibniz Institute provided some camera traps. A warm thank you to our Bornean Banteng Programme's field assistants, Ruslee Rahman, Rusdi Saibin and Siti Hadijah Abdul Rasyak.

Funding
Yayasan Sime Darby, Malaysian Palm Oil Council, Mohamed bin Zayed Species Conservation Fund, Houston Zoo and Woodland Park Zoo funded this project but had no role in the study design, data collection, preparation of the manuscript, or the decision to publish.

Authors' contributions
NP carried out the background research, formulated the methods including creating the body condition scoring system for banteng, identifying the individuals and carrying out the statistical analysis. PG facilitated research and conducted surveys and data acquisition, provided advice and ideas on the methodology and carried out extensive editing of the manuscript. JS provided much assistance with the data analysis and editing of the results section. JG conducted surveys and data acquisition. LA was a local counterpart, provided research permits and facilitated research. BG facilitated research and gave extensive advice and edits on the manuscript. All authors read and approved the final manuscript.

Competing interests
The authors declare that they have no competing interests.

Author details
[1]Danau Girang Field Centre, c/o Sabah Wildlife Department, Wisma Muis, 88100 Kota Kinabalu, Sabah, Malaysia. [2]Organisms and Environment Division, School of Biosciences, Cardiff University, Sir Martin Evans Building, Museum Avenue, Cardiff CF10 3AX, UK. [3]Eco-explore Public Interest Company, Cardiff School of Biosciences, Cardiff University, Cardiff CF10 3AX, UK. [4]Sabah Wildlife Department, Wisma Muis, 88100 Kota Kinabalu, Sabah, Malaysia. [5]Sustainable Places Research Institute, Cardiff University, 33 Park Place, Cardiff CF10 3BA, UK.

References
1. Ezenwa VO, Jolles AE, O'Brien MP. A reliable body condition scoring technique for estimating condition in African buffalo. Afr J Ecol. 2009;47:476–81.
2. Soares FS, Dryden GM. A body condition scoring system for Bali cattle. Asian-Australasian J Anim Sci. 2011;24:1587–94.
3. Domecq JJ, Skidmore AL, Lloyd JW, Kaneene JB. Validation of Body Condition Scores with Ultrasound Measurements. J Dairy Sci. 1995;78:2308–13.
4. Ramesh T, Sankar K, Qureshi Q, Kalle R. Assessment of wild Asiatic elephant (Elephas maximus indicus) body condition by simple scoring method in a tropical deciduous forest of Western Ghats, Southern India. Wildl Biol Pract. 2011;7:47–54.
5. Marshal JP, Krausman PR, Bleich VC. Body condition of mule deer in the Sonoran Desert is related to rainfall. Southwest Nat. 2008;53:311–8.
6. Fernando P, Janaka HK, Ekanayaka SKK, Nishantha HG, Pastorini J. A simple method for assessing elephant body condition. Gajah. 2009;31:29–31.
7. Audigé L, Wilson PR, Morris RS. A body condition score system and its use for farmed red deer hinds. N Z J Agric Res. 1998;41:545–53.
8. Caslini C. Wild Red Deer (Cervus elaphus, Linnaeus, 1758) Populations Status Assessment: Novel Methods using Hair. PhD. Milan: Università Degli Studi di Milano; 2012.
9. Wemmer C, Krishnamurthy V, Shrestha S, Hayek L-A, Thant M, Nanjappa KA. Assessment of body condition in Asian elephants (Elephas maximus). Zoo Biology. 2006;25:187–200.
10. Cabezas S, Calvete C, Moreno S. Vaccination success and body condition in the European wild rabbit: applications for conservation strategies. J Wildl Manag. 2006;70:1125–31.
11. Choquenot D. Density-dependent growth, body condition, and demography in feral donkeys: testing the food hypothesis. Ecology. 1991;72:805–13.
12. Bérubé CH, Festa-Bianchet M, Jorgenson JT. Individual differences, longevity, and reproductive senescence in bighorn ewes. Ecology. 2014;80:2555–65.
13. Stevenson RD, Woods WA. Condition indices for conservation: new uses for evolving tools. Integr Comp Biol. 2006;46:1169–90.
14. Bos javanicus [http://dx.doi.org/10.2305/IUCN.UK.2008.RLTS.T2888A9490684.en].
15. Gardner PC, Pudyatmoko S, Bhumpakphan N, Yindee M, Ambu LN, Goossens B. Species accounts: Banteng Bos javanicus d'Alton, 1823. In: Melletti M, Burton J, editors. Ecology, Evolution and Behaviour of Wild Cattle. Cambridge: Cambridge University Press; 2014. p. 216–30.
16. Purwantara B, Noor RR, Andersson G, Rodriguez-Martinez H. Banteng and Bali Cattle in Indonesia: Status and Forecasts. Reprod Domest Anim. 2012;47:2–6.
17. Pedrono M, Tuan HM, Chouteau P, Vallejo F. Status and distribution of the endangered banteng Bos javanicus birmanicus in Vietnam: a conservation tragedy. Oryx. 2009;43:618–25.
18. Meijaard E, Sheil D, Nasi R, Augeri D, Rosenbaum B, Iskandar D, Setyawati T, Lammertink M, Rachmatika I, Wong A, et al. Life After Logging: Reconciling Wildlife Conservation and Production Forestry in Indonesian Borneo. Jakarta: CIFOR and UNESCO; 2005.
19. Meijaard E, Sheil D. The persistence and conservation of Borneo's mammals in lowland rain forests managed for timber: observations, overviews and opportunities. Ecol Res. 2008;23:21–34.
20. Ancrenaz M, Ambu L, Sunjoto I, Ahmad E, Manokaran K, Meijaard E, Lackman I. Recent surveys in the forests of Ulu Segama Malua, Sabah, Malaysia, show that orang-utans (P. p. morio) can be maintained in slightly logged forests. PLoS ONE. 2010;5:e11510.
21. Lagan P, Sam M, Matsubayashi H. Sustainable use of tropical forests by reduced-impact logging in Dermakot Forest Reserve, Sabah, Malaysia. Ecol Res. 2007;22:414–21.
22. Bryan JE, Shearman PL, Asner GP, Knapp DE, Aoro G, Lokes B. Extreme differences in forest degradation in Borneo: comparing practices in Sarawak, Sabah, and Brunei. PLoS ONE. 2013;8:e69679.
23. General Climate of Malaysia [http://www.met.gov.my/web/metmalaysia/climate/generalinformation/malaysia?p_p_id=56_INSTANCE_zMn7KdXJhAGe&p_p_lifecycle=0&p_p_state=normal&p_p_mode=view&p_p_col_id=column-1&p_p_col_pos=1&p_p_col_count=2&_56_INSTANCE_zMn7KdXJhAGe_page=1].
24. Gardner PC. The natural history, non-invasive sampling, activity patterns and population genetic structure of the Bornean banteng Bos javanicus lowi in Sabah. Cardiff: Cardiff University, School of Biosciences; 2015.
25. Edwards DP, Woodcock P, Edwards FA, Larsen TH, Hsu WW, Benedick S, Wilcove DS. Reduced-impact logging and biodiversity conservation: a case study from Borneo. Ecol Appl. 2012;22:561–71.
26. R Development Core Team. R: A language and environment for statistical computing. Vienna, Austria: R Foundation for Statistical Computing; 2013.
27. Christensen RHB. Ordinal-Regression Models for Ordinal Data R package version 2013.9-30. 2013.
28. Putz FE, Blate GM, Redford KH, Fimbel R, Robinson J. Tropical forest management and conservation of biodiversity: an overview. Conserv Biol. 2001;15:7–20.
29. Aoyagi R, Imai N, Kitayama K. Ecological significance of the patches dominated by pioneer trees for the regeneration of dipterocarps in a Bornean logged-over secondary forest. Forest Ecology and Management. For Ecol Manag. 2013;289:378–84.
30. Bischoff W, Newbery DM, Lingenfelder M, Schnaeckel R, Petol GH, Madani L, Ridsdale CE. Secondary succession and dipterocarp recruitment in Bornean rain forest after logging. For Ecol Manag. 2005;218:174–92.
31. Haworth J. Life after Logging: the impacts of commercial timber extraction in tropical rainforests. London: Friends of The Earth; 1999.
32. Class 7: Tabin Wildlife Reserve [http://www.sabah.gov.my/htan_caims/Level 2 frame pgs/Class 7 Frames/tabin_fr.htm].
33. New Forests Ltd. Malua Forest Reserve Conservation Management Plan. 2008.
34. Sabah Forestry Department. Maliau Basin Forest Reserve. 2005.
35. Sabah Forest Industries. Sabah Forest Industries Sdn. Bhd. (SFI). 2011.

17

Active parental care, reproductive performance, and a novel egg predator affecting reproductive investment in the Caribbean spiny lobster *Panulirus argus*

J. Antonio Baeza[1,2,3*], Lunden Simpson[1], Louis J. Ambrosio[1], Nathalia Mora[4], Rodrigo Guéron[5] and Michael J. Childress[1]

Abstract

Background: We used the Caribbean spiny lobster *Panulirus argus*, one of the largest brooding invertebrates in the Western Atlantic, to test for the presence/absence of active parental care and to explore reproductive performance in large brooding marine organisms. Given [i] the compact and large embryo masses produced by *P. argus*, [ii] the expected disproportional increase in brooding costs with increasing embryo mass size, and [iii] the theoretical allometry of egg production with increasing body size, we predicted that parental females in this large species will engage in active brood care. We also predicted that larger broods from larger lobsters should suffer higher mortality and brood loss than smaller broods from smaller lobsters if parental care was minimal or absent. Lastly, we expected smaller females to allocate disproportionably more resources to egg production than larger females in the case of minimal parental care.

Results: Females brooding early and late embryos were collected from different reefs in the Florida Keys Reef tract, transported to the laboratory, and maintained in separate aquaria to describe and quantify active parental care during day and night. A second set of females was retrieved from the field and their carapace length, fecundity, egg size, reproductive output and presence/absence of brood-dwelling pathogens was recorded. Laboratory experiments demonstrated that brooding females of *P. argus* engaged in active brood care. Females likely use some of the observed behaviors (e.g., abdominal flapping, pleopod beating) to provide oxygen to their brood mass. In *Panulirus argus*, females did not suffer brood loss during embryo development. Also, reproductive output increased more than proportionally with a unit increase in lobster body weight.

Conclusions: Our results agree with the view that large brooding marine invertebrates can produce large embryo masses if they engage in active parental care and that the latter behavior greatly diminishes reproductive performance costs associated with producing large embryo masses. Lastly, we report on a nemertean worm that, we show, negatively impacts female reproductive performance.

Background

The degree of parental care varies broadly among marine invertebrates, even within monophyletic clades [1, 2]. At one extreme, some groups do not provide any form of parental care, spawning small unfertilized or fertilized eggs into the pelagic environment in which development of embryos and/or larvae takes place (various bivalves and sea urchins [3, 4]). At another extreme, some species with abbreviated or direct development produce large yolky eggs that are brooded and hatch as advanced larval stages or juveniles. These early ontogenetic stages might remain in the parental brood chamber and/or dwelling for long periods of time and can be fed, defended, and groomed by females (various seastars, isopods, and amphipods, among others [5–7]). In between extremes, many species exhibit indirect development,

* Correspondence: jbaezam@clemson.edu
[1]Department of Biological Sciences, 132 Long Hall, Clemson University, Clemson, SC 29634, USA
[3]Departamento de Biología Marina, Facultad de Ciencias del Mar, Universidad Católica del Norte, Larrondo 1281, Coquimbo, Chile
Full list of author information is available at the end of the article

and parental care is restricted to the protection of embryos incubated in bodily chambers of varying complexity (e.g., many crustaceans, including spiny lobsters [8]). Explaining the evolution and adaptive value of parental care is one of the most relevant yet still not completely understood problems in evolutionary biology [1, 2].

In marine brooding invertebrates, benefits to brooded offspring include, but are not limited to, protection against predators [5, 9, 10], protection from adverse abiotic conditions [5, 9, 10], and/or physiological provisioning [11]. Still, brooding embryos in bodily chambers is not exempt of costs. For instance, reproductive performance (i.e., fecundity, reproductive output) of parental females might decrease due to brood loss, in turn, driven by increases in embryo volume during development, embryo crowding and loss from the abdominal chamber [12]. Embryo masses accumulate sediment, detritus, bacteria, algae, fungi, and many other epizootic organisms (e.g., ciliates) that might further impact parental reproductive performance [13–16]. Egg predators are known to destroy embryo masses when experiencing population outbreaks, impacting not only female fecundity but also the host population health [17–19]. Perhaps more importantly, large densely packed embryo masses can be considered living tissue but without a circulatory system [20]. Most embryo masses are larger than the theoretical 1 mm limiting thickness that allows sufficient oxygen supply by diffusion [21] and oxygen limitation does occur at their centers [20, 22, 23]. Oxygen depletion at the interior of embryo masses has been shown to be severe, even early during embryo development, when respiration rates of early embryos are much lower than those of late embryos [20]. Hypoxic conditions do impact embryo development, often driving asynchronous development within embryo masses (e.g., periphery versus center) [9, 24, 25] and even embryo mortality [26].

Likely, parental behaviors exclusively directed toward embryo masses, as reported for various invertebrates, have evolved as a mechanism to retard or prevent fouling, repel egg-predators, and improve oxygen availability to developing embryos [14, 20, 27]. Ultimately, active parental care is expected to, and has been shown to, improve reproductive performance. For instance, grooming of the brood mass by parental females increases embryo survival and hatching rates [14, 27]. Similarly, ventilation of brooded embryos by parental females (e.g., abdominal flapping or pleopod fanning [23, 25]) is known to increase oxygen levels at the center and periphery of embryo masses and appears to speed up embryo development [25]. Our understanding of what constitutes 'active parental care' has improved considerably during the last decade thanks to studies focused on a few marine invertebrates [25, 27–30]. Nonetheless, whether or not active parental care is the rule rather than the exception in brooding marine invertebrates still remains an open question.

In marine brooding invertebrates, brooding costs should be considerable in large females from large species, as they have the potential to produce massive embryo masses and also suffer the putatively heavy costs of brooding [31, 32]. Theoretical considerations suggest that the costs associated with brooding (i.e., oxygen provision, grooming) increase non-linearly with increases in brood mass, potentially resulting in the inability of large adults to successfully rear all brooded embryos [23–25]. Furthermore, with increases in body size, the capacity for egg production is also expected to scale at a pace greater than the space available for brooding, and thus, larger adults might be capable of producing more gametes than can be successfully brooded [33, 34] (Fig. 1a).

Overall, large species of brooding marine invertebrates should suffer exacerbated brooding costs, resulting in poor reproductive performance, unless these large parental individuals allocate considerable time and energy to attending their broods [23]. Indeed, brood mass size and parental behaviors are likely interlinked; large parental individuals with large broods either allocate a considerable amount of energy and time to brood their embryos, or suffer the costs in terms of reproductive performance, when producing large embryo masses. Various studies during the last decades have demonstrated that physiological costs (i.e., increased metabolic rate) are considerable in females engaging in active parental care [20, 23, 35]. Yet, whether brooding costs also result in diminished reproductive performance in large species of brooding marine invertebrates has been poorly explored. Studies on large species are particularly relevant as it will help to test whether or not excessive physiological costs associated to brooding large embryo masses limit the evolution of brooding in large marine invertebrates [34].

In this study, we are particularly interested in testing whether or not large marine invertebrates engage in active parental care, and whether brooding costs, measured in terms of reproductive performance, are suffered by these brooding species. For this purpose, we used the Caribbean spiny lobster Panulirus argus as a model organism, one of the largest crustaceans in the Atlantic, attaining up to 45 cm in total length [36, 37] (Fig. 1b). As reported for all members of the Pleocyemata, a species-rich clade of crustaceans to which spiny lobsters belong, P. argus produces a large number of embryos that are carried by females underneath the abdomen immediately after spawning and are maintained until they hatch as larvae [38, 39]. The problem of oxygen limitation seems particularly critical in this lobster, which produces compact, large, semi-spherical masses of embryos

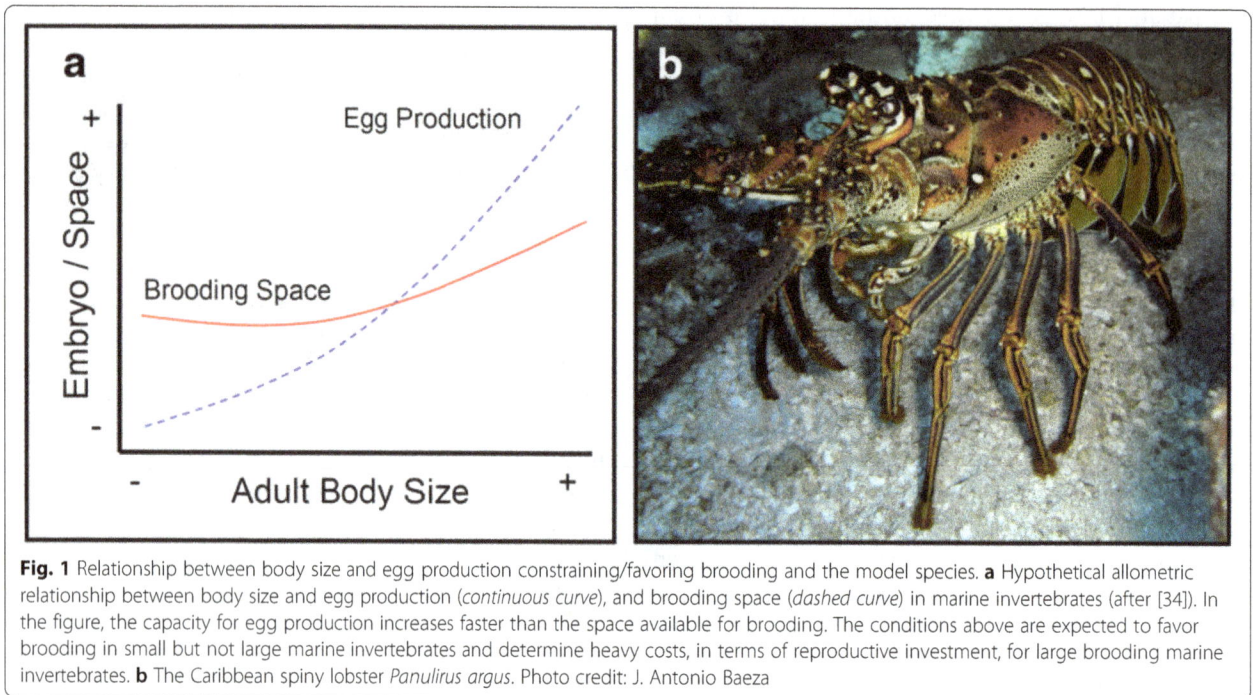

Fig. 1 Relationship between body size and egg production constraining/favoring brooding and the model species. **a** Hypothetical allometric relationship between body size and egg production (*continuous curve*), and brooding space (*dashed curve*) in marine invertebrates (after [34]). In the figure, the capacity for egg production increases faster than the space available for brooding. The conditions above are expected to favor brooding in small but not large marine invertebrates and determine heavy costs, in terms of reproductive investment, for large brooding marine invertebrates. **b** The Caribbean spiny lobster *Panulirus argus*. Photo credit: J. Antonio Baeza

that frequently exceed 8–10 cm in maximum diameter (JAB, pers. obs). In agreement with the notion above, a recent study has demonstrated asynchronous development of the embryo mass of *P. argus*: embryos at the periphery develop more quickly (24–48 h ahead) than those at the center [40]. We predict that ovigerous females of *P. argus* should engage in active brood care given the large embryo masses produced by this species. If the above is correct, then costs in terms of reproductive performance should be low or absent in this large marine invertebrate. By contrast, if *P. argus* does not engage in active parental care, then we expect considerable costs in terms of reproductive performance, including severe brood loss, decreased reproductive output, as well as negative allometric scaling of brood loss and reproductive output with body size. While studying reproductive performance in *P. argus*, we noticed the presence of an egg mass-dwelling nemertean belonging to the genus *Carcinonemertes*. We provide information about the prevalence and impact of this worm on the brood masses of infected lobsters.

Results

Active parental care in *Panulirus argus*

Six behaviors were observed and recorded in ovigerous and non-ovigerous (control) females of *P. argus*. Four of these behaviors were classified as events (i.e. abdominal flapping, pleopod fanning, 4th and 5th pereopod probing), while the remaining two behaviors were considered as states (i.e. standing, abdomen extension) (Fig. 2).

Only one (i.e., 4th pereopod probing, Fig. 2) out of the six recorded behaviors were found to be exclusive to ovigerous females in *P. argus*. However, the remaining of the recorded behaviors were observed at extremely low frequencies in non-ovigerous females. For instance, abdominal flapping and pleopod fanning were observed for a very short time period only three times in all non-ovigerous females.

Ovigerous females were observed to either exclusively or predominantly perform certain behaviors in a specific sequence. Females first raised their body from the bottom of the aquarium (stand up) by extending the pereopods and, immediately or after a few seconds of raising their bodies, females expanded the abdomen (abdomen extension). Next, females flapped the abdomen (abdominal flapping), fanned the pleopods back and forth (pleopod fanning) or performed both behaviors at the same time repetitively from a few seconds to up to an hour (maximum time period for analysis of behavior) (Additional file 1: Video 1). While standing up and with the abdomen extended, females were also observed to probe their embryos either with the 4th or 5th (chelae bearing) pereopods. The latter two behaviors were performed while females were fanning their embryos or during short time periods in which females stopped flapping and/or fanning the abdomen and/or pleopods. The sequence of the behaviors above lasted from a few seconds to an hour (maximum time period used for measuring female behaviors). After performing the behavioral sequence above for a few seconds, minutes or hours, females either walked around or crouched again at the bottom of the aquarium (Fig. 2).

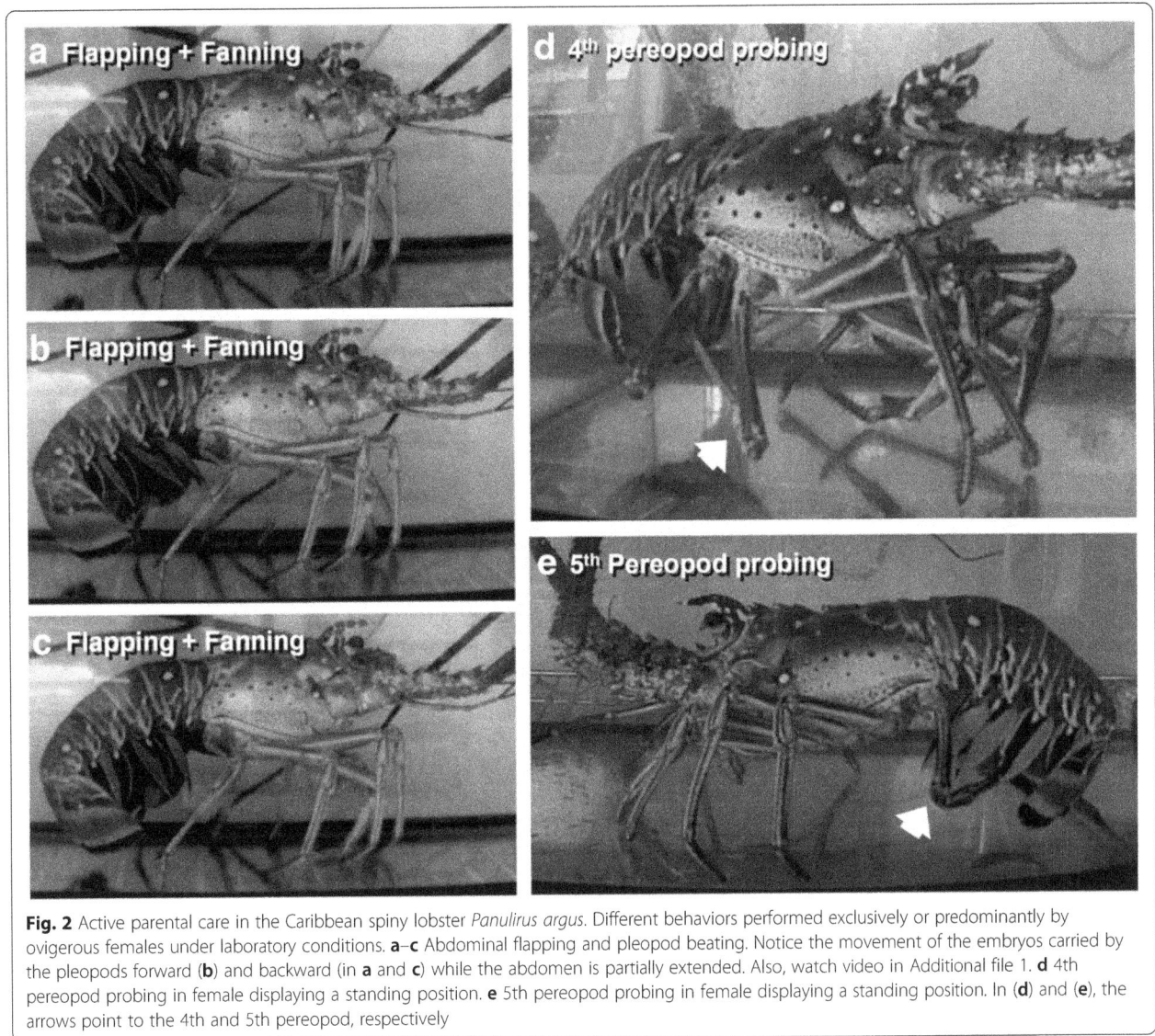

Fig. 2 Active parental care in the Caribbean spiny lobster *Panulirus argus*. Different behaviors performed exclusively or predominantly by ovigerous females under laboratory conditions. **a–c** Abdominal flapping and pleopod beating. Notice the movement of the embryos carried by the pleopods forward (**b**) and backward (in **a** and **c**) while the abdomen is partially extended. Also, watch video in Additional file 1. **d** 4th pereopod probing in female displaying a standing position. **e** 5th pereopod probing in female displaying a standing position. In (**d**) and (**e**), the arrows point to the 4th and 5th pereopod, respectively

The frequency or duration of the behaviors above did change depending on embryo developmental stage and/or diel cycle. The frequency of grooming bouts by the 4th pereopods, pleopod fanning / abdominal flapping, and the total time spent in a standing position were greater in females carrying late than early stage embryos (mixed nested ANOVAs: $P < 0.05$ in all cases; Table 1 and Fig. 3). Also, grooming bouts by the 4th pereopod, and pleopod fanning / abdominal flapping bouts were more frequently performed during night than during day hours (mixed nested ANOVA: $P < 0.05$ in all cases; Table 1 and Fig. 3). The remainder of the studied behaviors did not vary statistically between females carrying early or late embryos during day or night.

Reproductive performance in *Panulirus argus*

A total of 19, 18, 20, and 11 female lobsters carrying embryos in stages I, II, III, and IV, respectively were sampled. The mean ($\pm$ standard deviation, SD) CL was 76.06 ± 8.95 mm and ranged from 51.3 to 100.4 mm. No significant differences in body size (CL, carapace length) were detected between females carrying embryos in different developmental stages (ANOVA: F = 2.41, $df = 3,64$, $P = 0.0750$). A total of five females were found to host nemertean parasites in their brood masses. These parasitized females were not considered for the initial analysis of reproductive investment (but, see below).

Average ($\pm$ SD, range) fecundity varied between 299 328 ($\pm$162 537, 105 858–757 278) embryos lobster^{-1} and 347 001 ($\pm$131 652, 205 569–674 041) embryos lobster^{-1} in females carrying early and late stage embryos, respectively. An ANCOVA did not detect any effect of embryo stage (I versus IV) on fecundity ($F = 0.8$, d.f. = 1, 29, $P = 0.3784$). On the other hand, female body size (CL) did affect fecundity; large females carried more embryos

Table 1 Mixed model nested ANOVAs to test the effect of embryo developmental stage (early *versus* late or control *versus* early *versus* late) and diel cycle (day *versus* night) in the active brood care of the Caribbean spiny lobster *Panulirus argus*

Source	$N_{parameters}$	d.f.	d.f._denominator	F	P	Variance (%)
4th pereopod grooming bouts (early *versus* late)						
Embryo Stage	1	1	4	0.61	0.4791	
Day Time (Stage, Female ID)	6	6	12	7.19	**0.0020**	
Female ID						63.32
5th pereopod grooming bouts (control *versus* early *versus* late)						
Embryo Stage	2	2	6	5.19	**0.0491**	
Day Time (Stage, Female ID)	9	9	18	2.74	**0.0331**	
Female ID						53.73
Pleopod fanning + Abdomen flapping (control *versus* early *versus* late)						
Embryo Stage	2	2	6	15.13	**0.0045**	
Day Time (Stage, Female ID)	9	9	18	3.38	**0.0134**	
Female ID						67.83
Abdomen extension (control *versus* early *versus* late)						
Embryo Stage	2	2	6	4.08	0.0759	
Day Time (Stage, Female ID)	9	9	18	1.77	0.1442	
Female ID						<1.00
Standing (early *versus* late)						
Embryo Stage	1	1	4	2.85	0.1665	
Day Time (Stage, Female ID)	6	6	12	0.26	0.9470	
Female ID						<1.00
Abdominal extension (control *versus* early *versus* late)						
Embryo Stage	2	2	6	5.4443	**0.0448**	
Day Time (Stage, Female ID)	9	9	18	0.4	0.919	
Female ID						7.24

In the different analyses, developmental stage and diel cycle were considered fixed effects while female identity was considered a random effect. The proportion of the variance in the dataset explained by female identity is shown for each analysis. Numbers in bold indicate significant p-values

than small females ($F = 58.04$, d.f. = 1, 29, $P < 0.0001$). The interaction term of the ANCOVA was not significant ($F = 0.1972$, d.f. = 1, 29, $P = 0.6603$) (Fig. 4a). Thus, females, either large or small, did not loose embryos significantly throughout embryo development.

Embryo volume varied between 0.0834 (±0.0204, 0.122 - 0.054) mm^3 and 0.1008 (±0.0142, 0.1200 - 0.083) mm^3 in females carrying early and late stage embryos, respectively. An ANCOVA demonstrated an effect of embryo stage on egg volume ($F = 6.55$, d.f. = 1, 29, $P = 0.0160$). In turn, CL did not affect embryo volume ($F = 0.20$, d.f. = 1, 29, $P = 0.6570$). The interaction term of the ANCOVA was not significant ($F = 0.349$, d.f. = 1, 29, $P = 0.5589$; Fig. 4b). In general, embryo volume in *P. argus* increased by 17.2 % from early to late stage and females from all body sizes carried similarly sized embryos.

Reproductive output varied between 31.30 and 61.46 % and represented a mean ± SD of 49.21 % (±8.17) of lobster body dry weight. Reproductive output increased

with lobster body weight, i.e., RO exhibited positive allometry, as the slope ($b = 1.46$, SE$_b = 0.14$) of the line describing the relationship between these two variables (after log-log transformation) was significantly greater than unity (*t*-test: $t = 3.29$, $df = 1$, 14, $P = 0.0026$; Fig. 4c).

Carcinonemertes sp. in brood masses of *Panulirus argus*

We noticed the presence of parasitic nemertean worms from the genus *Carcinonemertes* in five (7.4 %) out of the 68 sampled ovigerous females and we estimated density of *Carcinonemertes* and embryo mortality in four of these infested females (one each carrying embryos in stages I, II, III, and IV, respectively). Density of *Carcinonemertes* varied between one worm per 1262 embryos (or 0.08 worms per 100 embryos) in the female carrying Stage I embryos (78.45 mm CL) and one worm per 193 embryos (or 0.51 worms per 100 embryos) in the female carrying Stage IV eggs (73.78 mm CL). The mean ± (SD) density of *Carcinonemertes* in these parasitized females

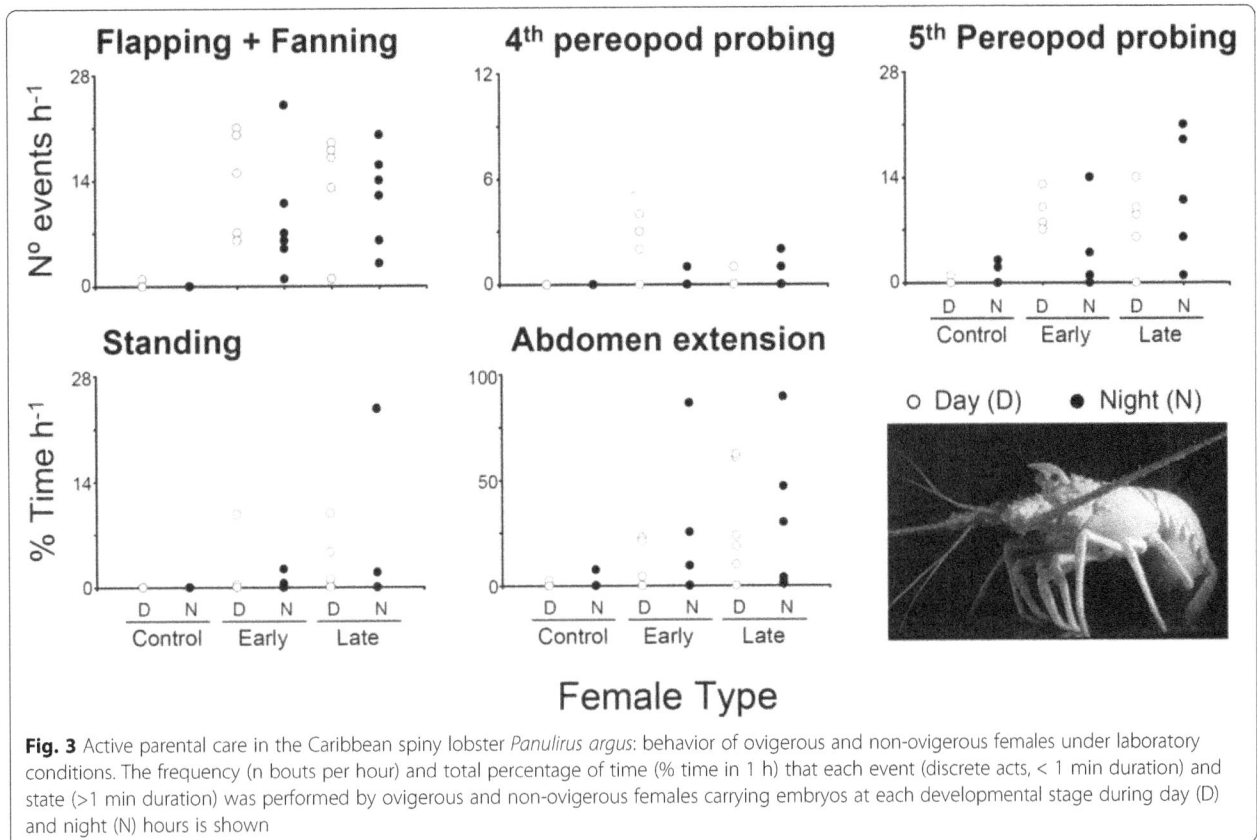

Fig. 3 Active parental care in the Caribbean spiny lobster *Panulirus argus*: behavior of ovigerous and non-ovigerous females under laboratory conditions. The frequency (n bouts per hour) and total percentage of time (% time in 1 h) that each event (discrete acts, < 1 min duration) and state (>1 min duration) was performed by ovigerous and non-ovigerous females carrying embryos at each developmental stage during day (D) and night (N) hours is shown

was one worm per 667 ± (460) embryos or 0.24 ± (0.19) worms per 100 embryos (Fig. 5).

The mean proportion of empty capsules in the infected females was 12.7 ± (10.8) % and varied between 0.8 % in the female carrying Stage I embryos (78.45 mm CL) and 25 % in the female carrying Stage III embryos (76.1 mm CL). In turn, the number of dead embryos varied between 0 % in the female carrying Stage I embryos (78.45 mm CL) and 23 % in the female carrying Stage III embryos (76.1 mm CL). The mean number of dead embryos in these infected females was 7.2 ± 10.7 %. No empty capsules or dead embryos were noticed in non-infected females (Fig. 5).

We compared fecundity, egg size and reproductive output among infected and non-infected females. Visual examination of the data indicated that fecundity in one of the infected females carrying stage I embryos and the infected female carrying stage II embryos was not dissimilar from that found in non-infected females of similar body size (arrows in Fig. 4a). By contrast, fecundity in the second infected female carrying stage I embryos and in infected females carrying stage III, and IV embryos was much smaller than that expected for non-infected females of similar body size (see encircled data points in Fig. 4a). Indeed, these last three infected females

represented data outliers whose fecundity was statistically lower than that expected for a non-infected female of equal body size and carrying embryos in the same stage (Fig. 4a). *Carcinonemertes* sp. did not affect embryo volume of infected females, regardless of embryo developmental stage. Lastly, reproductive output in one of the infected females carrying stage I embryos was much lower than that expected for non-infected females of the same body size (lower arrow in Fig. 4c).

Discussion

Active parental care in *Panulirus argus*

We expected the Caribbean spiny lobster *Panulirus argus* to engage in active brood care, considering the putatively high costs, in terms of reproductive performance, that large brooding marine invertebrates should face [20, 23–25]. Our results agree with the notion above; laboratory observations have clearly shown that ovigerous females do direct precise behaviors toward their embryo masses. These behaviors were performed either exclusively (i.e., 4th pereopod probing, standing) or predominantly by ovigerous females (i.e., abdominal flapping, pleopod fanning, 5th pereopod probing). Therefore, abdominal flapping, pereopod fanning, standing, and pereopod probing, among a few others, can be

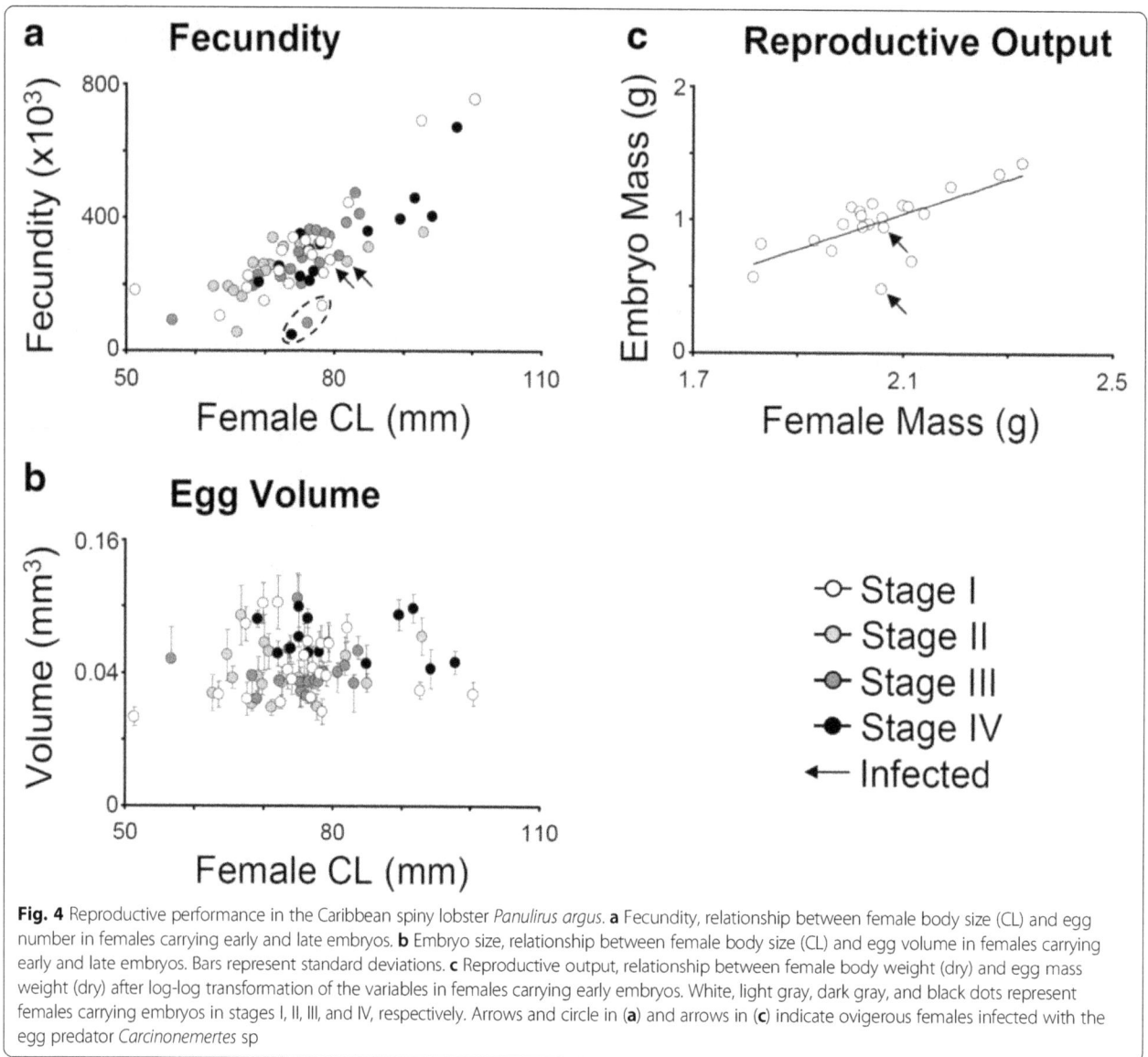

Fig. 4 Reproductive performance in the Caribbean spiny lobster *Panulirus argus*. **a** Fecundity, relationship between female body size (CL) and egg number in females carrying early and late embryos. **b** Embryo size, relationship between female body size (CL) and egg volume in females carrying early and late embryos. Bars represent standard deviations. **c** Reproductive output, relationship between female body weight (dry) and egg mass weight (dry) after log-log transformation of the variables in females carrying early embryos. White, light gray, dark gray, and black dots represent females carrying embryos in stages I, II, III, and IV, respectively. Arrows and circle in (**a**) and arrows in (**c**) indicate ovigerous females infected with the egg predator *Carcinonemertes* sp

considered 'active brood care' behaviors. We believe that some of the observed behaviors are used by parental females to provide oxygen to their brood mass while others are related to the detection of oxygen conditions and/or foreign objects (e.g., detritus, egg predators) that have the potential to impact reproductive performance [41, 42].

Ovigerous females of *P. argus* ventilate their embryo masses (i.e., likely increasing oxygen ability at its center) using abdominal flapping and fanning of the pleopods. Abdominal flapping has been demonstrated to be the specific behavior used by ovigerous females to increase oxygen concentration at the center of the brood mass in brachyuran crabs [20, 23, 25]. Abdominal flapping and a second behavior, pleopod fanning, have also been observed in various other groups of crustaceans that engage in active brood care and their role in oxygen delivery to embryos was proposed earlier in brachyuran

crabs [15, 16, 22, 43] and clawed lobsters [26], and most recently in caridean shrimps [44]. In turn, the standing position, another behavior frequently observed in ovigerous females of *P. argus*, probably improves the efficiency of oxygen delivery to embryos in a manner similar to that previously reported for crabs [20]. Experiments measuring particular behaviors concomitantly with changes in oxygen availability at the center of the embryo mass are required to demonstrate that abdominal flapping and pleopod fanning are indeed delivering oxygenated water to the center of the embryo mass in *P. argus*.

In contrast to abdominal flapping and pleopod fanning, 4th pereopod probing might be used in sensing the environment at the periphery or within the embryo mass of ovigerous lobsters. In decapod crustaceans, including lobsters, the dactyls of pereopods do bear setae that serve as mechanoreceptors or contact chemoreceptors

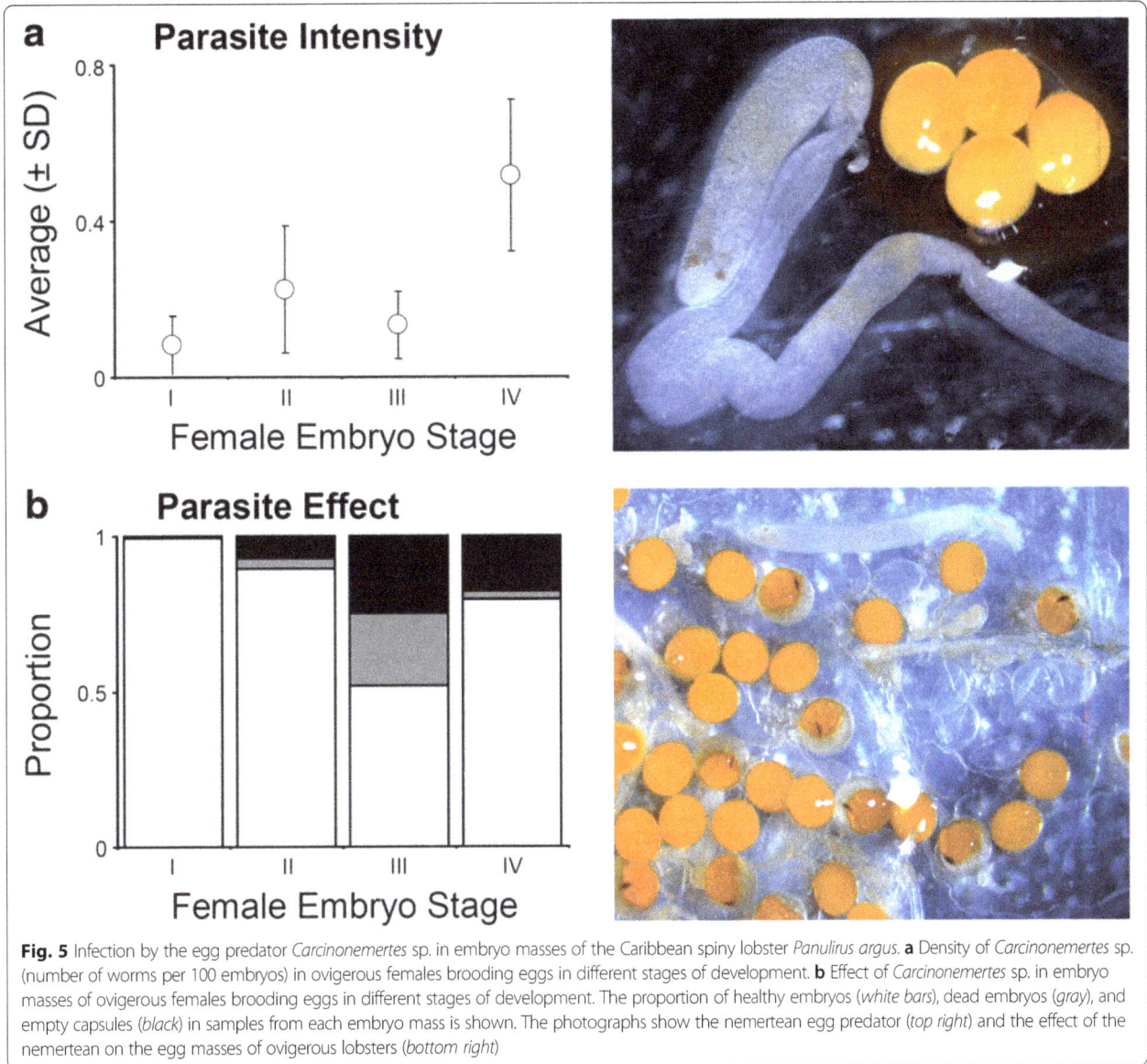

Fig. 5 Infection by the egg predator *Carcinonemertes* sp. in embryo masses of the Caribbean spiny lobster *Panulirus argus*. **a** Density of *Carcinonemertes* sp. (number of worms per 100 embryos) in ovigerous females brooding eggs in different stages of development. **b** Effect of *Carcinonemertes* sp. in embryo masses of ovigerous females brooding eggs in different stages of development. The proportion of healthy embryos (*white bars*), dead embryos (*gray*), and empty capsules (*black*) in samples from each embryo mass is shown. The photographs show the nemertean egg predator (*top right*) and the effect of the nemertean on the egg masses of ovigerous lobsters (*bottom right*)

[45–51]. Information gathering from the brood mass via pheromones has been previously reported for brooding female crabs and spiny lobsters hatching their embryos [41, 42, 52–56] and most recently in caridean shrimps [44]. Ovigerous females of *P. argus* might use their 4th pereopods to detect compounds produced by embryos and/or foreign objects (e.g., detritus, parasites) and respond accordingly, either by providing oxygen or removing particles, respectively. The 5th pereopods, each bearing a small but well developed claw [36], might play a function similar to that of the 4th pereopods, helping both with (chemical) information gathering and removal of exogenous particles that might potentially impact reproductive performance, as shown before in other crustaceans with minute but ornamented claws on their last pair of pereopods (in the anomuran crab *Petrolisthes*

violaceus [27]). Although 4th and 5th pereopod probing was frequently performed by ovigerous females carrying both early and late stage embryos, its putative function might not be entirely efficient as suggested by the presence of predatory worms in the brood mass of a few ovigerous females (see below).

The Caribbean spiny lobster *P. argus* exhibits active parental care and the intensity with which ovigerous females performed different behaviors differed with embryo developmental stage and time of the day. In particular, abdominal flapping and pereopod fanning were more often performed by females carrying late compared to early embryos, in agreement with preliminary observations (in [41, 42]). Since the frequency of abdominal flapping increases throughout the brooding period, an increase in the amount of oxygen provided to the developing embryos is likely to occur [20]. Oxygen consumption in embryos has been shown to increase dramatically during development in every species of crustacean in which this physiological function has been studied [20, 22, 23, 25, 29, 57–59]. In general, the increased brood care by brooding females throughout embryo development probably serves to avoid problems related to oxygen limitation for the embryos (e.g. increased mortality, low development rate, small size at hatching [9, 24]). Studies on oxygen consumption of embryos through development in *P. argus* are needed to corroborate the assertion above.

Panulirus argus females are also 'full time mothers' as they engage in parental care both during day and night [60]. Nonetheless, abdominal flapping and pereopod fanning, two behaviors that likely deliver oxygen to the embryos, together with embryo grooming using the 4th pereopods, a third behavior likely involved in information gathering from the brood mass, were more often performed during the day than night. Previous studies examining daily locomotor activity rhythm in *P. argus* have shown that females are more active at dusk and night than during the day [61, 62]. The information above suggests that ovigerous females of *P. argus* allocate time and energy differentially during the circadian cycle to different functions likely to optimize resource acquisition and use.

Lastly, previous studies in brachyuran crabs have demonstrated that active parental care does result in metabolic costs to ovigerous females [20, 58]. Herein, we have focused on costs associated with reproductive performance (see below). Nonetheless, studying metabolic costs associated with active brood care in *P. argus* remains to be addressed. These brooding costs (and others, e.g., lost feeding opportunities) are expected to be considerable in a large brooding marine invertebrate like *P. argus* as well as in other spiny lobsters [20].

Reproductive performance in *Panulirus argus*
We expected *P. argus* to suffer high brooding costs (e.g., elevated brood loss during development), if ovigerous females did not exhibit parental care. By contrast, if parental females engaged in active parental care, then we expected ovigerous females to suffer low brooding costs. Our results agree with the rationale above; we have shown that parental care is considerable in this species and concomitantly, costs measured in terms of reproductive performance were low or absent.

In the Caribbean spiny lobster, fecundity increased with body size as reported for every species of spiny lobster in which this reproductive parameter has been studied [63–71]. Also, average fecundity reported herein for *P. argus* is within the range expected for a spiny lobster of its body size range. For instance, fecundity varied

between 109,865 and 590,530 in *P. marginatus* from Hawaii, a species in which ovigerous females vary between 54.3 and 105.4 mm CL [72]. Perhaps more importantly, differences in fecundity between females carrying early and late embryos were negligible (i.e., non-statistically significant), and thus, brood loss (if any) did not increase disproportionately with increases in brood size. Our results agree with those reported for the Indo-Pacific ornate rock lobster *P. ornatus* and the Western Australian rock lobster *P. cygnus*, two species in which no significant brood loss was observed during incubation [73, 74]. In the few other species of spiny lobsters in which brood loss has been estimated, this has been rather low (7.9 % in *P. gracilis* [75], 14–17 % in *P. gilchristi* [76], 10–16 % in *P. delagoae* [77], 7.2 % in *P. inflatus* [78]). *Palinurus elephas* from the western Mediterranean can be considered an exception as brood loss reaches 26 % [70]. Brood loss appears to increase with incubation time in lobsters [79]. Interestingly, brood loss can be considerable in crustaceans that attain much smaller body sizes than spiny lobsters. For instance, in fiddler crabs from the genus *Uca* that grow no larger than 20 mm in carapace width, brood loss varies and can represent up to 43 % of initial fecundity [80]. In turn, in caridean shrimps, in which ovigerous females attain body sizes no larger than 25 cm total length, brood loss varies between 12 % and can reach up to 74 % (Table 6 in [81]). The information above, albeit limited, suggests an inverse relationship between brood loss and body size within one of many clades of brooding marine invertebrates. This relationship deserves further attention. As yet, that brood loss was negligible in *P. argus* does agree with the notion that large brooding species that allocate considerable attention to their brood masses experience reduced costs in terms of reproductive performance.

In *P. argus*, embryo volume did increase, but not considerably, during development. The difference between late and early embryos was, on average, only ~20 %, slightly larger than that reported for other spiny lobsters (egg size increase: 8.3 % in *P. delagoae* [77]). The modest increase in volume throughout development in embryos of *P. argus* might alleviate or ease, to some extent, the costs associated with brooding. Studies on oxygen consumption of eggs might help reveal whether or not small changes in embryo size throughout development do represent an adaptation to alleviate brooding costs in large marine invertebrates that produce large embryo masses such as *P. argus*.

Reproductive output (RO) in *P. argus* was high (49.21 % ± 8.17 of body dry weight) compared to that of other crustaceans. For instance, in most of brachyuran crabs, RO is limited to about 10 % (range: 3–22 %) of body weight ([82–84], but see [85] for exceptions). In caridean shrimp, with a body shape similar to that of lobsters, RO is slightly greater, varying between ~12 and 21 % of body weight [86, 87]. Unfortunately, to the best of our knowledge, no previous studies have estimated reproductive output in other species of spiny lobsters. Importantly, the scaling of reproductive output with female body weight in *P. argus* was positively allometric; egg mass did increase more than proportionally with a unit of increase in female body weight. By contrast, brood weight most often exhibits an isometric or nearly isometric constraint with increasing female body size in most brachyuran crabs [85]. A limitation on space available for yolk accumulation in the cephalothorax appears to be the main factor constraining RO and its scaling with body size in other brooding crustaceans [85]. Unlike crabs, but resembling caridean shrimps, the abdomen of lobsters is massive and their ovaries extend out of the cephalothorax into the first abdominal segments ([88], and JAB, pers. obs.). An elongated abdomen most likely allows distension of the organs (including gonads), and subsequent increases in reproductive output, likely explaining the positive allometric increases in RO with body size in *P. argus*. We argue in favor of additional studies examining constraints driven by body shape in determining RO in marine invertebrates. As yet, the positive allometric scaling of reproductive output with body size in *P. argus* supports the notion that RO and other reproductive parameters related to reproductive performance do not decrease in large brooding invertebrates that do actively provide intensive brood care.

Carcinonemertes sp. in brood masses of *Panulirus argus*

Our study reports for the first time an embryo predator from the brood mass of the Caribbean spiny lobster *P. argus*. Preliminary observations on preserved specimens have confirmed that the observed worms represent a previously undescribed species from the genus *Carcinonemertes* and its description is underway. The genus *Carcinomertes* can and does have a major impact on the reproductive performance of crustaceans, and has been implicated in the decline of king crab fisheries [18, 19, 89–91]. A few species of *Carcinonemertes* have been reported before in spiny lobsters from the Pacific (see Table 1 in [92]). However, the impact of this worm on the reproductive performance of spiny lobsters is unknown [19]. This study demonstrates for the first time the negative impact of *Carcinonemertes* on the reproductive performance of a spiny lobster. One infected lobster carrying early embryos (Stages I) experienced a significant reduction in fecundity and reproductive output. In turn, lobsters with late stage (III and IV) embryos suffered considerable reductions in fecundity and reproductive output. The above pattern suggests that the effect of the predator on the embryo mass progresses quickly during a single brooding event in *P. argus*.

Prevalence of *Carcinonemertes*. sp on *P. argus* was relatively low (7.4 %) when compared to reports in other infected populations (*Callinectes arcuatus* – 35 % [93], *Libinia spinosa* – 69.2 % [94], *Randallia ornata* – 70 % [95], *Panulirus interruptus* – 42 % [96]). Importantly, in some crustacean host species, *Carcinonemertes* might experience population outbreaks, destroying embryo masses of their crustacean hosts, and, sometimes, affecting host population health, and even resulting in decreased crab landings [18, 19, 91]. In the Caribbean, landings of *P. argus* have declined during the last decades, and various conditions are thought to play a role in such population declines, including hurricanes, overfishing, pollution, poor water quality, hypoxia, temperature extremes, viral disease, or a combination of the conditions above [19, 97]. We argue in favor of future studies focusing on the population biology of *Carcinonemertes* sp., given the potential of this embryo predator to cause extensive damage to natural populations of the commercially and ecologically important spiny lobster *P. argus*.

Conclusions

Parental care behaviors have rarely been reported for marine invertebrates (for exceptions see [6, 20, 43, 59], including crustaceans, a species-rich clade with great anatomical, physiological, reproductive, and ecological disparity. We have shown that the Caribbean spiny lobster *Panulirus argus*, a large marine invertebrate that broods bulky embryo masses, engages in active parental care. Costs, measured in terms of reproductive performance, were absent in this species, reinforcing the idea that brooding can be achieved in marine invertebrates with relatively large body sizes, if parental individuals play a major role in attending to brooded embryos. Active parental care should result in considerable metabolic costs to parental individuals [20]. Future studies in *P. argus* examining these putative metabolic costs are warranted as they will help to illustrate the relationship between mode of development (i.e., brooding *versus* broadcasting) and body size, recognized for marine invertebrates [9, 25, 34, 98].

Lastly, our study reports for the first time a nemertean egg predator belonging to the genus *Carcinonemertes* from the embryo mass of the Caribbean spiny lobster *P. argus*. We argue in favor of studies aimed at revealing the relative importance of this and other pathogens in explaining declines in the landing of *P. argus* throughout the wider Caribbean during recent decades.

Methods
Model organism: the Caribbean spiny lobster *Panulirus argus*

Panulirus argus supports a multimillion-dollar fishery in the Greater Caribbean and Gulf of Mexico, with landings from 1984 to 1998 surpassing 330,187 metric tons [99].

The above explains in great part the large body of literature on its anatomy, life history, ecology, behavior and physiology accumulated during the last decades [36]. Still, despite the commercial value and ecological importance of this large brooding invertebrate, no studies on active brood care have been conducted (for exceptions see [41, 42]) and limited information exists on individual-level reproductive performance [38, 100].

Mating in *Panulirus argus* may occur year round but peaks in the early spring and summer with an increasing photoperiod and warmer temperatures [101, 102]. Generally, males occupy and defend dens in shallow water and attract females who are ready to spawn [39, 103]. Males transfer a spermatophore to the ventral surface of the female [104]. Larger males have the potential to produce larger spermatophores, but males can scale their spermatophore to the size of the female [105, 106]. Female mating receptivity appears to decrease upon receipt of the spermatophore which is used to fertilize a single clutch of eggs [107]. Clutch size increases with female size [39] and the developing embryos are incubated for approximately 3–4 weeks [108], during which time females may migrate to deep reefs before the embryos hatch [109, 110]. Female can produce 2–4 clutches of eggs per year with larger, older females reproducing earlier and having more clutches per year [111]. Fishing pressure has a strong influence on reproduction, which slows growth [112] and decreases the size (but not the age) at first reproduction [111]. Protection from fishing not only increases the average size of females [113] but may also increase the proportion of reproductively mature females in each size class [114]. Previous studies of commercially exploited marine fishes have suggested that larger females may produce higher quality larvae [115] but in *P. argus* female size was found to be unrelated to egg size, egg C:N ratio, larvae size, or larvae mortality [106].

Sampling of *Panulirus argus*

Ovigerous females of *P. argus* were collected between June 22nd and July 25th, 2015 from various coral reefs (6–20 m depth) along the Florida Reef Tract, 2 to 8 km off Long Key (N 24° 49'26", W 80°48'48"), Florida Keys, USA. At each locality, lobsters were gently captured by hand (with the aid of a tickle stick and hand net) while scuba diving and transported alive to the laboratory. Lobsters were euthanized in coolers full of ice and maintained therein until measurements on reproductive investment were taken.

In the laboratory, the carapace length (CL) of each lobster was measured using a dial caliper (precision = 0.1 mm). Next, the embryos carried by brooding females were gently removed with micro-forceps and classified according to four different categories. Stage I embryos

displayed uniformly distributed yolk; stage II embryos showed cell differentiation and yolk clusters; stage III embryos had well-developed eyes and some visible chromatophores but no obvious development of abdomen and thoracic appendages; stage IV embryos possessed well-developed eyes and chromatophores, free abdomens, and thoracic appendages.

Lastly, ovigerous lobsters used to test whether or not *P. argus* engages in active parental care (see below) were collected during the same time period and from the same coral reefs above and maintained in 3.8 L transparent glass aquaria ($51 \times 51 \times 30$ cm^3) filled with aerated sea water (25–31 °C and 35–38 ppt) in an outdoor wet lab with natural photoperiod (day : night = ~ 13: 11 h) before being used for experiments.

Active parental care in *Panulirus argus*

To test whether ovigerous females of *P. argus* engaged in active parental care (i.e. displaying behaviors likely resulting in the provision of oxygen to and maintenance of brooded embryos) female behavior was recorded under laboratory conditions. Females with embryos at different stages of development (I and IV, $n = 3$ per stage) were placed individually in a 3.8 L transparent glass aquarium ($51 \times 51 \times 30$ cm^3) filled with aerated sea water (25–31 °C and 35–38 ppt) in an outdoor wet lab with natural photoperiod (day : night = ~ 13: 11 h). The behavior of each female was videotaped continuously over a 24-h period using a Brinno High Dynamic Range (HDR) Time Lapse Camera - TLC200 Pro, starting 1 h after the lobster was introduced to the tank. Infrared illuminators (Model IRLamp6, 40° beam angle, Wildlife Engineering) were used to continuously record female behavior throughout the night. Behavior of ovigerous females during the experimental period was classified and quantified as events or states, according to their relative duration [20]. Behaviors classified as states occurred over relatively long time periods (more than 1 min, e.g. standing) while discrete acts of relatively short duration (less than 1 min, e.g. pleopod beating) were classified as events most often occurring in bouts [116]. The frequency of the occurrence of each behavioral event (n° event bouts h^{-1}) and the proportion of time that females spent in each behavioral state (% time h^{-1}) were recorded during two haphazardly selected time blocks of 1 h each, for females carrying embryos at the two developmental stages. Time blocks were randomly selected when females were facing laterally, diagonally, or frontally to the camera, and their behavior could be recorded. Also, to identify brooding (i.e., active parental care) from non-brooding behaviors, non-ovigerous females ($n = 3$) were recorded and their behavior was measured as described above.

To test whether behaviors related to brood care varied during embryonic development, the frequency or percentage of time that each behavior was performed per unit of time (1 h) was compared between day and night (daytime effect) among brooding females carrying embryos at different developmental stages (embryo stage effect: early *vs* late or non-brooding *vs* early eggs *vs* late eggs depending upon the dependent variable, see Table 1) using independent mixed nested ANOVAs [117]. In the different ANOVAs, daytime and embryo stage were treated as fixed effects while female identity was treated as a random effect [117]. The behavior of non-brooding females was included in these ANOVAs only when these females performed the events or states analyzed more than once. Whenever, a specific event or state was observed only once or not observed in each one of the three non-brooding females, the behavior was not considered part of the behavioral repertoire of these females, and was not included in the mixed nested ANOVAs. Data were square root or log-transformed in order to meet the assumptions of the statistical model [117].

Reproductive performance in *Panulirus argus*

We estimated three different individual-level reproductive performance parameters in brooding females of *P. argus*: fecundity, embryo size, and reproductive output. For this purpose, we gently detached with forceps the totality of the embryos carried by females underneath the abdomen. Then, ten embryos from each mass were subsampled randomly, and the lengths along the short and long axis of each embryo were measured under the microscope (Leica S8AP0) to a precision of 0.001 mm. Embryo volume was estimated with the formula for the volume of an ellipsoid [118], EV = 1 / 6(L S2 π), where L = long axis and S = short axis. The effect of female body size (CL, covariate) and egg stage (Stage I vs IV, main factor) on egg volume was tested using an ANCOVA [117].

Next, four sub-samples of 100 embryos each were isolated from the brood mass of each female and dried with the respective remaining embryo mass and female body for at least 72 h at 70 °C and weighed to the nearest 0.01 mg with an analytical balance (Sartorius; ± 0.1 mg). Fecundity was calculated with the equation F = EmMass / (Ess1 + Ess2 + Ess3 + Ess4) * 400 + 400, where F = total number of embryos, and EmMass = dry weight of the remaining embryo mass after the four egg sub-samples (Ess1-4) have been taken. The effect of female body size (CL) and egg stage (Stage I vs IV) on fecundity was tested using an ANCOVA [117].

Lastly, reproductive output was estimated as the ratio between dry weight of embryos and dry weight of the females carrying early embryos (Stage I). This latter parameter represents the amount of resources (biomass) that females invest in reproduction [119]. We tested whether reproductive output increased linearly (isometrically) with female body size. The relationship between egg dry mass and female body dry mass was examined using the allometric model y = a*x^b [120]. The slope b of the log-log least-squares linear regression represents the rate of power increase ($b > 1$) or decrease ($b < 1$) of the estimate of reproductive allocation with a unit of increase in lobster dry mass. To determine whether the relationship deviated from linearity, a *t*-test was used to test if the estimated slope b deviated from the expected slope of unity. Before conducting the test above, assumptions of normality and homogeneity of variances were checked and found to be satisfactory [121]. Previous studied in *P. argus* has calculated fecundity using the wet weight of the embryo and female mass [39]. No previous studies have estimated egg size and reproductive output in this species as well as fecundity using the dry weight of the embryo and females mass, known to provide more accurate estimates [39, 108].

Carcinonemertes sp. in brood masses of *Panulirus argus*

We noticed the presence of parasitic nemertean worms from the genus *Carcinonemertes* in five (7.4 %) out of the 68 studied ovigerous females (two lobsters carrying stage I embryos and one each carrying stages II, III, and IV embryos). We examined four out of these five infected lobsters to describe the density of *Carcinonemertes* as well as the impact that the parasite had on brood mortality and individual-level reproductive parameters (see above). First, we estimated embryo mortality by haphazardly collecting five sub-samples, each comprised 100 eggs from the embryo mass of each female and counting the number of live embryos, dead embryos, and empty capsules (i.e., embryos putatively consumed by the parasite [19]). These embryo samples were then returned to the egg-mass of each female to be included in future fecundity measurements. Embryo mortality was calculated as the average proportion (%) of empty capsules and dead eggs present in our samples. Next, the density of *Carcinonemertes* sp. was estimated by haphazardly removing five sub-samples of 500 to 900 embryos each from the embryo mass of each infected female and counting the total number of worms in all sub-samples. These embryos were also returned to the egg-mass of each infected female. Density of *Carcinonemertes* was calculated as the number of worms found per 100 embryos. All examinations and counts were done under a Leica S8AP0 Stereoscope and a Wild M5-97874 dissecting scope.

Abbreviations
ANCOVA, analysis of covariance; ANOVA, analysis of variance; CL, carapace length; RO, reproductive output; SD, standard deviation

Acknowledgements
We sincerely thank Kylie Smith and Cindy Lewis for logistic and material assistance that helped make possible this study. We also appreciate the research opportunity provided by the Clemson University Creative Inquiry program to Mr. Dallas Owen who gained research experience at the Baeza-LAB during the summer, 2015 field season. This is Smithsonian Marine Station at Fort Pierce contribution number 1040.

Funding
Funding provided by Clemson University to JAB.

Authors' contribution
Conceived and designed the experiments: JAB LJA. Performed the experiments: JAB LJA LS NM RG. Analyzed the data: JAB LJA. Contributed reagents/materials/analysis tools: JAB. Wrote the paper: JAB LJA MC. All authors read and approved the final manuscript.

Authors' information
JAB is an ecologist specialized in evolutionary, behavioral and molecular ecology. JAB's research goals focus on testing sex allocation and mating systems theories using marine invertebrates (mostly crustaceans) as model systems. He is also interested in understanding the role of the environment in the evolution of social behaviors and other behavioral traits (i.e., alternative mating tactics, territoriality, and symbiosis). Currently, JAB is an assistant professor at the Department of Biological Sciences, Clemson University, USA, and an associate researcher at the Smithsonian Marine Station, Fort Pierce, Florida, USA. He has published more than 80 original papers on the behavioral ecology and life history of crustaceans.

Competing interests
The authors declare that they have no competing interests.

Author details
[1]Department of Biological Sciences, 132 Long Hall, Clemson University, Clemson, SC 29634, USA. [2]Smithsonian Marine Station at Fort Pierce, 701 Seaway Drive, Fort Pierce, FL 34949, USA. [3]Departamento de Biología Marina, Facultad de Ciencias del Mar, Universidad Católica del Norte, Larrondo 1281, Coquimbo, Chile. [4]Departamento de Biología, Facultad de Ciencias, Universidad del Valle, Cali 760032, Colombia. [5]Instituto Federal de Educação, Ciência e Tecnologia do Espírito Santo - Campus Alegre, Alegre, Espírito Santo, Brazil.

References
1. Clutton-Brock TH. The evolution of parental care. Princeton: Princeton University Press; 1991.
2. Royle NJ, Smiseth PT, Kolliker M. The evolution of parental care. Cambridge: Oxford University Press; 2012.
3. Evans JP, Marshall DJ. Male-by-female interactions influence fertilization success and mediate the benefits of polyandry in the sea urchin Heliocidaris erythrogramma. Evolution. 2005;59:106–12.
4. Neo ML, Vicentuan K, Teo SLM, Erftemeijer PLA, Todd PA. Larval ecology of the fluted giant clam, Tridacna squamosa, and its potential effects on dispersal models. J Exp Mar Biol Ecol. 2015;469:76–82.
5. Thiel M. Extended parental care in marine amphipods II. Maternal protection of juveniles from predation. J Exp Mar Biol Ecol. 1999;234:235–53.
6. Thiel M. Extended parental care in crustaceans – an update. Rev Chil Hist Nat. 2003;76:205–18.
7. Byrne M, Hart MW, Cerra A, Cisternas P. Reproduction and larval morphology of broadcasting and viviparous species in the Cryptasterina species complex. Biol Bull. 2003;205:285–94.
8. Phillips B, Kittaka J. Spiny lobsters: fisheries and culture 2nd edn. Oxford: Blackwell Science; 2000.
9. Chaffee C, Strathmann RR. Constraints on egg masses. I. Retarded development within thick egg masses. J Exp Mar Biol Ecol. 1984;84:73–83.
10. Strathmann RR. Feeding and nonfeeding larval development and life-history evolution in marine invertebrates. Ann Rev Ecol Syst. 1985;16:339–61.
11. Morritt D, Spicer JI. The physiological ecology of talitrid amphipods: an update. Can J Zool. 1988;76:1965–82.
12. Ramirez-Llodra E. Fecundity and life-history strategies in marine invertebrates. Adv Mar Biol. 2002;43:87–170.
13. Fisher TR. Oxygen uptake of the solitary tunicate Styela plicata. Biol Bull. 1976;1976(151):297–305.
14. Bauer RT. Adaptive modification of appendages for grooming (cleaning; antifouling) and reproduction in the Crustacea. In: Thiel M, Watling L, editors. Functional Morphology of Crustacea, vol. 1. New York: Oxford University Press; 2013. p. 337–75.
15. Silva P, Luppi TA, Spivak ED. Limb autotomy, epibiosis on embryos, and brooding care in the crab Crytograpsus angulatus (Brachyura: Varunidae). J Mar Biol Ass UK. 2003;83:1015–22.
16. Silva P, Luppi TA, Spivak ED. Epibiosis on eggs and brooding care in the burrowing crab Chasmagnathus granulatus (Brachyura: Varunidae): comparison between mudflats and salt marshes. J Mar Biol Ass UK. 2007;87:893–901.
17. Aiken DE, Waddy SL, Uhazy LS. Aspects of the biology of Pseudocarcinonemrtes homari and its association with the American Lobster, Homarus americanus. Can J Fish Aquat Sci. 1985;42:351–6.
18. Kuris AM. Crustacean Egg Production. Boca Raton: CRC; 1990.
19. Shields JD. The impact of pathogens on exploited populations of decapod crustaceans. J Invertebr Pathol. 2012;110:211–24.
20. Baeza JA, Fernández M. Active brood care in Cancer setosus (Crustacea: Decapoda): the relationship between female behavior, embryo oxygen consumption and the cost of brooding. Funct Ecol. 2002;16:241–51.
21. Prosser CL. Animal models for biomedical-research. Invertebrates – Introduction. Fed Proc. 1973;32:2177–8.
22. Naylor JK, Taylor EW, Bennett DB. Oxygen uptake of developing eggs of Cancer pagurus (Crustacea: Decapoda: Cancridae) and consequent behavior of the ovigerous female. J Mar Biol Ass UK. 1999;79:305–15.
23. Fernández M, Bock C, Pörtner HO. The cost of being a caring mother: the ignored factor in the reproduction of marine invertebrates. Ecol Lett. 2000;3:487–94.
24. Strathmann RR, Strathmann MF. Oxygen supply and limits on aggregation of embryos. J Mar Biol Ass UK. 1995;75:413–28.
25. Fernández M, Pardo LM, Baeza JA. Patterns of oxygen supply in embryo masses of brachyuran crabs throughout development: the effect of oxygen availability and chemical cues in determining female brooding behavior. Mar Ecol Prog Ser. 2002;245:181–90.
26. Eriksson SP, Nabbing M, Sjöman E. Is brood care in Nephrops norvegicus during hypoxia adaptive or a waste of energy? Funct Ecol. 2006;20:1097–104.
27. Förster C, Baeza JA. Active brood care in the anomuran crab Petrolisthes violaceus (Decapoda: Anomura: Porcellanidae): Grooming of brooded embryos by the fifth pereiopods. J Crustac Biol. 2001;21:606–15.
28. Sarvesan R. Some observations on parental care in Octopus dollfusi Robson (Cephalopoda: Octopodidae). J Mar Biol Ass India. 1969;11:203–5.
29. Dick JT, Faloon SE, Elwood RW. Active brood care in an amphipod: influences of embryonic development, temperature, and oxygen. Anim Behav. 1998;56:663–72.

30. Burris ZP. Costs of exclusive male parental care in the sea spider *Achelia simplissima* (Arthropoda: Pycnogonida). Mar Biol. 2011;158:381–90.

31. Gardner C. Effect of size on reproductive output of giant crabs *Pseudocarcinus gigas* (Lamarck): Oziidae. Mar Freshw Res. 1997;48:1323–650.

32. McClain CR, et al. Sizing ocean giants: patterns of intraspecific size variation in marine megafauna. PeerJ. 2015;3, e715.

33. Heath DJ. Simultaneous hermaphroditism: cost and benefit. J Theor Biol. 1977;64:363–73.

34. Strathmann RR, Strathmann MF. The relationship between adult size and brooding in marine invertebrates. Am Nat. 1982;119:91–101.

35. Arundell KL, Wedell N, Dunn AM. The impact of predation risk and of parasitic infection on parental care in brooding crustaceans. Anim Behav. 2014;96:97–105.

36. Holthuis LB. Marine lobsters of the world. Rome: FAO; 1991.

37. Carpenter KE, De Angelis N. The living marine resources of the eastern central Atlantic. Rome: FAO; 2014.

38. Fonseca-Larios ME, Briones-Fourzán P. Fecundity of the spiny lobster *Panulirus argus* (Latreille, 1804) in the Caribbean coast of Mexico. Bull Mar Sci. 1998;63:21–32.

39. Bertelsen RD, Matthews TR. Fecundity dynamics of female spiny lobster (*Panulirus argus*) in a south Florida fishery and Dry Tortugas National Park lobster sanctuary. Mar Freshw Res. 2001;52:1559–65.

40. Ziegler TA. Larval release rhythms and larval behavior of palinurid lobsters: a comparative study. (Doctoral dissertation). Nicholas School of the Environment and Earth Sciences Duke University. Durham, Duke University; 2007.

41. Ziegler TA, Forward RB. Control of larval release in the Caribbean spiny lobster, *Panulirus argus*: role of chemical cues. Mar Biol. 2007;152:589–97.

42. Ziegler TA, Forward RB. Larval release behaviors in the Caribbean spiny lobster, *Panulirus argus*: Role of peptide pheromones. J Chem Ecol. 2007;33:1795–805.

43. Wheatly MG. The provision of oxygen to developing eggs by female shore crabs (*Carcinus maenas*). J Mar Biol Ass UK. 1981;61:117–28.

44. Reinsel KA, et al. Egg mass ventilation by caridean shrimp: similarities to other decapods and insight into pheromone receptor location. J Mar Biol Ass UK. 2014;94:1009–17.

45. Ache BW, Derby CD. Functional organization of olfaction in crustaceans. Trends Neurosci. 1985;8:356–60.

46. Altner I, Hatt H, Altner H. Structural properties of bimodal chemosensitive and mechanosensitive setae on the pereiopod chelae of the crayfish, *Austropotamobius torrentium*. Cell Tissue Res. 1983;228:357–74.

47. Schmidt M, Gnatzy W. Are the funnel-canal organs the campaniform sensilla of the shore crab, *Carcinus maenas* (Decapoda, Crustacea) 2. Ultrastructure. Cell Tissue Res. 1984;237:81–93.

48. Ache BW, McClintock TS. The lobster olfactory receptor cell as a neurobiological model: the action of histamine. In: Wiese K, Krenz WD, Tautz J, Reichert H, Mulloney B, editors. Advances in Life Sciences: Frontiers in Crustacean Neurobiology. Basel: Verlag; 1990. p. 33–9.

49. Schmidt M, Ache BW. Processing of antennular input in the brain of the spiny lobster, *Panulirus argus*. Non-olfactory chemosensory and mechanosensory pathway of the lateral and median antennular neuropils. J Comp Physiol A. 1996;178:579–604.

50. Derby CD. Learning from spiny lobsters about chemosensenory coding of mixtures. Physiol Behav. 2000;69:203–9.

51. Rittschof D. Chemosensation in the daily life of crabs. Am Zool. 1992; 32:363–9.

52. Rittschof D, Forward RB, Mott DD. Larval release in the crab *Rhithropanopeus harrisii* (Gould) – chemical cues from hatching eggs. Chem Senses. 1985;10:567–77.

53. Forward RB. Larval release rhythms of decapod crustaceans – an overview. Bull Mar Sci. 1987;41:165–76.

54. De Vries MC, Forward RB. Control of egg-hatching time in crabs from different tidal heights. J Crustac Biol. 1991;11:29–39.

55. Rittschof D. Body odors and neutral-basic peptide mimics – a review of responses by marine organisms. Am Zool. 1993;33:487–93.

56. Rittschof D, Cohen JH. Crustacean peptide and peptide-like pheromones and kairomones. Peptides. 2004;25:1503–16.

57. Tankersley RA, Bullock TM, Forward RB, Rittschof D. Larval release behaviors in the blue crab *Callinectes sapidus*: role of chemical cues. J Exp Mar Biol Ecol. 2002;273:1–14.

58. Fernández M, Brante A. Brood care in brachyuran crabs: the effect of oxygen provision on reproductive costs. Rev Chil Hist Nat. 2003;76:157–68.

59. Phillips G. Incubation of English prawn *Palaemon serratus*. J Mar Biol Ass UK. 1971;51:43–8.

60. Ruiz-Tagle N, Fernández M, Pörtner HO. Full time mothers: daily rhythms in brooding and nonbrooding behaviors of brachyuran crabs. J Exp Mar Biol Ecol. 2002;276:31–47.

61. Kanciruk P, Herrnkind WF. Preliminary investigation of the daily and seasonal locomotor activity rhythms of the spiny lobster, *Panulirus argus*. Mar Freshw Behav Phy. 1972;1:351–9.

62. Lipcius RN, Herrnkind WF. Molt cycle alterations in behavior, feeding, and diel rhythms of a decapod crustacean, the spiny lobster *Panulirus argus*. Mar Biol. 1982;68:241–52.

63. Arana E, Dupre M, Gaete M. Ciclo reproductivo, talla de primera madurez sexual y fecundidad de la langosta de Juan Fernandez (*Jasus frontalis*). Investig Mar. 2000;28:165–74.

64. Annala JH, Bycrofft BL. Fecundity of the New Zealand red rock lobster *Jasus edwardsii*. New Zeal J Mar Fresh Res. 1987;21:591–7.

65. Kagwade PV. Fecundity in the spiny lobster *Panulirus polyphagus* (Herbst). J Mar Biol Ass India. 1988;30:114–20.

66. MacDiarmid AB. Size at onset of maturity and size-dependent reproductive output of female and male spiny lobsters *Jasus edwardsii* (Hutton) (Decapoda, Palinuridae) in northern New Zealand. J Exp Mar Biol Ecol. 1989; 127:229–43.

67. DeMartini EE, Ellis DM, Honda VA. Comparisons of spiny lobster *Panulirus marginatus* fecundity, egg size, and spawning frequency before and after exploitation. Fish Bull. 1992;91:1–7.

68. Melville-Smith R, Goosen PC, Stewart TJ. The spiny lobster *Jasus lalandii* (H. Milne Edwards, 1837) off the South African coast: inter-annual variations in male growth and female fecundity. Crustaceana. 1995;68:174–83.

69. Hogarth PJ, Barratt LA. Size distribution, maturity and fecundity of the spiny lobster *Panulirus penicillatus* (Oliver 1791) in the Red Sea. Trop Zool. 1996;9: 399–408.

70. Goñi R, Quetglas A, Reñones O. Size at maturity, fecundity and reproductive potential of a protected population of the spiny lobster *Palinurus elephas* (Fabricius, 1787) from the western Mediterranean. Mar Biol. 2003;143:583–92.

71. Green BS, Gardner C, Kennedy RB. Generalized linear modeling of fecundity at length in southern rock lobsters, *Jasus edwardsii*. Mar Biol. 2009;156:1941–7.

72. DeMartini EE, DiNardo GT, Williams HA. Temporal changes in population density, fecundity, and egg size of the Hawaiian spiny lobster (*Panulirus marginatus*) at Necker Bank, Northwestern Hawaiian Islands. Fish Bull. 2003; 101:22–31.

73. Morgan GR. Fecundity in western rock lobster *Panulirus longipes-cygnus* (George) (Crustacea, Decapoda, Palinuridae). Aust J Mar Fresh Res. 1972; 23:133–41.

74. MacFarlane JW, Moore R. Reproduction of the ornate rock lobster *Panulirus ornatus* (Fabricius), in Papua New Guinea. Aust J Mar Freshw Res. 1986;37:55–65.

75. Pérez-González R, Puga-Lopez D, Castro-Longoria R. Ovarian development and size at sexual maturity of the Mexican spiny lobster *Panulirus inflatus*. New Zeal J Mar Freshw Res. 2009;43:163–72.

76. Groenveld JC. Fecundity of spiny lobster *Palinurus gilchristi* (Decapoda: Palinuridae) off South Africa. Af J Mar Sci. 2005;27:231–8.

77. Groeneveld JC, Greengrass CL, Branch GM, McCue SA. Fecundity of the deep-water spiny lobster *Palinurus delagoae*. W I O J Mar Sci. 2005;4:135–43.

78. Gracia GA. Seasonal variation of the fecundity of the lobster *Panulirus inflatus* (Bouvier, 1895) (Crustacea: Decapoda: Palinuridae). Cienc. 1985; 11:7–27.

79. Pérez-González R, Valadez LM, Rodríguez-Domínguez G, Aragón-Noriega EA. Seasonal variation in brood size of the spiny lobster *Panulirus gracilis* (Decapoda: Palinuridae) in Mexican waters of the Gulf of California. J Shellfish Res. 2012;31:935–40.

80. Torres P, Penha-Lopez G, Narciso L, Macia A, Paula J. Fecundity and brood loss in four species of fiddler crabs, genus *Uca* (Brachyura: Ocypodidae), in the mangroves of Inhaca Island, Mozambique. J Mar Biol Assoc UK. 2009;89:371–8.

81. Oh CW, Hartnoll RG. Brood loss during incubation in *Philocheras trispinosus* (Decapoda) in Port Erin Bay, Isle of Man. J Crustac Biol. 1999;19:467–76.

82. Einum S, Hendry AP, Fleming IA. Egg-size evolution in aquatic environments: does oxygen availability constrain size? Proc Roy Soc B Biol Sci. 2002;269:2325–30.

83. Fernández M, Calderón R, Cifuentes M, Pappalardo P. Brooding behaviors and cost of brooding in small body size brachyuran crabs. Mar Ecol Prog Ser. 2006;309:213–20.

84. Hines AH. Allometric constraints and variables of reproductive effort in brachyuran crabs. Mar Biol. 1982;69:309–20.

85. Hines AH. Constraint on reproductive output in brachyuran crabs: Pinnotherids test the rule. Am Zool. 1992;32:503–11.

86. Clarke A. Temperature, latitude, and reproductive effort. Mar Ecol Prog Ser. 1987;38:89–99.

87. Anger K, Moreira GS. Morphometric and reproduction traits of tropical caridean shrimps. J Crustac Biol. 1998;18:823–38.

88. Anderson JR, Spadaro AJ, Baeza JA, Behringer DC. Ontogenetic shifts in resource allocation: colour change and allometric growth of defensive and reproductive structures in the Caribbean spiny lobster *Panulirus argus*. Biol J Linn Soc. 2013;108:87–98.

89. Wickham DE. *Carcinomertes errans* and the fouling and mortality of eggs of the Dungeness crab, *Cancer magister*. J Fish Res Board Can. 1979;36:1319–24.

90. Wickham DE. Epizootic infestation by nemertean brood parasites on commercially important crustaceans. Can J Fish Aquat Sci. 1986;43: 2295–302.

91. Shields JD, Kuris AM. Temporal variation in abundance of the egg predator *Carcinonemertes epialti* (Nemertea) and its effect on egg mortality of its host, the shore crab, *Hemigrapsus oregonensis*. Hydrobiologia. 1988;156:31–8.

92. Shields JD. Diseases of spiny lobsters: a review. J Invertebr Pathol. 2011; 106:79–91.

93. Okazaki RK, Wehrtmann IS. Preliminary survey of a nemertean crab egg predator, *Carcinonemertes*, on its host crab, *Callinectes arcuatus* (Decapoda, Portunidae) from Golfo de Nicoya, Pacific Costa Rica. ZooKeys. 2014;457: 367–75.

94. Santos C, Bueno SLS, Norenburg JL. Infestation by *Carcinonemertes divae* (Nemertea: Carcinonemertidae) in *Libinia spinosa* (Decapoda: Pisidae) from São Sebastião Island, SP, Brazil. J Nat Hist. 2006;40:999–1005.

95. Sadeghian PS, Kuris AM. Distribution and abundance of a nemertean egg predator (*Carcinonemertes* sp.) on a leucosiid crab, *Randallia ornate*. Hydrobiologia. 2001;456:59–63.

96. Shields JD, Kuris AM. *Carcinonemertes wickhami* n. sp. (Nemertea), a symbiotic egg predator from the spiny lobster *Panulirus interruptus* in Southern California, with remarks on symbiont-host adaptations. Fish Bull. 1990;88:279–87.

97. Shields JD, Behringer DC. A new pathogenic virus in the Caribbean spiny lobster *Panulirus argus* from the Florida Keys. Dis Aquat Org. 2004;59:109–18.

98. Lee CE, Strathmann RR. Scaling of gelatinous clutches: Effects of siblings' competition for oxygen on clutch size and parental investment per offspring. Am Nat. 1998;151:293–310.

99. Tavares M. Lobsters. In: Carpenter KE, De Angelis N, editors. The living marine resources of the eastern central Atlantic. Rome: FAO; 2014. p. 294–310.

100. Cruz R, Bertelsen RD. The spiny lobster (*Panulirus argus*) in the wider Caribbean: A review of life cycle dynamics and implications for responsible fisheries management. Proc 61st Gulf Caribbean Fish Inst. 2008;61:10–4.

101. Lipcius RN, Herrnkind WF. Photoperiodic regulation and daily timing of spiny lobster mating behavior. J Exp Mar Biol Ecol. 1987;89:191–204.

102. Butler MJ, et al. Patterns of spiny lobster (*Panulirus argus*) postlarval recruitment in the Caribbean: a CRTR Project. Proc 62nd Gulf Caribbean Fish Inst. 2010;62:361–9.

103. Butler MJ, Bertelsen R, MacDiarmid A. Mate choice in temperate and tropical spiny lobsters with contrasting reproductive systems. ICES J Mar Sci. 2015;72 Suppl 1:i101–14.

104. Lipcius RN, Edwards ML, Herrnkind WF, Waterman SA. In situ mating behaviour of the spiny lobster *Panulirus argus*. J Crustac Biol. 1983;3:217–22.

105. MacDiarmid AB, Butler MJ. Sperm economy and limitation in spiny lobsters. Behav Ecol Sociobiol. 1999;46:14–24.

106. Butler MJ, MacDiarmid A, Gnanalingam G. The effect of parental size on spermatophore production, egg quality, fertilization success, and larval characteristics in the Caribbean spiny lobster, *Panulirus argus*. ICES J Mar Sci. 2015;72 Suppl 1:i115–23.

107. Butler MJ, Heisig-Mitchell J, MacDiarmid AB, Swanson RJ. The effect of male size and spermatophore characteristics on reproduction in the Caribbean spiny lobster, *Panulirus argus*. New Front Crustac Biol. 2011;15:69–84.

108. Saul S. A review of the literature and life history study of the Caribbean spiny lobster, *Panulirus argus*. Caribbean Southeast Data Assessment Review Workshop Report, SEDAR-DW-05. Sustainable Fisheries Division Contribution No. SFD- 2004–048. North Charleston: SouthEast Data, Assessment, and Review; 2004.

109. Bertelsen RD, Hornbeck J. Using acoustic tagging to determine adult spiny lobster (*Panulirus argus*) movement patterns in the Western Sambo Ecological Reserve (Florida, United States). New Zeal J Mar Freshw Res. 2009; 43:35–46.

110. Bertelsen RD. Characterizing daily movements, nomadic movements, and reproductive migrations of *Panulirus argus* around the Western Sambo Ecological Reserve (Florida, USA) using acoustic telemetry. Fish Res. 2013; 144:91–102.

111. Maxwell KE, Matthews TR, Bertelsen RD, Derby CD. Using age to evaluate reproduction in Caribbean spiny lobster, *Panulirus argus*, in the Florida Keys and Dry Tortugas, United States. New Zeal J Mar Freshw Res. 2009;43:139–49.

112. Matthews TR, Maxwell KE, Bertelsen RD, Derby CD. Use of neurolipofuscin to determine age structure and growth rates of Caribbean spiny lobster *Panulirus argus* in Florida, United States. New Zeal J Mar Fresh. 2009;43:125–37.

113. Cox C, Hunt JH. Change in size and abundance of Caribbean spiny lobsters *Panulirus argus* in a marine reserve in the Florida Keys National Marine Sanctuary, USA. Mar Eco Prog Ser. 2005;294:227–39.

114. Maxwell KE, Matthews TR, Bertelsen RD, Derby CD. Age and size structure of Caribbean spiny lobster, *Panulirus argus*, in a no-take marine reserve in the Florida Keys, USA. Fish Res. 2013;144:84–90.

115. Birkeland C, Dayton PK. The importance in fishery management of leaving the big ones. Trends Ecol Evol. 2005;20:356–8.

116. Martin P, Bateson P. Measuring Behaviour: An Introductory Guide. Cambridge: Cambridge University Press; 1986.

117. JMP Pro, Version 10, SAS Institute Inc., Cary, N.C., 1989–2007

118. Turner RL, Lawrence JM. In: Stancyk SE, editor. Volume and composition of echinoderm eggs: implications for the use of egg size in life history models. Columbia: University of South Carolina Press; 1979. p. 25–40.

119. Baeza JA. Testing three models on the adaptive significance of protandric simultaneous hermaphroditism in marine shrimp. Evolution. 2006;60:1840–50.

120. Hartnoll RG. Growth. In: Abele LG, editor. The Biology of Crustacea 2. New York: Academic; 1982. p. 111–96.

121. Zar JH. Biostatistical analysis. Upper Saddle River: Prentice-Hall; 1996.

The cellular basis of bioadhesion of the freshwater polyp *Hydra*

Marcelo Rodrigues[1]* ⓘ, Philippe Leclère[2], Patrick Flammang[3], Michael W. Hess[4], Willi Salvenmoser[1], Bert Hobmayer[1] and Peter Ladurner[1]*

Abstract

Background: The freshwater cnidarian *Hydra* temporarily binds itself to numerous natural substrates encountered underwater, such as stones, leafs, etc. This adhesion is mediated by secreted material from specialized ectodermal modified cells at the aboral end of the animal. The means by which *Hydra* polyps attach to surface remain unresolved, despite the fact that Hydra is a classic model in developmental and stem cell biology.

Results: Here, we present novel observations on the attachment mechanism of *Hydra* using high pressure transmission electron microscopy, scanning electron microscopy, atomic force microscopy, super-resolution microscopy, and enzyme histochemistry. We analyzed the morphology of ectodermal basal disc cells, studied the secreted material, and its adhesive nature. By electron microscopy we identified four morphologically distinct secretory granules occurring in a single cell type. All the secretory granules contained glycans with different distribution patterns among the granule types. Footprints of the polyps were visualized under dry conditions by atomic force microscopy and found to consist of a meshwork with nanopores occurring in the interstices. Two antibodies AE03 and 3G11, previously used in cell differentiation studies, labelled both, basal disc cells and footprints. Our data suggest that the adhesive components of *Hydra* are produced, stored and delivered by a single cell type. Video microscopy analysis corroborates a role of muscle contractions for the detachment process.

Conclusion: We clearly demonstrated that bioadhesion of *Hydra* relies on the secreted material. Our data suggest that glycans and/or glycoproteins represent an important fraction of the secreted material. Detachment seems to be initiated by mechanical forces by muscular contractions. Taken together, our study represents the characterization of an unique temporary adhesive system not known in aquatic organisms from other metazoan phyla.

Keywords: Hydra, Basal disc, Biological adhesion, Adhesion

Background

Aquatic organisms, freshwater and marine, have evolved a myriad of effective solutions for underwater adhesion. Cases range from microscopic organisms, such as bacteria, through much larger and complex marine algae, invertebrates, and vertebrates. Examples include the permanent attachments of sessile mussels [1] and barnacles [2, 3], the temporary attachment of starfish and flatworms during locomotion [4, 5], the construction of protective shelters by sandcastle worms [6], and the defence against predators by the Cuverian tubules of sea cucumbers [7]. All these bioadhesives were adapted by natural selection for specific roles in the organism's life style. Likewise, the way they attach are also remarkably complex and involve a large range of interactions and components with different functions [8, 9]. Generally, these multicomponent adhesives are composed of protein, carbohydrates, and inorganic components. The amount of each component is highly variable in different organisms. For instance, in sea stars 21 % are proteins, 8 % carbohydrates, and 40 % inorganic material [10]. On the contrary, in barnacles 90 % of the adhesive is made out of proteins with the remainder being 1 % carbohydrates, 1 % lipids, and 4 % inorganic material [11]. Mussels have one of the best studied bioadhesive systems [12, 13]. These marine molluscs routinely stick to all kinds of surfaces underwater using their so-called byssus. This structure consists of a protein complex

* Correspondence: marcelo.rodrigues@uibk.ac.at; peter.ladurner@uibk.ac.at
[1]Institute of Zoology and Center for Molecular Biosciences Innsbruck, University of Innsbruck, Innsbruck, Austria
Full list of author information is available at the end of the article

secreted as a fluid that spreads spontaneously and exhibits strong reversible interfacial bonding and tunable cross-linking. Similarly, the sabelariid polychaete sandcastle worm *Phragmatopoma californica* [6, 14] secretes micro-droplets of adhesive to build a tube-like burrow from sand grains and other particles. Alternatively, the marine flat-worm *Macrostomum lignano*, possesses a duo-gland adhesive system in the tail plate [5] that allows it to adhere and release from the substrate multiple times within seconds. The different characteristics of all these adhesives are often derived from the physico-chemical properties of the adhesive proteins, and in particular, from their post-translational modifications (PTM), such as hydroxylation, phosphorylation, and glycosylation [15, 16]. Some organisms contain glycans associated with the adhesive proteins, but it is unknown whether they are covalently attached to the proteins [10, 17].

The freshwater cnidarian *Hydra* (Fig. 1a, b) is a solitary polyp inhabiting any unpolluted body of shallow freshwater all year round. During its whole lifecycle it lives temporarily attached underwater, being able to attach and detach repetitively, but only detaching when looking for better living conditions [18]. Reproduction is sexual and/ or by asexual budding which dominates when food is plentiful. The fertilized egg develops into an embryo that grows to an adult polyp, or as late blastula enters a resting stage surrounded by a chitinous covering. The animal has a single axis and consists of only two layers of epithelial cells: the endoderm and the ectoderm [19]. At the oral end is the hypostome (mouth opening), which is surrounded by a ring of tentacles, and at the aboral end is an adhesive disc called "basal disc" [20]. It is well known that basal disc cells are derived from ectodermal cells of the lower gastric column [21, 22], therefore consisting of modified ectodermal cells that secrete an adhesive material by which *Hydra* can attach strongly to a number of surfaces underwater, i.e., stones, wood sticks, leaves, and other submerged parts of plants.

Hydra is a classical model organism in axial pattern, regeneration, and stem cell biology [22–26]. In contrast to the existing detailed information about differentiation and regeneration of the ectodermal basal disc cells, only few studies have addressed *Hydra*'s astonishing attachment ability. The ultrastructure of ectodermal basal disc cells was studied by Chaet [27] and Philpott et al. [28] who provided a first description of *Hydra*'s secretory granules. They placed these granules in three categories, two of them representing the same type of granule in different stages of matureness, while the third granule is of a different type. Further, Davis [29] proposed that basal disc cells produce, by themselves or jointly with other cells, at least six types of granules, and a seventh one that originates from the neighbouring digestive cells. The only constituent identified inside the basal disc cells

is the presence of hyaluronic acid as seen after an Alcian Blue [28, 29] and PAS staining [28]. At the ultrastructural level, granules between 0.5 and 1.5 µm in diameter are known to be peroxidase positive [30]. The mode of action and the components of the secreted adhesives are not understood. Enlarged cytoplasmic protrusions which were named as filopodia were observed during *Hydra* attachment by Pan et al. [31].

In this study, our goal was to characterize the cellular components responsible for *Hydra*'s underwater adhesion. Light-and electron microscopic techniques were utilised for a comprehensive description of basal disc morphology of free and attached polyps. Additionally, morphology and adhesiveness of the secreted material (footprint) was analysed using atomic force microscopy. To investigate the components involved in adhesion, we first used energy electron loss spectroscopy to identify nitrogen and phosphorus atom distribution. Second, the periodic acid-Schiff reaction was carried out to verify the presence of glycans, and diamino benzidine was used to attest peroxidase activity. The localisation of two antigens labelled by two antibodies was confirmed to be present in the footprints. Finally, we showed that lipids most probably do not play a role in *Hydra* adhesion. We suggest that detachment is driven by muscular activity. After investigating all these components, we paved the way for *Hydra* as a model organism for bioadhesion research using molecular approaches [32]. It provides the basis for our current efforts to uncover the biochemical basis of the glue of *Hydra*.

Results

Basal disc cell morphology and secretion of adhesive granules

Using interference contrast microscopy of squeezed live animals the basal disc cells can be seen at the aboral end of the animal (Fig. 1c and d). A longitudinal section through a Carnoy-fixed unattached polyp showed an overview of the basal disc cells, which constitute the adhesive system (Fig. 1e and f). The external morphology of the basal disc consisted of a cylindrical peduncle covered by a flat disc (Fig. 2a). Between the basal disc cells there were uniformly distributed pores (Fig. 2b). Secreted material was visible on the basal disc (Fig. 2c). In a *Hydra* polyp attached to the substrate (Fig. 3a and b), the basal disc cells appeared directly in contact with the surface. When it detached, it left behind a footprint made up of the secreted material on the surface (Fig. 3c). The basal disc cells left behind an imprint outlining their apical cell-to-cell contact sites. The flat zones between the rims outlining the basal disc cells also contain a thin film of adhesive material (see below). In summary, the basal disc is a specialized secretory tissue allowing *Hydra* to attach to the surface.

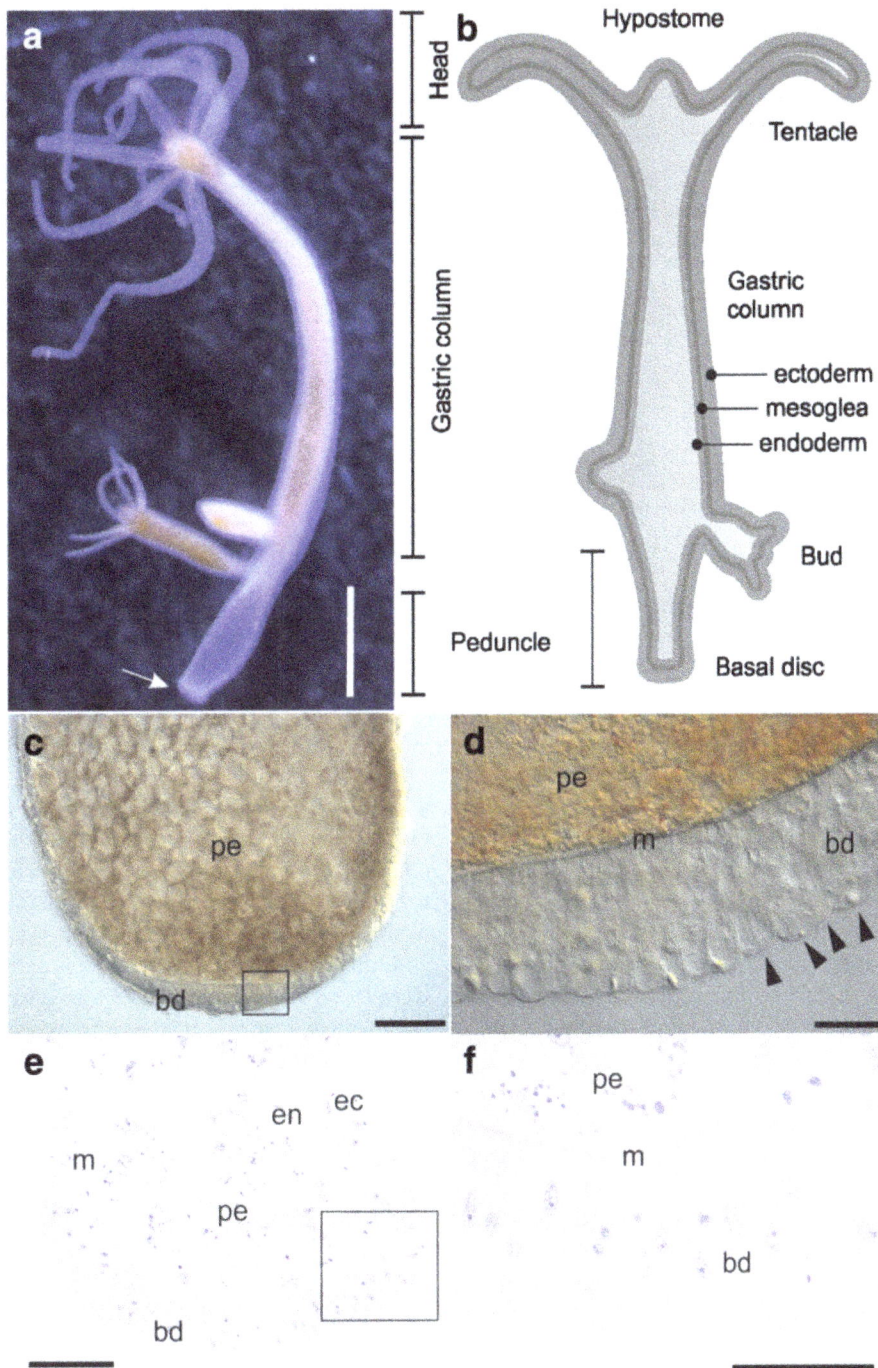

Fig. 1 Light microscopy images of live and fixed *Hydra magnipapillata* strain 105. **a** An adult polyp. The arrow indicates the basal disc. **b** Scheme of an adult polyp indicating details of the animal morphology. **c** Squeezed preparation of the foot region. The square indicates the area magnified in figure **d**. **d** The arrowheads point at individual basal disc cells. **e** Longitudinal section stained with hematoxylin eosin showing a general morphology of the basal disc. Inset shows the zone magnified in figure **f**. **f** Organization of basal disc cells with stained nuclei. Scale *bars* 1 mm (**a**), 100 μm (**c**), 50 μm (**d-f**). *Abbreviations*: ec, ectoderm; m, mesoglea; en, endoderm; pe, peduncle; bd, basal disc

The detachment process was recorded with video microscopy (Additional file 1). In the beginning, the full basal disc was attached. The polyp started the detachment from the outer rim towards the center of the basal disc. When the detachment got closer to the center, *Hydra* suddenly detached the last cells. A footprint was left behind which was transparent underwater (Additional file 1). Phalloidin staining revealed the actin filament distribution

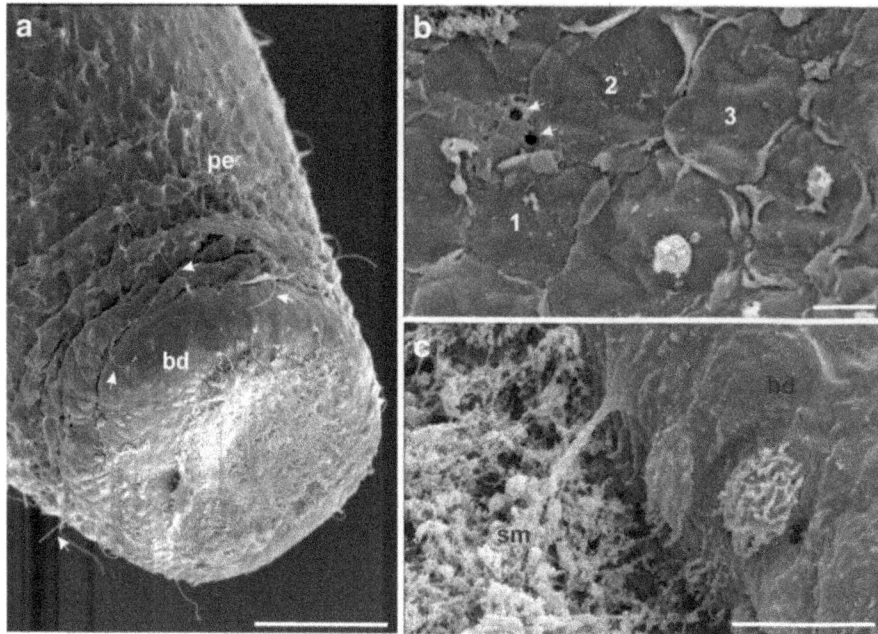

Fig. 2 Scanning Electron Microscopy. **a** Outer aspect of the basal disc. Arrows point at discharged nematocysts. **b** Basal disc surface. The numbers 1, 2, 3, mark three different basal disc cells. Arrows point at pores between the cells. **c** basal disc and substrate covered with secretory material. *Scale bars* 100 μm (**a**), 2 μm (**b**), 10 μm (**c**). *Abbreviations*: pe, peduncle; bd, basal disc; sm, secretory material

in the basal disc (Additional file 2). Within the ectodermal layer the myonemes were radially distributed (Additional file 2: Figure S2 a and b). From the radial myonemes a branch of myonemes goes perpendicularly towards the apical side of ectodermal basal disc cells (Additional file 2: Figure S2 c). This distribution of myonemes was corroborated by transmission electron microscopy experiments (Additional file 3: Figure S3 a). Within the endodermal layer they were circular (Additional file 2 Figure S2 a and b). Based on these observations, we propose that muscle contractions of the two cell layers (ectoderm and endoderm) of the basal disc were involved in the detachment of the animals.

In side view at the ultrastructural level, the basal disc cells had an irregular rectangular-like shape with a planar diameter ranging from 10 to 17 μm at the apical end of the cell and an apical-basal diameter about 51 to 60 μm (Fig. 4a). Several water containing vacuoles were seen as major constituents of the cells (Fig. 4 and Additional file 3: Figure S3 b). Their most apical region, which actually gets in contact with the surface when attached (Fig. 4c and d), beared an array of protruded cytoplasmic extensions with a diameter of about 0.4–1 μm. Filopodia-like cytoplasmic extensions were observed to be in contact with the surface after basal disc attachment (Fig. 4d, and Additional file 3: Figure S3 b). Basal disc cells presented many morphological characteristics of gland cells. The cells cytoplasm was rich in endoplasmic reticulum (ER), golgi fields, and mitochondria around the stored secretory

granules. Based on morphology and size, four types of granules located close to the secretory cell membrane could be discriminated (Fig. 4b and Additional file 3: Figure S3 b). First, *Hydra* secretory granules I (HSGI) are likely precursor (maturing granules) of the granules HSGII (Additional file 3: Figure S3 c). These maturing granules were closely associated with ER, suggesting that these organelles are involved in the synthesis and/or maturation of the secretory granule contents. HSGI measured between 1.90 and 2.10 μm in diameter, and their content were electron lucent. Second, *Hydra* secretory granules II (HSGII) represented electron dense granules with a diameter ranging from 0.40 to 1.10 μm. *Hydra* secretory granules III (HSGIII) measured between 0.45 and 0.49 μm in diameter, and appeared less electron dense than the *Hydra* secretory granules IV (HSGIV). HSGIV were by far the smallest granules with a diameter of 0.18 to 0.21 μm (Fig. 4b). In thick sections of 350 nm, secretory granules can be seen overlapping each other, corroborating that these are true vesicles (Additional file 3: Figure S3 b). Besides the granules membrane, there was no compartmentation of adhesive components inside the cell at any stage of the secretory process, and there was no intracellular drainage system. Once the cell surface established contact with the substratum (Fig. 4c), the basal disc cells secreted the adhesive material by exocytosis. The secreted material was deposited as a thin film filling any space between the cells and the substratum (Fig. 4d and see also AFM Fig. 5). This film gives rise to the footprint after detachment.

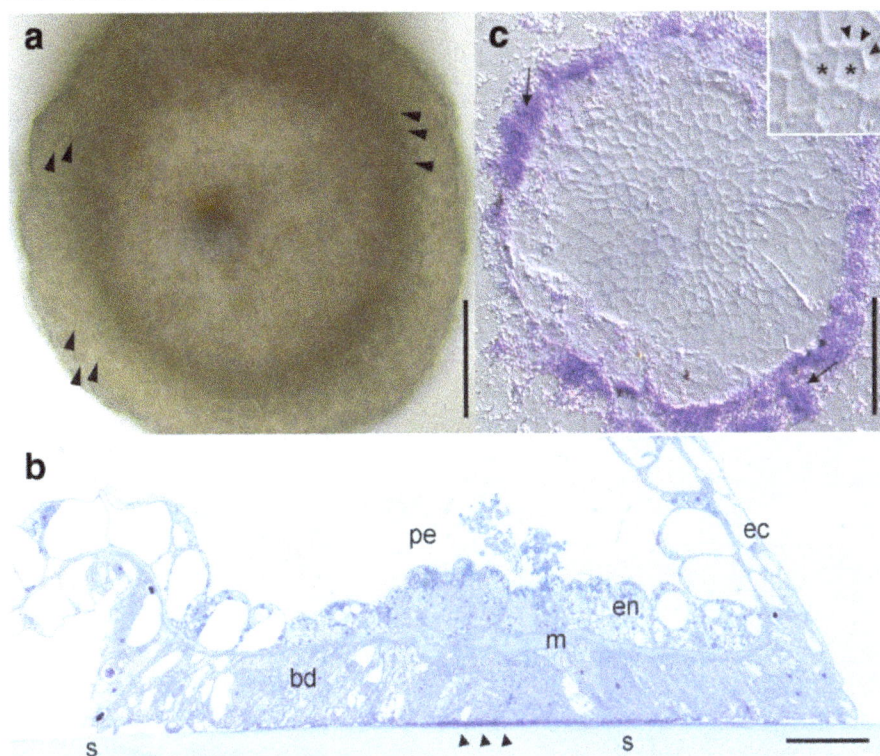

Fig. 3 Light microscopy images of live (**a**) and fixed-attached *Hydra* polyp (**b**), and secreted footprint (**c**). **a** Overview of the basal disc of a live attached polyp taken through inverted microscopy. Individual basal disc cells (arrowheads) are in focus along the outer margin. **b** Longitudinal semi-thin section stained with the basic dye methylene blue and Azur II. The polyp was chemically-fixed when still attached to a substratum. *Arrowheads* indicate the interface between substratum and polyp where the adhesive is secreted. **c** Footprint deposited by the basal disc of a *Hydra* polyp stained with crystal violet. Imprints derived from individual cells are visible in the centre of the footprint while at the periphery fungi can be observed [*arrows*]. Inset is a magnification of a central area of the footprint where the flat area [*asterisks*] and the rims [*arrowheads*] can be seen in a better detail. The fungi are contamination and/or symbionts of *Hydra* culture. *Scale bars* 100 μm (**a**, **c**), 50 μm (**b**). *Abbreviations*: s, substratum; bd, basal disc; ec, ectoderm; m, mesoglea; en, endoderm; pe, peduncle

Overall, our results confirm that the basal disc cells produce secretory granules and that their contents play the core role in *Hydra* adhesion.

The air dried footprints were easily located using phase contrast and could be precisely positioned beneath the AFM cantilever. Care was taken to select a scan site near to the rim of the footprints and to avoid regions that were too thick to image. When visualized through AFM, the footprints of *Hydra* were found to be a meshwork with nanopores occurring in the interstices of the deposited material (Fig. 5a). The secreted material seems homogeneous, traces of individual secretory granules cannot be detected. Several pores of about 0.5–1 μm where present at the surface of the footprint whose diameter correlated to the one of the protruded cytoplasmic extensions. The adhesion profile (Fig. 5b) showed that adhesive forces were higher in the deeper (thinner) areas of the footprint, reaching up to 66.4 nN. In summary, the AFM results corroborate the adhesive nature of the secreted material.

Chemistry of granules

To determine whether granules contained vicinal diol-containing glycans, Periodic Acid Schiff (PAS) cytochemistry was carried out in both light-and electron microscopy. The PAS method positively stained basal disc cells showing a gradient towards the apical end of the cells. The strongest reaction was observed most apically where the granule secretion takes place (Fig. 6a and b). Figure 6c showed that HSGII were fully positive to PAS, with a strongest reaction near their surrounding membrane, while negative control did not show any staining (Fig. 6d). HSGIII and IV, positively reacted throughout the whole vesicle. Alcian blue staining was negative for ectodermal basal disc cells in *H. magnipapillata* at pH-value 2.5. As positive control reaction for AB we successfully stained nematocysts in tentacles sections (Additional file 4), validating that the method used was functional. Our results showed that secretory granules from *H. magnipapillata* contained neutral glycans instead of the acidic compounds found previously in other species.

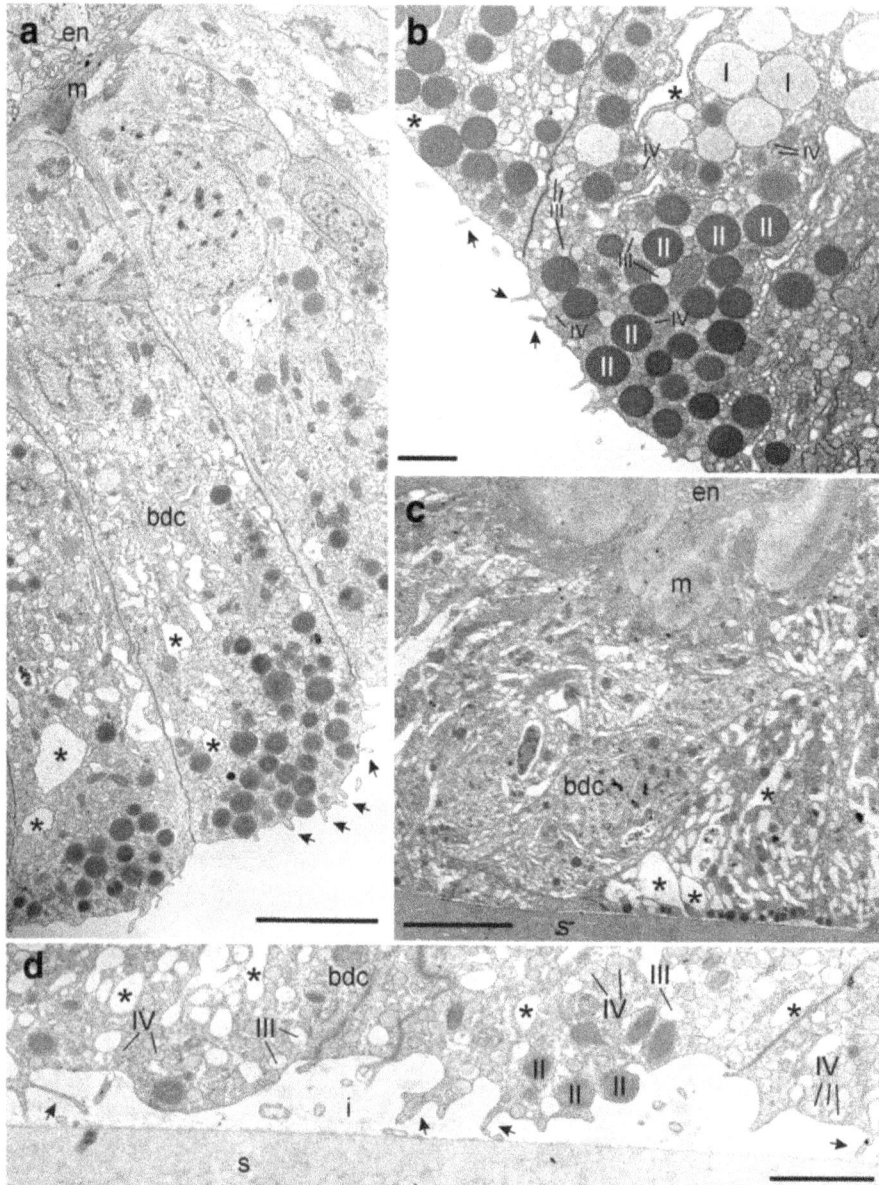

Fig. 4 Transmission electron microscopy images of chemically and cryo-fixed *Hydra* polyps. **a** Longitudinal section through a chemically-fixed basal disc showing individual basal disc cells. Electron dense granules tend to accumulate at the aboral end of the cell, which is the area attaching to the substratum. At the aboral end, several cytoplasmic extensions are present [*arrows*]. **b** Cryo-fixed basal disc reveals the fine structure of secretory granules. **c** Chemically-fixed *Hydra* polyp right after attachment to the substratum. **d** Interface between an attached basal disc and the substratum. Note protruded cytoplasmic extensions are in contact with the substratum. *Scale bars* 10 μm (**a**, **c**), 2 μm (**b**), 5 μm (**d**). *Abbreviations*: en, endoderm; m, mesoglea; bdc, basal disc cell; I, *Hydra* secretory granule I; II, *Hydra* secretory granule II; III, *Hydra* secretory granule III; IV, *Hydra* secretory granule IV; s, substratum; *arrows*, indicate cytoplasmic extensions; *asterisks*, indicate vacuoles of water

Diamino benzidine treated polyps allowed to visualize peroxidase activity in the basal disc of *Hydra* (Fig. 6e and f). The reaction was stronger close to where exocytosis takes place. The peroxidase reaction was associated with HSGI (Fig. 6g and h), but not for all the granules. Peroxidase activity was also detected at the level of cytoplasmic components surrounding HSGII.

Based on information from other model organisms, in which lipids played an important role in adhesion [33, 34], the occurrence of lipids was investigated in the present study using Nile Red on *Hydra* whole mounts. When exposing stained specimens to both blue (450–500 nm) and green (550 nm) exciting light, dispersed droplets of lipids measuring between 1 and 1.8 μm were visible in basal disc cells (Additional file 5), but apparently not associated to

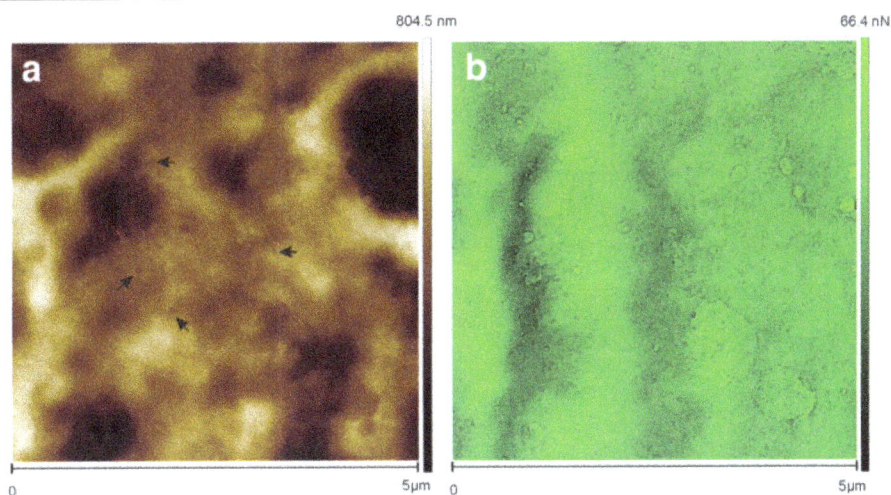

Fig. 5 Atomic force microscopy images collected in air on dried footprints from *Hydra*. **a** Height profile revealing the meshwork structure. *Arrows* point at pores presumably corresponding to cytoplasmic before polyp detachment. **b** Adhesion profile

secretory granules. Lipid droplets were also not observed in the footprints (data not shown). Therefore, these results support a view that lipids do not play a critical role in the adhesion process of *Hydra*.

Electron energy loss spectroscopy (EELS) and electron spectroscopic imaging (ESI) experiments allowed to determine the distribution of nitrogen and phosphorus on high-pressure freezing ultrathin resin sections from cryofixed specimens. Among the secretory granules, high contents of N atoms were found in HSGI and II (Additional file 6), demonstrating their protein nature. P atoms were observed in HSGI and II (Additional file 6) although at a lower density when compared to N atoms. In summary, EELS and ESI showed the presence of proteins and potentially post-translationally modified by phosphorylation, in some granules of basal disc.

Two monoclonal antibodies had been previously applied to study nematocyte development [35] and basal disc cell differentiation [36, 37]. Notably, both antibodies also stained secreted material, however, this aspect was not further pursued in the earlier studies. Therefore, we applied immunohistochemistry with antibodies AE03 [35, 36], and 3G11 [37] to label basal disc cells in whole mount preparations or macerated single cells, and of footprints. Both antibodies were confirmed to label the basal disc of polyps as well as the secreted material (Fig. 7a-h). AE03 also recognized nematoblasts and mature nematocytes [35], but this immunoreactivity did not impede the present study which focused at basal disc cells and their secretion. In super-resolution microscopy images, AE03 showed a clear specificity to an inner ring of granular structures (Fig. 7c) whereas 3G11 also recognized granular structures as well as cytoplasmic constituents throughout the whole cell (Fig. 7g). However, we cannot infer which granule type was labelled with the

antibodies. Although in macerated cells the staining patterns of the two antibodies were different, the pattern appeared very similar in the footprint (Fig. 7d and h). The stained pattern of secreted material with a densely stained rim outlining the cell margins and a weaker stained flat centre, corroborates our earlier observation of a crystal violet stained footprint (Fig. 3c). Therefore, our results confirm that the granule material is in fact secreted by basal disc cells.

Discussion

Hydra is considered to represent the most basal animals with a defined body plan and organized epithelia. A common feature within the cnidarian group is their ability of producing bioadhesives either for attaching permanently or temporarily to surfaces through specialized ectodermal cells, and their ability for food capture through specialized nematocysts. Here we investigated the ability of temporary adhesion in *Hydra* through ectodermal basal disc cells (Fig. 8) which represents one of the most ancient metazoan ways of cell-to-surface adhesion.

In organisms well-known for their capacity to adhere temporarily, such as free living flatworms [5] and sea stars [10] the adhesive mechanism relies on a duo-gland system where two or more secretory cell types collaborate in a way that allows attachment and detachment by using adhesive and de-adhesive components. The number of cell types are variable, free living flatworms enclose one adhesive and one de-adhesive cell, while sea stars have two adhesive and one de-adhesive cell type. Some parasitic flatworms possess two different glands producing dissimilar adhesive components, while cells producing de-adhesive constituents are absent [38]. In contrast, substrate detachment in *Hydra* is mediated by differentiated epithelia muscle cells of the ectoderm of

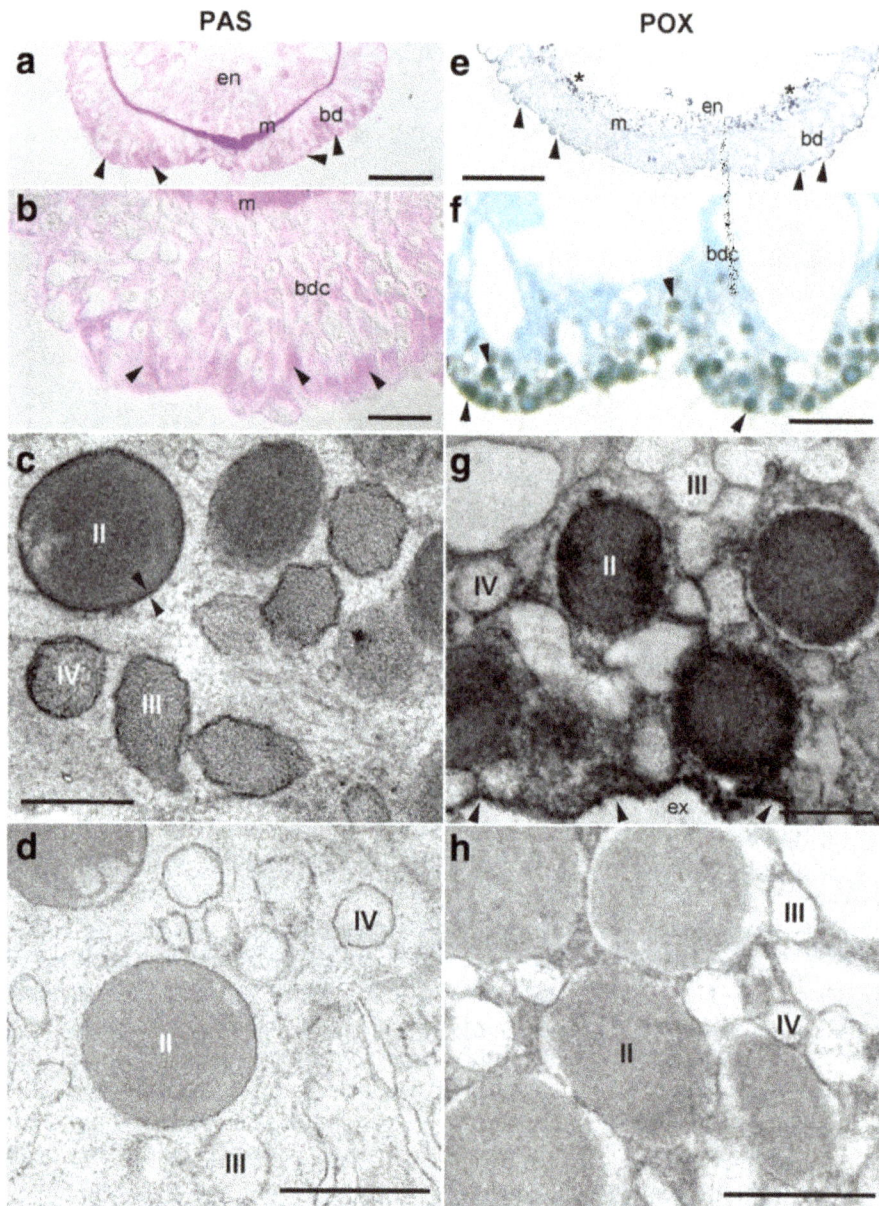

Fig. 6 Cytochemistry of *Hydra* basal disc. **a** Overview of semi-thin Epon section stained for glycans with PAS method. The reaction is stronger in the mesoglea and at the aboral tip [*arrows*] of the cell. **b** Magnification of **a**. *Arrows* indicate strong PAS reactions. Note the reaction gets stronger towards the aboral end of basal disc cells. **c** TEM- Periodic acid-thiocarbohydrazide-silver proteinate reaction (the EM-correlate PAS staining) was performed on cryo-fixed basal disc revealing the glycan distribution in the *Hydra* secretory granules. HSGII react positively in and close proximities of the granule membrane, while HSGIII and IV reacts quite uniformly. **d** Basal disc section not exposed to Periodic acid but regular thiocarbohydrazide-silver proteinate staining. Note secretory granules do not react to glycan staining. **e** Longitudinal semi-thin section through the basal disc stained with diamino benzidine and oxidized with Osmiun tetroxide. Endogenous peroxidase activity is strong at the aboral end of the basal disc [*arrows*] and endodermal lipid granules are dark from osmium fixation [*asterisks*]. **f** Longitudinal section of basal disc cells reacted for peroxidase after diamino bencidine staining counterstained with methylene blue and Azur II. *Arrows* point at some reacted secretory granules located at the most-aboral end of the cells. **g** TEM-peroxidase staining in chemically fixed basal disc is positive for some *Hydra* secretory granules II and their neighbouring cytoplasm. Arrows point at the most aboral end cell membrane, region with highly peroxidase activity. **h** Section from a negative control not exposed to diamino benzidine. *Scale bars* 100 µm (**a**), 50 µm (**b**, **e**), 20 µm (**f**), 1 µm (**c**, **d**, **g**, **h**). *Abbreviations*: PAS, periodic acid Schiff; POX, peroxidase; en, endoderm; m, mesoderm; bd, basal disc; bdc, basal disc cell; I, *Hydra* secretory granule I; II, *Hydra* secretory granule II; III, *Hydra* secretory granule III, IV, *Hydra* secretory granule IV; ex, exterior of the cell

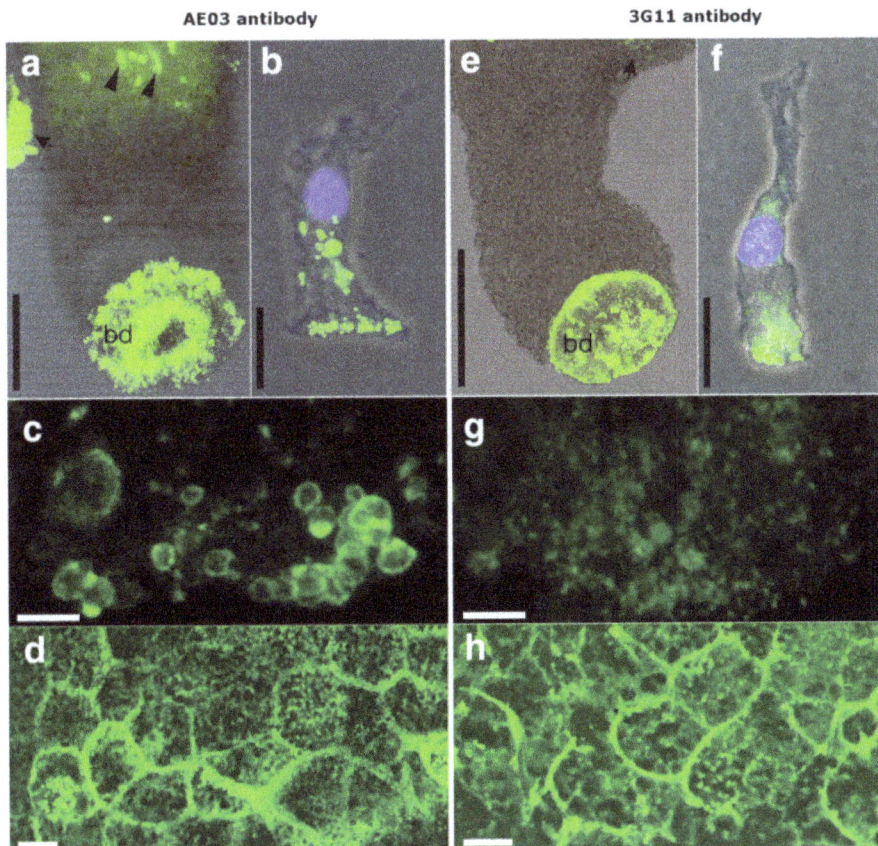

Fig. 7 Immunofluorescence staining of whole mount (merged bright field and fluorescence), macerated cell (merged phase contrast and fluorescence), and footprints with anti-AE03 and 3G11 antibodies. **a** Whole mount, **b** Macerated basal disc cell, **c** apical side of macerated cell imaged with super-resolution microscopy, and **d** footprint stained with AE03 antibody. **e** Whole mount, **f** macerated basal disc cell, **g** apical side of macerated cell imaged with super-resolution microscopy and **h** footprint stained with 3G11 antibody. Both AE03 and 3G11 have affinity to basal disc cells. Their corresponding antigens are eventually secreted and become a component of the adhesive anchoring *Hydra* polyps underwater. *Scale bars* 200 μm (**a**), 10 μm (**b, f**), 1 μm (**c, g**), 15 μm (**d, h**), 400 μm (**e**). Abbreviations: bd, basal disc; *arrows*, indicate bud; *arrowheads*, indicate nematocysts

basal disc. EM observations on secretory granules morphology in these cells [28, 29] with additional information on the molecules present. We have identified four morphologically distinct secretory granules (HSGs) as a major component of the basal disc cells that could be involved in adhesion although it remains unclear to which degree each granule type contributes to the adhesion process. We classified the HSGII as a mature granule derived from the HSGI, because these two granules share biochemical features, they are the only ones of protein nature, and contain phosphor. We also observed possible transition stages from HSGI to HSGII. Another feature is that HSGI is never seen close to the most-apical side of the cell. However, we have no direct proof for granule maturation. Therefore, this process must be further addressed in order to understand granule development.

The component that is actually labeled by the antibodies was not precisely identified. However, super-resolution microscopy experiments (Fig. 7d and h)

showed that both antibodies stained granular structures that are eventually secreted and become a constituent of the secreted material. The footprint of *Hydra* is a meshwork of quite homogenous strands. The starfish and sea cucumber footprint have a similar meshwork morphology [7, 39], but contain globular nanostructures which are absent in *Hydra*. In the case of *Hydra*, all HSG have spherical morphology before release and are not recognizable in the footprint. The contents of the granules seem to have merged or fused into larger aggregates forming the strands. In the footprints of sea cucumber [7], adhesive forces of 17nN were measured under dry conditions, and for the rootlets of the English ivy [40] 298 nN. Our measurements of *Hydra* footprints showed an adhesion force as high as 66 nN. Measurements in natural *Hydra* living conditions (i.e., under freshwater) would be necessary for a better understanding of its footprint adhesion forces. Yet, despite repeated attempts, the topographical visualization of the footprints under

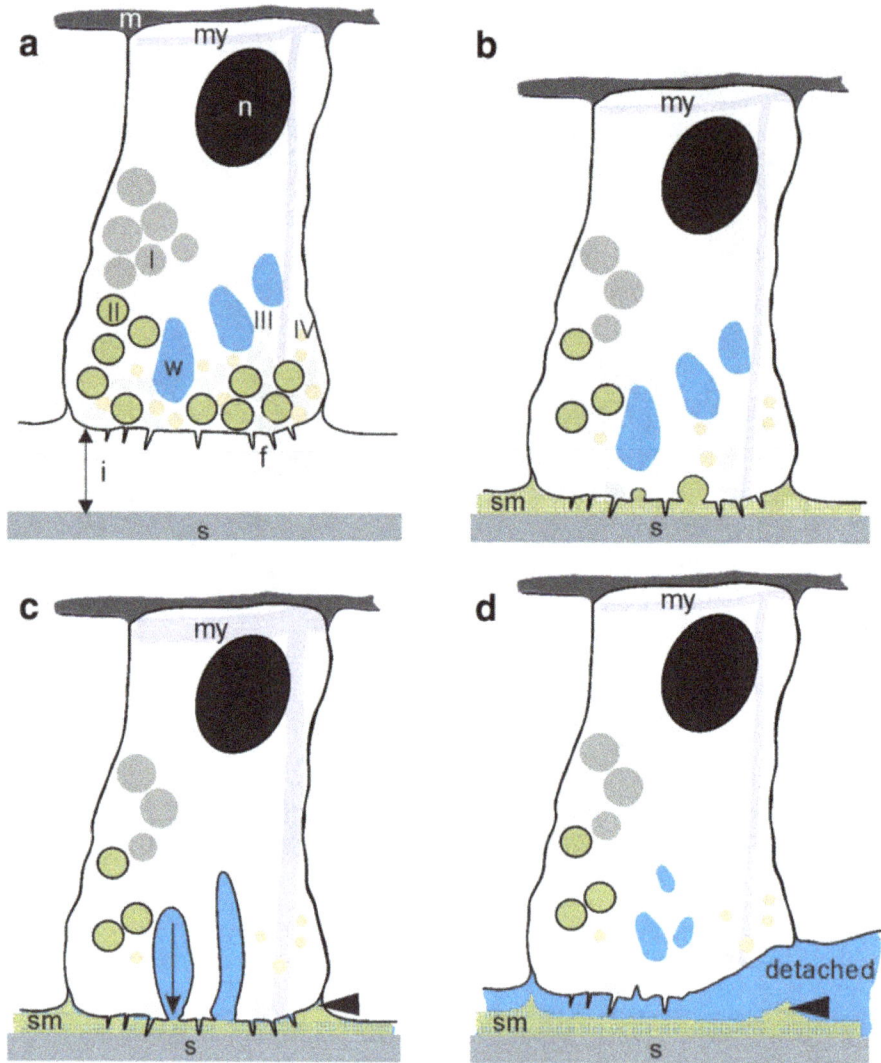

Fig. 8 Schematic representation of attachment (**a-c**) and detachment (**d**) of *Hydra*. Arrowheads indicate rims of footprints. Abbreviations: m, mesoglea; n, nucleus; my, myoneme; w, vacuoles of water; I-IV, *Hydra* secretory granule types; i, interface; s, substratum; f, cytoplasmic extensions; sm, secreted material; fo, footprint

native conditions in culture medium was not possible, due to the transparent nature of the footprints. New technical developments would be needed to provide an appropriate contrast for identifying the footprint underwater.

Glycans are considered important components for temporary adhesives, e.g., in cephalopods, gastropods, and echinoderms [10, 17, 41–43]. Permanent glues seems to be generally composed mainly of protein. Glycans have not been considered as an important moiety, though reported from mussels, stalked and acorn barnacles, sandcastle worms, and caddisfly larva [2, 11, 44, 45]. Although previous reports showed alcian blue staining of basal disc cells of other *Hydra* species [28, 29], our alcian blue experiments performed at pH-value 2.5, staining was negative for *H. magnipapillata*. In *Hydra*, EM cytochemistry showed

that glycans occur in all the HSG, with differential distributions (the present study). The presence of phosphor (P) in HSGI and II raises the question about possible protein phosphorylation of the *Hydra* adhesive. Phosphorus in the form of phosphate group may indeed be involved in phosphorylation, a PTM which confer additional functionalities to proteins, including adhesive proteins [16]. Phosphoproteins have been identified in a number of aquatic adhesives, e.g., in adults and larvae of barnacles [3, 34], in the caddisfly larvae [46], in mussels [47], in sandcastle worms [48], in sea cucumbers [16], and in kelp spores [49]. Although, covalently attached phosphate groups are usually present at substoichiometric levels (less than 5 % of the protein is modified) in intracellular proteins [50, 51], their proportion is usually much higher in extracellular structural proteins such as adhesive proteins. For instance, in Pc3 an adhesive protein from the sandcastle worm, up to 90 mol% (residues per 100 residues) of the serine residues are modified by phosphorylation [48]. The P concentration observed within HSGII (Additional file 6) is lower than that of N (Additional file 6). Both, glycosylation and phosphorylation, could be important in the adhesion process of *Hydra*.

Peroxidase activity in the basal disc of *Hydra* [52, 53] has been used as an important biochemical marker for tracking the reappearance of basal disc cells during polyp regeneration [53]. The *Hydra* genome [54] encodes five isoforms of putative peroxidases. Their functional significance for the animal is still under debate [55, 56], but most likely they play several roles. The obtained results are similar to the ones described earlier for other *Hydra* species [30]. Although peroxidase-like enzymes are highly concentrated in basal disc cells, it is not the adhesive *per se*, and its role in *Hydra's* adhesion remains to be elucidated. The peroxidase-like enzyme could catalyse the crosslinking of other components to post-draw the secreted adhesive as it occurs in the freshwater caddisfly silk [57]. Furthermore, it is well known that peroxidase-like enzymes possess antimicrobial features [58], and can function to either foster beneficial relationships or control pathogenesis. During the attachment process, the basal disc secretes a protein-rich adhesive material which might serves as a nutrient source for bacteria and the peroxidase may have a protective role, reducing bacterial biodegradation over the time protein needs for curing. The combination of both functions, i.e., curing and antimicrobial, may occur in *Hydra* adhesion, though the functional significance of this is yet to be examined.

In various organisms, detachment from temporary adhesion is controlled by additional secretory products, enzymes, or by creating forces. However, the detachment processes are largely not well understood. The observations on *Hydra* presented here support the hypothesis that its detachment is mechanically induced (Additional files 1 and 2). If chemical detachment were the case, the basal disc would detach at once, and individual basal disc cells would not be seen detaching individually. When detachment is necessary, myonemes within both ectoderm (radial myoneme) and endoderm (circular myoneme) composing the basal disc contract, leading to an expansion of both radial and circular myonemes. This retracts the attached cells from the surface that in combination with the longitudinal myonemes contracting the vacuoles of water which expels water into the footprint meshwork would pull the polyp off the cured adhesive.

Conclusion

This study showed that cnidarian *Hydra* polyps secrete elaborate adhesive composites underwater (freshwater) to temporarily anchor themselves to substrate surfaces. The adhesive system used by *Hydra* exhibits unique features among metazoans. Glue based adhesion is the main component of the system: basal disc cells release their adhesive vesicles whose contents would have the ability to spread over the surface, displace water, and create a proper environment for curing the secreted glue. Projecting structures were observed, but the way they functions is enigmatic. The adhesive components of *Hydra* are produced, stored and delivered by a single cell type, the ectodermal basal disc cells. Our results revealed that the secretory granules contained glycans and phosphorus, which are important components in other bioadhesive systems. A Peroxidase-like enzyme was associated with secretory granules and could play a role in *Hydra* adhesion. This work was intended to offer a first overview of the *Hydra* adhesive system. The characteristics presented here provide a basis for an ongoing project aimed at unravelling the molecular components of *Hydra'* glue.

Methods
Animals

Hydra magnipapillata strain 105 was used for all the experiments carried out in this study in compliance with animal welfare laws and policies (Austrian Law for animal experiments, TVG 2012, §1). Permanent mass cultures were bred and kept at 18 °C in growth chambers, and day/night light cycle at the Institute of Zoology, University of Innsbruck. *Hydra* cultures were fed five times per week with freshly hatched *Artemia* nauplii as previously described [59]. Under these conditions, animals remained asexual and reproduced by budding. We selected animals that had at least one bud. Animals were starved for 24 h before experiments. Before fixation, animals were relaxed in 2 % urethane in culture medium for 2 min.

Light microscopy

For bright field or differential interference contrast visualization, processed samples were examined with a Leica DM5000. Images were taken with a Leica DFC495 digital camera and a Leica LAS software.

Footprint: To collect *Hydra's* footprints, polyps were placed onto glass slides and allowed attach for 30 min. After this period, polyps were gently detached with the help of a glass pipet. Glass slides bearing footprints were rinsed three times with *Hydra* culture medium before staining. Fresh footprints were stained using a 0.05 % solution (in culture medium) of Crystal Violet, and rinsed in culture medium.

Squeezing preparations: Living polyps were anesthetized in a 2:1 mixture of 2 % Urethane and culture medium, transferred in a drop onto a slide and slightly squeezed under a coverslip. The specimens were observed with interference contrast under the same microscopy as mentioned above.

Histology: Adult *Hydra* polyps were fixed in Carnoy's fixative (ethanol, chloroform, glacial acetic acid, 6 + 3 + 1 respectively), Bouin's fluid (saturated picric acid, 36 % formaldehyde, and glacial acid, 15 + 3 + 1 respectively), 4 % paraformaldehyde (PFA) in 0.1 M Phosphate Buffer (PBS)) and/or in Flemmings fixative (1 % chromium (VI) oxide, 2 % osmium tetroxide and glacial acetic acid, 15 + 4 + 1 respectively), dehydrated and embedded into paraplast or in Technovit 7100 resin. Paraplast sections (7 μm) and resin sections (3 μm) were produced with a Reichert Autocut 2030 (Reichert, Austria) and stained with hematoxylin and eosin (HE), periodic acid Schiff (PAS), or alcian blue (AB) pH 2.5.

Enzymehistochemistry: For peroxidase activity, *Hydra* polyps were fixed with 4 % PFA in 0.1 M PBS, stained with diamino benzidine (DAB + CHROMOGEN, Dako) post fixed either with 2.5 % glutaraldehyde or 1 % osmium tetroxide, dehydrated and embedded in PolyBed 812 resin. Semi thin sections (350 nm to 500 nm) were cut with a Leica ultra-microtome UCT (Leica, Austria) and stained according to Richardson et al. [60].

Lipid staining: PFA fixed polyps were stained with the fluorescence Nile Red method as whole mounts for lipid detection following method used by Gohad et al. [34]. Negative controls were performed by exposing specimens to ETOH washes. Whole mounts were visualized under a Leica SP5 II confocal laser scanning microscope.

Antibody staining

Two antibodies labelling basal disc cells were used: AE03 [35, 36], and 3G11 [37] were kindly provided by the corresponding authors. The antibody staining method was slightly modified from the original protocols. Experiments were performed on whole mount, macerated cells and footprints. Samples were mounted in Vectashield (Vector), and visualized with a Leica DM5000, or a Leica SP5 II confocal scanning microscope. For super resolution microscopy, macerated cells samples were mounted in Mowiol and examined with a Leica TCS SP8 gSTED microscope system. Obtained super-resolution images were deconvoluted using the Huygens software from Scientific Volume Imaging implemented in the TCS SP8.

Whole mount preparation: For AE03 labelling, whole polyps were fixed in Zamboni's fixative (2 % PFA, 0.2 % picric acid in 0.1 M PBS pH 7.2). For 3G11 labelling, whole polyps were fixed in 4 % PFA. Both fixations were done at 4 °C overnight. The following steps were applied to both antibodies: After three washes with PBS, the polyps were permeabilized with 0.5 % Triton in PBS for 30 min, and incubated with 0.5 % Triton, 1 % bovine serum albumin (BSA, w/v) in PBS with primary antibody (dilutions = AE03 1:5, and 3G11 1:1000) overnight at 4 °C. After this period, polyps were washed three times in PBS, and incubated for 2 h with fluorescein isotthiocyanate-

conjugated (FITC) antimouse lgG (Dako) secondary antibody (1:200). Polyps were washed again three times in PBS and mounted.

Macerated cells: Basal discs (from approx. 150 polyps) were excised and incubated in 200 μl maceration medium (acetic acid, glycerol, and distilled water, 1:1:7) for 2 h at 30 °C. Basal discs were then mechanically disrupted by shearing them through the opening (roughly 1 mm diameter) of a pipette. The same amount of fixative, either Zamboni or PFA, were added to the medium containing cells and gently mixed. 50 μl of the sample were spread onto gelatine-coated slides and allowed dry for 20 min at RT. Steps for antibody staining were as for whole mount. Differences were a Triton concentration of 0.1 %, and an incubation time for the secondary antibody of 4 h. Slides were additionally counterstained with the DNA-specific fluorochrome, Hoechst 33342 (Life Technologies; 1 μg/ml). Samples examined with super-resolution microscopy were incubated with antimouse abberior STAR 488 (Abberior) secondary antibody diluted 1:100.

Footprint: Secreted material was collected on glass slides as described for light microscopy purposes. Immunofluorescence staining with AE03 and 3G11 was carried out as described above.

We tried to elucidate the subcellular binding of AE03 and 3G11, but several different immunogold approaches failed: i) post-embedded immunogold on both, cryo and chemically fixed material [61], ii) post-embedded immunogold on thawed cryosections according Tokuyasu [62], and iii) pre-embedded with horseradish labelled streptavidin conjugated antibodies.

Phalloidin staining

Configuration of actin filaments in the basal disc were detected by phalloidin staining. Experiments were performed using amputated basal discs. Animals were let to attach to a glass slide for approximately 1 h and were then amputated. Basal discs were fixed with 4 % PFA for 1 h at room temperature, then washed three times for 10 min in PBS-0.5 % Triton, and then incubated in Alexa 488 phalloidin (Invitrogen) in a concentration of 1:400 for 1 h at RT in the dark. Afterwards, they were washed three times for 10 min with PBS. Samples were mounted in Vectashield (Vector), and visualized with a Leica SP5 II confocal scanning microscope.

Scanning electron microscopy

Hydra polyps were fixed in 4 % PFA for 24 h. They were dehydrated in graded ETOH, dried by the critical point method (with CO_2 as transition fluid), mounted on aluminium stubs, coated with gold in a sputter coater, and observed with a JEOL JSM-6100 scanning electron microscope.

Transmission electron microscopy

Conventional chemical fixation and cryofixation were performed basically as described by Holstein et al. [61]. *Chemical fixation*: Hydra polyps were allowed to attach on a dialysis membrane, relaxed in 2 % urethane for 3 min and immediately fixed with a combined 2.5 % glutaraldehyde and 1 % osmium tetroxide fixative, dehydrated in a graded acetone series and embedded in Polybed 812 resin.

High pressure freezing (HPF) and freeze substitution

Basal discs were dissected and frozen with a HPM-010 (HPF apparatus from BAL-TEC, Baltzers, Liechtenstein), freeze substituted with acetone containing osmium tetroxide and uranyl acetate, and embedded into PolyBed 812 as previously described [61]. Thick sections (350 nm) and ultrathin sections (70 nm) were cut with a Leica ultramicrotome UCT (Leica, Austria), mounted on copper grids and stained with uranyl acetate and lead citrate and examined with a Zeiss Libra 120 energy filter transmission electron microscope using zero loss electrons. Images were taken with a TRS 2048 high speed camera (Tröndle, Germany) and visualized through Olympus SiS iTEM 5.0 software.

Cytochemical detection of PAS-positive 1–2 vicinal diols was carried out according to Thiery [63]. Sections from cryofixed samples were mounted on gold grids, exposed to periodic acid, thiocarbohydrazide, and silver proteinate. Negative control included omission of periodic acid treatment.

Peroxidase activity was detected in *Hydra* polyps fixed with 4 % PFA in 0.1 M PBS. They were stained with diamino benzidine (DAB + CHROMOGEN, Dako), and post fixation with 2.5 % glutaraldehyde/1 % osmium tetroxide, dehydration and embedding into Polybed 812. Images were taken without any further section post-staining. Controls included the inhibition of peroxidase activity by incubating samples in 3 % hydrogen peroxide for 20 min.

Electron energy loss spectroscopy (EELS) and electron spectroscopic imaging (ESI) were performed on ultrathin sections in order to detect element distribution in structures rich in nitrogen (N) and phosphorus (P) within the basal disc cells. The EELS charts and ESI images were collected using the software iTEM 5.0$^{©}$. Distribution of N and P was measured according to a three-window power law difference ESI model with an energy slit width of 15 eV. Here, two background images and one image at the ionization edge of the appropriate element, K edge 397 eV for N and $L_{2,3}$ edge at 129 eV for P were taken. Maximum element distribution was finally mix mapped with an inverted high contrast image taken at 250 eV.

The window-one was placed at 382 eV, the window-two at 350 eV, and window-three at 410 eV. The difference of the three windows coincide with the onset of the N –K ionization edge at 397 eV. The background image was set with the subtraction model obtained from the EELS analyses default, which ensures that only the energy loss from the ion under examination is mapped. Likewise, for constructing the distribution map of P, the window-one was placed at 121 eV, window-two at 110, and window-three at 153 eV. The energy-loss contribution of three windows coincide with the onset of P –$L_{2,3}$ ionization edge of 129 eV.

Atomic force microscopy

Footprints from *Hydra* were collected on glass slides. The samples were let dry at room temperature and analyzed with Atomic Force Microscopy (AFM- Peak Force Tapping™, PFT). Data collection was achieved by applying controlled, low forces on the tip of the cantilever during imaging, which allows a direct comparison between the morphology and the adhesive properties at the nanometer (nm) scale. With the PFT method, an adhesion profile can be obtained by evaluating one force-distance curve for each pixel of the obtained image. Thereof, the adhesion force is defined as the maximum force needed to pull off the cantilever tip. The probe (silicon tip on silicon nitride cantilever—SNL, Bruker, Santa Barbara, CA, USA, k¼0.12 N/m) was calibrated on a stiff surface prior to the experiment for the measurements of the mechanical properties, in order to quantify the tip sample force.

Additional files

Additional file 1: *Hydra magnipapillata* dettaching from a glass slide in real time. (MP4 3304 kb)

Additional file 2: Phalloidin staining of actin filaments visualized with confocal microscopy. a overview of basal disc myoneme organization. The square indicates the area magnified in figure b. b detail of actin filaments in the basal disc. c lateral view of basal disc showing actin filaments which branched perpendicularly towards the apical side. Arrowheads point at actin filaments from basal disc cells, and arrows indicate the ones from endodermal cells. *Scale bars* 50 μm (a), 15 μm (b, c). *Abbreviation*: ap, aboral pore; m, mesoglea; bdc, ectodermal basal disc cells. (TIF 18280 kb)

Additional file 3: Transmission electron microscopy images of chemically fixed basal discs from *Hydra magnipapillata*. a 350 nm sections from longitudinal section of apical side of basal disc cell. Note that given the thickness of sections, overlapping secretory granules can be seen. b 350 nm sections from longitudinal section of basal side of basal disc cell. Myoneme branching towards the apical side are seen. c 70 nm section longitudinal section of basal disc showing possible HSGII developing stages. *Scale bars* 2 μm (a), 1 μm (b, c). *Abbreviations*: I, *Hydra* secretory granule I; II, *Hydra* secretory granule II; I/II, possible developing transition between HSGI and HSGII; III, *Hydra* secretory granule III; IV, *Hydra* secretory granule IV; ex, exterior of the cell; m, mesoglea; my, myoneme; er, endoplasmic reticulum; arrows, indicate cytoplasmic extensions; asterisks, indicate vacuoles of water; arrowheads, indicate myoneme branching towards apical side of basal disc cell. (TIF 9364 kb)

Additional file 4: Cytochemistry of *Hydra* basal disc. a longitudinal section of a basal disc stained with alcian blue. Basal disc cells did not react for acidic mucopolysaccharides. b Longitudinal section through a *Hydra* tentacle. Nematocytes were alcian blue positive (arrows) corroborating the method used was correct. *Scale bars* 100 μm (a), 50 μm (b). *Abbreviations*: bd, basal disc; m, mesoglea; en, endoderm (TIF 3667 kb)

Additional file 5: Lipidaceous granules distribution within *Hydra* basal disc. Whole mounts were stained with Nile Red. a-a''' *Hydra* polyps contain few scattered lipid droplets within basal disc cells. Unlike basal disc cells, the endoderm is rich in lipids. b-b''' negative control was established after a series of ETOH washes. *Scale bars* 30 μm. *Abbreviations*: bd, basal disc; m, mesoglea; en, endoderm (TIF 6150 kb)

Additional file 6: Electron spectral imaging (ESI), and electron energy loss spectroscopy (EELS) performed in sections of cryo-fixed basal disc. a Merged ESI micrographs revealing the nitrogen atoms profile –green dots. Nitrogen atoms are densely distributed in *Hydra* secretory granule II, covering its full surface. b Merged ESI micrographs depicting P atoms distribution –green dots. Note P atoms are found in the same secretory granules as N but in much lower density. *Scale bars* 1 μm. *Abbreviations*: ex, exterior of the cell. (TIF 2992 kb)

Abbreviations
AB, alcian blue; EELS, electron energy loss spectroscopy; EM, electron microscopy; ER, endoplasmic reticulum; ESI, electron spectroscopic imaging; HPF, high pressure freezing; HSG I-IV, Hydra secretory granule one to four; PAS, periodic acid Schiff

Acknowledgements
The authors are most grateful to Karin Gutleben and Thi Chinh Ngo for excellent assistance for high pressure freezing/freeze substitution, and atomic force microscopy, respectively. We are indebted to Marina P. Samoylovich, Yoshitaka Kobayakawa, and Alexander V. Klimovich for kindly providing the antibodies used in this study. We also thank Martin Offterdinger for support during SP8 experiments.

Funding
MR was supported by Marie-Curie FP7-PEOPLE-2013-IEF 626525 fellowship, and COST Action TD0906. The project was also supported by Austrian Science Fund (FWF) grant 25404-B25.

Authors' contributions
MR and PLa conceived the study, performed experiments and wrote the paper. PLe performed and supervised AFM experiments. PF contributed to electron microscopy. MWH contributed to cryo-based electron microscopy and (immune) cytochemistry. WS performed histological and ultrastructural experiments. BH contributed on light microscopy, and cell macerations and Hydra biology. All authors have read and approved the manuscript.

Competing interests
The authors declare that they have no competing interests.

Author details
[1]Institute of Zoology and Center for Molecular Biosciences Innsbruck, University of Innsbruck, Innsbruck, Austria. [2]Laboratory for Chemistry of Novel Materials, Research Institute for Material Sciences and Engineering, Center of Innovation and Research in Materials and Polymers, University of Mons, Mons, Belgium. [3]Biology of Marine Organisms and Biomimetics, Research Institute for Biosciences, University of Mons, Mons, Belgium. [4]Division of Histology and Embryology, Innsbruck Medical University, Innsbruck, Austria.

References
1. Waite JH, Andersen NH, Jewhurst S, Sun C. Mussel adhesion: finding the tricks worth mimicking. J Adhes. 2005;81:297–317.
2. Jonker J-L, von Byern J, Flammang P, Waltraud K, Power AM. Unusual adhesive production system in the barnacle *Lepas anatifera*: an Ultrastructural and histochemical investigation. J Morphol. 2012;273:1377–91.
3. Zheden V, Klepal W, von Byern J, Bogner FR, Thiel K, Kowalik T, Grunwald I. Biochemical analyses of the cement float of the goose barnacle *Dosima fascicularis* – a preliminary study. Biofouling. 2014;30:949–63.
4. Hennebert E, Wattiez R, Demeuldre M, Ladurner P, Hwang DS, Waite JH, Flammang P. Sea star tenacity mediated by a protein that fragments, then aggregates. Proc Natl Acad Sci U S A. 2014;111:6317–22.
5. Lengerer B, Pjeta R, Wunderer J, Rodrigues M, Arbore R, Schärer L, Berezikov E, Hess MW, Pfaller K, Egger B, Obwegeser S, Salvenmoser W, Ladurner P. Biological adhesion of the flatworm *Macrostomum lignano* relies on a duo-gland system and is mediated by a cell type-specific intermediate filament protein. Front Zool. 2014;11:12.
6. Wang CS, Stewart RJ. Localization of the bioadhesive precursors of the sandcastle worm, *Phragmatopoma californica* (Fewkes). J Exp Biol. 2012; 215(Pt 2):351–61.
7. Demeuldre M, Chinh Ngo T, Hennebert E, Wattiez R, Leclère P, Flammang P. Instantaneous adhesion of Cuvierian tubules in the sea cucumber *Holothuria forskali*. Biointerphases. 2014;9:029016.
8. Smith AM, Callow JA. Biological Adhesives. Berlin: Springer Berlin Heidelberg; 2006.
9. von Byern J, Grunwald I. Biological Adhesive Systems. Vienna: Springer Vienna; 2010.
10. Flammang P, Michel A, Cauwenberge A, Alexandre H, Jangoux M. A study of the temporary adhesion of the podia in the sea star *Asterias rubens* (Echinodermata, asteroidea) through their footprints. J Exp Biol. 1998;201(Pt 16):2383–95.
11. Walker G. The biochemical composition of the cement of two barnacle species, *balanus hameri* and *balanus crenatus*. J Mar Biol Assoc U K. 1972;52: 429–35.
12. Bandara N, Zeng H, Wu J. Marine mussel adhesion: biochemistry, mechanisms, and biomimetics. J Adhes Sci Technol. 2013;27:2139–62.
13. Lee H, Dellatore SM, Miller WM, Messersmith PB. Mussel-inspired surface chemistry for multifunctional coatings. Science. 2007;318:426–30.
14. Becker PT, Lambert A, Lejeune A, Lanterbecq D, Flammang P. Identification, characterization, and expression levels of putative adhesive proteins from the tube-dwelling polychaete *Sabellaria alveolata*. Biol Bull. 2012;223:217–25.
15. Sagert J, Sun C, Waite JH. Chemical Subtleties of Mussel and Polychaete Holdfasts. In: Smith DAM, Callow DJA, editors. Biological Adhesives. Berlin: Springer Berlin Heidelberg; 2006. p. 125–43.
16. Flammang P, Lambert A, Bailly P, Hennebert E. Polyphosphoprotein-containing marine adhesives. J Adhes. 2009;85:447–64.
17. Hennebert E, Wattiez R, Flammang P. Characterisation of the carbohydrate fraction of the temporary adhesive secreted by the tube feet of the sea star *Asterias rubens*. Mar Biotechnol N Y N. 2011;13:484–95.
18. Quinn B, Gagné F, Blaise C. *Hydra*, a model system for environmental studies. Int J Dev Biol. 2012;56:613–25.
19. Steele RE. Developmental signaling in *hydra*: what does It take to build a "simple" animal? Dev Biol. 2002;248:199–219.
20. Lentz T. The Cell Biology of Hydra. Amsterdam: North-Holland publishing company; 1966.
21. Dübel S, Hoffmeister SA, Schaller HC. Differentiation pathways of ectodermal epithelial cells in *Hydra*. Differ Res Biol Divers. 1987;35:181–9.
22. Hobmayer B, Jenewein M, Eder D, Eder M-K, Glasauer S, Gufler S, Hartl M, Salvenmoser W. Stemness in *Hydra* - a current perspective. Int J Dev Biol. 2012;56:509–17.
23. Bode HR. Axial patterning in *hydra*. Cold Spring Harb Perspect Biol. 2009;1: a000463.
24. Holstein TW, Hobmayer E, Technau U. Cnidarians: an evolutionarily conserved model system for regeneration? Dev Dyn. 2003;226:257–67.
25. Bosch TCG. *Hydra* and the evolution of stem cells. BioEssays. 2009;31:478–86.
26. Bosch TCG. Why polyps regenerate and we don't: towards a cellular and molecular framework for *hydra* regeneration. Dev Biol. 2007;303:421–33.
27. Chaet AB. Invertebrate adhering surfaces: secretions of the starfish, *Asterias forbesi*, and the coelenterate, *Hydra pirardi*. Ann N Y Acad Sci. 1965;118:921–9.
28. Philpott DE, Chaet AB, Burnett AL. A study of the secretory granules of the basal disk of *Hydra*. J Ultrastruct Res. 1966;14:74–84.
29. Davis LE. Histological and ultrastructural studies of the basal disk of *Hydra*. Z Für Zellforsch Mikrosk Anat. 1973;139:1–27.

30. Hoffmeister-Ullerich SAH, Herrmann D, Kielholz J, Schweizer M, Schaller HC. Isolation of a putative peroxidase, a target for factors controlling foot-formation in the coelenterate Hydra. Eur J Biochem. 2002;269:4597–606.

31. Pan HC, Yang HQ, Zhao FX, Qian XC. Molecular cloning, sequence analysis, prokaryotic expression, and function prediction of foot-specific peroxidase in Hydra magnipapillata Chinese strain. Genet Mol Res. 2014;13:6610–22.

32. Rodrigues M, Lengerer B, Ostermann T, Ladurner P. Molecular biology approaches in bioadhesion research. Beilstein J Nanotechnol. 2014;5:983–93.

33. Alibardi L, Edward D, Patil L, Bouhenni R, Dhinojwala A, Niewiarowski P. Histochemical and ultrastructural analyses of adhesive setae of lizards indicate that they contain lipids in addition to keratins. J Morphol. 2011;272:758–68.

34. Gohad NV, Aldred N, Hartshorn CM, Jong Lee Y, Cicerone MT, Orihuela B, Clare AS, Rittschof D, Mount AS. Synergistic roles for lipids and proteins in the permanent adhesive of barnacle larvae. Nat Commun. 2014;5:4414.

35. Amano H, Koizumi O, Kobayakawa Y. Morphogenesis of the atrichous isorhiza, a type of nematocyst, in Hydra observed with a monoclonal antibody. Dev Genes Evol. 1997;207:413–6.

36. Kobayakawa Y, Kodama R. Foot formation in Hydra: commitment of the basal disk cells in the lower peduncle. Dev Growth Differ. 2002;44:517–26.

37. Shirokova VN, Begas OS, Knyazev NA, Samoilovich MP. Antigenic marker of differentiated cells of a Hydra basal disc. Cell Tissue Biol. 2009;3:84–92.

38. Whittington ID, Armstrong WD, Cribb BW. Mechanism of adhesion and detachment at the anterior end of Neoheterocotyle rhinobatidis and Troglocephalus rhinobatidis (Monogenea: Monopisthocotylea: Monocotylidae). Parasitol Res. 2004;94:91–5.

39. Hennebert E, Viville P, Lazzaroni R, Flammang P. Micro- and nanostructure of the adhesive material secreted by the tube feet of the sea star Asterias rubens. J Struct Biol. 2008;164:108–18.

40. Xia L, Lenaghan SC, Zhang M, Wu Y, Zhao X, Burris Jr JN, Stewart Jr CN. Characterization of English ivy (Hedera helix) adhesion force and imaging using atomic force microscopy. J Nanoparticle Res. 2010;13:1029–37.

41. von Byern J, Klepal W. Adhesive mechanisms in cephalopods: a review. Biofouling. 2006;22:329–38.

42. Flammang P. Adhesive Secretions in Echinoderms: An Overview. In: Smith DAM, Callow DJA, editors. Biological Adhesives. Berlin: Springer Berlin Heidelberg; 2006. p. 183–206.

43. Smith AM. Gastropod Secretory Glands and Adhesive Gels. In: von Byern DD-BJ, Grunwald DD-BI, editors. Biological Adhesive Systems. Vienna: Springer; 2010. p. 41–51.

44. Silverman HG, Roberto FF. Understanding marine mussel adhesion. Mar Biotechnol N Y N. 2007;9:661–81.

45. Stewart RJ, Ransom TC, Hlady V. Natural underwater adhesives. J Polym Sci Part B Polym Phys. 2011;49:757–71.

46. Stewart RJ, Wang CS. Adaptation of caddisfly larval silks to aquatic habitats by phosphorylation of H-fibroin serines. Biomacromolecules. 2010;11:969–74.

47. Waite JH, Qin X. Polyphosphoprotein from the adhesive pads of Mytilus edulis†. Biochemistry (Mosc). 2001;40:2887–93.

48. Zhao H, Sun C, Stewart RJ, Waite JH. Cement proteins of the tube-building polychaete phragmatopoma californica. J Biol Chem. 2005;280:42938–44.

49. Petrone L, Easingwood R, Barker MF, McQuillan AJ. In situ ATR-IR spectroscopic and electron microscopic analyses of settlement secretions of Undaria pinnatifida kelp spores. J R Soc Interface. 2011;8:410–22.

50. Engholm-Keller K, Larsen MR. Technologies and challenges in large-scale phosphoproteomics. Proteomics. 2013;13:910–31.

51. Jensen ON. Interpreting the protein language using proteomics. Nat Rev Mol Cell Biol. 2006;7:391–403.

52. Hand AR. Ultrastructural localization of catalase and L-alpha-hydroxy acid oxidase in microperoxisomes of Hydra. J Histochem Cytochem. 1976;24:915–25.

53. Hoffmeister S, Schaller HC. A new biochemical marker for foot-specific cell differentiation in Hydra. Wilhelm Rouxs Arch Dev Biol. 1985;194:453–61.

54. Chapman JA, Kirkness EF, Simakov O, Hampson SE, Mitros T, Weinmaier T, Rattei T, Balasubramanian PG, Borman J, Busam D, Disbennett K, Pfannkoch C, Sumin N, Sutton GG, Viswanathan LD, Walenz B, Goodstein DM, Hellsten U, Kawashima T, Prochnik SE, Putnam NH, Shu S, Blumberg B, Dana CE, Gee L, Kibler DF, Law L, Lindgens D, Martinez DE, Peng J, et al. The dynamic genome of Hydra. Nature. 2010;464:592–6.

55. Böttger A, Doxey AC, Hess MW, Pfaller K, Salvenmoser W, Deutzmann R, Geissner A, Pauly B, Altstätter J, Münder S, Heim A, Gabius H-J, McConkey BJ, David CN. Horizontal gene transfer contributed to the evolution of extracellular surface structures: the freshwater polyp hydra is covered by a complex fibrous cuticle containing glycosaminoglycans and proteins of the PPOD and SWT (sweet tooth) families. PLoS ONE. 2012;7.

56. Technau U, Miller MA, Bridge D, Steele RE. Arrested apoptosis of nurse cells during Hydra oogenesis and embryogenesis. Dev Biol. 2003;260:191–206.

57. Wang C-S, Ashton NN, Weiss RB, Stewart RJ. Peroxinectin catalyzed dityrosine crosslinking in the adhesive underwater silk of a casemaker caddisfly larvae, Hysperophylax occidentalis. Insect Biochem Mol Biol. 2014;54:69–79.

58. Weis VM, Small AL, McFall-Ngai MJ. A peroxidase related to the mammalian antimicrobial protein myeloperoxidase in the Euprymna–Vibrio mutualism. Proc Natl Acad Sci. 1996;93:13683–8.

59. Hobmayer B, Holstein TW, David CN. Stimulation of tentacle and bud formation by the neuropeptide head activator in Hydra magnipapillata. Dev Biol. 1997;183:1–8.

60. Richardson KC, Jarett L, Finke EH. Embedding in epoxy resins for ultrathin sectioning in electron microscopy. Biotech Histochem. 1960;35:313–23.

61. Holstein TW, Hess MW, Salvenmoser W. Preparation techniques for transmission electron microscopy of Hydra. Methods Cell Biol. 2010;96:285–306.

62. Tokuyasu KT. A technique for ultracryotomy of cell suspensions and tissues. J Cell Biol. 1973;57:551–65.

63. Thiéry J. Mise en évidence des polysaccharides sur coupes fines en microscopie électronique. J Microsc. 1967;6:987–1018.

Permissions

The contributors of this book come from diverse backgrounds, making this book a truly international effort. This book will bring forth new frontiers with its revolutionizing research information and detailed analysis of the nascent developments around the world.

We would like to thank all the contributing authors for lending their expertise to make the book truly unique. They have played a crucial role in the development of this book. Without their invaluable contributions this book wouldn't have been possible. They have made vital efforts to compile up to date information on the varied aspects of this subject to make this book a valuable addition to the collection of many professionals and students.

This book was conceptualized with the vision of imparting up-to-date information and advanced data in this field. To ensure the same, a matchless editorial board was set up. Every individual on the board went through rigorous rounds of assessment to prove their worth. After which they invested a large part of their time researching and compiling the most relevant data for our readers.

The editorial board has been involved in producing this book since its inception. They have spent rigorous hours researching and exploring the diverse topics which have resulted in the successful publishing of this book. They have passed on their knowledge of decades through this book. To expedite this challenging task, the publisher supported the team at every step. A small team of assistant editors was also appointed to further simplify the editing procedure and attain best results for the readers.

Apart from the editorial board, the designing team has also invested a significant amount of their time in understanding the subject and creating the most relevant covers. They scrutinized every image to scout for the most suitable representation of the subject and create an appropriate cover for the book.

The publishing team has been an ardent support to the editorial, designing and production team. Their endless efforts to recruit the best for this project, has resulted in the accomplishment of this book. They are a veteran in the field of academics and their pool of knowledge is as vast as their experience in printing. Their expertise and guidance has proved useful at every step. Their uncompromising quality standards have made this book an exceptional effort. Their encouragement from time to time has been an inspiration for everyone.

The publisher and the editorial board hope that this book will prove to be a valuable piece of knowledge for researchers, students, practitioners and scholars across the globe.

List of Contributors

Naim Saglam
Department of Aquaculture and Fish Diseases, Fisheries Faculty, Firat University, 23119 Elazig, Turkey

Ralph Saunders, Daniel H. Shain
Biology Department, Rutgers The State University of New Jersey, 315 Penn Street, Camden, NJ 08102, USA

Shirley A. Lang
Rowan University Graduate School of Biomedical Sciences at SOM, Stratford, NJ 08084, USA

Vesa Selonen
Department of Biology, Section of Ecology, FI-20014 University of Turku, Turku, Finland

Ralf Wistbacka
Department of Biology, FI-90014 University of Oulu, Oulu, Finland

Andrea Santangeli
The Helsinki Lab of Ornithology, Finnish Museum of Natural History, University of Helsinki, Helsinki, Finland

David Crouse
Department of Computer Science and Engineering, Michigan State University, East Lansing, MI, USA
Present address: Samsung Semiconductor, Inc., San Jose, CA, USA

Rachel L. Jacobs
Department of Anthropology, Center for the Advanced Study of Human Paleobiology, The George Washington University, Washington, DC, USA

Zach Richardson
Department of Computer Science and Engineering, Michigan State University, East Lansing, MI, USA

Scott Klum
Department of Computer Science and Engineering, Michigan State University, East Lansing, MI, USA
Present address:Noblis, Inc., Annandale, VA, USA

Anil Jain
Department of Computer Science and Engineering, Michigan State University, East Lansing, MI, USA

Andrea L. Baden
Department of Anthropology, Hunter College, City University of New York, New York, NY, USA
The Graduate Center of City University of New York, New York, NY, USA
The New York Consortium in Evolutionary Primatology (NYCEP), New York, NY, USA

Stacey R. Tecot
School of Anthropology, The University of Arizona, Tucson, AZ, USA

Dieter Slos, Wim Bert
Department of Biology, Nematology Research Unit, Ghent University, K.L. Ledeganckstraat 35, 9000 Ghent, Belgium

Walter Sudhaus
Institut für Biologie/Zoologie,Freie Universität Berlin, Königin-Luise-Str. 1-3, 14195 Berlin, German

Lewis Stevens, Mark Blaxter
Institute of Evolutionary Biology, University of Edinburgh, Edinburgh EH9 3FL, UK

Philina A. English, David J. Green
Department of Biological Sciences, Simon Fraser University, Burnaby, BC V5A 1S6, Canada

Alexander M. Mills
Department of Biology, York University, Toronto, ON M3J 1P3, Canada

Michael D. Cadman
Canadian Wildlife Service, Environment and Climate Change Canada, Burlington, ON L7S 1A1, Canada

Audrey E. Heagy
Bird Studies Canada, Front Street, Port Rowan, ON N0E 1M0, Canada

Greg J. Rand
Canadian Museum of Nature, 1740 Pink Road, Gatineau, QC J9J 3N7, Canada

Joseph J. Nocera
Faculty of Forestry and Environmental Management, University of New Brunswick, Fredericton, NB E3B 5A3, Canada.

Suvi Viranta
Department of Anatomy, Faculty of Medicine, University of Helsinki, Helsinki, Finland

Anagaw Atickem
Department of Biosciences, Centre for Ecological and Evolutionary Synthesis (CEES), University of Oslo, Blindern, N-0316 Oslo, Norway
Cognitive Ethology Laboratory, German Primate Center, Kellnerweg
Department of Zoological Sciences, Addis Ababa University, Addis Ababa, Ethiopia.

Lars Werdelin
Department of Palaeobiology, Swedish Museum of Natural History, S-10405 Stockholm, Sweden.

Nils Chr. Stenseth
Department of Biosciences, Centre for Ecological and Evolutionary Synthesis (CEES), University of Oslo, Blindern, N-0316 Oslo, Norway
Department of Zoological Sciences, Addis Ababa University, Addis Ababa, Ethiopia.

Pushpinder S. Jamwal, Pankaj Chandan, Rohit Rattan
Western Himalayas Landscape, WWF-India, New Delhi 110003, India

Anupam Anand
Global Land Cover Facility, University of Maryland, College Park, MD 20742, USA

Prameek M. Kannan
Department of Biology, Pace University, 861 Bedford Road, Pleasantville, NY 10570, USA

Michael H. Parsons
Department of Biology, Hofstra University, Hempstead, NY 11549, USA
Department of Biological Sciences, Fordham University, Bronx, NY 10458, USA

Claire L. McAroe
School of Biological Sciences, Queen's University Belfast, Medical Biology Centre, 97 Lisburn Road, Belfast BT9 7BL, Northern Ireland
School of Psychology, Queen's University Belfast, University Road, Belfast BT7 1NN, Northern Ireland

Cathy M. Craig
School of Psychology, Queen's University Belfast, University Road, Belfast BT7 1NN, Northern Ireland.

Richard A. Holland
School of Biological Sciences, Queen's University Belfast, Medical Biology Centre, 97 Lisburn Road, Belfast BT9 7BL, Northern Ireland
Current address, School of Biological Sciences, Bangor University, Deiniol Road, Bangor LL57 2UW, UK

Kevin A. M. Sullivan, Roy N. Platt II, David A. Ray
Department of Biological Sciences, Texas Tech University, Lubbock, TX 79409, USA

Robert D. Bradley
Department of Biological Sciences, Texas Tech University, Lubbock, TX
79409, USA
Museum of Texas Tech University, Lubbock, TX 79409, USA

Riley F. Bernard
Department of Ecology and Evolutionary Biology, University of Tennessee, Knoxville, TN, USA
Department of Ecosystem Science and Management, Pennsylvania State University, University Park, Pennsylvania, USA

Emma V. Willcox
Department of Forestry, Wildlife, and Fisheries, University of Tennessee, Knoxville, TN, USA

Katy L. Parise, Jeffrey T. Foster
Center for Microbial Genetics and Genomics, Northern Arizona University, Flagstaff, AZ, USA
Department of Molecular, Cellular, and Biomedical Sciences, University of New Hampshire, Durham, NH, USA

Gary F. McCracken
Department of Ecology and Evolutionary Biology, University of Tennessee, Knoxville, TN, USA

Uthpala A. Jayawardena
Postgraduate Institute of Science, University of Peradeniyai, Peradeniya, Sri Lanka
Department of Zoology, University of Peradeniya, Peradeniya, Sri Lanka

Jason R. Rohr
Department of Integrative Biology, University of South Florida, Tampa, FL, USA

Priyanie H. Amerasinghe
International Water Management Institute, C/o
ICRISAT,Patancheru – 502, Hyderabad, Andhra
Pradesh 324, India

Ayanthi N. Navaratne
Department of Chemistry, University of Peradeniya,
Peradeniya, Sri Lanka

Rupika S. Rajakaruna
Department of Zoology, University of Peradeniya,
Peradeniya, Sri Lanka

**Ryo Nozu , Rui Matsumoto, Makio Yanagisawa,
Keiichi Ueda, Keiichi Sato**
Okinawa Churashima Research Center, Okinawa
Churashima Foundation, 888 Ishikawa, Motobu,
Okinawa 905-0206, Japan
Okinawa Churaumi Aquarium, 424 Ishikawa,
Motobu, Okinawa 905-0206, Japan

**Kiyomi Murakumo, Yosuke Matsumoto, Nagisa
Yano**
2Okinawa Churaumi Aquarium, 424 Ishikawa,
Motobu, Okinawa 905-0206, Japan

Masaru Nakamura
Okinawa Churashima Research Center, Okinawa
Churashima Foundation, 888 Ishikawa, Motobu,
Okinawa 905-0206, Japan

Christoph Gertler
Friedrich-Loeffler-Institut,   Federal   Research
Institute for Animal Health, Institute of Novel and
Emerging Infectious Diseases, Greifswald-Insel
Riems, Greifswald, Germany
Present Address: RWTH Aachen, Institute for
Biotechnology, Aachen, Germany.

Mathias Schlegel
Friedrich-Loeffler-Institut,   Federal   Research
Institute for Animal Health, Institute of Novel and
Emerging Infectious Diseases, Greifswald-Insel
Riems, Greifswald, Germany
Present Address: Seramun Diagnostica GmbH,
Heidesee, Germany

Miriam Linnenbrink, Diethard Tautz
Max Planck Institute of Evolutionary Biology, Plön,
Germany

Rainer Hutterer
Stiftung   Zoologisches   Forschungsmuseum
Alexander Koenig, ZFMK, Bonn, Germany

Patricia König
Friedrich-Loeffler-Institut, Federal Research Institute
for Animal Health, Institute of Diagnostic Virology,
Greifswald-Insel Riems, Greifswald, Germany

Bernhard Ehlers
Robert   Koch-Institut,   Division   12   "Measles,
Mumps,   Rubella,   and   Viruses   Affecting
Immunocompromised Patients", Berlin, Germany

Kerstin Fischer, René Ryll, Kathrin Baumann
Friedrich-Loeffler-Institut,   Federal   Research
Institute for Animal Health, Institute of Novel and
Emerging Infectious Diseases, Greifswald-Insel
Riems, Greifswald, Germany

Jens Lewitzki
Landratsamt   Weilheim-Schongau   Veterinäramt,
Weilheim i. OB, Germany

Sabine Sauer
Bundeswehr Medical Academy, Military Medical
Research and Development, Division E, Munich,
Germany

Angele Breithaupt, Jens P. Teifke
Friedrich-Loeffler-Institut, Federal Research Institute
for Animal Health, Department of Experimental
Animal   Facilities   and   Biorisk   Management,
Greifswald-Insel Riems, Greifswald, Germany

Michael Faulde
Zentrales   Institut   des   Sanitätsdienstes   der
Bundeswehr   Koblenz,   Abteilung   I   Medizin,
Koblenz, Germany

Rainer G. Ulrich
Friedrich-Loeffler-Institut,   Federal   Research
Institute for Animal Health, Institute of Novel and
Emerging Infectious Diseases, Greifswald-Insel
Riems, Greifswald, Germany
German Center for Infection Research (DZIF),
Partner Site Hamburg-Luebeck-Borstel-Insel Riems,
Greifswald-Insel Riems, Greifswald, Germany

Ayan Sadhu, Yadvendradev Vikramsinh Jhala
Department of Animal Ecology and Conservation
Biology, Wildlife Institute
of India, Chandrabani, Dehra Dun, Uttarakhand

Peter Prem Chakravarthi Jayam
Department of Animal Ecology and Conservation
Biology, Wildlife Institute
of India, Chandrabani, Dehra Dun, Uttarakhand

Tamilnadu Forest Department, Udumalpet, Tamil Nadu 642126, India

Qamar Qureshi
Department of Population Management, Capture and Rehabilitation, Wildlife Institute of India, Chandrabani, Dehra Dun, Uttarakhand 248001, India

Raghuvir Singh Shekhawat, Sudarshan Sharma
Rajasthan Forest Department, Jaipur, Rajasthan 302004, India

Jan T. Lifjeld , Jarl Andreas Anmarkrud, Terje Laskemoen, Gunnhild Marthinsen,Even Stensrud, Lars Erik Johannessen, Arild Johnsen
Natural History Museum, University of Oslo, Blindern, 0318 Oslo, Norway

Pascual Calabuig
Wildlife Recovery Center "Tafira", Vivero Forestal, Cabildo de Gran Canaria, 35017 Las Palmas de Gran Canaria, Canary Islands, Spain

Joseph E. J. Cooper, Robert F. Lachlan
3Department of Biological and Experimental Psychology, Queen Mary University of London, London E1 4NS, UK

Anna M. Kearns
Natural History Museum, University of Oslo, Blindern, 0318 Oslo, Norway
Department of Biological Sciences, University of Maryland, Baltimore County, 1000 Hilltop Circle, Baltimore, MD 21250, USA

Eduardo Garcia-del-Rey
Macaronesian Institute of Field Ornithology, C/ Enrique Wolfson 11-3, 38004 Santa Cruz de Tenerife, Canary Islands, Spain

Naomi S. Prosser, Penny C. Gardner
Danau Girang Field Centre, c/o Sabah Wildlife Department, Wisma Muis, 88100 Kota Kinabalu, Sabah, Malaysia
Organisms and Environment Division, School of Biosciences, Cardiff University, Sir Martin Evans Building, Museum Avenue, Cardiff CF10 3AX, UK

Jeremy A. Smith
Organisms and Environment Division, School of Biosciences, Cardiff University, Sir Martin Evans Building, Museum Avenue, Cardiff CF10 3AX, UK

Eco-explore Public Interest Company, Cardiff School of Biosciences, Cardiff University, Cardiff CF10 3AX, UK

Jocelyn Goon Ee Wern
Danau Girang Field Centre, c/o Sabah Wildlife Department, Wisma Muis, 88100 Kota Kinabalu, Sabah, Malaysia

Laurentius N. Ambu
Sabah Wildlife Department, Wisma Muis, 88100 Kota Kinabalu, Sabah, Malaysia

Benoit Goossens
Danau Girang Field Centre, c/o Sabah Wildlife Department, Wisma Muis, 88100 Kota Kinabalu, Sabah, Malaysia
Organisms and Environment Division, School of Biosciences, Cardiff University, Sir Martin Evans Building, Museum Avenue, Cardiff CF10 3AX, UK
Sabah Wildlife Department, Wisma Muis, 88100 Kota Kinabalu, Sabah, Malaysia
SustainablePlaces Research Institute, Cardiff University, 33 Park Place, Cardiff CF10 3BA,UK.

J. Antonio Baeza
Department of Biological Sciences, 132 Long Hall, Clemson University, Clemson, SC 29634, USA
Smithsonian Marine Station at Fort Pierce, 701 Seaway Drive, Fort Pierce, FL 34949, USA
Departamento de Biología Marina, Facultad de Ciencias del Mar, Universidad Católica del Norte, Larrondo 1281,Coquimbo, Chile

Lunden Simpson, Louis J. Ambrosio, and Michael J. Childress
Department of Biological Sciences, 132 Long Hall, Clemson University,Clemson, SC 29634, USA.

Nathalia Mora
Departamento de Biología, Facultad de Ciencias, Universidad del Valle, Cali 760032, Colombia

Rodrigo Guéron
Instituto Federal de Educação, Ciência e Tecnologia do Espírito Santo - Campus Alegre, Alegre, Espírito Santo, Brazil

Marcelo Rodrigues, Willi Salvenmoser, Bert Hobmayer, Peter Ladurner
Institute of Zoology and Center for Molecular Biosciences Innsbruck, University of Innsbruck, Innsbruck, Austria.

Philippe Leclère
Laboratory for Chemistry of Novel Materials, Research Institute for Material Sciences and Engineering, Center of Innovation and Research in Materials and Polymers, University of Mons,Mons, Belgium

Patrick Flammang
Biology of Marine Organisms and Biomimetics, ResearchInstitute for Biosciences, University of Mons, Mons, Belgium

Michael W. Hess
Division of Histology and Embryology, Innsbruck Medical University, Innsbruck, Austria

Index

A

African Wolf, 63-64, 67-70
Animal Biometrics, 23-24, 36
Antrostomus Vociferous, 52-53

B

Bats, 17, 21-22, 80, 89, 104-114, 145
Bayesian Inference (BI), 1, 3
Biodiversity, 36, 51, 64, 70, 72, 80-81, 89, 104-105, 112, 134, 182, 185-186
Biological Indicators, 72, 81
Body Condition, 21, 62, 104, 109, 114, 179-186
Body Condition Scoring, 179, 182, 184-186

C

Caenorhabditis, 37-51
Camera Trap, 147, 150-153, 155, 158, 179-184
Camera Traps, 147-151, 179, 181-182, 184-185
Canidae, 63, 65, 67, 69-71
Canis Aureus, 63-65, 67, 69-71, 149
Canis Lupaster, 63-67, 69-70
Conservation, 1, 10-11, 23, 34, 36, 53, 60-61, 63-64, 70-89, 97, 105, 112, 123, 132, 147-148, 155, 157-159, 176-178, 180-183, 185-186

D

Danio Rerio, 90-91, 97
Dispersal, 7, 12, 15, 22, 24, 35, 70, 73, 80, 82, 89, 147, 150-157, 159, 173, 199

E

Eulemur Rubriventer, 23, 25-26, 36

F

Feeding Guilds, 72-73, 79, 81-82, 88
Female Defence, 13, 17-18

G

Geolocator, 52-54, 56, 58, 60-61

H

Habitat Degradation, 179, 184
Hibernation, 104-105, 107-109, 111-114
Hormones, 125, 127, 129-132

I

In-captivity, 125
Insectivore, 52-53, 61
Integrative Taxonomy, 160, 177-178
Inter-birth Interval, 147, 153

K

Known Fate, 147, 155, 158

L

Leapfrog, 52, 58
Linear Discriminant Analysis, 23, 29-30, 35
Litter Size, 140, 147, 151, 153, 155, 157, 159
Local Binary Pattern (LBP), 29

M

Mammal, 14, 21-23, 36, 102-103, 113-114, 133-138, 140, 144-145, 159, 179, 181, 185
Mating Behaviors, 125, 129
Maximum Likelihood (ML), 1, 100
Medicinal Leeches, 1-2, 5-7, 10-12
Migration, 52-62, 159
Mitochondrial Genome, 98-99, 102-103, 125, 163-164, 166, 177-178
Mitogenomes, 98-102, 160, 162, 166-168, 173, 176
Mobula Alfredi, 125
Morphology, 2, 11, 22, 37, 47, 49-50, 63-64, 125, 161, 176-178, 199, 202-205, 210, 214
Mortality, 19, 35, 52, 60-61, 105, 111-119, 122-123, 147, 151-154, 156, 159, 180, 187-188, 191, 195, 197-198, 201
Multiscale Local Binary Pattern, 23, 26, 35

N

Navigation, 90-91, 93, 95-97
Nightjar, 52-53, 55, 60-61

P

Paraphyletic, 45, 51, 98, 100
Pelage, 23-25, 29, 34, 64-65, 67
Peromyscus, 98-103
Photograph, 23, 27, 36, 127, 150
Phylogenetics, 11, 98, 102-103

Phylogeny, 3, 10-12, 22, 37, 50-51, 70, 98, 101-103, 132, 169, 177

Plumage Colour, 160, 162, 164, 172, 175-176

Primate, 23, 34-36, 70

R

Radio-telemetry, 147, 149

Ranthambhore, 147-150, 154-159

Recapture Rate, 52

Reduced-impact Logging, 179-180, 184-186

Reef Manta Ray, 125, 132

Relative Abundance, 72-73, 81-82

Reproductive Barriers, 160

Residential Status, 72, 77, 81, 86

Ribosomal Rna (RRNA), 1

S

Sabah, 179-186

Scramble Competition Mating System, 13-14, 17-18

Sex Steroid, 125

Sex-differential Migration, 52

Sexual Maturation, 125, 130-131

Sexual Size Dimorphism, 13, 17, 21-22

Shoal, 90-97

Siberian Flying Squirrel, 13-14, 18, 21-22

Spatial Cognition, 90, 96-97

Spatial Memory, 90-91, 96

Spatially Explicit Capture-recapture, 147, 159

Speciation, 1, 8, 10, 103, 145, 160, 177

Sperm Size, 176, 178

Stopover, 52, 55, 57, 59-61, 73, 80, 82, 89

Survival, 22, 35-36, 52, 54, 56, 60, 73, 80, 82, 89, 105, 111, 114-116, 118-119, 121-124, 147-148, 152, 154-159, 162, 188

T

Taxonomy, 11-12, 37, 47, 50, 63, 67, 70, 102, 160-162, 171, 174, 177-178

Tennessee, 104-109, 111-114

Trans-gulf, 52

Tropical Forest, 61, 179, 186

W

Wetland Conservation, 72, 74-79, 81, 83-88

Whip-poor-will, 52-53, 55, 60-61

White-nose Syndrome, 104-105, 112-114

Z

Zebrafish, 81-88